Signal Processing: Theory and Implementation

Signal Processing: Theory and Implementation

Edited by
George Pilato

WILLFORD PRESS

www.willfordpress.com

Published by Willford Press,
118-35 Queens Blvd., Suite 400,
Forest Hills, NY 11375, USA

ISBN: 978-1-68285-335-1

Cataloging-in-publication Data

Signal processing : theory and implementation / edited by George Pilato.
    p. cm.
Includes bibliographical references and index.
ISBN 978-1-68285-335-1
1. Signal processing. 2. Signal processing--Digital techniques. 3. Adaptive signal processing.
4. Signal theory (Telecommunication). I. Pilato, George.
TK5102.9 .S54 2017
621.382 2--dc23

For information on all Willford Press publications
visit our website at www.willfordpress.com

Printed in the United States of America.

Contents

Permissions

List of Contributors

Index

Preface

This book outlines the various theories, techniques and implementation that are of utmost importance to signal processing. Technology that incorporates the basic theory, algorithms, and implementations of transferring data stored in varied spaces like physical or abstract format is known as signal processing. The aim of this book is to present researches that have transformed this discipline and aided its technological advancement. While understanding the long-term perspectives of the topics, the book makes an effort in highlighting their impact as a growing field. It aims to elucidate the varied applications of this technology like audio signal processing, video processing and image processing to name few. Researchers and students in this field will be assisted by this book. As it outline the various processes and techniques that are required to make the technology work efficiently.

This book is a comprehensive compilation of works of different researchers from varied parts of the world. It includes valuable experiences of the researchers with the sole objective of providing the readers (learners) with a proper knowledge of the concerned field. This book will be beneficial in evoking inspiration and enhancing the knowledge of the interested readers.

In the end, I would like to extend my heartiest thanks to the authors who worked with great determination on their chapters. I also appreciate the publisher's support in the course of the book. I would also like to deeply acknowledge my family who stood by me as a source of inspiration during the project.

Editor

Efficiently sphere-decodable physical layer transmission schemes for wireless storage networks

Hsiao-feng (Francis) Lu[1], Amaro Barreal[2], David Karpuk[2] and Camilla Hollanti[2*]

Abstract

Three transmission schemes over a new type of multiple-access channel (MAC) model with inter-source communication links are proposed and investigated in this paper. This new channel model is well motivated by, e.g., wireless distributed storage networks, where communication to repair a lost node takes place from helper nodes to a repairing node over a wireless channel. Since in many wireless networks nodes can come and go in an arbitrary manner, there must be an inherent capability of inter-node communication between every pair of nodes. Assuming that communication is possible between every pair of helper nodes, the newly proposed schemes are based on various smart time-sharing and relaying strategies. In other words, certain helper nodes will be regarded as relays, thereby converting the conventional uncooperative multiple-access channel to a multiple-access relay channel (MARC). The diversity-multiplexing gain tradeoff (DMT) of the system together with efficient sphere-decodability and low structural complexity in terms of the number of antennas required at each end is used as the main design objectives. While the optimal DMT for the new channel model is fully open, it is shown that the proposed schemes outperform the DMT of the simple time-sharing protocol and, in some cases, even the optimal uncooperative MAC DMT.

While using a wireless distributed storage network as a motivating example throughout the paper, the MAC transmission techniques proposed here are completely general and as such applicable to any MAC communication with inter-source communication links.

Keywords: Distributed communications, Distributed storage systems, Diversity-multiplexing gain tradeoff, MIMO, Multiple-access channel, Relay channel, Sphere decoding, Wireless networks

1 Introduction

The amount of data in cloud storage systems and worldwide data traffic have reached incredible numbers. It was estimated that in 2011, $1.8 \cdot 10^{21}$ bytes of data needed to be stored worldwide [1], a number that grew to an astonishing $4.4 \cdot 10^{21}$ bytes in 2013, and which is further expected to grow tenfold by 2020 [2]. The availability of such an astronomical amount of data and rapid progress in (wireless) communications engineering explain the observed growth of mobile data traffic, which increased by 69 % in 2014, reaching $2.5 \cdot 10^{18}$ bytes per month at the end of the year. This amount of data traffic, which is expected to increase tenfold until 2019, corresponds to nearly 30 times the size of the entire internet in 2000. In addition, about 497 million mobile devices and connections were added globally in 2014, of which smart phones account for 88 %, so that—as foreseen—the number of mobile-connected devices exceeded the number of people on earth by the end of 2014 [3].

The massive amount of available data demands that data no longer be stored on a single device, but rather distributed among several storage nodes in a network, hence usually referred to as *distributed storage systems* (DSSs) (see [4] for a nice introduction). One of the main advantages of storing information in a distributed manner is that the storage system can be made robust against failures by introducing some level of redundancy. Some examples of

*Correspondence: camilla.hollanti@aalto.fi
[2]Department of Mathematics and Systems Analysis, Aalto University, P.O. Box 11100, FI-00076 AALTO (Espoo), Finland
Full list of author information is available at the end of the article

real-life distributed storage systems are Apache Cassandra [5], which is a DSS initially developed at Facebook, and Windows Azure [6], created by Microsoft.

More formally, a DSS consists of n storage nodes over which a file is stored in a redundant manner by dividing it into fragments and distributing the fragments among n nodes using, for instance, a (n, k) maximum distance separable (MDS) erasure code [7]. MDS codes satisfy the Singleton bound and have the convenient property of being able to reconstruct the file by contacting any k of the nodes. Another key feature of a DSS is the ability to repair, meaning that when a node fails (that is a device breaks down or leaves the network), the failed node can be repaired or replaced. If an (n, k) storage code further satisfies the condition that any failed node can be repaired or replaced by contacting any K of the remaining nodes, termed *helpers*, the code is called an (n, k, K) storage code, and the node replacing the failed one is called a *repairing node*, or a *newcomer* [8], if it is not one of the already existing nodes in the network.

Sophisticated storage protocols have been developed, always giving a tradeoff between the amount of data that needs to be stored in any of the storage nodes, and the amount of data that needs to be retrieved for repairing a lost node, also called repair bandwidth (see e.g. [8, 9]. for details), and codes lying on the storage-repair bandwidth tradeoff curve [8, 10] are called *regenerating codes*. Explicit, tradeoff achieving regenerating codes can be found in the literature, see [9, 11], among others.

One important aspect of future DSSs lies in the ability to communicate over wireless channels, making it possible to store or retrieve a file using a wireless connection, even if the storage cloud itself might be wired. This is a feature related to the more general concept of *wireless edge* [12–15]. The mobility of a user has become crucial in everyday life, and wireless channels are used for data transmission for increased flexibility. However, it is well-known that communicating over a fading channel in a wireless DSS [16, 17] makes repair transmissions prone to physical layer errors.

Consider the wireless repair transmission of a DSS, that is, the case of repairing a failed/lost node and replacing it with a repairing node by contacting any K of remaining storage (helper) nodes via wireless links. We assume that both types of nodes may be equipped with multiple antennas. Then, the transmission from the K helpers to the repairing node can be regarded as wireless multiple-input multiple-output (MIMO) multiple-access communication [18–20] with an additional feature of inter-helper communication among the K helpers. To see this, note that the MIMO multiple-access channel (MAC) studied in classical information theory [21] assumes only the existence of communication links from the helpers to the repairing node, or equivalently from sources to destination. Yet

in many wireless distributed storage networks[1], there are often more nodes present and connected than those storing data, i.e., there are *blank nodes* in addition to the actual storage nodes. The total number of nodes can be dynamic even though the number of storage nodes would be fixed, see Fig. 1 for illustration. Now, the loss of a node can happen to any of the n storage nodes, and the K helpers can be any subset of the remaining storage nodes. The role of a repairing node can be taken by any of the blank nodes (also a new node entering the system will be blank in the beginning). This implies that an inherent communication link exists between the blank nodes and the storage nodes. After repair, a node that was previously blank becomes a storage node, and after this can assume the role of a helper node. Storage nodes may also erase their stored data and become blank, after which they can also assume the role of a repairing node, etc. This means that an inter-node communication capability actually exists between every pair of helper nodes, therefore calling for the design of efficient transmission schemes when the sources are further allowed to communicate with each other in a wireless MIMO-MAC.

Yet another example to motivate such inter-helper links is as follows. Notice that each of the n wireless storage nodes in the network consists of a wireless component and a memory component. In case of the wireless component of a node failing, we simply replace it with a new wireless component, and there is no need to contact the helpers to reconstruct the data. On the other hand, if the memory component fails, the "repair" of the node (hence the name of repairing node) happens by repairing the memory component of the node with the aid of helper nodes. That is, the node is not replaced by a completely new node. In this sense, the failed node is still one of the original n storage nodes, and the repairing process is done by contacting any K of the remaining nodes. In other words, the failed node and the repairing node are the same node with the same wireless component. This justifies the requirement that the inter-helper link must exist between every pair of nodes.

When communicating over a wireless channel between terminals equipped with multiple antennas, *space-time codes* [22–26] are often employed to protect the transmitted information from adverse channel effects such as fading and noise. The asymptotic error-performance of space-time codes is commonly dictated by the diversity-multiplexing gain tradeoff (DMT) [27]. Assume each of the K helper nodes has n_t transmit antennas and transmits simultaneously at the same rate of $R = r \log_2 \text{SNR}$ in bits per channel use to the repairing node with n_r receive antennas, where SNR is the signal-to-noise power ratio, and r is commonly referred to as the *multiplexing gain* [27]. The optimal MIMO-MAC DMT was given by Tse et al. in [28] and characterizes the maximal diversity gain,

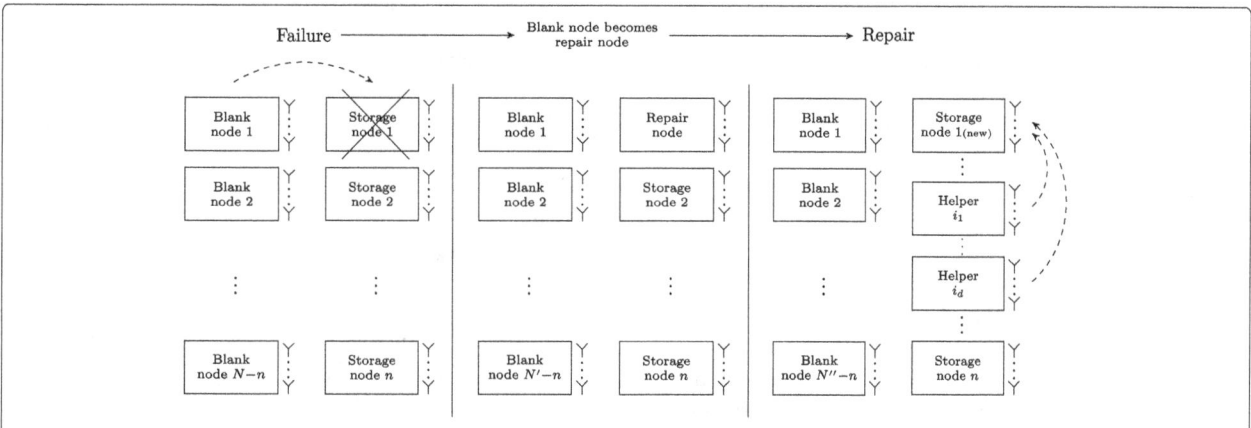

Fig. 1 Wireless storage network. A dynamic network with a varying number of nodes, out of which a fixed number of n nodes are storing data and the rest are inactive (*blank*)

or equivalently the smallest error probability, that can possibly be achieved by any space-time code. A general construction of MIMO-MAC DMT optimal space-time codes was proposed in [26] for any triple (n_t, n_r, K) and multiplexing gain r. These MIMO-MAC codes are constructed from cyclic division algebras [23, 29] and have a linear-dispersion form [30]. Therefore, they can be decoded in the maximal-likelihood (ML) sense by a sphere decoder [31]. While sphere decoding is known to be an efficient implementation of ML decoding, it is unfortunate that when $Kn_t > n_r$, decoding these DMT-optimal codes requires at least partial brute-force decoding before the sphere decoder begins to function [32], or alternatively one has to allow for suboptimal decoding methods, incurring degraded performance [33]. The former approach, though achieving the optimal performance, results in an exponential increase of decoding complexity when Kn_t becomes large and n_t remains fixed [32].

Motivated by these realistic problems in wireless DSSs, new transmission schemes based on various strategies are proposed in this paper. The DMT, together with efficient sphere-decodability and low transmitter and receiver structural complexity in terms of the number of antennas required at each end, are used as the main design objectives, thus naturally establishing a DMT-complexity tradeoff.

1.1 Contributions and related work

In most of the storage and network-coding related research the focus is on the (logical) network layer, while the physical layer functionality is usually ignored or assumed perfect. An exception is [16], where a so-called partial downloading scheme is proposed, which allows for data reconstruction with limited bandwidth by downloading only parts of the content of helper nodes. This is relevant in the fading channel scenario, and the idea can potentially be combined with the present work. In

[17], optimal storage codes are constructed for the error-and-erasure scenario, but fading is not addressed. Isolated from the storage point of view, a lot of research has been carried out in physical layer wireless communications, see e.g. [34] and the references therein.

Remark 1. *An obvious, but naïve attempt would be to try to simply combine an optimal storage code on the network layer and an optimal space-time code on the physical layer. The reason for avoiding this approach is both its structural (many antennas) and computational (decoding subject to partial brute-force) complexity.*

In this paper[2], we propose a class of transmission schemes for MIMO-MAC when communication links among the sources (helper nodes) do exist, which is generally true in many wireless storage networks. The proposed schemes allow for the design of efficiently sphere-decodable space-time codes[3] with only one or two receive antennas. This is in contrast to the state-of-the-art MIMO-MAC codes [26] that have extremely good performance but require Kn_t receive antennas at the repairing node to enable efficient sphere decoding. This is of course unacceptable even for a relatively small value of K, since wireless networks are often heterogenous and might include nodes with only few or even just one antenna. At the moment, to the best of the authors' knowledge, no such scheme exists for large value of K when the receiver has only 1 or 2 antennas, except for the trivial scheme of time-sharing among K helper nodes.

This paper is organized as follows. In Section 2, we will present the channel model for DSS repair transmission, which can be seen as a MIMO-MAC in the presence of communication links among helper nodes. A brief introduction on DMT will also be given therein. Section 2.2 briefly reviews the notion of complexity exponent, which was established by Jaldén and Elia [35] for measuring the minimal computational complexity required by sphere decoders in order to achieve a certain diversity performance. The complexity exponent of existing MIMO-MAC DMT optimal codes [26, 32, 36] is also given to further motivate our design objectives that are presented in Section 3 and to serve as a baseline for comparing the complexity of the first proposed scheme given in Section 4. The first scheme is based on a simple time sharing among *pairs of helpers*, when $n_t = 1$, $n_r = 2$ and $K \geq 2$. The resulting DMT falls between the simple time sharing DMT and optimal MIMO-MAC DMT [28].

We then present two more elaborate schemes extending the first one, achieving a higher DMT by taking advantage of the inter-helper communication links and transforming the overall DSS network into a series of relay networks, where conventional half-duplex[4] cooperative-communication protocols such as the non-orthogonal amplify-and-forward (NAF) strategy [37, 38] will be used. In particular, it will be seen that these schemes can outperform the MIMO-MAC DMT at certain multiplexing gains, simply due to the use of inter-helper communications in the DSS. Moreover, our results on the DMTs for NAF-based relay networks not only improve, but also extend the ones presented in related works, such as [39, 40].

2 Transmission model and preliminaries

Consider a wireless DSS with K helper nodes, equipped with n_t transmit antennas each, and a repairing node with n_r receive antennas. Let $H_i \in \mathbb{C}^{n_r \times n_t}$ be the channel matrix, and $X_i \in \mathbb{C}^{n_t \times T}$ the code matrix associated with the ith helper node, where T is the number of channel uses needed for transmitting X_i. The received signal matrix at the repairing node is given by

$$Y = \sum_{i=1}^{K} H_i X_i + W, \qquad (1)$$

where $W \in \mathbb{C}^{n_r \times T}$ is a matrix modeling complex additive white Gaussian noise (AWGN). The entries of H_i and W_i are independent and identically distributed (i.i.d.) circularly symmetric complex Gaussian random variables with zero mean and unit variance, a distribution which we henceforth denote as $\mathbb{CN}(0, 1)$. The code matrices X_i are required to satisfy the average power constraint $\mathbb{E}||X_i||^2 \leq T \cdot \text{SNR}$. It is also assumed throughout the paper that the repairing node has a complete knowledge of channel state information $\{H_i : i = 1, \ldots, K\}$.

Due to the nature of the DSS, the helper nodes can communicate with each other, a feature not seen in classical MIMO-MAC. Focusing on the ith helper node, let $G_{i,j} \in \mathbb{C}^{n_t \times n_t}$ be the channel matrix and $S_j \in \mathbb{C}^{n_t \times T'}$ be the

code matrix sent by the jth helper node, $j \neq i$; then, the signal matrix received at the ith helper node is given by

$$Y_i = \sum_{\substack{j=1 \\ j \neq i}}^{K} G_{i,j} S_j + Z_i, \qquad (2)$$

where the entries of $G_{i,j}$ and Z_i are again modeled as i.i.d. $\mathbb{CN}(0, 1)$ random variables, and the signal matrices S_j satisfy $\mathbb{E}||S_j||^2 \leq T' \cdot \text{SNR}$. A complete knowledge of $\{G_{i,j}\}$ is assumed to be available at the ith node. Finally, it is assumed throughout the paper that all communication links are half-duplex. A pictorial description of the above channel model is given in Fig. 2.

2.1 The DMT

One of the design objectives in this paper is to provide high performance transmission schemes for wireless DSS repair transmissions. The performance of each scheme will be measured by the DMT [27, 34]. In order to simplify the discussion of DMT, let us ignore the existence of the inter-helper channels for the moment and focus only on the channel input-output relation (1), where only the direct channels from the K helper nodes to the repairing node are of concern. Assuming each helper node transmits at the same multiplexing gain r to the repairing node, we say a scheme achieves diversity gain $d(r)$ if its outage probability $P_{\text{out}}(r)$, which is defined as the probability of mutual information $I\left(X_{i_1}, \ldots, X_{i_s}; Y | H_1, \ldots, H_K\right)$ being strictly less than $s \cdot r \log_2 \text{SNR}$ for some $\{i_1, \ldots, i_s\} \subseteq \{1, \ldots, K\}$, satisfies

$$-\lim_{\text{SNR} \to \infty} \frac{\log P_{\text{out}}(r)}{\log \text{SNR}} = d(r), \qquad (3)$$

and we will write the above as

$$P_{\text{out}}(r) \doteq \text{SNR}^{-d(r)}. \qquad (4)$$

The outage probability $P_{\text{out}}(r)$ is an asymptotic lower bound on the error probability of the scheme [27, 34]

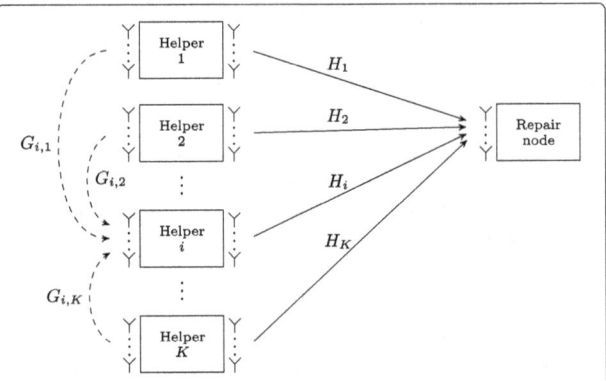

Fig. 2 DSS repair transmission. Complete channel model for DSS repair transmission with K helper nodes, each having n_t transmit antennas, and with n_r receiver antennas at the repairing node

when SNR is large and the multiplexing gain r remains fixed.

For objectivity, we will compare the performance of the proposed schemes with the following optimal DMT [28] for MIMO-MAC[5]

$$d^*_{n_t,n_r,K}(r) = \min\left\{d^*_{n_t,n_r}(r), d^*_{Kn_t,n_r}(Kr)\right\}, \quad (5)$$

where $d^*_{m,n}(r)$ is the optimal DMT for an ($n_t \times n_r$) point-to-point MIMO channel and is given by the piecewise linear function connecting the points $(r, (n_t - r)(n_r - r))$ for $r = 0, 1, \ldots, \min\{n_t, n_r\}$ [27]. We must emphasize that (5) is known to be optimal for the non-cooperative MIMO-MAC, that is, the case when the K helper nodes share no common information, meaning that the inter-helper channels (2) do not exist, and the file fragments stored at the K helper nodes are all statistically independent[6]. Therefore, with a properly designed scheme, it is expected that a higher DMT performance than (5) can be achieved in DSS repair transmission.

2.2 Sphere decoding complexity of state-of-the-art MIMO-MAC codes

A general construction of MIMO-MAC space-time codes was proposed in [26] and was shown to achieve the optimal DMT (5) for any triple (n_t, n_r, K) and multiplexing gain r. More specifically, given n_t and K, the component code C_i of the ith helper node is taken from an algebraic lattice of lattice rank[7] $2n_t K_o^2$ in [26], and C_i consists of $(n_t \times T)$ matrices with $|C_i| \doteq \text{SNR}^{rT}$ and $T = n_t K_o$, where K_o is the smallest odd integer $\geq K$.

To estimate the complexity of decoding the overall code $C_1 \times \cdots \times C_K$ using a joint sphere decoder, we follow [35] by using the notion of *complexity exponent* as a complexity measure.

Definition 1 (Complexity exponent [35]). *Given the multiplexing gain r, let $C_{r,k}$ be a lattice code consisting of $(n_t \times T)$ codeword matrices with $|C_{r,k}| \doteq SNR^{rT}$, $k = 1, \ldots, K$. Let D_r be a decoder for the overall code $C_r = C_{r,1} \times \cdots \times C_{r,K}$, subject to a computational constraint $N_{\max}(r)$, in floating point operations (flops) per T channel uses, in the sense that after $N_{\max}(r)$ flops, the decoder D_r must simply terminate, potentially prematurely and before completing the task, thus declaring an error. We then say D_r achieves diversity order $d(r)$ with complexity exponent $c(r)$ if D_r achieves error probability $P_e \doteq SNR^{-d(r)}$ using at most $N_{\max}(r) \doteq SNR^{c(r)}$ (cf. (3),(4)) flops of computational reserves.* □

The above definition means that in order to decode the code C_r using a joint sphere decoder, one does not have to decode every received signal matrix, especially when the communication channel is deeply faded. Instead, one can enforce a complexity constraint (also called a *halting*

policy) at the sphere decoder, say at most $N_{\max}(r)$ flops of computational reserves. By choosing $N_{\max}(r)$ large enough such that the probability of any premature termination of the sphere decoder is asymptotically no larger than $\text{SNR}^{-d^*_{n_t,n_r,K}(r)}$, the overall error probability at most $2 \cdot \text{SNR}^{-d^*_{n_t,n_r,K}(r)}$, thereby achieving the same diversity $d^*_{n_t,n_r,K}(r)$.

It was shown in [32, 36] that the complexity exponent for decoding the DMT optimal code [26] is given by

$$c_{n_t,n_r,K}(r) = K_o r(Kn_t - n_r) \cdot \mathbf{1}(Kn_t > n_r)$$
$$+ \sup_{\underline{\mu} \in \mathcal{B}(r)} K_o n_t \sum_{i=1}^{v} \left(\frac{r}{n_t} - (1 - \mu_i)^+\right)^+, \quad (6)$$

where $\mathbf{1}(\cdot)$ is the usual indicator function, $v = \min\{Kn_t, n_r\}$, $(x)^+ := \max\{x, 0\}$ and

$$\mathcal{B}(r)$$
$$= \left\{\underline{\mu} = [\mu_1 \cdots \mu_v]^\top \in \mathbb{R}^v : \begin{array}{l} \mu_1 \geq \cdots \geq \mu_v \geq 0, \\ \sum_{i=1}^{v}(|Kn_t - n_r| + 2i - 1)\mu_i \leq d^*_{n_t,n_r,K}(r) \end{array}\right\}.$$
$$(7)$$

There is an intuitive explanation for the term $K_o r(Kn_t - n_r)$ in (6) when $Kn_t > n_r$. Recall that the component code C_i is taken from a certain subset of an algebraic lattice Λ_i of rank $2n_t^2 K_o$. This means that each codeword matrix X_i of C_i is of the form $X_i = \sum_{\ell=1}^{n_t^2 K_o} x_{i,\ell} C_{i,\ell}$, where $\{C_{i,\ell} : \ell = 1, \ldots, n_t^2 K_o\}$ is a basis for Λ_i, and the $x_{i,\ell}$ are independent QAM symbols taken from a certain set $\mathcal{A} \subset \mathbb{Z}[\iota]$ of size $\text{SNR}^{\frac{r}{n_t}}$, $\iota = \sqrt{-1}$. Thus, we can rewrite (1) as

$$Y = \sum_{i=1}^{K} \sum_{\ell=1}^{n_t^2 K_o} H_i C_{i,\ell} x_{i,\ell} + W, \quad (8)$$

or equivalently in a vector form

$$\underline{y} = H\underline{x} + \underline{w}, \quad (9)$$

where $\underline{x} = \left[x_{1,1}, \ldots x_{1,n_t^2 K_o}, \ldots x_{K,n_t^2 K_o}\right]^\top$, $\underline{y}$ is the vectorization of the matrix Y, and H is the corresponding matrix of size $\left(n_r K_o n_t \times K n_t^2 K_o\right)$ by (8). When decoding (9) using a sphere decoder, one first performs a QR-decomposition of the matrix H, say $H = QR$. If $Kn_t > n_r$, the matrix R is no longer upper triangular; it is a trapezoidal matrix with

$$Kn_t^2 K_o - n_r K_o n_t + 1 = K_o n_t (Kn_t - n_r) + 1$$

nonzero entries in the bottom row. Hence, any sphere decoder for (9) must first resolve – perhaps by brute-force – the $|\mathcal{A}|^{K_o n_t(Kn_t - n_r)} = \text{SNR}^{K_o r(Kn_t - n_r)}$ ambiguities before processing the root of the sphere decoding tree. The number of ambiguities then forms the first term in (6).

Remark 2. *A different definition of complexity exponent has appeared in [41], where Damen et al. studied the* *number of flops required by a sphere decoder to decode a fixed-rate space-time code at various finite SNR values. In particular, they defined the complexity exponent as the logarithm to base m of the number of flops required by a sphere decoder to complete its task, where m is the length of vector $\underline{x}$ defined in (9). Below we highlight some of the major differences between Damens' definition of complexity exponent and the one considered in this paper (cf. Definition 1).*

- *Damens' definition focuses on a code with a fixed rate, and Definition 1 concerns more with the theoretical asymptote at high SNR regime when the rate scales linearly with $\log_2$ SNR.*
- *Definition 1 considers the possibility of having a halting policy, while Damens' definition requires the sphere decoder to complete its task at all channel realizations.*

Remark 3. *In [42] Damen et al. proposed to decode (9) by using GDFE-MMSE preprocessing followed by the sphere decoder when $Kn_t > n_r$, in hope of making the matrix R upper-triangular and avoiding the need of resolving the ambiguities. However, it can be seen from [32, 36] that at high SNR regime the matrix R – after MMSE-GDFE preprocessing – is ill-conditioned with $K_0 n_t(Kn_t - n_r)$ number of singular values arbitrarily close to zero. This also explains the appearance of the first term in (6).*

On the other hand, when the code has a fixed rate and operates in the low or moderate SNR regime, the MMSE-GDFE approach does offer a certain complexity reduction with a negligible performance loss, as the singular values of R are numerically well-behaved in general. Other approaches for further complexity reductions under such premises are also available in the literature. For instance, Barbero and Thompson [43] proposed a fixed-complexity sphere decoder, where the number of candidates to be searched at the i-th level of sphere decoding tree is at most n_i, thereby yielding a constant complexity $\prod_i n_i$. Another way to reduce complexity is through the various orderings of singular values of R. A comprehensive study in this direction can be found for example in [44]. We shall emphasize that the complexity exponents simulated in [43, 44] are both based on Damens' definition [41] (cf. Remark 2) because of the aforementioned premises.

In Fig. 3, we plot the complexity exponents for the sphere decoding of MAC DMT optimal codes $\mathcal{C}_r$ [26] when $n_t = 2$, $K = 5$, and $n_r = 2, 10, 100$, respectively. It can be seen that these codes can be efficiently decoded by sphere decoders only when $n_r \gg Kn_t$. Such a requirement is often impossible in practice, particularly in heterogeneous storage networks, where nodes may have only a small number of antennas in use.

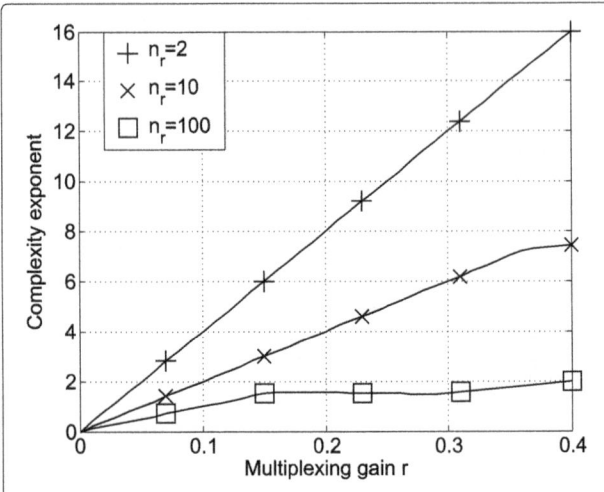

Fig. 3 Complexity exponent comparison. Complexity exponents for the sphere decoding of MAC-DMT optimal codes [26] when $n_t = 2$, $K = 5$ and $n_r = 2, 10, 100$

Remark 4. *In case of $n_r \geq Kn_t$, it has been shown [45, 46] that the DMT optimal MIMO-MAC lattice codes can be decoded with sub-exponential complexity, i.e., having a complexity exponent asymptotically equal to 0, using the Lenstra-Lenstra-Lovász-based lattice reduction aided regularized lattice decoder. The decoder is a combination of GDFE-MMSE, lattice reduction and sphere decoding, and it has a vanishing gap of performance loss to the exact ML decoding as SNR approaches infinity.*

3 Objectives for the design of transmission schemes

In Section 2, we have seen that there is a fundamental difference between the channel for DSS repair transmission and the classical MIMO-MAC, in the sense that the former includes additional inter-helper communication links. Thus, the MIMO-MAC DMT (5) and the MIMO-MAC codes [26] are no longer optimal in scenarios such as DSS repair transmission. Moreover, due to these additional inter-helper channels, it is expected that the DSS repair transmission can have a higher optimal DMT than (5). This then calls for the design of new transmission schemes with good DMT performance for DSS repairing, which is the first design objective considered in this paper.

The second design objective comes from the observation of high decoding complexity of MIMO-MAC codes [26] in Fig. 3 when $Kn_t > n_r$. In a DSS, it is often possible that K is large, and n_r is relatively small and fixed. This then calls for the design of new transmission schemes that can yield efficiently sphere-decodable space-time codes avoiding the need to process the ambiguities by brute-force. Potentially, such an aim could be achieved by reducing the number of "active" helper nodes, i.e., reducing the effective value of K in (1), such that the average number of independent QAM symbols received by the repairing node at each channel use be no larger than n_r, as observed from (6).

In the subsequent sections, we will focus on the case of $n_t = 1$ and $K \gg n_r$, and we will provide three transmission schemes, each for a different configuration of the wireless DSS network and for a different design objective. The first scheme is given in Section 4 for the case of two receive antennas and an arbitrary number of helper nodes, each having one transmit antenna. It is based on a simple time sharing among pairs of helpers and is aimed at having a low sphere-decoding complexity at a cost of certain DMT performance-loss due to its neglect of existing inter-helper links. The DMT for this scheme falls between the simple time sharing DMT and optimal MIMO-MAC DMT (5).

Two more elaborate schemes will be presented in Sections 5 and 6, respectively, where we aim to improve the DMT performance at the possible cost[8] of higher decoding complexity. These schemes take advantage of inter-helper channels and transform the overall DSS network into a series of relay networks, where the conventional half-duplex NAF protocol [37, 38] will be used. In particular, we will see that these schemes can outperform the MIMO-MAC DMT (5) at certain multiplexing gains, simply by exploiting inter-helper communications in the DSS.

4 Scheme 1: $n_t = 1$, $n_r = 2$, and K helper nodes

We have seen in Section 2.2 that the existing state-of-the-art MIMO-MAC space-time codes [26] could incur an extremely high decoding complexity when the repairing node has only a few number of antennas. Thus, our major aim in this section is to provide a new transmission scheme that can yield space-time codes with reduced decoding complexity. In particular, we would like these potential codes to be efficiently sphere-decodable, by which we mean that the H matrix, when writing the channel input-output relation in a vector form (cf. (9)), has linearly independent columns with probability one.

Besides the desired property of being efficiently sphere-decodable, the complexity of the transmission schemes should also be considered. In other words, if we ignore the existence of inter-helper links (2), then the schemes for DSS repair transmission can be made relatively simple. These are the main objectives of Scheme 1.

Let $\mathcal{K} = \{1, 2, \ldots, K\}$ denote the set of K helper nodes, and let $\mathcal{U}$ be a collection of two subsets sof $\mathcal{K}$, defined as below

$$\mathcal{U} := \begin{cases} \{\{1,2\},\{3,4,\},\ldots,\{K-1,K\}\}, & \text{if } K \text{ even,} \\ \{\{1,2\},\ldots,\{K-2,K-1\},\{K,1\},\{2,3\},\ldots,\{K-1,K\}\}, & \text{if } K \text{ odd.} \end{cases}$$

With the above, the proposed scheme is the following. For each $U = \{u_1, u_2\} \in \mathcal{U}$, only helper nodes u_1 and

u_2 are allowed to transmit during the active period of U. This implies that the probability of helper node k transmitting equals $\frac{2}{K}$ for every $k \in \mathcal{K}$. In order to achieve an average multiplexing gain r, each helper node k, when chosen according to U, i.e. $k \in U$, should actually transmit at a higher multiplexing gain $\frac{Kr}{2}$. We summarize the above scheme below, and a pictorial description of Scheme 1 is given in Fig. 4.

Scheme 1

1: **for** each $U = \{u_1, u_2\} \in \mathcal{U}$ **do**

2:   Helper-nodes u_1 and u_2 transmit using the MIMO-MAC code given in [26, Eq. (20)] for $n_t = 1$, two users and multiplexing gain $\frac{Kr}{2}$.

3: **end for**

The following theorem is a straightforward consequence of [34].

Theorem 1. *The DMT performance achieved by Scheme 1 is*

$$d_1(r) = \min\left\{ d_{1,2}^*\left(\frac{Kr}{2}\right), d_{2,2}^*(Kr) \right\} \qquad (10)$$

In Fig. 5, we consider the case $n_t = 1$, $n_r = 2$ and $K = 10$, and compare $d_1(r)$ to $d_{1,2,10}^*(r)$, which is the DMT corresponding to all 10 helper nodes transmitting simultaneously. The function $d_0(r)$ is the DMT for the time-division multiple-access (TDMA)-based scheme, by which we mean that each helper node takes turns in an orthogonal manner to transmit information to the repairing node at multiplexing gain Kr. It can be seen that the first proposed scheme outperforms the TDMA scheme in terms of the DMT, and there is a considerable gap between $d_1(r)$ and $d_{1,2,10}^*(r)$. However, the comparison is unfair in the sense that in order to achieve $d_{1,2,10}^*(r)$ the codes in [26] would require exponentially large computational reserves, or equivalently an exponentially long time, for decoding.

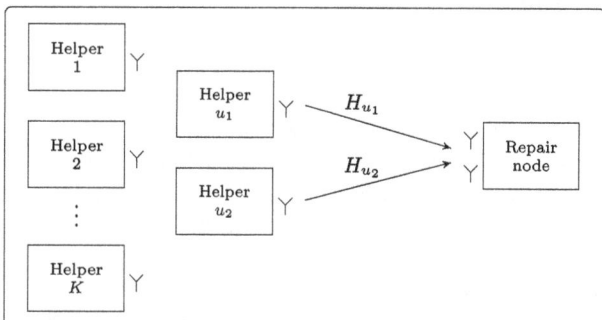

Fig. 4 Channel model of Scheme 1. Channel model for Scheme 1 at the Uth step, $U = \{u_1, u_2\} \subset \{1, 2, \ldots, K\}$

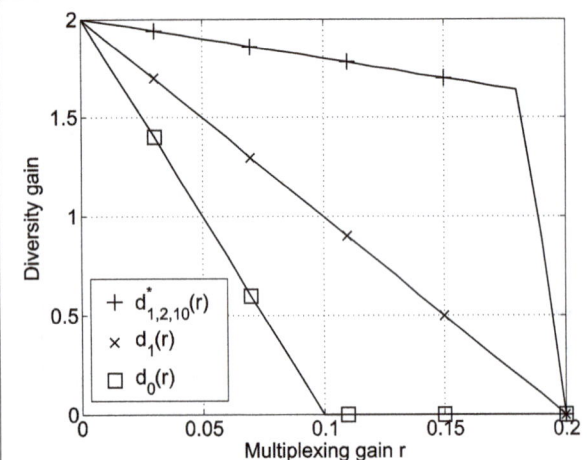

Fig. 5 Scheme 1: DMT comparison. DMT performances achieved by MIMO-MAC, Scheme 1 and time-sharing scheme for $K = 10$ helper nodes, $n_t = 1$, and $n_r = 2$

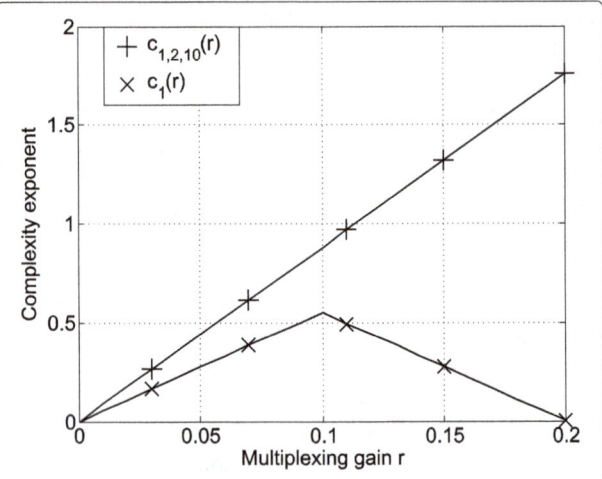

Fig. 6 Complexity exponent comparison. Complexity exponents for the sphere decoding of the MIMO-MAC code given in [26] ($c_{1,2,10}(r)$) and the proposed code ($c_1(r)$) based on Scheme 1 for the case of $n_t = 1$, $n_r = 2$ and $K = 10$

Continuing the example of $n_t = 1$, $n_r = 2$ and $K = 10$, by modifying the two-user MIMO-MAC code given in [26] according to Scheme 1, it can be directly seen from (6) that the resulting code achieves DMT $d_1(r)$ with complexity exponent

$$c_1(r) := 3 \sup_{\underline{\mu} \in \mathcal{B}_1(r)} \sum_{i=1}^{2} \left[\min\left\{ \frac{rK}{2}, \frac{Kr}{2} + \mu_i - 1 \right\} \right]^+, \quad (11)$$

where

$$\mathcal{B}_1(r) = \left\{ \underline{\mu} = [\mu_1 \ \mu_2]^\top \in \mathbb{R}^2 : \mu_1 \geq \mu_2 \geq 0, \mu_1 + 3\mu_2 \leq d_1(r) \right\}, \quad (12)$$

when it is decoded using a sphere decoder with halting policies. In Fig. 6, we compare $c_1(r)$ to the complexity exponent $c_{1,2,10}(r)$ of the MIMO-MAC code given in [26] for the case $n_t = 1$, $n_r = 2$ and $K = 10$. It can be clearly seen that the proposed scheme can yield a code with with a much lower decoding complexity.

5 Scheme 2: $n_t = 1$, general n_r, and K helper nodes

The aim of Scheme 1 presented in the previous section was to have a small decoding complexity, at a cost of certain DMT performance loss due to the neglect of inter-helper links in DSS repair-communication. In this section as well as the next, we will shift our focus to designing transmission schemes that take into account these inter-helper links and beat the DMT performance $d^*_{n_t, n_r, K}(r)$.

Consider a DSS repair channel with K helpers, each having $n_t = 1$ transmit antenna, and a repairing node with n_r receive antennas. To make good use of the inter-helper links, we interpret in Scheme 2 some of the links as links

of a relay channel. More specifically, in this scheme each of the K helper nodes will take turns acting as the source in a cooperative relay network [37], while the remaining $K - 1$ helper nodes play the role of relays helping the source to send information to the repairing node.

With the above, the proposed scheme is a modification of the NAF protocol [37, 38] for a cooperative relay network with $K - 1$ relays. It consists of K phases, and each phase requires at least $2(K - 1)$ channel uses. Thus, the total number of channel uses required by Scheme 2 is at least $2K(K - 1)$.

Let $\mathcal{K} = \{1, 2, \ldots, K\}$ denote the set of K helper nodes. Given $k \in \mathcal{K}$, the scheme is at the kth phase, and helper node k acts as the source of a relay network. The remaining helper nodes $\mathcal{R}_k := \mathcal{K} \setminus \{k\} = \{u_1, \ldots, u_{K-1}\}$ are the relays. At the tth channel use of the kth phase, $t = 1, 2, \ldots, 2(K - 1)$, node k broadcasts a signal $x_{k,t}$, subject to the power constraint $\mathbb{E}|x_{k,t}|^2 \leq \text{SNR}$, to all nodes in $\mathcal{R}_k$ as well as to the repairing node. Due to the half-duplex assumption in Section 2, the nodes in $\mathcal{R}_k$ can either receive or transmit, but not both at the same time. Therefore, the behavior of each node $u_i \in \mathcal{R}_k$ is set such that it receives the signal from node k when $t = 2i - 1$ and transmits to the repairing node when $t = 2i$. More specifically, the signal received by node u_i at $t = 2i - 1$ is given by

$$r_{u_i, 2i-1} = g_{u_i, k} x_{k, 2i-1} + z_{u_i, k, 2i-1}, \quad (13)$$

where $g_{u_i, k}$ and $z_{u_i, k, 2i-1}$ are i.i.d. $\mathbb{CN}(0, 1)$ random variables representing the channel gain from node k to node u_i and the additive noise, respectively, as defined in (2). Node

u_i then amplifies the signal $r_{u_i,2i-1}$ with an amplification factor $a_{u_i,k}$ set such that

$$\mathbb{E}|a_{u_i,k}r_{u_i,2i-1}|^2 \leq \text{SNR}, \tag{14}$$

where the expectation is taken with respect to $x_{k,2i-1}$ and $z_{u_i,k,2i-1}$, since $g_{u_i,k}$ is already known to node u_i. Equivalently, we have

$$|a_{u_i,k}|^2 \leq \frac{\text{SNR}}{1+\text{SNR}|g_{u_i,k}|^2}. \tag{15}$$

Then, at channel use $t = 2i$, node u_i joins node k and sends the amplified signal $a_{u_i,k}r_{u_i,2i-1}$ to the repairing node.

Since each helper node k is allowed to transmit its own message to the repairing node during the kth phase, its multiplexing gain must be increased to $K \cdot r$ in order to achieve the desired average multiplexing gain r. We now summarize the steps of Scheme 2 below. A pictorial description of Scheme 2 is given in Fig. 7.

Scheme 2

1: **for** each $k = 1, 2, \ldots, K$ **do**
2:     Set $\mathcal{K} \setminus \{k\} = \{u_1, \ldots, u_{K-1}\}$
3:     **for** $i = 1, 2, \ldots, (K-1)$ **do**
4:         Node k broadcasts a signal $x_{k,2i-1}$ at multiplexing gain Kr to all nodes at channel use $t = 2i-1$. The signals received by node u_i and the repairing node are respectively given by

$$r_{u_i,2i-1} = g_{u_i,k}x_{k,2i-1} + z_{u_i,k,2i-1}, \tag{16}$$

$$\underline{y}_{k,2i-1} = \underline{h}_k x_{k,2i-1} + \underline{w}_{k,2i-1}. \tag{17}$$

5:         Node k broadcasts a signal $x_{k,2i}$ to all nodes at channel use $t = 2i$, and node u_i simultaneously sends $a_{u_i,k}r_{u_i,2i-1}$. The signal received by the repairing node when $t = 2i$ is

$$\underline{y}_{k,2i} = \underline{h}_k x_{k,2i} + \underline{h}_{u_i}a_{u_i,k}r_{u_i,2i-1} + \underline{w}_{k,2i}. \tag{18}$$

6:     **end for**
7: **end for**

5.1 DMT achieved by Scheme 2

Note firstly that by the symmetry among the phases of Scheme 2, it suffices to analyze the DMT achieved within the first phase, i.e., for $k = 1$, where the helper node 1 acts as the source, and the remaining helper nodes are relays. Thus, for notational convenience, we will henceforth drop the subindex k.

Set $N = 2(K-1)$, and let x_t be a $\mathbb{CN}(0, \text{SNR})$ random variable, representing the signal sent by helper node 1 at time instance t for $t = 1, 2, \ldots, N$. Then, the signal received by the repairing node at the tth channel use is

$$\underline{y}_t = \begin{cases} \underline{h}_1 x_t + \underline{w}_t, & t \text{ odd,} \\ \underline{h}_1 x_t + a_i \underline{h}_i(g_i x_{t-1} + z_i) + \underline{w}_t, & t \text{ even and } i = \frac{t}{2}+1, \end{cases} \tag{19}$$

where g_i and z_i's are i.i.d. $\mathbb{CN}(0,1)$ random variables obtained by re-indexing the corresponding variables in (16) for notational convenience. The amplification factor $a_i \in \mathbb{R}^+$, $i = 2, \ldots, K$, is set such that

$$|a_i|^2 \leq \frac{\text{SNR}}{1+\text{SNR}|g_i|^2}.$$

We can equivalently reformulate the received vectors $\underline{y}_t$ in (19) in matrix form, as

$$\underline{y} = \begin{bmatrix} \underline{y}_1 & \underline{y}_2 & \cdots & \underline{y}_N \end{bmatrix}^\top$$

$$= \underbrace{\begin{bmatrix} \underline{h}_1 & 0 & \cdots & 0 & 0 \\ a_2 g_2 \underline{h}_2 & \underline{h}_1 & \cdots & 0 & 0 \\ \vdots & \vdots & \ddots & \vdots & \vdots \\ 0 & 0 & \cdots & \underline{h}_1 & 0 \\ 0 & 0 & \cdots & a_K g_K \underline{h}_K & \underline{h}_1 \end{bmatrix}}_{H} \underbrace{\begin{bmatrix} x_1 \\ x_2 \\ \vdots \\ x_{N-1} \\ x_N \end{bmatrix}}_{\underline{x}} + \underbrace{\begin{bmatrix} \underline{w}_1 \\ a_2 z_2 \underline{h}_2 + \underline{w}_2 \\ \underline{w}_3 \\ a_3 z_3 \underline{h}_3 + \underline{w}_4 \\ \vdots \\ \underline{w}_{N-1} \\ a_K z_K \underline{h}_K + \underline{w}_N \end{bmatrix}}_{\underline{v}}. \tag{20}$$

Given H, the instantaneous mutual information between the transmitted signal $\underline{x}$ and the received signal $\underline{y}$ is

$$I(\underline{x};\underline{y} \mid H) = \log_2 \det\left(K_v + \text{SNR}HH^\dagger\right) - \log_2 \det(K_v)$$

$$= \sum_{i=2}^{K} \log_2 \det\left(I_2 + \text{SNR}H_i^\dagger K_i^{-1}H_i\right), \tag{21}$$

where

$$K_v = \mathbb{E}(\underline{v}\underline{v}^\dagger), \quad H_i = \begin{bmatrix} \underline{h}_1 & 0 \\ a_i g_i \underline{h}_i & \underline{h}_1 \end{bmatrix}, \text{ and } K_i = \begin{bmatrix} I_{n_r} & \\ & I_{n_r} + |a_i|^2 \underline{h}_i \underline{h}_i^\dagger \end{bmatrix}. \tag{22}$$

Thus, the outage probability for Scheme 2 is given by

$$\Pr\left\{ H: \sup_{|a_i|^2 \leq \frac{\text{SNR}}{1+\text{SNR}|g_i|^2}} I\left(\underline{x};\underline{y} \mid H\right) < 2K(K-1)r\log_2 \text{SNR} \right\} \doteq \text{SNR}^{-d_2(r)}, \tag{23}$$

where the target information rate $2K(K-1)r\log_2 \text{SNR}$ arises from the facts that

(i)   the scheme takes K phases to complete, and
(ii) each phase requires $2(K-1)$ channel uses.

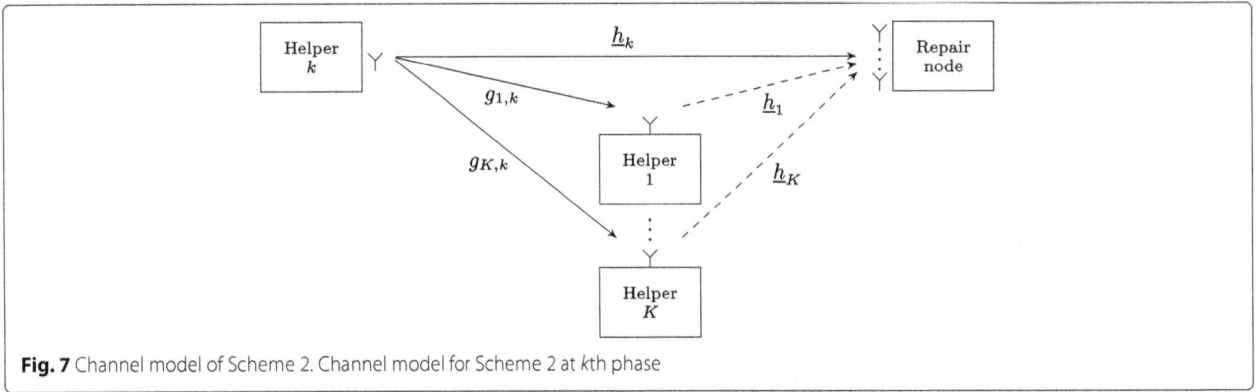

Fig. 7 Channel model of Scheme 2. Channel model for Scheme 2 at kth phase

5.2 DMT achieved by Scheme 2 when $n_r = 1$

When $n_r = 1$, it can be seen that the DMT achieved by Scheme 2 is exactly the DMT for the NAF protocol derived by Azarian et al. [38] with $K - 1$ relays and multiplexing gain Kr. Hence, the following result is immediate from [38].

Theorem 2. *The DMT achieved by Scheme 2 when $n_r = 1$ is the following*

$$d_2(r)\bigg|_{n_r=1} = (1 - Kr)^+ + (K - 1)(1 - 2Kr)^+. \quad (24)$$

In Fig. 8, we plot the DMT performance achieved by this scheme for the case of $K = 10$ helper nodes. We also include the base-line TDMA scheme for comparison. It can be seen that the proposed scheme has a better DMT performance than $d^*_{n_t,n_r,K}(r)$ for $r \leq \frac{1}{2K+1} = \frac{1}{21}$, due to the use of additional inter-helper links.

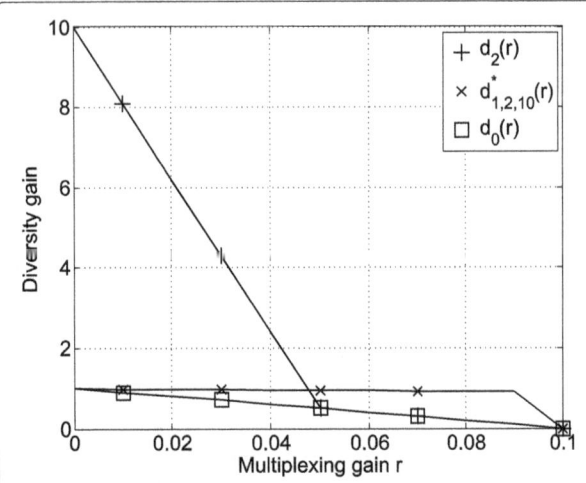

Fig. 8 Scheme 2: DMT comparison. DMT performances achieved by Scheme 2, MIMO-MAC and time-sharing scheme for $K = 10$ helper nodes, $n_t = 1$, and $n_r = 1$

5.3 Upper and lower bounds on $d_2(r)$ with general n_r

Analyzing the outage probability (23) turns out to be very challenging in general when the repairing node has multiple antennas, i.e., $n_r \geq 2$. Almost all existing works such as [38, 47] consider only the case $n_r = 1$. In [39] Yang and Belfiore investigated the DMT for the MIMO-NAF protocol and provided a lower bound for such DMT. Their result can be modified to yield a lower bound for $d_2(r)$. We will comment more on that particular lower bound at the end of this subsection.

To provide bounds on the DMT $d_2(r)$ for general values of n_r, let U be an $(n_r \times n_r)$ unitary matrix such that $U\underline{h}_1 = [\,\|\underline{h}_1\|\ 0 \cdots 0\,]^\top := \underline{h}$.

For H_i defined in (22), $i = 2, \ldots, K$, we get

$$\mathrm{diag}(U, U)H_i = \begin{bmatrix} U & \\ & U \end{bmatrix} H_i = \begin{bmatrix} \underline{h} & 0 \\ a_i g_i \underline{\ell}_i & \underline{h} \end{bmatrix} = S_i, \quad (25)$$

where $\underline{\ell}_i = U\underline{h}_i$ has the same probability density function as $\underline{h}_i$, $i = 2, \ldots, K$. Let $\Sigma_i := I_{n_r} + |a_i|^2 \underline{h}_i \underline{h}_i^\dagger$. Clearly, we have the following partial ordering for positive-definite matrices,

$$I_{n_r} \prec \Sigma_i \prec \left(1 + |a_i|^2\|\underline{h}_i\|^2\right)I_{n_r} = \left(1 + |a_i|^2\|\underline{\ell}_i\|^2\right)I_{n_r},$$

which in turn implies $\frac{1}{1+|a_i|^2\|\underline{\ell}_i\|^2}I_{n_r} \prec \Sigma_i^{-1} \prec I_{n_r}$. With the above, $I(\underline{x};\underline{y} \mid H)$ can be upper bounded by

$$I(\underline{x};\underline{y} \mid H)$$
$$\leq \sum_{i=2}^{K} \log_2 \det\left(I_2 + \mathrm{SNR}H_i^\dagger H_i\right) \quad (26)$$
$$= \sum_{i=2}^{K} \log_2\left[\left(1 + \mathrm{SNR}\|\underline{h}\|^2\right)^2 + \mathrm{SNR}|a_i g_i|^2\|\underline{\ell}_i\|^2\right.$$
$$\left. + \mathrm{SNR}^2|a_i g_i|^2\|\underline{h}\|^2 \sum_{j=2}^{n_r} |\ell_{i,j}|^2\right]. \quad (27)$$

Similarly, set $c_i = \frac{1}{1+|a_i|^2||\ell_i||^2}$, and $I(\underline{x};\underline{y} \mid H)$ is lower bounded by

$$I(\underline{x};\underline{y} \mid H)$$

$$\geq \sum_{i=2}^{K} \log_2 \det\left(I_2 + \mathrm{SNR}S_i^\dagger \begin{bmatrix} I_{n_r} & \\ & c_i I_{n_r} \end{bmatrix} S_i\right) \quad (28)$$

$$= \sum_{i=2}^{K} \log_2\Big[1 + (1+c_i)\mathrm{SNR}||\underline{h}||^2 + c_i|a_ig_i|^2||\underline{\ell}_i||^2\mathrm{SNR}$$

$$+ c_i\mathrm{SNR}^2||\underline{h}||^4 + c_i^2|a_ig_i|^2\mathrm{SNR}^2||\underline{h}||^2\sum_{j=2}^{n_r}|\ell_{i,j}|^2\Big]. \quad (29)$$

Equations (27) and (29) then yield the following theorem for bounding the DMT $d_2(r)$ for Scheme 2.

Theorem 3. *The DMT $d_2(r)$ of Scheme 2 for a general number $n_r \geq 1$ of receive antennas at the repairing node has the following upper bound $d_{2,U}(r)$ and lower bound $d_{2,L}(r)$:*

$$d_{2,U}(r) := \inf_{g} \sup_{b \leq g} \inf_{(\alpha,\beta_1,\beta_2) \in \mathcal{A}_U(r,b,g)} n_r\alpha + (K-1)\beta_1$$

$$+ (n_r-1)(K-1)\beta_2 + (K-1)g$$

$$(30)$$

$$d_{2,L}(r) := \inf_{g} \sup_{b \leq g} \inf_{(\alpha,\beta_1,\beta_2) \in \mathcal{A}_L(r,b,g)} n_r\alpha + (K-1)\beta_1$$

$$+ (n_r-1)(K-1)\beta_2 + (K-1)g$$

$$(31)$$

where

$$\mathcal{A}_U(r,b,g)$$

$$= \left\{\alpha,\beta_1,\beta_2 \in [0,1]: \max\begin{cases} 2(1-\alpha), \\ 1+b-g-\min\{\beta_1,\beta_2\}, \\ 2+b-g-\beta_2 \end{cases} \leq 2Kr\right\}$$

$$(32)$$

and

$$\mathcal{A}_L(r,b,g)$$

$$= \left\{\alpha,\beta_1,\beta_2 \in [0,1]: \max\begin{cases} 1-\alpha, \\ 2-2\alpha-(b-\beta)^+, \\ 1-\beta+b-g-(b-\beta)^+, \\ 2-\alpha+b-g-\beta_2-2(b-\beta)^+ \end{cases} \leq 2Kr\right\}.$$

$$(33)$$

Proof. Note that the random variables g_i's are i.i.d., hence there is no need to distinguish them in (27) and (29) when deriving the DMT. The same holds also true for a_i, $\underline{\ell}_i$, and its elements $\ell_{i,j}$ for $i = 2,\ldots,K$. Thus, we set

$|a_i|^2 \doteq \mathrm{SNR}^b$, $|g_i|^2 \doteq \mathrm{SNR}^{-g}$, $||\underline{h}||^2 \doteq \mathrm{SNR}^{-\alpha}$, $||\underline{\ell}_i||^2 \doteq \mathrm{SNR}^{-\beta}$, and $|\ell_{i,j}|^2 \doteq \mathrm{SNR}^{-\beta_j}$ with $\beta = \min_{j=1,\ldots,n_r}\beta_j$. Moreover, we note that $||\underline{h}||^2$ is a χ^2 random variable with $2n_r$ degrees of freedom, hence it contributes the term $n_r\alpha$ to (27). Each $\underline{\ell}_i$ consists of n_r i.i.d. $\mathbb{CN}(0,1)$ complex random variables, and there is no need to distinguish $\ell_{i,j}$ for $i = 2,\ldots,K$ and for $j = 2,\ldots,n_r$ as can be seen from (27) and (29). Hence, we can set $|\ell_{i,j}|^2 \doteq \mathrm{SNR}^{-\beta_2}$ for $i = 2,\ldots,K$ and for $j = 2,\ldots,n_r$. Similarly, there is no need to distinguish $\ell_{i,1}$ for $i = 2,\ldots,K$, hence we set $|\ell_{i,1}|^2 \doteq \mathrm{SNR}^{-\beta_1}$ for $i = 2,\ldots,K$. Finally, note that $|g_i|^2$ and $|\ell_{i,j}|^2$ are i.i.d. χ^2 random variables with two degrees of freedom. Plugging the above into (27) and (29) and applying the Laplace principle as in [27] yield the desired upper and lower bounds (30) and (31).

$\square$

In Fig. 9, we plot the DMT bounds $d_{2,L}(r)$ and $d_{2,U}(r)$ of Scheme 2 as well as the DMT $d^*_{1,2,10}(r)$ with $K = 10$ helper nodes, $n_t = 1$ and $n_r = 2$. While there is a gap between bounds $d_{2,L}(r)$ and $d_{2,U}(r)$ when the multiplexing gain r is small, it can be clearly seen that Scheme 2 can offer a better DMT performance than $d^*_{1,2,10}(r)$ when r is small. Regarding the sharpness of $d_{2,L}(r)$ and $d_{2,U}(r)$, let us focus on the case when r is approaching zero from the right, i.e., when $r \downarrow 0$. Note that there are nine SISO channels from helper node 1 to the remaining helper nodes, and the channel between node 1 and the repairing node is a (1×2) SIMO channel. Therefore, the communication to the repairing node would be in outage if the nine SISO channels and the (1×2) SIMO channel are all in deep fade, thereby yielding a maximal diversity order of $9 + 2 = 11$. We therefore conclude that the upper bound $d_{2,U}(r)$ can be further improved.

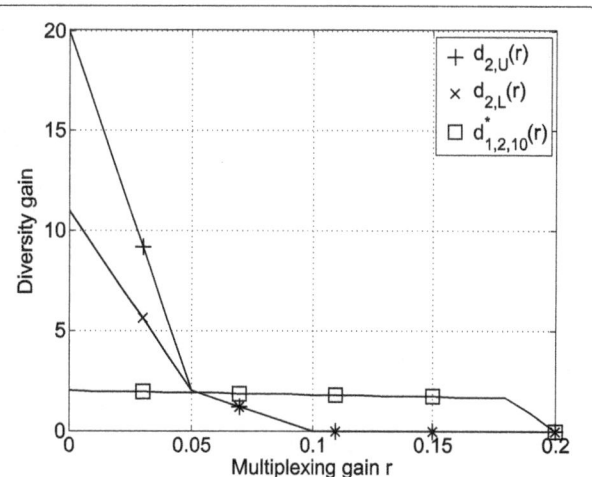

Fig. 9 Scheme 2: DMT bounds and comparison. DMT performances achieved by Scheme 2 (lower bound and the first upper bound) and MIMO-MAC for $K = 10$ helper nodes, $n_t = 1$, and $n_r = 2$

As mentioned earlier, Yang and Belfiore ([39], Theorem 2) provided a lower bound on the DMT for MIMO-NAF protocol. Their bound can be modified to become a lower bound for $d_2(r)$ and has the following form

$$d_{2,\mathrm{L,YB}}(r) = n_r \cdot (1 - Kr)^+ + (K-1) \cdot d_{\mathrm{RP}}(2Kr), \quad (34)$$

where $d_{\mathrm{RP}}(r)$ is the DMT for the Rayleigh product channel $\underline{h}_i \cdot g_i$, and an exact expression for $d_{\mathrm{RP}}(r)$ can be found in ([39], Proposition 1).

In Fig. 10, we compare our lower bound $d_{2,L}(r)$ to the lower bound $d_{2,\mathrm{L,YB}}(r)$ for the case $n_t = 1$, $n_r = 2$ and $K = 10$. It can be clearly seen that, in this case, our bound is shaper than the bound (34).

5.4 Another upper bound on $d_2(r)$ with general n_r

To obtain another upper bound on the instantaneous mutual information $I(\underline{x}; \underline{y} \mid H)$, we consider the situation that the repairing node has further knowledge of $r_{i,t-1} = g_i x_{t-1} + z_i$ when $t = 2, 4, \ldots, N$ and $i = \frac{t}{2} + 1$. In this case, define

$$\underline{y}'_t = \underline{h}_1 x_t + w_t, \quad t = 1, 2, \ldots, N. \quad (35)$$

Writing $\underline{y}_t = \underline{y}'_t + a_i \underline{h}_i r_{i,t-1}$ for $t = 2(i-1)$, it follows that

$$I\left(\underline{x}; \underline{y} \mid H\right) \le I\left(\underline{x}; \underline{y}'_1, \ldots, \underline{y}'_N, r_{2,1}, r_{3,3}, \ldots, r_{K,N-1} \mid H\right), \quad (36)$$

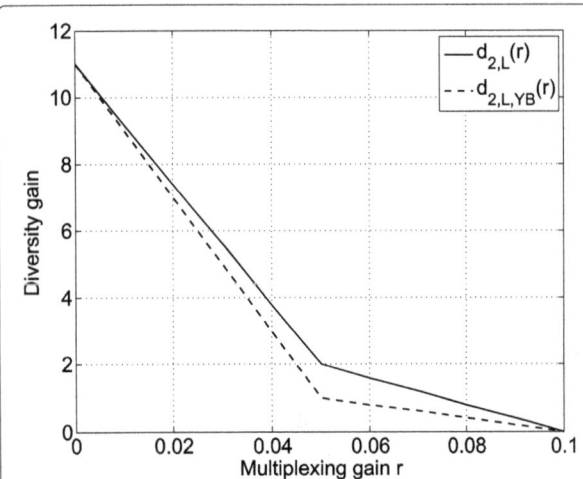

Fig. 10 Scheme 2: DMT bounds and comparison. A comparison between Yang-Belfiore lower bound $d_{2,L,YB}(r)$ [39] and our lower bound $d_{2,L}(r)$ in (31) for the DMT achieved by Scheme 2 when $n_t = 1$, $n_r = 2$ and $K = 10$

and the upper bound has a much simpler expression than $I\left(\underline{x}; \underline{y} \mid H\right)$. To see this, formulate the received vectors as

$$\underline{y}_U := \begin{bmatrix} \underline{y}'_1 \\ \underline{y}'_3 \\ \vdots \\ \underline{y}'_{N-1} \\ \underline{y}'_2 \\ \vdots \\ \underline{y}'_N \\ r_{2,1} \\ \vdots \\ r_{K,N-1} \end{bmatrix} = \underbrace{\begin{bmatrix} \underline{h}_1 & & & & \\ & \_h_1 & & & \\ & & \ddots & & \\ & & & \underline{h}_1 & \\ & & & & \underline{h}_1 \\ & & & & & \ddots \\ & & & & & & \underline{h}_1 \\ g_2 & & & & \\ & \ddots & & & \\ & & g_K & & \end{bmatrix}}_{H_U} \begin{bmatrix} x_1 \\ x_3 \\ \vdots \\ x_{N-1} \\ x_2 \\ x_4 \\ \vdots \\ x_N \end{bmatrix} + \begin{bmatrix} w_1 \\ w_3 \\ \vdots \\ w_{N-1} \\ w_2 \\ \vdots \\ w_N \\ z_2 \\ \vdots \\ z_K \end{bmatrix}; \quad (37)$$

then

$$H_U^\dagger H_U = \begin{bmatrix} \|\underline{h}_1\|^2 + |g_2|^2 & & & \\ & \ddots & & \\ & & \|\underline{h}_1\|^2 + |g_K|^2 & \\ & & & \|\underline{h}_1\|^2 I_{K-1} \end{bmatrix}. \quad (38)$$

This implies that

$$\begin{aligned} & I\left(\underline{x}; \underline{y} \mid H\right) \\ & \le I\left(\underline{x}; \underline{y}_U \mid H\right) \\ & = (K-1) \log_2(1 + \mathrm{SNR}\|\underline{h}_1\|^2) \\ & \quad + \sum_{i=2}^{K} \log_2\left(1 + \mathrm{SNR}\left(\|\underline{h}_1\|^2 + |g_i|^2\right)\right). \end{aligned} \quad (39)$$

Hence, the outage probability for the second scheme is lower bounded by

$$\Pr\left\{H: \sup_{|a_i|^2 \le \frac{\text{SNR}}{1+\text{SNR}|g_i|^2}} I\left(\underline{x};\underline{y}\mid H\right) < 2K(K-1)r\log_2 \text{SNR}\right\}$$

$$\ge \Pr\left\{H : I\left(\underline{x};\underline{y}_{\mathcal{U}}\mid H\right) < 2K(K-1)r\log_2 \text{SNR}\right\} \tag{40}$$

$$\doteq \text{SNR}^{-d_{2,\mathcal{U}'}(r)}. \tag{41}$$

Theorem 4. *The DMT $d_2(r)$ for Scheme 2 for a general number $n_r \ge 1$ of receive antennas at the repairing node is upper bounded by*

$$d_{2,\mathcal{U}'}(r) = \begin{cases} (n_r + K - 1)(1 - Kr)^+, & \text{if } n_r \ge K - 1, \\ 2n_r(1 - Kr)^+ + (K - 1 - n_r)(1 - 2Kr)^+, & \text{if } n_r \le K - 1. \end{cases} \tag{42}$$

Proof. Similar to the proof of Theorem 3, it is unnecessary to distinguish the random variables g_i in (39) for $i = 2, \ldots, K$ when calculating the DMT. Thus, let $||\underline{h}_1||^2 = \text{SNR}^{-\alpha}$ and $|g_i|^2 = \text{SNR}^{-\beta}$. Note $||\underline{h}_1||^2$ is a χ^2 random variable with $2n_r$ degrees of freedom and $|g_i|^2$ is a χ^2 random variable with 2 degrees of freedom. Plugging the above into (41) and applying the Laplace principle as in [27] gives

$$d_{2,\mathcal{U}'}(r) = \inf_{\mathcal{B}(r)} n_r\alpha + (K - 1)\beta,$$

where

$$\begin{aligned}\mathcal{B}(r) &= \{\alpha, \beta \in [0, 1] : 1 - \alpha + \max\{1 - \alpha, 1 - \beta\} \le 2Kr\} \\ &= \{\alpha, \beta \in [0, 1] : 2(1 - \alpha) + (\alpha - \beta)^+ \le 2Kr\}.\end{aligned}$$

Solving the above optimization problem gives the desired result. $\square$

In Fig. 11, we plot $d_{2,L}(r)$, $d_{2,U}(r)$, and $d_{2,U'}(r)$ for the second proposed scheme with $K = 10$ helper nodes, $n_t = 1$ and $n_r = 2$. It can be seen that $d_{2,L}(r) = d_{2,U'}(r)$ for all values of r, hence we have $d_2(r) = d_{2,L}(r) = d_{2,U'}(r)$ in this case.

5.5 Remarks on the complexity exponents of Scheme 2
Determining the complexity exponents of the second scheme requires much more effort than determining the DMT. At least two major difficulties must be resolved before any identification of complexity exponents is possible. Notice that the notion of complexity exponents resides in an actual construction of space-time codes for the scheme, and that the complexity exponents can vary from one code to another. Codes with a smaller complexity exponent are more favorable in practice, provided that the codes are optimal in the DMT sense, i.e., achieve the DMT $d_2(r)$. Therefore, we have to at least identify a space-time code for Scheme 2 first. In [39], Yang and Belfiore provided a systematic construction of space-time

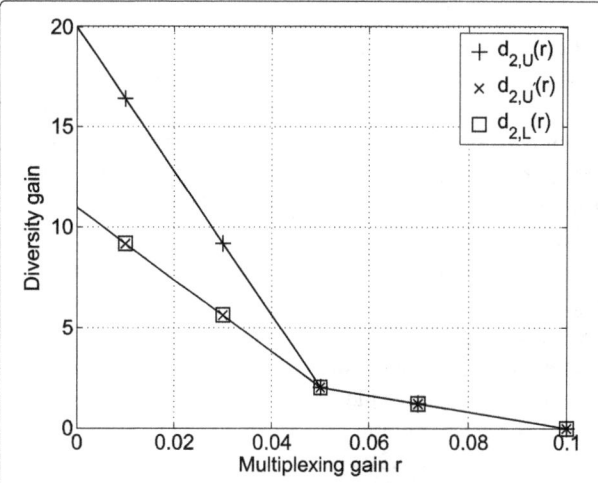

Fig. 11 Scheme 2: improved DMT bounds. DMT performances achieved by Scheme 2 (two upper bounds and a lower bound) for $K = 10$ helper nodes, $n_t = 1$, and $n_r = 2$

codes that is *approximately universal* [48] for NAF-based cooperative relay communications. It is certainly possible to adapt their construction to the transmission using Scheme 2.

The second issue complicating the investigation of complexity exponents arises from the need of an exact characterization of eigenvalues of the matrices $H_i^\dagger K_i^{-1} H_i$ for $i = 2, \ldots, K$, appearing in (21). Determining these eigenvalues is particularly difficult. It is in fact the main reason preventing us from obtaining an exact expression for $d_2(r)$ in previous subsections, and we are only able to provide bounds on $d_2(r)$ in this paper.

Nevertheless, it can be seen from (20) that the equivalent channel matrix H is of size $(Nn_r \times N)$ and has linearly independent columns with probability 1. This implies that when applying a sphere decoder to decode the codes—for instance, the code constructed by Yang and Belfiore [39]— transmitted using Scheme 2, the QR decomposition of the matrix H would result in an upper triangular matrix R; hence, there is no ambiguity to be resolved prior to processing the root of the sphere decoding tree. Therefore, the code must be efficiently sphere decodable.

6 Scheme 3: $n_t = 1$, $n_r \ge 2$, and K helper nodes
In the previous section, we presented a powerful scheme that makes a good use of the inter-helper links to improve the DMT performance of DSS repair transmission. The scheme allows one helper node to transmit information in each phase, and the remaining helper nodes are regarded as relays. Furthermore, we have introduced a novel technique that allows us to upper-bound the DMT for the NAF protocol in a cooperative relay network with multiple antennas at the repairing node. In this section, we will present our third scheme, which can be seen as an

enhancement of Schemes 1 and 2 and can provide a further improvement on the DMT performance.

The third proposed scheme concerns the case $n_t = 1$, $n_r \geq 2$ and K helper nodes. It allows L helper nodes, $L \leq \min\{n_r, K-1\}$, to transmit simultaneously and non-cooperatively to the repairing node as well as to the remaining $(K-L)$ helper nodes, which will function as relays[9] in the network. To achieve an average multiplexing gain r, each of the selected L helper nodes must transmit at a higher multiplexing gain of $\frac{K}{L}r$. In particular, we could later seek to improve the overall DMT performance by optimizing over the choices of L. Therefore, L can actually be a function of the multiplexing gain r.

Given L, the third scheme consists of $\binom{K}{L}$ phases, one for each possible L-subset $\mathcal{L} = \{i_1, \ldots, i_L\}$ of $\mathcal{K}$, where $\mathcal{K} = \{1, 2, \ldots, K\}$ is the set of helper nodes. The helper nodes in set $\mathcal{L}$ transmit simultaneously and non-cooperatively throughout the phase, which has a duration of $N = 2(K-L)$ channel uses. The remaining nodes in $\mathcal{K} \setminus \mathcal{L} = \{j_1, \ldots, j_{K-L}\}$ will function as relays following the NAF protocol. Details of this scheme are given as below, and a pictorial description of this scheme is given in Fig. 12.

Scheme 3

1: Let L be an integer with $1 \leq L \leq \min\{n_r, K-1\}$.
2: **for** each $\mathcal{L} = \{i_1, \ldots, i_L\} \subseteq \mathcal{K}$ **do**
3:     **for** $t = 1, 2, \ldots, N = 2(K-L)$ **do**
4:         At the tth channel use of the phase associated with $\mathcal{L}$, each helper node $i_\ell \in \mathcal{L}$ broadcasts a signal $x_{i_\ell, t}$, with $\mathbb{E}|x_{i_\ell, t}|^2 \leq \mathrm{SNR}$, to the nodes in $\mathcal{K} \setminus \mathcal{L}$ and to the repairing node
5:         Helper-node $j_s \in \mathcal{K} \setminus \mathcal{L}$, $s = 1, \ldots, (K-L)$, receives the following signal when t is odd

$$r_{j_s, t} = \sum_{\ell=1}^{L} g_{j_s, i_\ell} x_{i_\ell, t} + z_{j_s, t}, \qquad (43)$$

        where g_{j_s, i_ℓ} and $z_{j_s, t}$ are i.i.d. $\mathbb{CN}(0,1)$ random variables defined in (2). When $t = 2s$, node j_s broadcasts the signal $a_{j_s} r_{j_s, 2s-1}$, where a_{j_s} is chosen such that

$$|a_{j_s}|^2 \leq \frac{\mathrm{SNR}}{1 + \mathrm{SNR} \sum_{\ell=1}^{L} |g_{j_s, i_\ell}|^2}. \qquad (44)$$

6:         The signal received at the repairing node at the t-th channel use of the phase associated with $\mathcal{L}$ is

$$\underline{y}_t = \begin{cases} \sum_{\ell=1}^{L} \underline{h}_{i_\ell} x_{i_\ell, t} + \underline{w}_t, & t \text{ odd} \\ \sum_{\ell=1}^{L} \underline{h}_{i_\ell} x_{i_\ell, t} + \underline{h}_{j_s} a_{j_s} r_{j_s, t-1} + \underline{w}_t, & t = 2s \end{cases} \qquad (45)$$

7: **end for**
8: **end for**

6.1 DMT analysis for Scheme 3

The communication channel deduced from Scheme 3 resembles the *multiple-access relay channel* (MARC), which was first introduced by Kramer and van Wijngaarden [49]. The DMTs for the two-user and single-relay MARC—in terms of our notation this means $n_t = 1$, $n_r = 1$, $K = 3$ and $L = 2$—using various protocols have been studied in the past. For instance, Azarian et al. [50] investigated the DMT for such MARC using the dynamic-decode-and-forward (DDF) strategy, and Yuksel and Erkip [51] focused on the compress-forward (CF) protocol. Furthermore, a protocol similar to Scheme 3 was proposed in [40] and was termed *multiple-access amplify-and-forward* (MAF), which is a variation of the NAF protocol. It was found in [40] that the MAF outperforms the DDF in the high multiplexing gain regime and the CF protocol [51] in the low multiplexing gain regime when $n_t = 1$, $n_r = 1$, $K = 3$, and $L = 2$. The MAF thus provides a nice balance between complexity and performance.

Scheme 3 considers a much more complicated scenario than the one in [40], with $n_t = 1$, and general values of n_r, K and $L \leq \min\{n_r, K-1\}$. To the best of our knowledge, the DMT analysis for the MAF protocol has never been taken to such complexity level. On the other hand, our novel bounding technique employed in the proof of Theorem 4 is extremely powerful and enables us to analyze the DMT for general MARC using the MAF protocol.

To this end, for any subset $\mathcal{U} = \{u_1, \ldots, u_k\} \subseteq \mathcal{L}$ of the selected helper nodes, let $\mathcal{E}_{\mathcal{U}}$ denote the event that helper nodes $u_1, \ldots, u_k$ are in outage. The probability for $\mathcal{E}_{\mathcal{U}}$ is given by

$$\Pr\{\mathcal{E}_{\mathcal{U}}\}$$
$$= \Pr\Bigg\{ \{\underline{h}_i\}_{i=1}^{K}, \{g_{i,j}\}_{i,j=1}^{K} : \sup_{\substack{a_{j_s} \\ s=1,\ldots,K-L}} I\Big(\{x_{u,t} : u \in \mathcal{U}\}_{t=1}^{N} ; \underline{y}_1, \ldots, \underline{y}_N \Big|$$
$$\{x_{u,t} : u \in \mathcal{L} \setminus \mathcal{U}\}_{t=1}^{N}, \{\underline{h}_i\}_{i=1}^{K}, \{g_{i,j}\}_{i,j=1}^{K}\Big) < \frac{NK}{L} rk \log_2 \mathrm{SNR} \Bigg\}, \qquad (46)$$

where $N = 2(K-L)$. The overall outage probability for the third proposed scheme with given L is

$$P_{\mathrm{out},3}(L, r) := \Pr\Bigg\{ \bigcup_{\mathcal{U} \subseteq \mathcal{L}} \mathcal{E}_{\mathcal{U}} \Bigg\} \doteq \max_{\mathcal{U} \subseteq \mathcal{L}} \Pr\{\mathcal{E}_{\mathcal{U}}\} \doteq \mathrm{SNR}^{-d_3(L, r)}. \qquad (47)$$

The technique introduced in Section 5.4 can be applied to yield the following upper bound on $d_3(L, r)$.

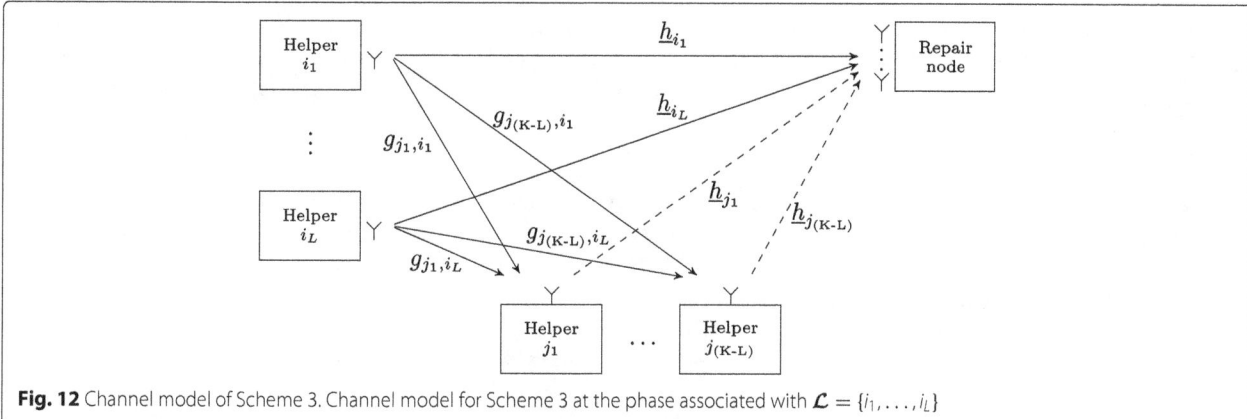

Fig. 12 Channel model of Scheme 3. Channel model for Scheme 3 at the phase associated with $\mathcal{L} = \{i_1, \ldots, i_L\}$

Theorem 5. *The DMT $d_3(L, r)$ can be upper bounded as*

therefore, it knows $\underline{y}'_t = H_{\mathcal{U}}\underline{x}_t + \underline{w}_t$ for $t = 1, 2, \ldots, N$. We then have

$$d_3(L, r) \leq d_{3,U}(L, r) := \min_{k=1,\ldots,L} \inf_{\mathcal{A}(L,k,r)} \sum_{i=1}^{k} [2i - 1 + (n_r - k)]\alpha_i$$
$$+ (K - L) \left[\sum_{j=1}^{k} \beta_j + (n_r - k)\beta_{k+1} \right], \tag{48}$$

where

$$\mathcal{A}(L, k, r)$$
$$:= \left\{ \begin{array}{l} \alpha_1, \cdots \alpha_k, \beta_1, \cdots \beta_{k+1} : \\ 1 \geq \alpha_1 \geq \alpha_2 \geq \cdots \geq \alpha_k \geq 0, \\ \alpha_i \geq \beta_i \geq 0, \quad i = 1, 2, \ldots, k, \\ \beta_{k+1} \geq 0 \, and \, \beta_{k+1} = 0 \, if \, n_r = k, \\ \sum_{i=1}^{k}(1 - \alpha_i) + \frac{1}{2} \max\{\alpha_1 - \beta_1, \ldots, \alpha_k - \beta_k, \beta_{k+1}\} < \frac{Krk}{L} \end{array} \right\}. \tag{49}$$

$$\begin{bmatrix} \underline{y}'_1 \\ \underline{y}'_3 \\ \vdots \\ \underline{y}'_{N-1} \\ \underline{y}'_2 \\ \vdots \\ \underline{y}'_N \\ r_1 \\ \vdots \\ r_{K-L} \end{bmatrix} = \begin{bmatrix} H_{\mathcal{U}} & & & & \\ & \ddots & & & \\ & & H_{\mathcal{U}} & & \\ & & & H_{\mathcal{U}} & \\ & & & & \ddots \\ & & & & & H_{\mathcal{U}} \\ \underline{g}_1^\top & & & & \\ & \ddots & & & \\ & & \underline{g}_{K-L}^\top & & \end{bmatrix} \begin{bmatrix} \underline{x}_1 \\ \underline{x}_3 \\ \vdots \\ \underline{x}_{N-1} \\ \underline{x}_2 \\ \underline{x}_4 \\ \vdots \\ \underline{x}_N \end{bmatrix} + \begin{bmatrix} \underline{w}_1 \\ \underline{w}_3 \\ \vdots \\ \underline{w}_{N-1} \\ \underline{w}_2 \\ \vdots \\ \underline{w}_N \\ z_1 \\ \vdots \\ z_{K-L} \end{bmatrix}.$$
$$\underbrace{}_{\underline{y}_{\mathcal{U}}} \qquad \underbrace{}_{H_{eq}} \qquad \underbrace{}_{:=\underline{x}}$$
$$\tag{50}$$

It follows that

$$I\left(\{x_{u,t} : u \in \mathcal{U}\}_{t=1}^{N}; \underline{y}_1, \ldots, \underline{y}_N \,\Big|\, \{x_{u,t} : u \in \mathcal{K} \setminus \mathcal{U}\}_{t=1}^{N}, \right.$$
$$\left. \{\underline{h}_i\}_{i=1}^{K}, \{g_{i,j}\}_{i,j=1}^{K} \right)$$
$$\leq I\left(\underline{x}; \underline{y}_{\mathcal{U}} \,\big|\, H_{eq} \right)$$
$$= \frac{N}{2} \log_2 \det\left(I_k + \text{SNR} H_{\mathcal{U}}^\dagger H_{\mathcal{U}} \right)$$
$$+ \sum_{s=1}^{\frac{N}{2}} \log_2 \det\left(I_k + \text{SNR} H_{\mathcal{U}}^\dagger H_{\mathcal{U}} + \text{SNR} \underline{g}_s^* \underline{g}_s^\top \right). \tag{51}$$

Proof. Given any $\mathcal{U} = \{u_1, \ldots, u_k\} \subseteq \mathcal{L}$ of selected helper nodes, we first reformulate the channel input-output relations (45) in matrix form. For the sake of notational convenience, we set $r_s = r_{j_s, 2s-1}$, $z_s = z_{j_s, 2s-1}$,

$$\underline{x}_t := \begin{bmatrix} x_{u_1, t} \\ \vdots \\ x_{u_k, t} \end{bmatrix}, \; H_{\mathcal{U}} = \begin{bmatrix} \underline{h}_{u_1} & \cdots & \underline{h}_{u_k} \end{bmatrix}, \text{and } \underline{g}_s = \begin{bmatrix} g_{j_s, u_1} \\ \vdots \\ g_{j_s, u_k} \end{bmatrix},$$

for $j_s \in \mathcal{K} \setminus \mathcal{L}$ and $s = 1, \ldots, (K - L)$. Following the same approach as in Section 5.4, we assume the repairing node has further knowledge of r_s for $s = 1, \ldots, (K - L)$;

Let $H_{\mathcal{U}}^\dagger H_{\mathcal{U}} = E\Lambda E^\dagger$ be the eigen-decomposition of $H_{\mathcal{U}}^\dagger H_{\mathcal{U}}$, where E is a $(k \times k)$ unitary matrix, $\Lambda = \text{diag}(\lambda_1, \cdots, \lambda_k)$, and $0 < \lambda_1 \leq \cdots \leq \lambda_k$ are the nonzero ordered eigenvalues of $H_\mathcal{U}^\dagger H_\mathcal{U}$, since rank$(H_\mathcal{U}^\dagger H_\mathcal{U}) = k$ with probability one. The instantaneous mutual information $I\left(\underline{x};\underline{y}_\mathcal{U} \mid H_{eq}\right)$ can be further simplified to

$$
\begin{aligned}
&I\left(\underline{x};\underline{y}_\mathcal{U} \mid H_{eq}\right) \\
&= N\sum_{s=1}^{k}\log_2(1+\mathrm{SNR}\lambda_s) \\
&\quad + \sum_{s=1}^{\frac{N}{2}}\log_2\left(I_k + (I_k + \mathrm{SNR}\Lambda)^{-1}\,\mathrm{SNR}E^\dagger\underline{g}_s^*\underline{g}_s^\top E\right) \\
&= N\sum_{s=1}^{k}\log_2(1+\mathrm{SNR}\lambda_s) \\
&\quad + \sum_{s=1}^{\frac{N}{2}}\log_2\left(1+\sum_{j=1}^{k}\frac{\mathrm{SNR}}{1+\mathrm{SNR}\,\lambda_j}|v_{s,j}|^2+\sum_{j=k+1}^{n_r}\mathrm{SNR}|v_{s,j}|^2\right),
\end{aligned}
\tag{52}
$$

where we have set $\underline{v}_s = E^\dagger \underline{g}_s^*$, which is a length-$n_r$ random vector with i.i.d. $\mathbb{CN}(0,1)$ entries. It follows that

$$
\Pr\{\mathcal{E}_\mathcal{U}\}\geq\Pr\left\{I\left(\underline{x};\underline{y}_\mathcal{U}\mid H_{eq}\right)<\frac{NK}{L}rk\log_2\mathrm{SNR}\right\}\doteq\mathrm{SNR}^{-d_\mathcal{U}(r)}.
\tag{53}
$$

We set

- $\lambda_s = \mathrm{SNR}^{-\alpha_s}$ for $s = 1,\ldots,\frac{N}{2}$, with each α_s contributing the term $(2s - 1 + (n_r - k))\alpha_s$ to the overall diversity order.
- $|v_{s,j}|^2 = \mathrm{SNR}^{-\beta_{s,j}} = \mathrm{SNR}^{-\beta_j}$ for $s = 1,\ldots,\frac{N}{2}$ and $j = 1,\ldots,k$, since there is no need to distinguish $v_{s,j}$ in these cases when applying the Laplace principle to (53). Each $\beta_j, j = 1,\ldots,k$, contributes the term $\frac{N}{2}\beta_j$ to the overall diversity order.
- $|v_{s,j}|^2 = \mathrm{SNR}^{-\beta_{s,j}} = \mathrm{SNR}^{-\beta_{k+1}}$ for $s = 1,\ldots,\frac{N}{2}$ and $j = k + 1,\ldots,n_r$, for the same reason. The factor β_{k+1} contributes the term $\frac{N}{2}(n_r - k)\beta_{k+1}$ to the overall diversity order.

It follows from the above that

$$
\begin{aligned}
d_\mathcal{U}(r) := \inf_{\mathcal{A}'(L,k,r)}\sum_{i=1}^{k}[\,2i - 1 + (n_r - k)]\alpha_i + (K - L)\sum_{j=1}^{k}\beta_j \\
+ (K - L)(n_r - k)\beta_{k+1},
\end{aligned}
\tag{54}
$$

where

$$
\mathcal{A}'(L,k,r) :=
$$

$$
\left\{
\begin{aligned}
&\alpha_1,\ldots,\alpha_k,\beta_1,\ldots,\beta_{k+1}\in\mathbb{R}: \\
&\quad \alpha_1 \geq \alpha_2 \geq \cdots \geq \alpha_k \geq 0,\ \beta_1,\beta_2,\ldots,\beta_{k+1}\geq 0, \\
&\quad \beta_{k+1} = 0 \text{ if } n_r = k, \\
&\quad \sum_{i=1}^{k}(1-\alpha_i)^+ + \frac{1}{2}\max\{(1 - \beta_j - (1 - \alpha_j)^+)^+,\ldots, \\
&\qquad (1 - \beta_k - (1 - \alpha_k)^+)^+,\beta_{k+1}\}<\frac{Krk}{L}
\end{aligned}
\right\}
\tag{55}
$$

Finally, the upper bound $d_{3,\mathcal{U}}(L,r)$ is obtained after minimizing $d_\mathcal{U}(r)$ for all possible subsets $\mathcal{U}\subseteq\mathcal{L}$ (or equivalently for all $k = 1,\ldots,L$) and after simplifying the constraints in (55). $\qquad\square$

By optimizing over all possible $L = 1, 2,\ldots,\min\{n_r, K - 1\}$ for $d_{3,\mathcal{U}}(r)$, we obtain an upper bound on the DMT performance for the third scheme.

Corollary 6. *The DMT performance for Scheme 3 is upper bounded by*

$$
d_{3,\mathcal{U}}(r) := \max_{L=1,\ldots,\min\{K-1,n_r\}} d_{3,\mathcal{U}}(L,r).
\tag{56}
$$

In Fig. 13, we illustrate the overall picture for the case of $n_t = 1$, $n_r = 2$, and $K = 10$. While we do not yet have a lower bound or a tight DMT result, we believe that Scheme 3 is indeed likely to be superior[10] to all other schemes presented in this paper, namely the TDMA scheme, Schemes 1 and 2.

In particular, we note that the DMT upper bound for Scheme 3 achieves the maximal possible multiplexing gain of $\frac{2}{10} = 0.2$, which is the same as the TDMA scheme and the MIMO-MAC. Such possibility for the optimality of Scheme 3 turns out to be generally true, at least from the viewpoint of the upper bound (51). To see this, note that by (51), we have

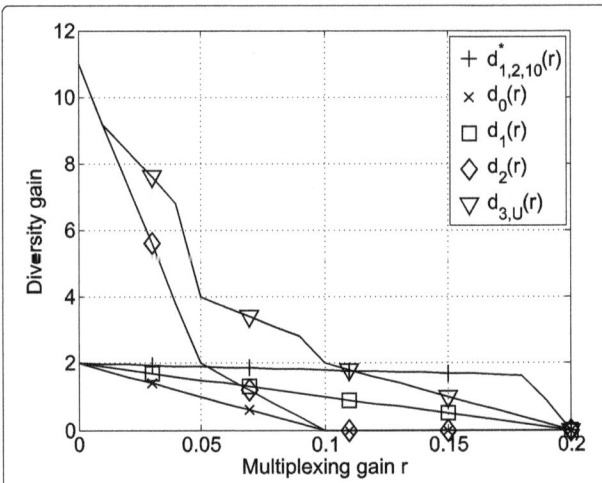

Fig. 13 Overall DMT comparison. DMT performances achieved by the MIMO-MAC, time-sharing scheme and Schemes 1,2, and 3, for $K = 10$ helper nodes, $n_t = 1$, and $n_r = 2$

$$\mathbb{E}\left[\frac{I\left(\underline{x};\underline{y}_{\mathcal{L}}|H_{\text{eq}}\right)}{N\log_2 \text{SNR}}\right]$$

$$=\frac{\mathbb{E}\frac{1}{2}\log_2 \det\left(I_L + \text{SNR}H_{\mathcal{L}}^{\dagger}H_{\mathcal{L}}\right) + \frac{1}{2}\mathbb{E}\log_2 \det\left(I_L + \text{SNR}H_{\mathcal{L}}^{\dagger}H_{\mathcal{L}} + \text{SNR}\underline{g}_1^*\underline{g}_1^{\top}\right)}{\log_2 \text{SNR}}$$

$$=\frac{1}{2}\min\{L, n_r\} + \frac{1}{2}\min\{L, n_r + 1\} + o(1), \tag{57}$$

as SNR $\rightarrow \infty$, where the last equality follows from the asymptotic analysis of the degrees-of-freedom (DoF) for the MIMO channel [27, 52] and from the fact that $H_{\mathcal{L}}$ is a channel matrix of size $(n_r \times L)$, and $H_{\mathcal{L}}^{\dagger}H_{\mathcal{L}} + \underline{g}_1^*\underline{g}_1^{\top} = FF^{\dagger}$ with $F = [H_{\mathcal{L}}^{\dagger}\ \underline{g}_1^*]$ is a matrix of size $(L \times (n_r + 1))$. Eq. 57 shows that the channel capacity resulting from Scheme 3 equals $L \cdot \log_2 \text{SNR} + O(\log_2 \text{SNR})$ in high SNR regime for $L \leq \min\{n_r, K - 1\}$, and such an amount of capacity is shared by the L selected helper nodes. In other words, each selected helper node gets $1 \cdot \log_2 \text{SNR} + O(\log_2 \text{SNR})$ bits per channel use as the maximal achievable transmission rate. Note that in Scheme 3 the selected helper node must transmit at a higher multiplexing gain $\frac{K}{L}r$ such that the average multiplexing gain equals r. This then implies

$$\frac{K}{L}r\log_2 \text{SNR} \leq 1 \cdot \log_2 \text{SNR} + o(\log_2 \text{SNR}), \tag{58}$$

i.e., $r \leq \frac{L}{K}$. Now, with $L = n_r < K$ we see that Scheme 3 achieves the maximal possible multiplexing gain of $\frac{n_r}{K}$ for each helper node, same as MIMO-MAC [28], where the maximal possible multiplexing gain is given by $\frac{\min\{Kn_t, n_r\}}{K} = \frac{n_r}{K}$.

7 Conclusions

The communications within a wireless storage network can be modeled as a multiple-access channel with additional inter-source communication links. Motivated by this observation, we have proposed three physical layer transmission schemes based on different time-sharing and relaying strategies that are suitable for the given channel model. In contrast to the state-of-the-art MAC DMT optimal algebraic space-time codes, our schemes are efficiently sphere-decodable with only one or two antennas. Their DMT performance reaches between the time-sharing DMT and the optimal MAC DMT—the one for conventional MIMO-MAC having no inter-source links—in the high-multiplexing gain regime. When the desired multiplexing gain is low, the schemes even outperform the optimal MAC DMT. Naturally, the schemes are also applicable to DSS file reconstruction, as well as to any MAC communications with inter-source links.

In the future, even small devices with very limited power may be equipped with several antennas thanks to massive MIMO at 60 Hz. However, implementation of a practical massive MIMO system still calls for a considerable amount of research efforts regarding pilot design, channel estimation, and code design. Before all that is realized, we believe that the proposed schemes provide a good and efficient alternative.

Endnotes

[1] Device-to-device (D2D) communication networks provide one such example, see e.g. [53, 54].

[2] Preliminary results related to this work were reported in the Global Wireless Summit 2014 GWS'14 [18] (invited abstract which is considered a preprint), 21st International Symposium on Mathematical Theory of Networks and Systems MTNS'14 [19] (short invited abstract, Scheme 1), and 2014 International Symposium on Information Theory and Its Applications (ISITA) [20] (Schemes 1–3, now combined to Scheme 2). We point out that the numbering of the schemes has been changed so that the schemes previously called 2 and 3 [20] have been combined to Scheme 2, and the new scheme is hence now called Scheme 3 and has not appeared anywhere before. This paper extends the results by additional proofs for the bounds related to Scheme 2, and with a completely new scheme, Scheme 3, that improves upon the other schemes.

[3] By efficiently sphere-decodable space-time code we mean that the code can be sphere-decoded without the need of performing an exhaustive search for part of the symbols before starting processing the root of a sphere-decoding tree. See discussions in Section 2.2.

[4] By half-duplex we mean each node can choose to either transmit or receive, but not both at the same time.

[5] Such a comparison might not seem fair to some readers as (5) assumes no inter-helper links. However, the DMT (5) is the best DMT result that can be found in the related literature.

[6] This latter condition might seem unrealistic in certain (logical) distributed storage codes. However, it would be extremely difficult to determine the mutual information between the helpers and the repairing node if one takes into account the shared information among helper nodes.

[7] A lattice is a discrete abelian subgroup of a real or complex vector space, and its rank is given by its rank as a module over $\mathbb{Z}$. By an algebraic lattice we refer to one constructed from a number field extension or a division algebra, see e.g. [29].

[8] It is unfortunate that measuring the exact complexity exponents for these schemes is extremely complicated, and we are unable to complete the task in this paper. Nevertheless, it can still be seen that these schemes can yield efficiently sphere-decodable space-time codes without the need of resolving ambiguities when processing the sphere-decoding tree.

[9] Here we have implicitly assumed $K - L \geq 1$ such that at least one helper node will function as a relay.

[10] Cf. the corresponding upper bound $d_{2,u'}(r) = d_2(r)$ for Scheme 3 that turned out to be tight.

Competing interests
The authors declare that they have no competing interests.

Acknowledgements
C. Hollanti is supported by the Academy of Finland grants #276031, #282938, and #283262, and by Magnus Ehrnrooth Foundation, Finland. D. Karpuk is supported by the Academy of Finland grant #268364. The support from the European Science Foundation under the ESF COST Action IC1104 is also gratefully acknowledged. The research of H. F. Lu was funded in part by Taiwan Ministry of Science and Technology under Grants MOST 101-2923-E-009-001-MY3 and MOST 103-2221-E-009-043-MY3.

Author details
[1] Department of Electrical and Computer Engineering, National Chiao Tung University, ED726, 1001 University Rd., 300 Hsinchu, Taiwan. [2] Department of Mathematics and Systems Analysis, Aalto University, P.O. Box 11100, FI-00076 AALTO (Espoo), Finland.

References
1. Extracting value from chaos (2011). www.emc.com/collateral/analyst-reports/idc-extracting-value-from-chaos-ar.pdf, digital Universe Study by the EMC corporation, Accessed 18.11.2015
2. The digital universe of opportunities (2014). http://www.emc.com/collateral/analyst-reports/idc-digital-universe-2014.pdf, digital Universe Study by the EMC corporation
3. Cisco visual networking index: Global mobile data traffic forecast update, 2014–2019, white paper (2015). http://www.cisco.com/c/en/us/solutions/collateral/service-provider/ip-ngn-ip-next-generation-network/white_paper_c11-481360.pdf
4. F Oggier, A Datta, Coding techniques for repairability in networked distributed storage systems. Found. Trends Commun. Inf. Theory. **9**(4), 383–466 (2013)
5. A Lakshman, P Malik, Cassandra: a decentralized structured storage system. SIGOPS Oper. Syst. Rev. **44**(2), 35–40 (2010)
6. R Jennings, Cloud computing with the Windows Azure platform. Wiley Publishing (2009)
7. FJ MacWilliams, NJA Sloane, The theory of error-correcting codes. North-Holland (1983)
8. AG Dimakis, B Godfrey, Y Wu, MJ Wainwright, K Ramchandran, Network coding for distributed storage systems. IEEE Trans. Inf. Theory. **56**(9), 4539–4551 (2010)
9. KV Rashmi, NB Shah, PV Kumar, Optimal exact-regenerating codes for distributed storage at the MSR and MBR points via a product-matrix construction. IEEE Trans. Inf. Theory. **57**(8), 5227–5239 (2011)
10. T Ernvall, SE Rouayheb, C Hollanti, HV Poor, Capacity and security of heterogeneous distributed storage systems. IEEE J. Sel. Areas Commun. **31**(12), 2701–2709 (2013)
11. SYE Rouayheb, K Ramchandran, in *Proc. 48th Annual Allerton Conference, Sept. 29 2010–Oct. 1 2010*. Fractional Repetition Codes for Repair in Distributed Storage Systems, (Monticello, IL, USA, 2010), pp. 1510–1517
12. J Rabaey, in *Keynote Address*. A brand new wireless day (ASPDAC, Seoul, 2008)
13. J Rabaey, Connectivity brokerage—enabling seamless cooperation in wireless networks. white paper (2010). https://faculty.ozyegin.edu.tr/aliercan/files/2012/10/Pub6.pdf
14. J Rabaey, in *Proc. Symp. VLSI Circuits, June 15-17, 2011*. The swarm at the edge of the cloud—a new perspective on wireless, (Honolulu, HI, USA, 2011), pp. 2158–5601
15. E Bastug, M Bennis, M Debbah, Living on the edge: The role of proactive caching in 5g wireless networks. IEEE Commun. Mag. **52**(8), 82–89 (2014)
16. C Gong, X Wang, On partial downloading for wireless distributed storage networks. IEEE Trans. Signal Process. **60**(6), 3278–3288 (2012)
17. KV Rashmi, NB Shah, K Ramchandran, PV Kumar, in *Proceedings of the 2012 IEEE International Symposium on Information Theory, ISIT 2012, July 16, 2012*. Regenerating codes for errors and erasures in distributed storage, (Cambridge, MA, USA, 2012), pp. 1202–1206
18. C Hollanti, DA Karpuk, A Barreal, HF Lu, Proc. 4th International Conference on Wireless Communications, Vehicular Technology, Information Theory and Aerospace & Electronic Systems (VITAE), May 11-14, 2014, Aalborg, Denmark. Space-time storage codes for wireless distributed storage systems, pp. 1–5 (2014)
19. A Barreal, C Hollanti, DA Karpuk, H Lu, in *Pre-proceedings of the 21st International Symposium on Mathematical Theory of Networks and Systems, MTNS 2014, July 7–11, 2014*. Algebraic Codes and a New Physical Layer Transmission Protocol for Wireless Distributed Storage Systems, (Groningen, The Netherlands, 2014)
20. C Hollanti, H Lu, DA Karpuk, A Barreal, in *Proc. 2014 IEEE International Sympos ium on Information Theory and its Applications, ISITA 2014, October 26–29, 2014*. New relay-based transmission protocols for wireless distributed storage systems, (Melbourne, VIC, Australia, 2014), pp. 585–589
21. TM Cover, JA Thomas, *Elements of Information Theory*, 2nd. edn. (John Wiley & Sons, New Jersey, 2006)
22. V Tarokh, N Seshadri, AR Calderbank, Space-time codes for high data rate wireless communication: performance criterion and code construction. IEEE Trans. Inf. Theory. **44**, 744–765 (1998)
23. P Elia, KR Kumar, SA Pawar, PV Kumar, HF Lu, Explicit space–time codes achieving the diversity–multiplexing gain tradeoff. IEEE Trans. Inf. Theory. **52**(9), 3869–3884 (2006)
24. C Hollanti, J Lahtonen, HF Lu, Maximal orders in the design of dense space-time lattice codes. IEEE Trans. Inf. Theory. **54**(10), 4493–4510 (2008)
25. JC Belfiore, G Rekaya, E Viterbo, The Golden code: a 2×2 full-rate space-time code with non-vanishing determinants. IEEE Trans. Inf. Theory. **51**(4), 1432–1436 (2005)
26. HF Lu, C Hollanti, R Vehkalahti, J Lahtonen, DMT optimal codes constructions for multiple-access MIMO channel. IEEE Trans. Inf. Theory. **57**(6), 3594–3617 (2011)
27. L Zheng, DNC Tse, Diversity and multiplexing: a fundamental tradeoff in multiple antenna channels. IEEE Trans. Inf. Theory. **49**(5), 1073–1096 (2003)
28. DNC Tse, P Viswanath, L Zheng, Diversity-multiplexing tradeoff in multiple-access channels. IEEE Trans. Inf. Theory. **50**(9), 1859–1874 (2004)
29. F Oggier, JC Belfiore, E Viterbo, *Cyclic Division Algebras: a Tool for Space-Time Coding*. (Publishers Inc. PO Box 1024 Hanover, MA 02339 USA, 2007)
30. B Hassibi, M Hochwald, High-rate codes that are linear in space and time. IEEE Trans. Inf. Theory. **48**(7), 1804–1824 (2002)
31. B Hassibi, H Vikalo, On the sphere-decoding algorithm I, expected complexity. IEEE Trans. Signal Process. **53**(8), 2806–2818 (2005)
32. T Tang, H Tien, HF Lu, Selection and rate-adaptation schemes for MIMO multiple-access channels with low-rate channel feedback. IEEE Trans. Inf. Theory. **61**(11), 5948–5975 (2015)
33. TW Tang, MK Chen, HF Lu, Improving the DMT performance for MIMO communication with linear receivers. IEEE Trans. Veh. Technol. **62**(3), 1189–1200 (2013)
34. D Tse, P Viswanath, *Fundamentals of Wireless Communication*. (Cambridge, University Press, Cambridge, UK, 2005)

35. J Jaldén, P Elia, Sphere decoding complexity exponent for decoding full-rate codes over the quasi-static MIMO channel. IEEE Trans. Inf. Theory. **58**(9), 5785–5803 (2012)
36. HF Lu, P Elia, A Singh, Performance-complexity analysis for MAC ML-based decoding with user selection. IEEE Trans. Signal Process. (2014). arxiv.1505.07725
37. JN Laneman, GW Wornell, Distributed space-time-coded protocols for exploiting cooperative diversity in wireless networks. IEEE Trans. Inf. Theory. **49**(10), 2415–2425 (2003)
38. K Azarian, H El Gamal, P Schniter, On the achievable diversity-multiplexing tradeoff in half-duplex cooperative channels. IEEE Trans. Inf. Theory. **51**(12), 4152–4172 (2005)
39. S Yang, JC Belfiore, Optimal space-time codes for the MIMO amplify-and-forward cooperative channel. IEEE Trans. Inf. Theory. **53**(2), 647–663 (2007)
40. D Chen, K Azarian, JN Laneman, A case for amplify-forward relaying in the block-fading multiple-access channel. IEEE Trans. Inf. Theory. **54**(8), 3728–3733 (2008)
41. MO Damen, HE Gamal, G Caire, On maximum-likelihood detection and the search for the closest lattice point. IEEE Trans. Inf. Theory. **49**(10), 2389–2402 (2003)
42. MO Damen, H El Gamal, G Caire, in *Proc. 2004 IEEE Int. Symp. Inf. Theory.* MMSE-GDFE Lattice Decoding for Solving Under-determined Linear Systems With Integer Unknowns, (Chicago, IL, 2004), p. 538
43. LG Barbero, T Ratnarajah, C Cowan, in *Proceedings of the IEEE International Conference on Acoustics, Speech, and Signal Processing, ICASSP 2008, March 30 - April 4, 2008, Caesars Palace, Las Vegas, Nevada, USA.* A low-complexity soft-mimo detector based on the fixed-complexity sphere decoder, (2008), pp. 2669–2672
44. K Su, Detection and decoding of signals transmitted over linear MIMO channels. PhD thesis, University of Cambridge (2005)
45. J Jaldén, P Elia, DMT optimality of LR-aided linear decoders for a general class of channels, lattice designs, and system models. IEEE Trans. Inf. Theory. **56**(10), 4765–4780 (2010)
46. AK Singh, J Jaldén, P Elia, Achieving a vanishing SNR gap to exact lattice decoding at a subexponential complexity. IEEE Trans. Inf. Theory. **58**(6), 3692–3707 (2012)
47. P Elia, K Vinodh, M Anand, PV Kumar, D-MG tradeoff and optimal codes for a class of AF and DF cooperative communication protocols. IEEE Trans. Inf. Theory. **55**(7), 3161–3185 (2009)
48. S Tavildar, P Viswanath, Approximately universal codes over slow fading channels. IEEE Trans. Inf. Theory. **52**(7), 3233–3258 (2006)
49. G Kramer, AJ van Wijngaarden, in *Proc. 2000 IEEE Int. Symp. Inf. Theory.* On the White Gaussian Multiple-acess Relay Channel, (Sorrento, Italy, 2000)
50. K Azarian, HE Gamal, P Schniter, On the optimality of the ARQ-DDF protocol. IEEE Trans. Inf. Theory. **54**(4), 1718–1724 (2008)
51. M Yuksel, E Erkip, Multiple-antenna cooperative wireless systems: A diversity–multiplexing tradeoff perspective. IEEE Trans. Inf. Theory. **53**(10), 3371–3393 (2007)
52. E Telatar, Capacity of multi-antenna Gaussian channels. Europ. Trans. Telecomm. **10**(6), 585–595 (1999)
53. J Pääkkönen, C Hollanti, O Tirkkonen, Device-to-device data storage for mobile cellular systems. Device-to-Device (D2D) Communication With and Without Infrastructure, Globecom 2013, arXiv:1309.6123 (2013)
54. J Pääkkönen, C Hollanti, O Tirkkonen, Device-to-device data storage with regenerating codes. 8th International Workshop on Multiple Access Communications (MACOM), arXiv:1411.1608 (2015)

2

Robust adaptive algorithm for active control of impulsive noise

Alina Mirza[*], Ayesha Zeb and Shahzad Amin Sheikh

Abstract

Active noise control (ANC) systems employing adaptive filters suffer from stability issues in the presence of impulsive noise. To overcome this limitation, new methods must be investigated. In this paper, we propose the filtered-x state-space recursive least square (FxSSRLS), an SSRLS-based practical and adaptive algorithm for ANC. Computer simulations are executed to verify the enhanced performance of the FxSSRLS algorithm. Symmetric α-stable (SαS) distributions are used to model impulsive noise. The results show that the proposed FxSSRLS algorithm is more robust in eliminating high-peaked impulses than the recently reported algorithms for ANC applications. Moreover, the suggested solution exhibits better stability and faster convergence, without jeopardizing the performance of the proposed solution in terms of residual noise suppression in the presence of impulses.

Keywords: Impulsive noise, Active noise control, Stable distribution, Step size, FxLMS, SSRLS, MNR

1 Introduction

Active noise control (ANC) has been extensively used by researchers over the last two decades, due to its superior performance in canceling low-frequency noise as compared to passive methods such as enclosures, barriers, and silencers [1]. Impulsive noise is non-Gaussian in nature, which means that it involves the frequent occurrence of amplitudes that are larger than those found in Gaussian noise. Hence, the information contained within the signal is altered significantly. Power line communication interference, underwater acoustic signals, low-frequency atmospheric noise, noise generated by punching machines, infusion pump sounds in hospitals, and all types of man-made noise can be classified as impulsive noise [2–4].

Figure 1 shows the basic block diagram of a single-channel feed-forward ANC system using an adaptive algorithm. The system consists of two microphones and a control system. The two microphones are used to obtain reference noise $x(n)$ and residual noise error signal $e(n)$ while the control system is used to generate an anti-noise signal $d(n)$. The output $y(n)$ of the adaptive control system drives the cancelation loudspeaker.

The biggest challenges incurred in the ANC of impulsive noise are convergence and stability of the noise reduction algorithms. Due to its simplicity and low computational complexity, the most widely used algorithm for ANC is filtered-x least mean square (FxLMS), which is designed to minimize the variance of error signal [5]. However, since the second-order moment does not exist in case of impulsive noise [6], it cannot be used for impulsive noise reduction. Sun et al. [7] proposed a modification in the FxLMS algorithm which ensures stability of the system. They applied fixed thresholds on the reference signal in order to eliminate the effect of large amplitudes of impulses. Instead of ignoring the samples as in [7], Akhtar and Mitsuhashi [8] improved the Sun algorithm by replacing impulses with new threshold values of error and reference signals to achieve faster convergence along with enhanced stability. The algorithms [5, 7, 8] are bound to update their threshold parameters during runtime operation which increases computational complexity. To reduce the complexity, another normalized step-size FxLMS (NSS-FxLMS) algorithm is reported in [9], which does not need modified reference or error signal. And in consequence to that, no selection of the threshold parameters are required. Wu et al. in [10] suggested a new technique (FxlogLMS) based on fair M-estimator that minimizes the squared logarithmic transformations of error signal to achieve robustness. However, the algorithm has the

---
* Correspondence: alina.mirza78@ceme.nust.edu.pk
Department of Electrical Engineering, College of Electrical and Mechanical Engineering, National University of Sciences and Technology, Islamabad, Pakistan

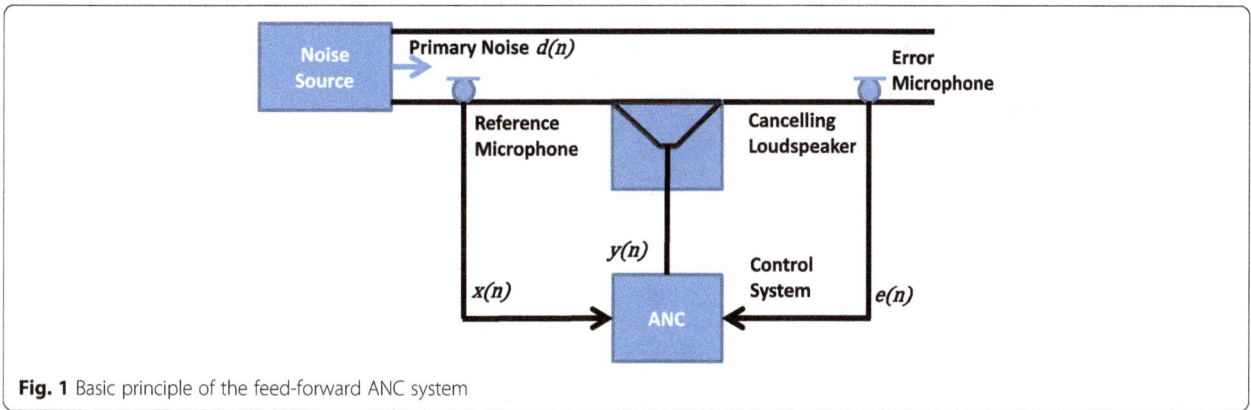

Fig. 1 Basic principle of the feed-forward ANC system

drawback of reaching a dead zone in the process of updating the filter coefficients. Bergamasco et al. [11] provided a solution based on online estimation of the secondary path for ANC applications. A modified Filtered x Least Mean M-estimator (FxLMM) algorithm [12], established on a two-part skewed triangular M-estimate, is presented to achieve stability in FxLMM when exposed to high-peaked impulses. Data reusing-based normalized step-size FxLMS (DR-NSSFxLMS) algorithm, recommended in [13] for active control of impulsive noise sources, improved the performance but at the expense of increased computational complexity. Similarly, the Filtered x Recursive Least Square (FxRLS) algorithm [14] is used for impulsive noise control, which gives faster convergence but at the cost of increased computational complexity. The Gauss-Siedel algorithm [15] and dichotomous coordinate descent (DCD) algorithm [16] are used for recursive least square adaptive filtering that gives reduced computational complexity. However, the main problem with recursive least square (RLS) family algorithms is the lack of robustness. To enhance the robustness

of the RLS algorithm, a modification, i.e., state-space RLS, is presented in [17, 18]. State-space recursive least square (SSRLS) exhibits excellent tracking performance due to its model-dependent state-space formulation but has not been tested in the ANC domain. Due to the presence of the secondary path $s(z)$ in Fig. 2, SSRLS cannot be used in its existing form. Therefore, in this paper, we have modified the SSRLS algorithm to track the filtered reference noise, making it suitable for ANC applications. The SSRLS algorithm is used in combination with filtered reference input and hence named as the filtered-x SSRLS algorithm.

The rest of the paper is organized as follows: Section 2 presents the proposed algorithm along with its associated schematics. Following this, the simulation results are shown in Section 3. Finally, the conclusions are drawn in Section 4.

2 Proposed algorithm

Consider the unforced discrete time system

$$r[n+1] = Ar[n] \tag{1}$$

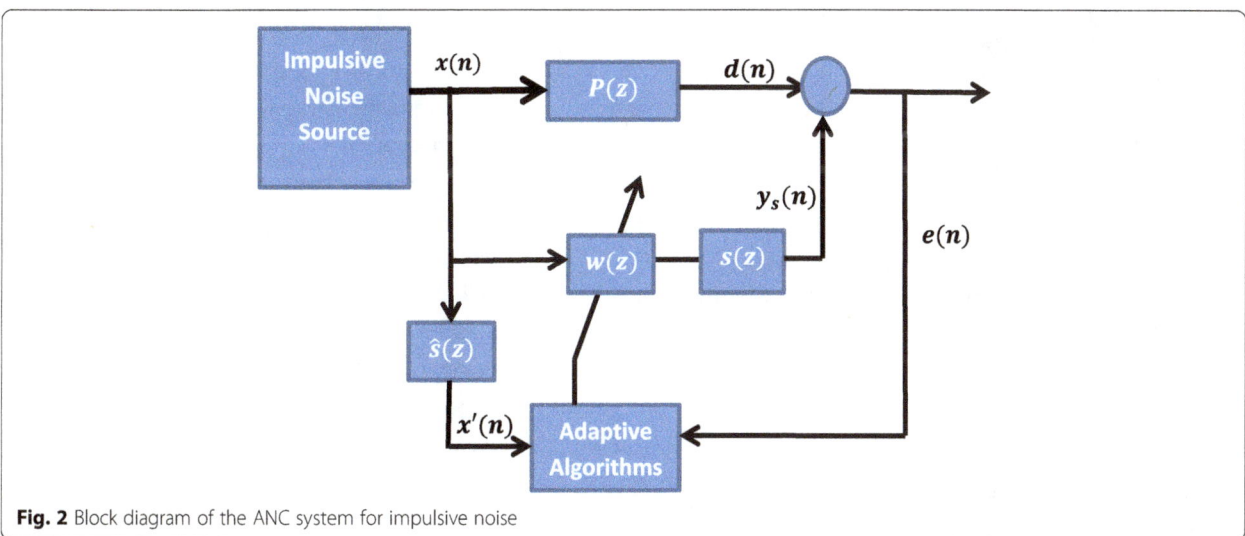

Fig. 2 Block diagram of the ANC system for impulsive noise

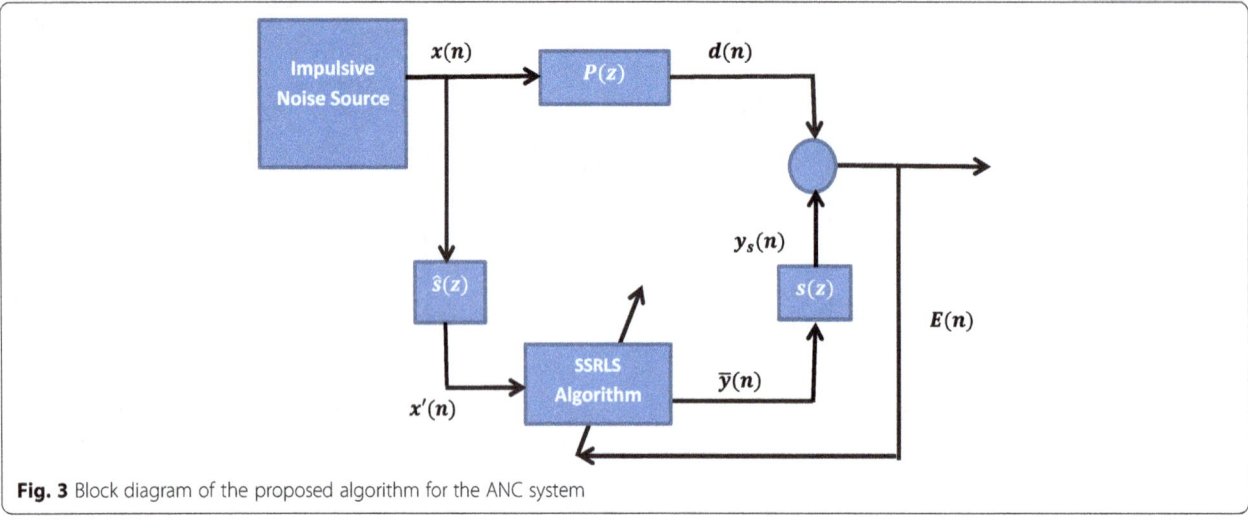

Fig. 3 Block diagram of the proposed algorithm for the ANC system

$$y[n] = Cr[n] \tag{2}$$

where r is the process states and y is the output, while A and C represent the system and the observation matrices, respectively. We assume that the pair (A, C) is L-step observable and A is invertible. The state-space formulation of SSRLS provides the designer with the freedom to choose an appropriate model for the underlying environment. However, the reasons that SSRLS cannot be used in its existing form for active noise control are as under:

1) SSRLS is designed for unforced system, i.e., system without input.
2) ANC applications have a secondary path $s(z)$ following the adaptive filter $w(z)$ as shown in Fig. 2.
3) Impulsive noise state-space model is unknown.

According to the required modifications in SSRLS for the ANC domain, the block diagram of the proposed algorithm is shown in Fig. 3.

In this figure, the reference noise signal vector is $x(n) = [x(n), x(n-1) \ldots \ldots x(n-L+1)]$ ', where L is the length of the reference noise. The desired signal $d(n)$ is calculated as

$$d(n) = P(n) * x(n) \quad \text{where opeartor} \ast \text{represents the convolution} \tag{3}$$

The filtered reference noise $x'(n) = [x'(n), x'(n-1) \ldots \ldots x'(n-L+1)]$ ' and the error signal $E(n)$ measured by the error microphone is

$$x'(n) = \hat{s}(n) * x(n) \tag{4}$$

$$E(n) = d(n) - y_s(n) \tag{5}$$

The $s(n)$ and $\hat{s}(n)$ are the impulse responses of the secondary path and its estimate, respectively.

Due to the presence of the secondary path following the adaptive filter, phase mismatch occurs between the desired signal and output of the filter as shown in Fig. 3, which consequently degrades the performance of the ANC system. Thus, for incorporating the effect of the secondary path, an identical filter is placed in the reference signal path leading to input of the filter. The modified output of the adaptive filter followed by the secondary path is given by

$$y_s(n) = s(n) * \bar{y}(n) \tag{6}$$

The filtered reference noise signal $x'(n)$ is passed to the SSRLS adaptive filter block which computes $\bar{y}^{\boxtimes}(n)$. The description of the parameters used for our modified filtered-x state-space recursive least square (FxSSRLS) algorithm are listed in Table 1.

Table 1 Detail of variables

Variables	Description
$P(z)$	Transfer function of the primary path
$S(z)$	Transfer function of the secondary path
$x(n)$	Reference impulsive noise
$S'(z)$	Estimated transfer function of the secondary path
$d(n)$	Desired signal
$x'(n)$	Filtered reference noise
$\hat{r}[n]$	Estimated states of the adaptive filter
$\bar{r}[n]$	Predicted states of the adaptive filter
$K[n]$	Gain of the adaptive filter
$\varepsilon[n]$	Prediction error of the adaptive algorithm
δ	Regularization parameter
λ	Forgetting factor
$\bar{y}(n)$	Predicted output of the adaptive filter
$y_s(n)$	Adaptive filter output followed by the secondary path
N	Total samples

Since the system given in (1) is for forced systems with $x'(n)$ being the input, the three special models of SSRLS in [17, 18] are modified in Table 2.

After modification of the SSRLS models, the next objective is to select the appropriate model for impulsive noise to get the best match of the underlying environment with the presumed model of SSRLS for achieving enhanced performance of SSRLS. The exact model for impulsive noise cannot be determined because of its random nature. As the higher order models can better approximate the abrupt changes in impulsive noise therefore, we have used an acceleration model in our application. The choice has been validated through extensive simulations. The proposed algorithm is summarized below.

2.1 Performance analysis and computational complexity

For adaptive filters, when a new algorithm is developed, it is important to carry out its performance analysis. Although the FxLMS algorithm has been widely used for the implementation of ANC applications, its convergence analysis is still an active area of research [19–21]. The inclusion of the secondary path in FxLMS makes its convergence analysis complex as compared to the standard LMS algorithm. Various attempts on derivation of theoretical convergence analysis for the FxLMS algorithm have been made with different simplified assumptions on inputs being single or multitonal, stationary or purely white, and secondary path being pure delay or moving average model, etc. [22, 23].

Proposed FxSSRLS Algorithm

Parameters:

Select A, C, N

Initialize δ, λ, T_s, $\bar{r}[0] = 0$, $\Phi[0] = \delta I + C^T C$, $K[0] = \Phi^{-1}(0)C^T$, $\varepsilon[0] = y[0] - \bar{y}[0]$

Computation:

   While {x(n)} available **do**

   $y(n) = p(n) * x(n)$

   $x'(n) = \hat{s}(n) * x(n)$

   $\hat{r}[n] = \bar{r}[n] + K[n]\varepsilon[n]$

   $\Phi[n] = \lambda\left(A^{-T}\Phi[n-1]\right)A^{-1} + C^T C$

   $K[n] = \Phi^{-1}[n]C^T$

   $\bar{r}[n] = A\hat{r}[n]$

   $\bar{y}[n] = C\bar{r}[n]$

   $y_s(n) = s(n) * \bar{y}(n)$

   $\varepsilon[n] = y[n] - y_s[n]$

   end while

Table 2 Modified state-space models of SSRLS

Sinusoidal model	Velocity model	Acceleration model
$A = \begin{bmatrix} cos(wT)sin(wT) \\ -sin(wT)cos(wT) \end{bmatrix}$	$A = \begin{bmatrix} 1 & T \\ 0 & 1 \end{bmatrix}$	$A = \begin{bmatrix} 1 & T & \frac{T^2}{2} \\ 0 & 1 & T \\ 0 & 0 & 1 \end{bmatrix}$
$C = [x'(n) \quad 0]$	$C = [x'(n) \quad 0]$	$C = [x'(n) \quad 0 \quad 0]$

The analysis of the SSRLS algorithm for the standard adaptive filter has been presented in [17, 18], which may be extended to perform theoretical analysis of the Fx version of the SSRLS algorithm for the ANC systems. This paper develops a modified SSRLS (FxSSRLS) algorithm for ANC of impulsive sources being modeled as symmetric α-stable (SαS) distributions. For stable distributions, the moments only exist for the order lesser than the characteristic exponent [6], i.e., for impulsive noise, second-order moments do not exist. The lower order moments are more difficult to compute than the second-order moments [24], which makes the theoretical analysis difficult, if not impossible. The non-Gaussian signal processing is in general much more complicated in terms of finding statistics than the Gaussian signal processing. This may be the reason that recent work on ANC of impulsive sources (being modeled as stable process) does not include the theoretical analysis, and in fact, the simulations have been used as a major tool to demonstrate the effectiveness of the proposal (see, for example, [7, 8, 11]). The interested reader may also look into the recent works on ANC [25–31]. Though simulations do not prove, they do demonstrate the effectiveness. In this paper, we have also used computer simulations as the evaluation tool and it is observed that the proposed algorithm outperforms the existing algorithms.

Moreover, computational complexity of an algorithm is usually of significant importance particularly in real-time applications. The complexity of individual equations of

the proposed FxSSRLS algorithm is given in Table 3, followed by the complexity analysis of other investigated algorithms in Tables 4, 5, 6, and 7.

Here, L represents the total number of states in the FxSSRLS algorithm while in other investigated algorithms, it represents the number of filter coefficients [17, 18]. M represents the secondary path, and N represents the data reuse order for the DR-NSSFxLMS algorithm. The computational complexities along with the memory requirements of the investigated algorithms are summarized in Table 8. The memory of the investigated algorithms is calculated using the method given in [32].

Figure 4 shows the plots for the computational complexities of the investigated algorithms. The proposed FxSSRLS algorithm has high computational complexity as compared to the FxLMS and FxRLS algorithm family, which makes it costly for few applications. Nevertheless, in the practical applications where stability and fast convergence is a matter of concern, the implementation of FxSSRLS in the ANC system can be easily handled by the latest DSPs.

3 Comparison with existing techniques and simulation results

The ANC system for impulsive noise is implemented using the MATLAB platform. The performance of the proposed algorithm is compared with that of the already reported adaptive algorithms in literature [5, 7–9, 13, 14]. The parameters used in simulating the ANC system are tabularized below.

In our simulation setup, the SαS distributions are used to model the statistical parameters of impulsive noise [6]. The analytical form for probability density functions (PDFs) of stable distributions does not exist, so they are normally expressed by their characteristic equations which is actually fourier transform of its PDF.

Table 3 Complexity analysis of the proposed algorithm

Equations	Operations	*	+/-	÷
1	$x'(n)_{1\times1} = s(n)_{1\times M} * x(n)_{M\times1}$	M	$M-1$	–
2	$r[n]_{L\times1} = \bar{r}[n]_{L\times1} + K[n]_{L\times1}\varepsilon[n]_{1\times1}$	L	L	–
3	$\Phi[n]_{L\times L} = \lambda(A^- T_{L\times L}\Phi[n-1]_{L\times L})A^{-1}_{L\times L} + CT_{L\times1}C_{1\times L}$	$2L^3 + 2L^2$	$2L^3 - 2L^2$	1
4	$K[n]_{L\times1} = \Phi^{-1}(n)_{L\times L}CT_{L\times1}$	L^2	$L^2 - L$	–
5	$\bar{r}[n]_{L\times1} = A_{L\times L}r[n]_{L\times1}$	L^2	$L^2 - L$	–
6	$\bar{y}[n]_{1\times1} = C_{1\times L}\bar{r}[n]_{L\times1}$	L	$L-1$	–
7	$y_s(n)_{M\times1} = s(n)_{M\times1} * \bar{y}(n)_{1\times1}$	M	–	–
8	$\varepsilon[n]_{1\times1} = y[n]_{1\times1} - y_s[n]_{1\times1}$	–	1	–
	Total	$2L^3 + 4L^2 + 2L + 2M$	$2L^3 + M$	1

Table 4 Complexity analysis of the FxLMS algorithm

Equations	Operations	*	+/-	÷
1	$x'(n)_{1\times1} = s\,(n)_{1\times M} * x(n)_{M\times1} * x(n)_{M\times1}$	M	$M-1$	–
2	$y(n)_{1\times1} = wT(n)_{1\times L} * x(n)_{L\times1}$	L	$L-1$	–
3	$w(n+1)_{L\times1} = w(n)_{L\times1} - \mu_{1\times1} * e(n)_{1\times1} * x'(n)_{1\times L}$	$L+1$	L	–
4	$e(n)_{1\times1} = d(n)_{1\times1} - y_s(n)_{1\times1}$	–	1	–
5	$y_s(n)_{1\times1} = s(n)_{1\times M} * y(n)_{M\times1}$	M	$M-1$	–
	Total	$2L+2M+1$	$2L+2M-2$	–

Table 5 Complexity analysis of the NSS-FxLMS algorithm

Equations	Operations	*	+/-	÷
1	$x'(n)_{1\times1} = s\,(n)_{1\times M} * x(n)_{M\times1} * x(n)_{M\times1}$	M	$M-1$	–
2	$y(n)_{1\times1} = wT(n)_{1\times L} * x(n)_{L\times1}$	L	$L-1$	–
3	$w(n+1)_{L\times1} = w(n)_{L\times1} - \mu(n)_{1\times1} * e(n)_{1\times1} * x'(n)_{1\times L}$	$L+1$	L	–
4	$\mu(n)_{1\times1} = \dfrac{\overline{\mu(n)_{1\times1}}}{\delta + x^T(n)_{1\times L} * x(n)_{L\times1} + E(n)}$	L	$L+1$	1
5	$e(n)_{1\times1} = d(n)_{1\times1} - y_s(n)_{1\times1}$	–	1	–
6	$y_s(n)_{1\times1} = s(n)_{1\times M} * y(n)_{M\times1}$	M	$M-1$	–
7	$E(n)_{1\times1} = \lambda E(n-1)_{1\times1} + (1-\lambda)E^2(n)_{1\times1}$	3	2	–
	Total	$3L+2M+4$	$3L+2M+1$	1

Table 6 Complexity analysis of the FxRLS algorithm

Equations	Operations	*	+/-	÷
1	$x'(n)_{1\times1} = s\,(n)_{1\times M} * x(n)_{M\times1} * x(n)_{M\times1}$	M	$M-1$	–
2	$y(n)_{1\times1} = wT(n)_{1\times L} * x(n)_{L\times1}$	L	$L-1$	–
3	$w(n+1)_{L\times1} = w(n)_{L\times1} + K(n)_{L\times L} * e(n)_{1\times1}$	L	L	–
4	$K(n)_{L\times1} = \dfrac{\pi(n)_{L\times1}}{\lambda + x'(n)_{L\times1} * \pi(n)_{L\times1}}$	$2L$	L	1
5	$\pi(n)L_{\times1} = p(n-1)_{L\times L} * x'(n)_{L\times1}$	L^2	L^2-L	–
6	$p(n)L_{\times}L = \lambda^{-1} * p(n-1)_{L\times L} - \lambda^{-1} * K(n)_{L\times1} * x'(n)_{1\times L} * p(n-1)_{L\times L}$	$3L^2$	$2L^2-L$	1
7	$e(n)_{1\times1} = d(n)_{1\times1} - y_s(n)_{1\times1}$	–	1	–
8	$ys(n)_{1\times1} = s(n)_{1\times M} * y(n)_{M\times1}$	M	$M-1$	–
	Total	$4L^2+4L+2M$	$3L^2+L+2M-2$	2

Table 7 Complexity analysis of the DR-NSSFxLMS algorithm

Equations	Operations	*	+/-	÷
1	$x'(n)_{1\times1} = s\,(n)_{1\times M} * x(n)_{M\times1} * x(n)_{M\times1}$	M	$M-1$	–
2	$y(n)_{1\times1} = w^T(n)_{1\times L} * x(n)_{L\times1}$	L	$L-1$	–
3	$d_1(n)_{1\times1} = e(n)_{1\times1} + s(n)_{1\times M} * y(n)_{M\times1}$	M	M	–
4	$e_1(n)_{1\times1} = d_1(n)_{1\times1} - w_1^T(n)_{1\times L} * x'(n)_{L\times1} * x'(n)_{L\times1}$	L	L	–
5	Compute $w_1(n+1)_{L\times1}$ using the DR algorithm in Table 2 from [25]	$N(3L+4)$	$N(3L+2)$	–
	Total	$2L+2M+N(3L+4)$	$2L+2M-2+N(3L+2)$	–

Table 8 Performance analysis of the investigated algorithms

Algorithm	Complexity		Memory
	Additions	Multiplications	
FxLMS	$2L + 2M - 2$	$2L + 2M + 1$	$2(L + M)$
NSS-FxLMS	$3L + 2M + 1$	$3L + 2M + 4$	$2(L + M)$
DR-NSSFxLMS	$2L + 2M - 2 + N(3L + 2)$	$2L + 2M + N(3L + 4)$	$(N + 2)L + 3M$
FxRLS	$3L^2 + L + 2M - 2$	$4L^2 + 4L + 2M$	$3L + 2M$
Proposed FxSSRLS	$2L^3 + M$	$2L^3 + 4L^2 + 2L + 2M$	$4L + 2M$

$$\varphi(t) = e^{-|t|^{\alpha}} \tag{7}$$

Some PDFs for SαS distributions are shown in Fig. 5. The SαS distributions have a characteristic exponent parameter α $(0 < \alpha < 2)$, which controls the spread of the PDF, i.e., a smaller value of α indicates that noise will be more impulsive with a heavier tail.

For the stable distributions, α ranges between 0 and 2. It is characterized as normal distribution for $\alpha = 2$, while the distribution is Cauchy for $\alpha = 1$. In Fig. 6, impulsive noise generated by the standard SαS process with $\alpha = 1.65$ is shown while the parameters used for simulating the impulsive noise are mentioned in Table 9.

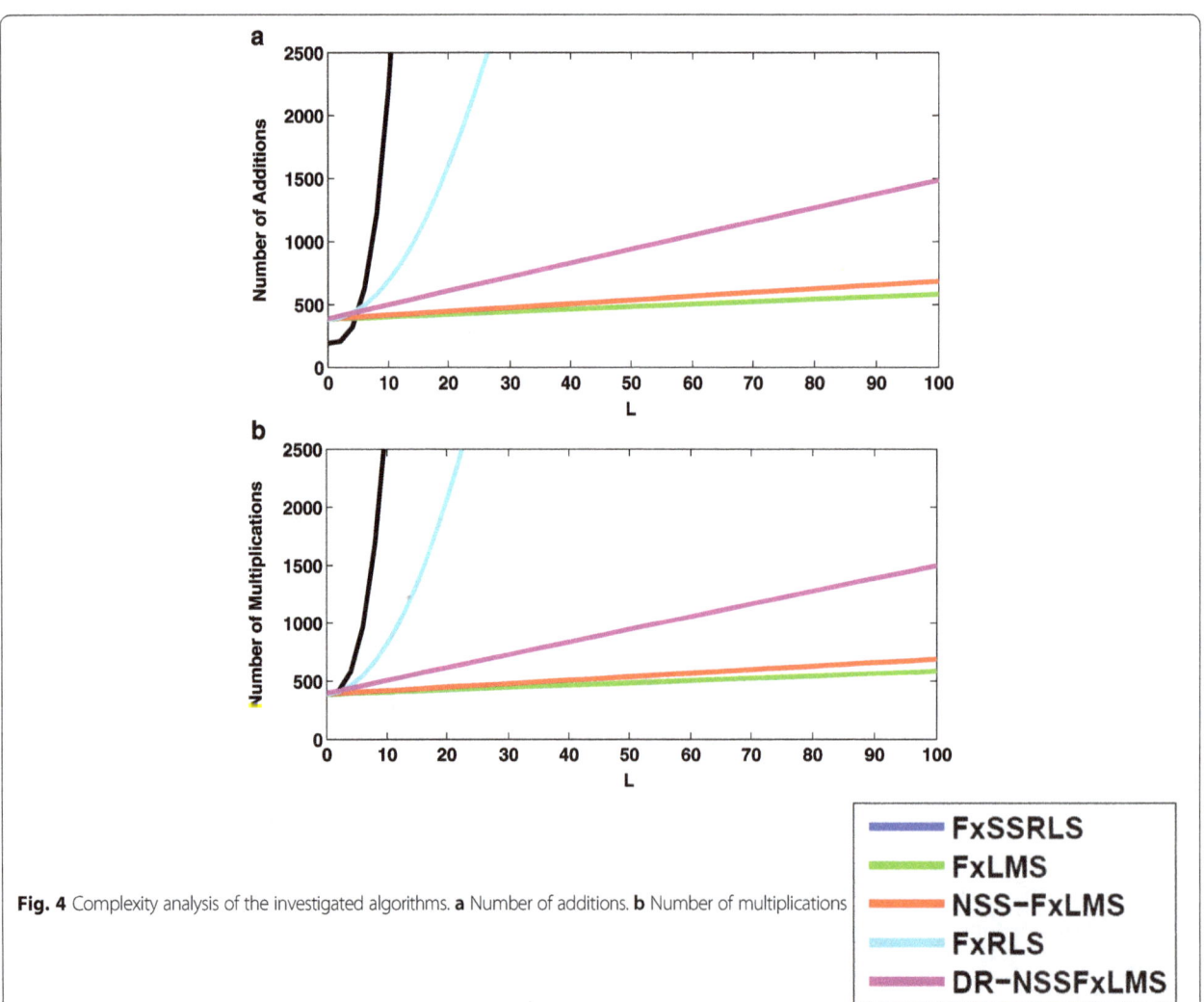

Fig. 4 Complexity analysis of the investigated algorithms. **a** Number of additions. **b** Number of multiplications

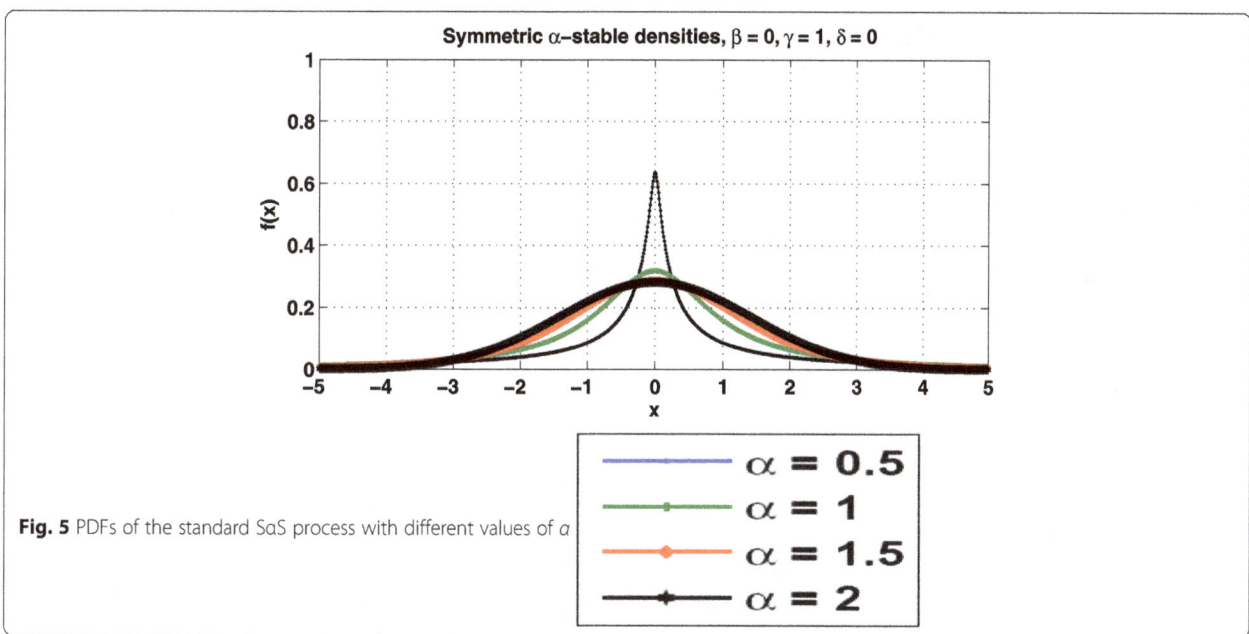

Fig. 5 PDFs of the standard SαS process with different values of α

The primary noise $d(n)$ for $\alpha = 1.65$ picked by the reference microphone is depicted in Fig. 7.

For the simplicity of our simulations, we have made an assumption that the estimated secondary path model $\hat{s}(z)$ is the same as $s(z)$ [13, 25, 30, 31]. The numeric values of the coefficients of the primary and secondary acoustic paths are taken from the data set given in [1]. The frequency response comprising of magnitude and phase of both path filters are depicted in Fig. 8.

The performance metric used in this research for comparison of the studied algorithms is mean noise reduction. It is calculated as

$$\text{MNR}(n) = E\left\{\frac{A_e(n)}{A_d(n)}\right\} \tag{8}$$

$$A_e(n) = \lambda A_e(n-1) + (1-\lambda)|e(n)| \tag{9}$$

$$A_d(n) = \lambda A_d(n-1) + (1-\lambda)|d(n)| \tag{10}$$

where $A_e(n)$ and $A_d(n)$ are the estimates of the absolute value of the residual error and disturbance signal, respectively.

In this section, we have validated the performance of our proposed algorithm for ANC of impulsive noise. The impulses for the research are generated by the symmetric alpha-stable model by considering $\alpha = 1.85$, $\alpha = 1.65$, and $\alpha = 1.45$, respectively, which corresponds to a small, mild, and heavy impulsiveness. Extensive simulations are carried out to find the optimum values of controlling the parameters of the discussed algorithms. The detailed simulation results for the step-size parameter of the NSS-FxLMS and DR-NSSFxLMS algorithms for $\alpha = 1.65$ are illustrated in

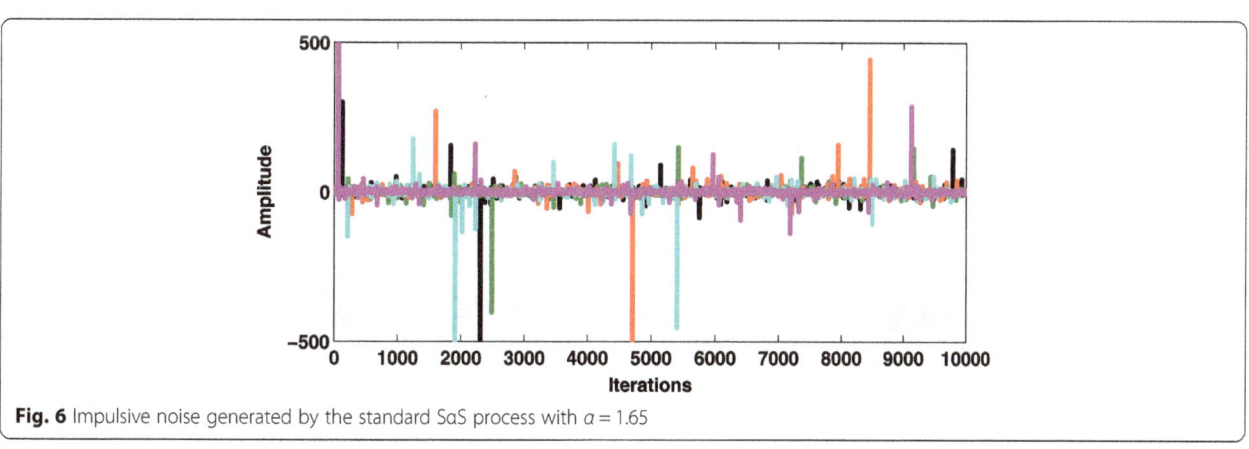

Fig. 6 Impulsive noise generated by the standard SαS process with $\alpha = 1.65$

Table 9 Parameter set for the proposed technique simulation

ANC system			Impulsive noise		
Parameters	Symbols	Values	Parameters	Symbols	Value
Primary path tap size	L	256	Total samples	N	10,000
Secondary path tap size	M	128	Total realizations	Avg	1
Adaptive filter tap size	L_w	192	Characteristic exponent	a	1.65
DR-NSSFxLMS algorithm step size	$\bar{\mu}$	5e−2	Scale parameter	γ	1
NSS-FxLMS algorithm step size	$\bar{\mu}$	5e−2	Location parameter	C	0
RLS forgetting factor	λ	0.99	Skewness parameter	δ	0
SSRLS forgetting factor	λ	1			

Fig. 9. It can be observed from Fig. 9 a, b that the optimum step-size value for both algorithms is 5e−2.

Similarly, the effect of the regularization parameter delta (δ) of the FxRLS algorithm is shown in Fig. 10 for $\alpha = 1.65$. The parameter δ depends on the signal-to-noise ratio (SNR) [5], i.e., the greater the value of the SNR, the smaller the value of delta is selected for better performance of algorithms and vice versa. The optimum value selected for further simulation is 100,000 for $\alpha = 1.65$.

Figure 11 depicts the convergence curves of the most widely used adaptive algorithms in the ANC domain for $\alpha = 1.65$. The optimum step sizes for the FxLMS, Sun, modified Sun, and Akhtar algorithms used in this simulation are 1e−3, 5e−6, 5e−5, and 5e−5, respectively. It can be seen that among the investigated algorithms of the LMS family, the NSS-FxLMS and DR-NSSFxLMS algorithms converge quickly after 1000 iterations and give good noise reduction by achieving the lowest mean noise reduction (MNR) as compared to the other investigated algorithms.

Similarly, the DR-NSSFxLMS algorithm is comparatively less affected by the occurrence of impulsive noise at different iterations and thus exhibits better stability. Therefore, we have selected the NSS-FxLMS and DR-NSSFxLMS algorithms for further comparison with our proposed FxSSRLS algorithm which can be visualized in Fig. 12.

It can be noticed from Fig. 12 that the NSS-FxLMS and DR-NSSFxLMS algorithms give slow convergence as compared to the FxRLS and FxSSRLS algorithms that achieve a steady state value at about 2000 and 500 iterations, respectively. The convergence curves of the FxSSRLS and FxRLS algorithms almost overlap after 3500 iterations. However, when an impulse is encountered at about 800 iterations, the FxRLS algorithm exhibits a sudden increase in MNR while the FxSSRLS algorithm is robust enough to remain unaffected.

Similarly, other simulation cases with SαS impulsive noises of $\alpha = 1.45$ and 1.85 were conducted to validate the effectiveness of the proposed FxSSRLS algorithm.

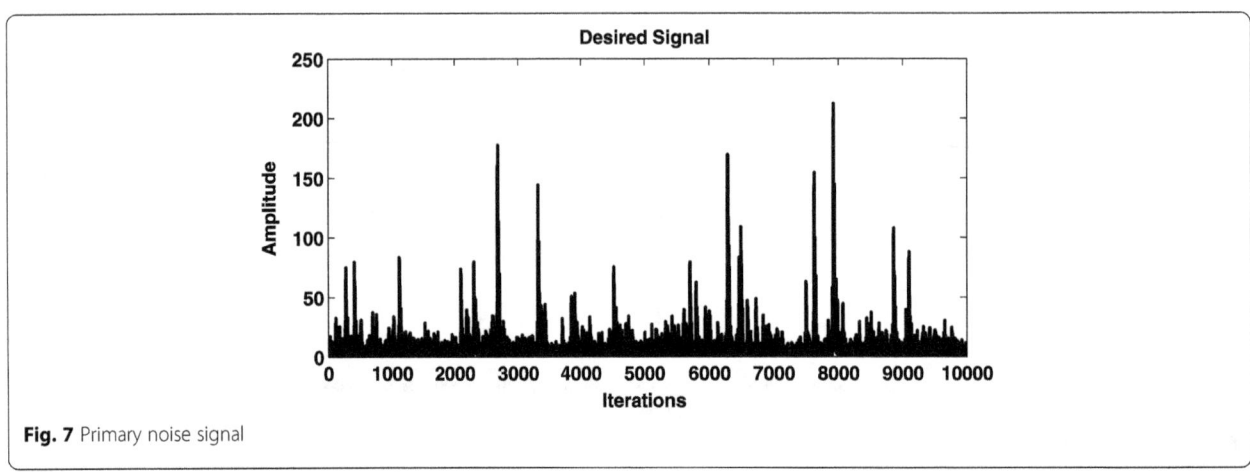

Fig. 7 Primary noise signal

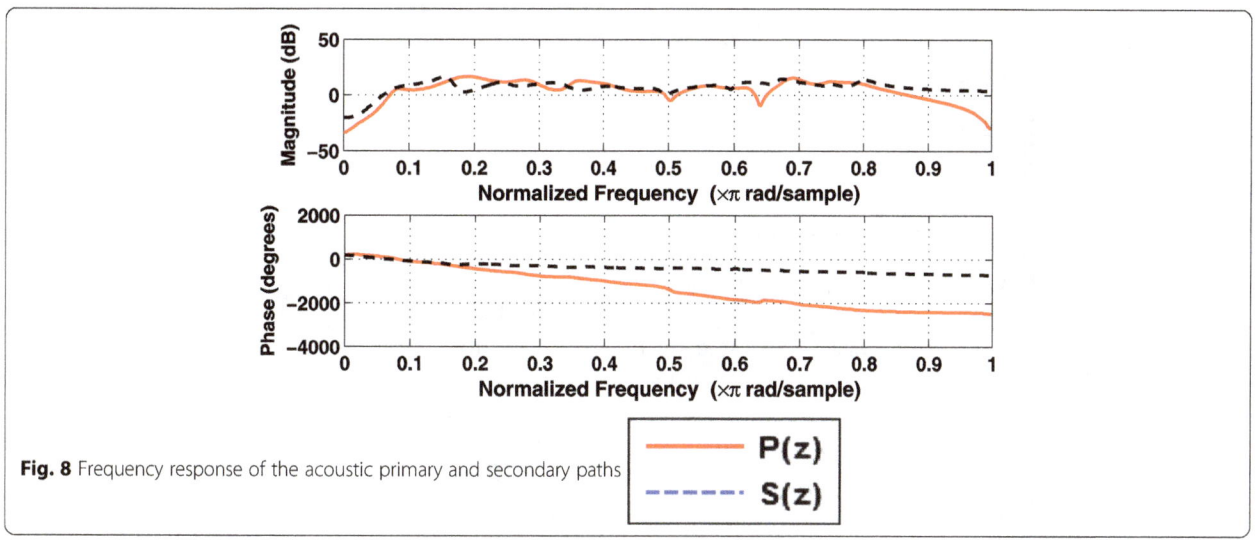

Fig. 8 Frequency response of the acoustic primary and secondary paths

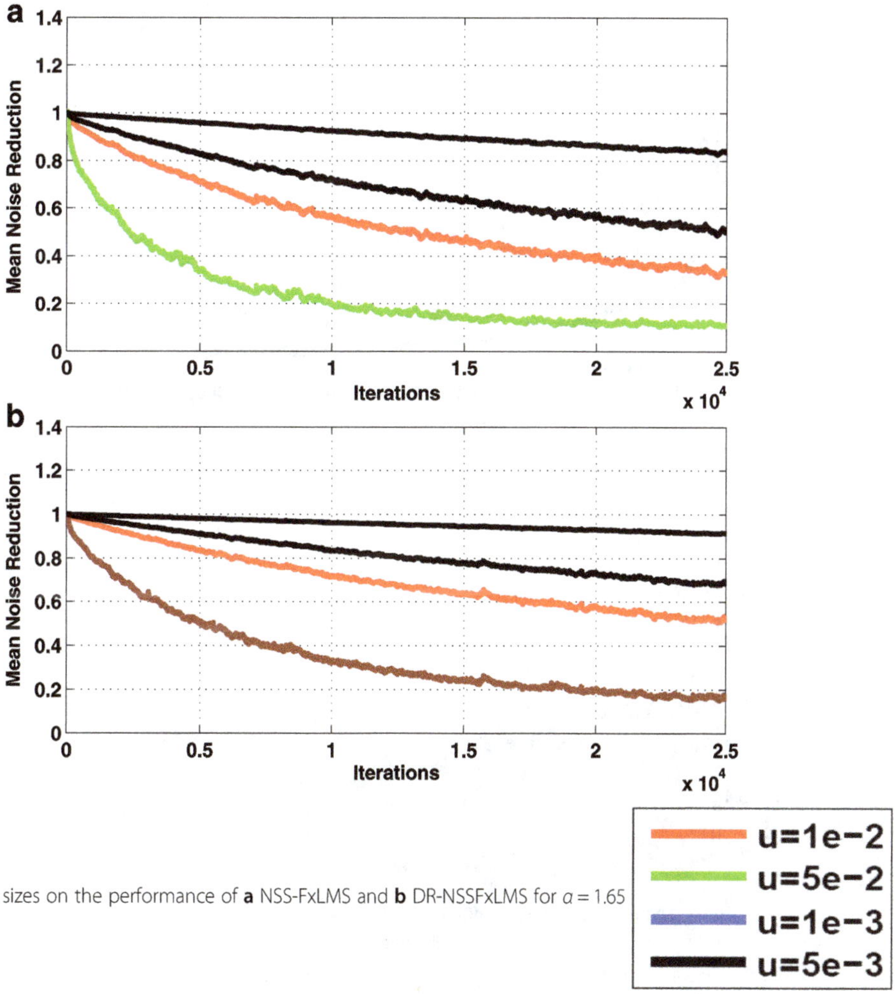

Fig. 9 Effect of varying step sizes on the performance of **a** NSS-FxLMS and **b** DR-NSSFxLMS for $a = 1.65$

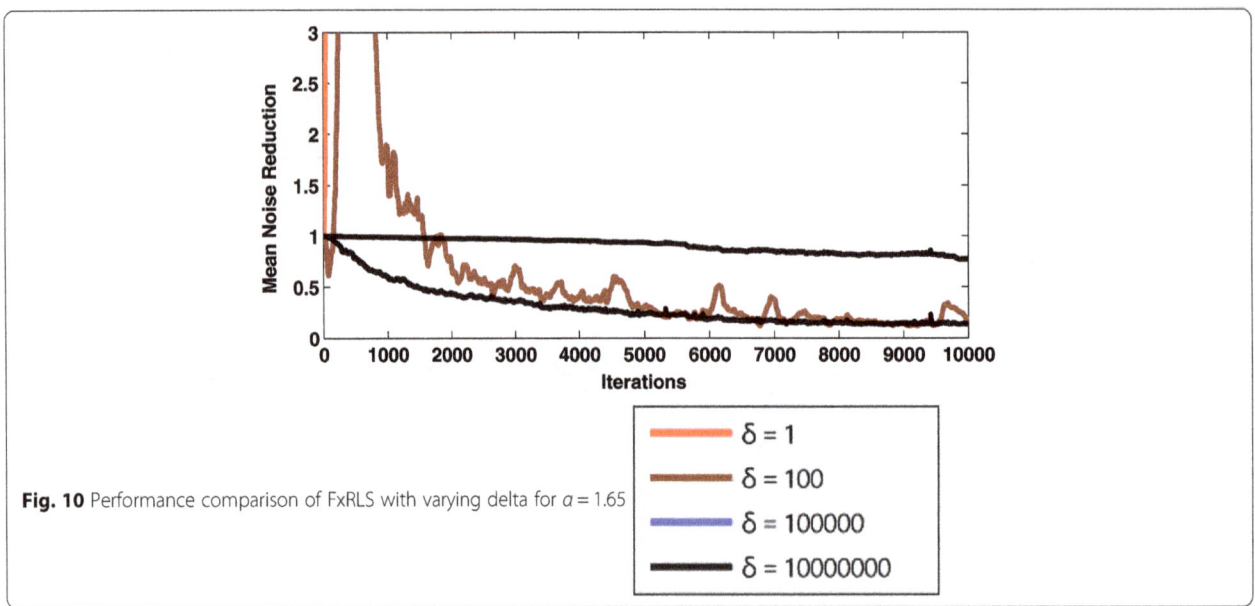

Fig. 10 Performance comparison of FxRLS with varying delta for $a = 1.65$

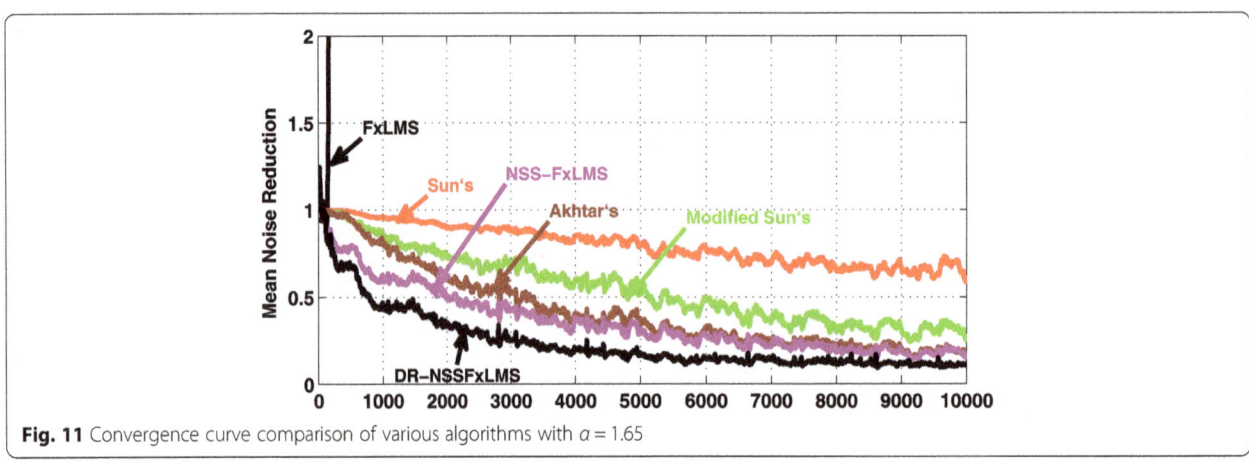

Fig. 11 Convergence curve comparison of various algorithms with $a = 1.65$

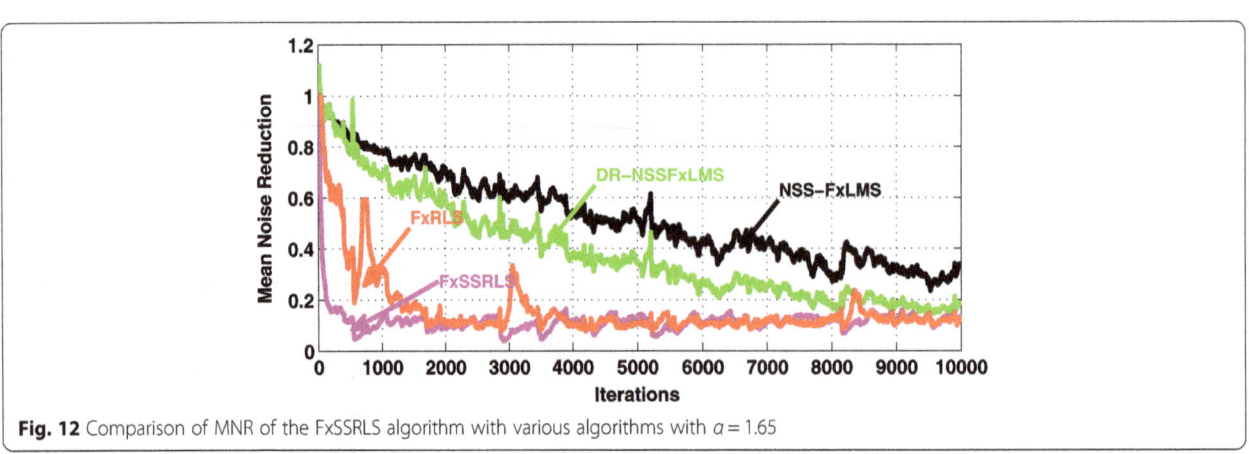

Fig. 12 Comparison of MNR of the FxSSRLS algorithm with various algorithms with $a = 1.65$

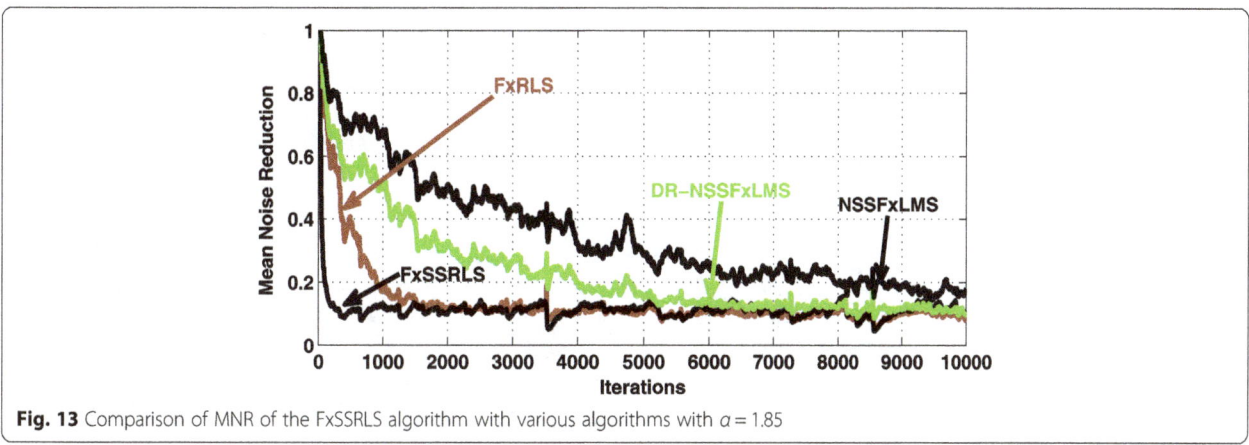

Fig. 13 Comparison of MNR of the FxSSRLS algorithm with various algorithms with $a = 1.85$

It is noticed that the proposed FxSSRLS algorithm demonstrates its improved performance for other selected values of impulsive noise over the investigated algorithms. As shown in Figs. 13 and 14, the steady state performance of the FxRLS and proposed FxSSRLS algorithms is better than that of the NSS-FxLMS and DR-NSSFxLMS algorithms. Also, the proposed FxSSRLS can yield improved convergence rate and robustness even in the presence of large impulses than that of the FxRLS, NSS-FxLMS, and DR-NSSFxLMS algorithms, thus making our proposed solution an excellent choice for mitigating the influence of impulses in ANC applications.

4 Conclusions

In this paper, we have analyzed non-Gaussian impulsive noise in the ANC domain. The adaptive algorithms employed in ANC applications become unstable and lack robustness in the presence of impulsive noise. To overcome this limitation in ANC applications, a new algorithm FxSSRLS has been developed and presented in this paper. Due to the recursive parameters of the proposed adaptive algorithm, the reduction in impulsive noise has been achieved, which has been further enhanced by the state-space formulation of the SSRLS models. To validate this improved performance of the newly suggested solution, extensive numerical simulations have been carried out. The results show that with the use of the presented algorithm for ANC, the large amplitude impulses have been significantly reduced. Moreover, the suggested algorithm for ANC applications outperforms the existing algorithms in terms of mean noise reduction, convergence, and stability. However, this improved performance has been achieved at the cost of slight increase in computational complexity. In applications where stability and fast convergence is a matter of concern, the little price paid in terms of computational complexity can be ignored.

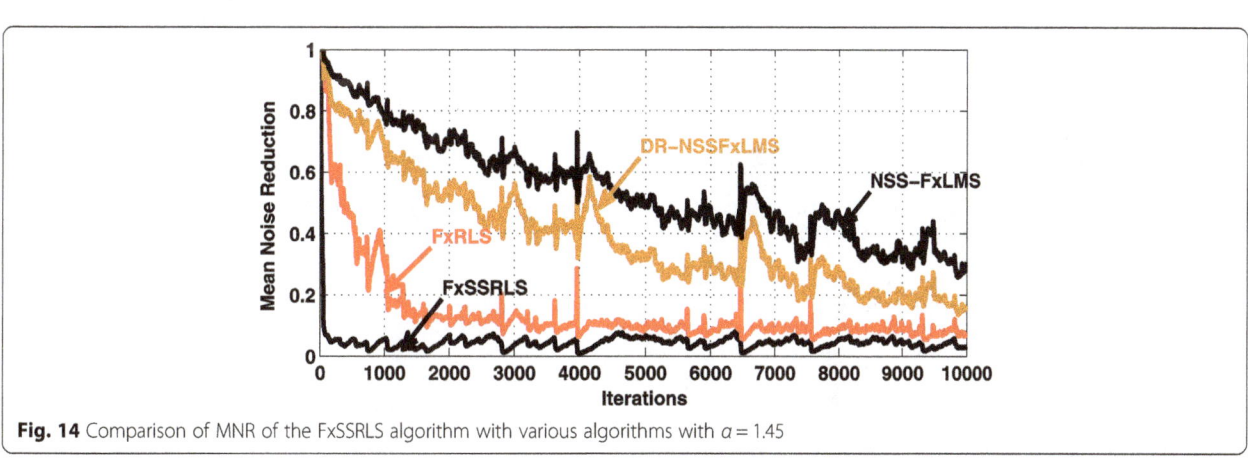

Fig. 14 Comparison of MNR of the FxSSRLS algorithm with various algorithms with $a = 1.45$

Abbreviations

ANC: active noise control; DR: data reuse; FxLMS: filtered-x least mean square; FxSSRLS: filtered-x state-space recursive least square; MNR: mean noise reduction; NSS-FxLMS: normalized step-size filtered-x least mean square; PDF: probability density function; RLS: recursive least square; SNR: signal-to-noise ratio; SSRLS: state-space recursive least square; SaS: symmetric α-stable.

Competing interests

The authors declare that they have no competing interests.

Acknowledgements

The authors would like to thank their colleague Qasim Umer Khan for improving the content of this paper. The authors would also like to thank the anonymous reviewers for their valuable suggestions and comments.

References

1. SM Kuo, DR Morgan, *Active Noise Control Systems—Algorithms and DSP Implementation* (Wiley, New York, 1996)
2. M Zimmermann, K Dostert, Analysis and modeling of impulsive noise in broad-band powerline communications. Electromag. Compat. IEEE Trans. **44**, 249–258 (2002). doi:10.1109/15.990732
3. L Lui, S Gujjula, P Thanigai, SM Kuo, Still in womb: intrauterine acoustic embedded active noise control for infant incubators. Adv. Acoust Vib (2008). doi:10.1155/2008/495317
4. YI Zhou, YX Yin, QZ Zhang, An optical repetitive control algorithm for periodic impulsive noise attenuation in a non-minimum phase ANC system. Appl. Acoust. **74**, 1175–1181 (2013). doi:10.1016/j.apacoust.2013.04.008
5. S Haykin, *Adaptive Filter Theory*, 4th edn. (Prentice-Hall, Englewood Cliffs, NJ, 2001)
6. CL Nikias, M Shao, *Signal Processing with Alpha-Stable Distribution and Applications* (Wiley, New York, USA, 1995)
7. X Sun, SM Kuo, M Guang, Adaptive algorithm for active noise control of impulsive noise. J. Sound Vib. **291**, 516–522 (2006). doi:10.1016/j.jsv.2005.06.011
8. MT Akhtar, W Mitsuhashi, Improving performance of FXLMS algorithm for active noise control of impulsive noise. J. Sound Vib. **327**, 647–656 (2009). doi:10.1016/j.jsv.2009.07.023
9. M.T. Akhtar, W. Mitsuhashi, A modified normalized FxLMS algorithm for active control of impulsive noise. Paper presented at the 18th European signal processing conference, Aalborg, Denmark, 23-27 August 2010
10. L Wu, H He, X Qiu, An active impulsive noise control algorithm with logarithmic transformation. IEEE Trans. Audio Speech Lang. Process. **19**, 1041–1044 (2011)
11. M Bergamasco, FD Rossa, L Piroddi, Active noise control with online estimation of non-Gaussian noise characteristics. J. Sound Vib. **331**, 27–40 (2012). doi:10.1016/j.jsv.2011.08.025
12. P Li, X Yu, Active noise cancellation algorithms for impulsive noise. Mech. Syst. Signal Process. **36**, 630–635 (2013). doi:10.1016/j.ymssp.2012.10.017
13. MT Akhtar, A Nishihara, Data reusing based filtered reference adaptive algorithms for active control of impulsive noise sources. Appl. Acoust. **92**, 18–26 (2015). doi:10.1016/j.apacoust.2015.01.006
14. A Zeb, A Mirza, SA Sheikh, FxRLS algorithm based ANC of impulsive noise. Paper presented at the 7th international conference on modelling, identification and control, Sousse, Tunisia, 18–20 December 2015
15. F Albu, Leading element dichotomous coordinate descent exponential recursive least squares algorithm for multichannel active noise control, Proc. of AAS Acoust. 21–23 November 2012
16. F Albu, C Paleologu, A recursive least square algorithm for active noise control based on the Gauss-Seidel method. Paper presented at the 15th IEEE international conference on electronics, circuits and systems, August 2008
17. MB Malik, State-space recursive least squares: part I. Signal Process. **84**, 1709–1718 (2004). doi:10.1016/j.sigpro.2004.05.022
18. MB Malik, State-space recursive least squares: part II. Signal Process. **84**, 1709–1718 (2004). doi:10.1016/j.sigpro.2004.05.021
19. LT Ardekani, WH Abdulla, Stochastic modelling and analysis of filtered x-least square adaptive algorithm. IET Signal Process. **7**, 486–496 (2013). doi:10.1049/iet-spr.2012.0090
20. G Sun, T Feng, M Li, TC Lim, Convergence analysis of filtered-x least mean squares algorithm for active control of repetitive impact noise. J. Acoust. Soc. Am. **134**, 4190–4190 (2013). doi:10.1121/1.4831368
21. G Sun, T Feng, M Li, TC Lim, Convergence analysis of FxLMS-based active noise control for repetitive impulses. Appl. Acoust. **89**, 178–187 (2015). doi:10.1016/j.apacoust.2014.09.026
22. LT Ardekani, WH Abdulla, Effects of imperfect secondary path modeling on adaptive active noise control systems. IEEE Trans. Control Syst. Technol. **20**, 1252–1262 (2012). doi:10.1109/TCST.2011.2161762
23. LT Ardekani, WH Abdulla, Theoretical convergence analysis of FxLMS algorithm. Signal Process. **90**, 3046–3055 (2010). doi:10.1016/j.sigpro.2010.05.009
24. M Shao, CL Nikias, Signal processing with fractional lower order moments: stable processes and their applications. Proc. IEEE **81**, 986–1010 (1993). doi:10.1109/5.231338
25. MT Akhtar, Binormalized data-reusing adaptive filtering algorithm for active control of impulsive sources. Digital Signal Process. **49**, 56–64 (2015). doi:10.1016/j.dsp.2015.11.002
26. Z Bo, C Sun, Y Xu, S Jiang, A variable momentum factor filtered-x weighted accumulated LMS algorithm for narrow band active noise control systems. Measurement **48**, 282–291 (2014). doi:10.1016/j.measurement.2013.11.010
27. SB Behera, DP Das, NK Rout, Nonlinear feedback active noise control for broad band chaotic noise. Appl. Soft Comput. **15**, 80–87 (2014). doi:10.1016/j.asoc.2013.10.025
28. L Wu, X Qui, IS Burnett, Y Guo, A recursive least square algorithm for Active control of mixed noise. J. Sound Vib. **339**, 1–10 (2015). doi:10.1016/j.jsv.2014.11.002
29. L Wu, X Qui, An M-estimator based algorithm for active impulse like noise control. Appl. Acoust. **74**, 407–412 (2013). doi:10.1016/j.apacoust.2012.06.019
30. Y Zhou, Q Zhang, Y Yin, Active control of impulsive noise with symmetric α-stable distribution based on an improved step-size normalized adaptive algorithm. Mech. Syst. Signal Process. **56**, 320–339 (2015). doi:10.1016/j.ymssp.2014.10.002
31. G Sun, M Li, TC Lim, A family of threshold based robust adaptive algorithms for active impulsive noise control. Appl. Acoust. **97**, 30–36 (2015). doi:10.1016/j.apacoust.2015.04.003
32. SM Kuo, I Panahi, KM Chung, T Horner, M Nadeski, J Chyan, *Design of Active Noise Control Systems with the TMS320 family* (Texas Instruments, USA, 1996)

3

Multi-camera object tracking using surprisal observations in visual sensor networks

Venkata Pathuri Bhuvana[1,2*], Melanie Schranz[1], Carlo S. Regazzoni[2], Bernhard Rinner[1], Andrea M. Tonello[1] and Mario Huemer[3]

Abstract

In this work, we propose a multi-camera object tracking method with surprisal observations based on the cubature information filter in visual sensor networks. In multi-camera object tracking approaches, multiple cameras observe an object and exchange the object's local information with each other to compute the global state of the object. The information exchange among the cameras suffers from certain bandwidth and energy constraints. Thus, allowing only a desired number of cameras with the most informative observations to participate in the information exchange is an efficient way to meet the stringent requirements of bandwidth and energy. In this paper, the concept of surprisal is used to calculate the amount of information associated with the observations of each camera. Furthermore, a surprisal selection mechanism is proposed to facilitate the cameras to take independent decision on whether their observations are informative or not. If the observations are informative, the cameras calculate the local information vector and matrix based on the cubature information filter and transmit them to the fusion center. These cameras are called as surprisal cameras. The fusion center computes the global state of the object by fusing the local information from the surprisal cameras. Moreover, the proposed scheme also ensures that on average, only a desired number of cameras participate in the information exchange. The proposed method shows a significant improvement in tracking accuracy over the multi-camera object tracking with randomly selected or fixed cameras for the same number of average transmissions to the fusion center.

Keywords: Kalman filters, Information filters, State estimation, Information entropy

1 Introduction

Object tracking is an extensively studied topic in visual sensor networks (VSN). A VSN is a network composed of smart cameras; they capture, process, and analyze the image data locally and exchange extracted information with each other [1]. The main applications of a VSN are indoor and/or outdoor surveillance, e.g., airports, massive waiting rooms, forests, deserts, inaccessible locations, and natural environments [2]. In general, the typical task of a VSN is to detect and track specific objects. The objects are usually described by a state that includes various characteristics of the objects such as position, velocity, appearance, behavior, shape, and color. These states can be used to detect and track the objects. Recursive state estimation algorithms are predominantly used to track objects in a VSN [3].

In [4–11], the authors presented several Kalman filter (KF)-based object tracking methods. Extended Kalman filter (EKF)-based object tracking method is proposed in [12]. The unscented Kalman filter (UKF) is applied for visual contour tracking in [13] and object tracking in [14]. In terms of object tracking in a VSN, the cubature Kalman filter (CKF) is primarily applied in our previous work [15]. In [16–24], the authors presented particle filter (PF)-based object tracking. The object tracking methods based on these conventional Bayesian filters have a varying degree of complexity and accuracy.

---

*Correspondence: venkata.pathuri@aau.at
[1] Institute of Networked and Embedded Systems, Alpen-Adria-Universität, Klagenfurt, Austria
[2] Department of Marine Engineering, Electrical, Electronics, and Telecommunications, University of Genova, Genova, Italy
Full list of author information is available at the end of the article

In general, the performance of the tracking algorithms suffers from different adverse effects such as distance or orientation of the camera, and occlusions. However, a VSN with overlapping field of views (FOVs) is capable of providing multiple observations of the same object simultaneously. The authors in [25] presented a distributed and collaborative sensing mechanism to improve the observability of the objects by dynamically changing the camera's pan, tilt, and zoom. Other examples of distributed object tracking methods are presented in [26] and [27].

Recently, information filters have emerged as suitable methods for multi-sensor state estimation [28]. In information filtering, the information vector and matrix are computed and propagated over time instead of the state vector and its error covariance. The information matrix is the inverse of the state error covariance matrix. The information vector is the product of the information matrix and state vector. The information filters have an inherent information fusion mechanism which makes them more suitable for multi-camera object tracking. A more detailed description of information filters is given in Section 3. The authors in [29] and [30] presented information weighted consensus-based distributed object tracking with an underlying KF or a distributed maximum likelihood estimation. In our work [31], we have presented a robust cubature information filter (CIF)-based distributed object tracking in VSNs. However, the limited processing, communication, and energy capabilities of the cameras in a VSN present a major challenge.

Nowadays, VSNs tend to evolve into large-scale networks with limited bandwidth and energy reservoirs. This allows a large number of cameras to observe a single object. In spite of the improved tracking accuracy, the information exchange of the large number of observations among the cameras increases the communication overhead and energy consumption. Hence, allowing only a desired number of cameras to participate in the information exchange is a way to meet the stringent requirements of bandwidth and energy.

Estimating an object's state with a selected set of cameras is a well-investigated topic. Several camera selection mechanisms have been proposed in literature to minimize and/or maximize different metrics such as estimation accuracy, monitoring area, number of transmissions, and amount of data transfer. In [32], the authors presented an object tracking method based on fuzzy automaton in handing over to expand the monitoring area. This method selects a single best camera to control and track the objects by comparing its rank with the neighboring cameras. This method fails to select multiple cameras, and cameras have to communicate with each other to select the best camera. In [33], the authors presented an efficient camera-tasking approach to minimize the visual hull area (maximal area that could be occupied by objects) for a given number of objects and cameras. They also presented several methods to select a subset of cameras based on the positions of the objects and cameras to minimize the visual hull area. If the objects are recognized in the vicinity of a certain location, then a subset of cameras that is best suited to observe this location performs the tracking. This method is capable of selecting multiple cameras but not the desired number of cameras on average. In [34], the authors presented a framework for dynamically selecting a subset of cameras to track people in a VSN with limited network resources to achieve the best possible tracking performance. However, the camera selection decision is made at the FC based on training data and the selection is broadcast to the cameras in the VSN. Hence, this selection process does not depend on the true observations.

The observations received by the cameras in the VSN are typically realizations of a random variable. Hence, they contain a varying degree of information about the state of the object. They can be broadly classified into informative and uninformative observations. The non-informative observations do not contribute significantly to the tracking accuracy. Hence, a camera selection strategy that allows only a desired number of cameras with most informative observations to participate in the information exchange and discards the cameras with non-informative observations is an efficient way to meet the requirements of bandwidth and energy.

In [35], the authors presented an entropy-based algorithm that dynamically selects multiple cameras to reduce transmission errors and subsequently communication bandwidth. In this work, the cameras in the VSN use the extended information filter (EIF) as the local filter and calculate the expected information gain (EIG) in the form of a logarithmic ratio of the expected and posterior information matrices. If the information gain is greater than the cost of transmissions, then the cameras participate in the information fusion. The calculated EIG in this method does not depend on the measurements directly, and the cluster head has to run an optimization step to select the best possible cameras at each step. Moreover, this method is not capable of selecting only a desired number of cameras on average. In [36], a camera set is selected based on an individual image quality metric (IQM) for spherical objects. The cameras that detect the spherical target are ranked in ascending order based on their value of the local IQM, and the required number of cameras with highest IQM are chosen. This approach is limited to spherical objects. However, it can be easily extended to non-spherical objects. The major disadvantage of this method is either all the cameras in the VSN or the FC should know IQM of all the other cameras in the VSN. Hence, this method does not ensure cameras to take independent decisions thus restricting the scalability.

In our work, a multi-camera object tracking method based on the CIF is proposed in which the cameras can take independent decisions on whether or not to participate in information exchange. Furthermore, the proposed method also ensures that on average, only a desired number of cameras participate in the information exchange to meet bandwidth requirements. We model the state of an object utilizing a dynamic state representation that includes its position and velocity on the ground plane. Further, we consider a VSN with overlapping FOVs; thus, multiple cameras can observe an object simultaneously. Each camera in the VSN has a local CIF on board. Hence, they can calculate the local information metrics (information contribution vector and matrix) based on their observations. The cameras that can observe a specific object form a cluster (observation cluster) with an elected fusion center (FC). In this paper, we consider the concept of surprisal [37] to evaluate the amount of information in the observations received by the cameras in the VSN. The surprisal of the measurement residual indicates the amount of new information received from the corresponding observation. The observations of a camera are informative only if the corresponding surprisal of the measurement residual is greater than a threshold. The threshold is calculated as a function of the ratio of the number of desirable cameras and the total number of cameras in the observation cluster. This ensures that on average, only the desired number of cameras are selected as the cameras with informative observations (surprisal cameras). The surprisal cameras calculate the local information metrics based on the CIF and transmit them to the FC. Then, the FC fuses the surprisal local information metrics

to achieve the global state by using the inherent fusion mechanism of the CIF. The proposed selection mechanism only requires the knowledge of the total number of cameras in the observation cluster and the desired number of cameras. Further, we compare the proposed multi-camera object tracking method with surprisal cameras with multi-camera object tracking with random and fixed cameras using simulated and experimental data.

The paper is organized as follows: Section 2 describes the considered VSN with motion and observation models. Section 3 presents theoretical concepts of information filtering. Section 4 describes the camera selection based on the surprisal of the measurement residual and the calculation of the surprisal threshold. Section 5 explains the proposed CIF-based multi-camera object tracking with surprisal cameras. Section 6 evaluates the proposed method based on simulation and experimental data. Finally, Section 7 presents the conclusions.

2 System model

In this work, we consider a VSN consisting of a fixed set of calibrated smart cameras c_i, where $i \in \{1, 2, \cdots, M\}$, with overlapping FOVs as illustrated in Fig. 1. The task of the cameras in the VSN is to monitor the given environment and to identify and track an object. As these cameras are calibrated, there exists a homography to calculate the object's position on the ground plane. The cameras c_i that can observe the object at time k form the observation cluster C_k. The state of the object comprises its position (x_k, y_k) and the velocity $(\dot{x}_k, \dot{y}_k)$ on the ground plane. Thus, the state at time k is described as $\mathbf{x}_k = \begin{bmatrix} x_k \ y_k \ \dot{x}_k \ \dot{y}_k \end{bmatrix}^T$. The motion model of the object at camera c_i at time k is given as

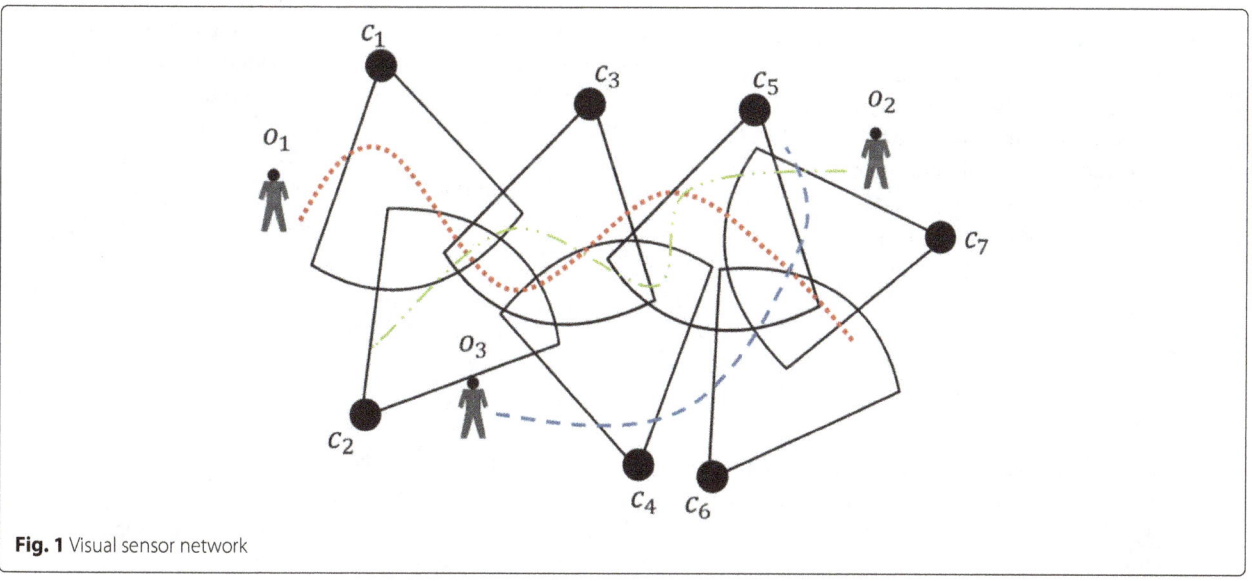

Fig. 1 Visual sensor network

$$\mathbf{x}_k = \mathbf{f}_{i,k}\left(\mathbf{x}_{k-1}, \mathbf{w}_{i,k}\right)$$

$$= \begin{bmatrix} x_{k-1} + \delta \dot{x}_{k-1} + \frac{\delta^2}{2}\ddot{x}_{i,k} \\ y_{k-1} + \delta \dot{y}_{k-1} + \frac{\delta^2}{2}\ddot{y}_{i,k} \\ \dot{x}_{k-1} + \delta \ddot{x}_{i,k} \\ \dot{x}_{k-1} + \delta \ddot{y}_{i,k} \end{bmatrix}, \qquad (1)$$

where $\ddot{x}$ and $\ddot{y}$ represent the acceleration of the object in x and y directions that are modeled by the independent and identically distributed (IID) white Gaussian noise vector $\mathbf{w}_{i,k} = [\ddot{x}_{i,k}\ \ddot{y}_{i,k}]^T$ with covariance $\mathbf{Q}_{i,k} = \mathrm{diag}\,(qx_i, qy_i)$. δ is time interval between two observations. The state transition model (1) can be further written as

$$\mathbf{x}_k = \begin{bmatrix} 1 & 0 & \delta & 0 \\ 0 & 1 & 0 & \delta \\ 0 & 0 & 1 & 0 \\ 0 & 0 & 0 & 1 \end{bmatrix} \mathbf{x}_{k-1} + \mathbf{w}_{i,k}^s, \qquad (2)$$

where $\mathbf{w}_{i,k}^s$ is IID white Gaussian noise vector with covariance

$$\mathbf{Q}_{i,k}^s = \begin{bmatrix} \frac{qx_i\delta^4}{4} & 0 & \frac{qx_i\delta^3}{2} & 0 \\ 0 & \frac{qy_i\delta^4}{4} & 0 & \frac{qy_i\delta^3}{2} \\ \frac{qx_i\delta^3}{3} & 0 & qx_i\delta^2 & 0 \\ 0 & \frac{qy_i\delta^3}{3} & 0 & qy_i\delta^2 \end{bmatrix}. \qquad (3)$$

The state of the object is estimated from observations taken at each time step k. The observation model of the object at camera c_i and time k is given as

$$\mathbf{z}_{i,k} = \mathbf{h}_{i,k}\left(\mathbf{x}_k\right) + \mathbf{v}_{i,k}, \qquad (4)$$

where $\mathbf{v}_{i,k}$ is an IID measurement noise vector with covariance $\mathbf{R}_{i,k}$. The measurement function $\mathbf{h}_{i,k}$ is the non-linear homography function which converts the object's coordinates from the ground to the image plane. The considered motion model (1) and measurement model (4) are adapted from [27].

3   Information filtering

The information filter is an alternative version of the Bayesian state estimation methods. In information filtering, the information vector and the information matrix are computed and propagated instead of the estimated state vector and the error covariance. The estimated global information matrix $\mathbf{Y}_{k-1|k-1}$ and information vector $\widehat{\mathbf{y}}_{k-1|k-1}$ at time $k-1$ are given as

$$\mathbf{Y}_{k-1|k-1} = \mathbf{P}_{k-1|k-1}^{-1}, \qquad (5)$$

$$\widehat{\mathbf{y}}_{k-1|k-1} = \mathbf{Y}_{k-1|k-1}\widehat{\mathbf{x}}_{k-1|k-1}, \qquad (6)$$

where $\widehat{\mathbf{x}}_{k-1|k-1}$ and $\mathbf{P}_{k-1|k-1}$ are the estimated global state vector and error covariance matrix at time $k-1$. At time k and camera c_i, the information filter has two steps: time and measurement update.

3.1   Time update

The information form of the predicted state and the corresponding information matrix are computed as

$$\mathbf{Y}_{i,k|k-1} = \mathbf{P}_{i,k|k-1}^{-1}, \qquad (7)$$

$$\widehat{\mathbf{y}}_{i,k|k-1} = \mathbf{Y}_{i,k|k-1}\widehat{\mathbf{x}}_{i,k|k-1}, \qquad (8)$$

where $\widehat{\mathbf{x}}_{i,k|k-1}$ and $\mathbf{P}_{i,k|k-1}$ are the predicted state vector and the error covariance matrix, respectively.

3.2   Measurement update

Upon receiving the measurement $\mathbf{z}_{i,k}$, the information contribution matrix $\mathbf{I}_{i,k}$ and information contribution vector $\mathbf{i}_{i,k}$ are computed as

$$\mathbf{I}_{i,k} = \mathbf{Y}_{i,k|k-1}\mathbf{P}_{\mathbf{xz},i,k}\mathbf{R}_{i,k}^{-1}\mathbf{P}_{\mathbf{xz},i,k}^T\mathbf{Y}_{i,k|k-1}^T, \qquad (9)$$

$$\mathbf{i}_{i,k} = \mathbf{Y}_{i,k|k-1}\mathbf{P}_{\mathbf{xz},i,k}\mathbf{R}_{i,k}^{-1} \\ \left(\mathbf{e}_{i,k} + \mathbf{P}_{\mathbf{xz},i,k}^T\widehat{\mathbf{y}}_{i,k|k-1}\right), \qquad (10)$$

where $\mathbf{P}_{\mathbf{xz},i,k}$, $\mathbf{R}_{i,k}$, and $\mathbf{e}_{i,k}$ are the cross-covariance of the state and measurement vector, the measurement noise variance, and the measurement residual, respectively. The measurement residual is defined as

$$\mathbf{e}_{i,k} = \mathbf{z}_{i,k} - \widehat{\mathbf{z}}_{i,k|k-1}, \qquad (11)$$

where $\widehat{\mathbf{z}}_{i,k|k-1}$ is the predicted measurement. In this work, the CIF is used at the cameras to track the objects locally. We refer to Appendices 1 and 2 and [38] for the CIF algorithm.

3.3   Information fusion

In multi-camera networks, multiple cameras have an overlapping FOV and thus can observe an object simultaneously. Hence, each camera c_i where $i \in C_k$ that observes the object computes its own information contribution vector $\mathbf{i}_{i,k}$ and information contribution matrix $\mathbf{I}_{i,k}$ as shown in (9) and (10), respectively. Let us consider that each camera sends their local information metrics to an elected FC, then the global information equivalents of the estimated state and error covariances at the FC c_o, where $o \in C_k$ are calculated as

$$\mathbf{Y}_{k|k} = \mathbf{Y}_{o,k|k-1} + \sum_{i=1}^{|C_k|} \mathbf{I}_{i,k}, \qquad (12)$$

$$\widehat{\mathbf{y}}_{k|k} = \widehat{\mathbf{y}}_{o,k|k-1} + \sum_{i=1}^{|C_k|} \mathbf{i}_{i,k}, \qquad (13)$$

where $\widehat{\mathbf{y}}_{o,k|k-1}$ and $\mathbf{Y}_{o,k|k-1}$ are the predicted information vector and matrix at the FC, respectively.

4 Surprisal camera selection

The VSNs usually have limited bandwidth and energy reservoirs. Therefore, it might be necessary that only a desired number of cameras (subset) transmit their local information to the FC. On the other hand, this can lead to decreased tracking accuracy. A better tracking accuracy can be achieved by selecting the cameras based on the information associated with their observations. This strategy improves the accuracy of the global state estimation under the given bandwidth and energy constraints. The information content associated with the observations can be calculated by applying the concept of self-information or surprisal.

4.1 Surprisal

The surprisal H is a measure of the information associated with the outcome x of a random variable. It is calculated as

$$H = -\log(\Pr(x)), \tag{14}$$

where $\Pr(x)$ is the probability of the outcome x and the base of the logarithm can be considered as 2, 10, or e. In this paper, the surprisal is calculated with the natural logarithm (base e) for the sake of mathematical simplification. The surprisal of the outcome of a random variable depends only on the probability of the corresponding outcome $\Pr(x)$. A highly probable outcome of a random variable is less surprising and vice versa.

4.2 Surprisal of measurement residual

In multi-camera object tracking, the local observations $\mathbf{z}_{i,k}$ of each camera c_i are random variables because of the additive Gaussian noise and the random initial state. Hence, they contain a varying degree of information about the state of the object. Within the framework of information filtering, the measurement residual $\mathbf{e}_{i,k}$ at camera c_i and time k is the disagreement between the predicted observation and the actual observation (see (11)). Hence, the surprisal of the measurement residual $\mathbf{e}_{i,k}$ gives the additional information associated with the received observations that is not available in the predicted observations through the predicted state. The surprisal of the measurement residual $\mathbf{e}_{i,k}$ at camera c_i and time k can be computed as[1]

$$H_{i,k} = -\log_e\left(p\left(\mathbf{e}_{i,k}\right)\right). \tag{15}$$

Under the assumptions of IID additive Gaussian observation noise, the measurement residual becomes approximately a Gaussian distributed variable with zero mean and the covariance $\mathbf{P}_{\mathbf{zz},i,k}$, called the innovation covariance

$$\mathbf{e}_{i,k} \sim \mathcal{N}\left(0, \mathbf{P}_{\mathbf{zz},i,k}\right). \tag{16}$$

By substituting (16) in (15), the surprisal of the measurement residual $\mathbf{e}_{i,k}$ becomes

$$H_{i,k} = -\log_e\left(\frac{\exp\left(-\frac{1}{2}\mathbf{e}_{i,k}^T\mathbf{P}_{\mathbf{zz},i,k}^{-1}\mathbf{e}_{i,k}\right)}{(2\pi)^{\frac{n_z}{2}}\det^{\frac{1}{2}}(\mathbf{P}_{\mathbf{zz},i,k})}\right)$$
$$= \alpha_{i,k} + \frac{1}{2}\mathbf{e}_{i,k}^T\mathbf{P}_{\mathbf{zz},i,k}^{-1}\mathbf{e}_{i,k}, \tag{17}$$

where $\alpha_{i,k}$ is

$$\alpha_{i,k} = \frac{n_z}{2}\log_e(2\pi) + \frac{1}{2}\log_e\left(\det(\mathbf{P}_{\mathbf{zz},i,k})\right), \tag{18}$$

and $n_{\mathbf{z}}$ is the length of the observation vector of camera c_i at time k. The observations of the camera c_i at time k are informative enough if the surprisal of the corresponding measurement residual $H_{i,k}$ is greater than a threshold

$$H_{i,k} \begin{cases} \geq \chi_k & \text{informative} \\ < \chi_k & \text{non-informative.} \end{cases} \tag{19}$$

The cameras with enough informative measurements are called surprisal cameras. The threshold χ_k has to be defined based on the bandwidth and energy constraints in such a way that at each time k, on average, only a given number of cameras are selected as surprisal cameras.

4.3 Surprisal threshold

Let $\mathbf{s}_k = \left(s_{1,k}, s_{2,k}, \cdots, s_{|C_k|,k}\right)$ be the indication vector at time k, where $|C_k|$ is the number of cameras in the observation cluster. Each element $s_{i,k}$ in the indication vector is either 1 or 0

$$s_{i,k} = \begin{cases} 1 & \text{surprisal camera} \\ 0 & \text{non-surprisal camera.} \end{cases} \tag{20}$$

From (17), (19), and (20), the average number of times a camera c_i becomes a surprisal camera is given as

$$\begin{aligned}
\mathbb{E}\left[s_{i,k}\right] &= \Pr\left(s_{i,k} = 1\right) \\
&= \Pr\left(H_{i,k} \geq \chi_k\right) \\
&= \Pr\left(\alpha_{i,k} + \frac{1}{2}\mathbf{e}_{i,k}^T\mathbf{P}_{\mathbf{zz},i,k}^{-1}\mathbf{e}_{i,k} \geq \chi_k\right) \\
&= \Pr\left(\mathbf{e}_{i,k}^T\mathbf{P}_{\mathbf{zz},i,k}^{-1}\mathbf{e}_{i,k} \geq 2\left(-\alpha_{i,k} + \chi_k\right)\right) \\
&= \Pr\left(\mathbf{e}_{i,k}^T\mathbf{P}_{\mathbf{zz},i,k}^{-1}\mathbf{e}_{i,k} \geq \beta_k\right),
\end{aligned} \tag{21}$$

where $\beta_k = 2\left(-\alpha_{i,k} + \chi_k\right)$. Since $\mathbf{e}_{i,k} \sim \mathcal{N}\left(0, \mathbf{P}_{\mathbf{zz},i,k}\right)$,

$$\mathbf{e}_{i,k}^T\mathbf{P}_{\mathbf{zz},i,k}^{-1}\mathbf{e}_{i,k} \sim \chi_{n_z}^2, \tag{22}$$

where $\chi_{n_{\mathbf{z}}}^2$ is a chi-square distribution with a degree of freedom of $n_{\mathbf{z}}$. The surprisal threshold β_k in (21) should be calculated in such a way that on average, $|l_k|$ cameras are selected as surprisal cameras. Thus,

$$\begin{aligned}
\mathbb{E}\left[\sum_{i=1}^{|C_k|} s_{i,k}\right] &= \sum_{i=1}^{|C_k|}\mathbb{E}\left[s_{i,k}\right] \\
&= |C_k|\,\mathbb{E}\left[s_{i,k}\right] = |l_k|.
\end{aligned} \tag{23}$$

From (21), (23), and (22), it is implied that

$$\Pr\left(\mathbf{e}_{i,k}^T \mathbf{P}_{\mathbf{zz},i,k}^{-1} \mathbf{e}_{i,k} \geq \beta_k\right) = \frac{|l_k|}{|C_k|}. \tag{24}$$

The surprisal threshold β_k can be calculated as the value for which the probability of chi-square distributed squared and normalized measurement residual $\chi_{n_\mathbf{z}}^2$ is greater than or equal to $|l_k| / |C_k|$ as

$$\beta_k = F_{\chi_{n_\mathbf{z}}^2}^{-1}\left(1 - \frac{|l_k|}{|C_k|}\right), \tag{25}$$

where $F\chi_{n_\mathbf{z}}^2$ is the cumulative distribution function of the chi-square distribution $\chi_{n_\mathbf{z}}^2$ with a degree of freedom of $n_\mathbf{z}$.

Hence, the surprisal threshold β_k at time k can be calculated by using the knowledge of the number of cameras in the observation cluster $|C_k|$ and the number of desirable surprisal cameras $|l_k|$. Thus, the cameras c_i in the cluster can independently decide whether their local observations are informative or not.

5 Multi-camera object tracking with surprisal cameras (MOTSC)

In the proposed scheme, the cameras c_i where $i \in \{1, 2, \cdots, M\}$ in the network that can observe an object at time k form a cluster (observation cluster) C_k with a FC $c_{o,k}$ as shown in the Fig. 2. The dynamic clustering can be achieved in several ways. One of such methods is pre-sented in [39]. Further, each camera in the VSN has an on-board CIF algorithm. At each time k, each camera in the observation cluster C_k except the FC independently decides whether it is a surprisal camera or not, as discussed in Section 4. All surprisal cameras in the cluster C_k transmit their information contribution vectors and matrices to the FC. Moreover, the FC also performs the local filtering based on the on-board CIF. The locally calculated and received information contribution metrics are then fused together to achieve the estimated global state of the object at time k.

The FC is initialized with the global initial information vector and matrix $(\widehat{\mathbf{y}}_{0|0}, \mathbf{Y}_{0|0})$. At each time step k, it has four main functions: surprisal threshold calculation, local filtering, information fusion, and global state dissemination as shown in Algorithm 1.

- *Surprisal threshold calculation*: The surprisal threshold can be calculated with the knowledge of the size $|C_k|$ of the observation cluster and desired size $|l_k|$ of the surprisal cluster as shown in (25). Hence, the FC which knows this information calculates and broadcasts the surprisal threshold whenever the observation and surprisal cluster sizes change.

- *Local filtering*: The FC performs the local estimation based on its measurement $\mathbf{z}_{o,k}$ by using the on-board CIF. Firstly, the FC predicts the information vector and matrix $(\widehat{\mathbf{y}}_{o,k|k-1}, \mathbf{Y}_{o,k|k-1})$ from the prior global information vector and matrix $(\widehat{\mathbf{y}}_{k-1|k-1}, \mathbf{Y}_{k-1|k-1})$

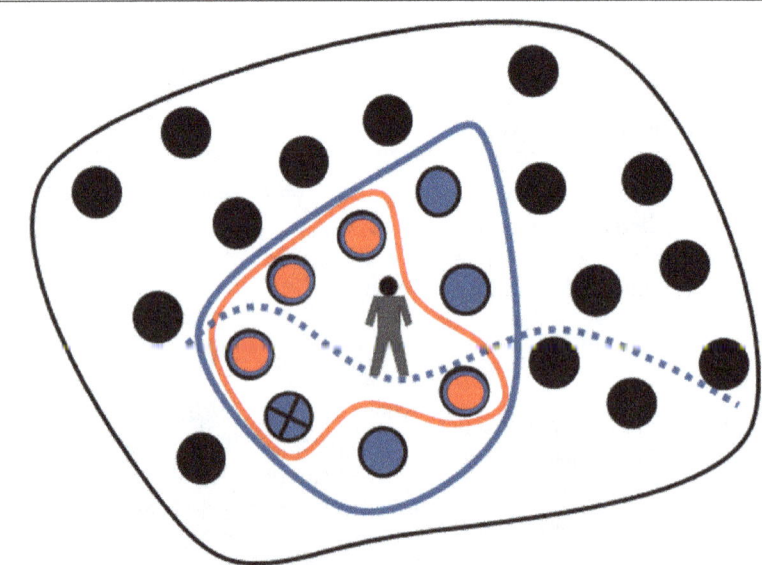

Fig. 2 The VSN with the observation and surprisal clusters with a fixed FC. All the cameras (*dots*) inside the *blue cluster* can observe the object at a given time and form the observation cluster. The cameras (*dots*) inside the *red cluster* are the surprisal cameras and form the surprisal cluster. The *blue dot with cross* represents the FC

as shown in Appendix 1. Then, it computes the information contribution vector and matrix $(\mathbf{i}_{o,k}, \mathbf{I}_{o,k})$ by using its own local observations $\mathbf{z}_{o,k}$ as shown in Appendix 2.

- *Information fusion*: The FC receives a set of information contribution metrics $(\mathbf{i}_{i,k}, \mathbf{I}_{i,k})$ where $i = 1, 2, \cdots, |l_k|$ from the surprisal cameras in the cluster. The global information vector and information matrix $(\widehat{\mathbf{y}}_{k|k}, \mathbf{Y}_{k|k})$ are obtained by fusing the received surprisal information contributions and its own information contributions $(\mathbf{i}_{f,k}, \mathbf{I}_{f,k})$ with the predicted information vector and matrix $(\widehat{\mathbf{y}}_{k|k-1}, \mathbf{Y}_{k|k-1})$.

- *Global state dissemination*: After the information vector and matrix $(\widehat{\mathbf{y}}_{k|k}, \mathbf{Y}_{k|k})$ are computed, the FC broadcasts it in the network. Hence, the cameras in the network have the global knowledge which can be used as prior information for the local filtering in the time step $k + 1$.

The cameras in the observation cluster C_k at time k have two main functions to perform: time update and surprisal update as shown in Algorithm 2. The cameras in the observation cluster know the prior global information of the object $(\widehat{\mathbf{y}}_{k-1|k-1}, \mathbf{Y}_{k-1|k-1})$. At each time step k, they perform the following:

- *Time update*: The camera predicts the information vector and matrix $(\widehat{\mathbf{y}}_{i,k|k-1}, \mathbf{Y}_{i,k|k-1})$ from the prior global information vector and matrix $(\widehat{\mathbf{y}}_{k-1|k-1}, \mathbf{Y}_{k-1|k-1})$ using the CIF time update as shown in Appendix 1.
- *Surprisal update*: Each camera receives the surprisal threshold β_k from the FC whenever the observation and/or surprisal cluster size changes. Upon receiving the measurement $\mathbf{z}_{i,k}$, each camera c_i calculates the corresponding measurement residual and innovation covariance $(\mathbf{e}_{i,k}, \mathbf{P}_{zz,i,k},)$. The proposed surprisal threshold rule in Section 4.3 is used to determine whether it is a surprisal camera or not. If the camera is a surprisal camera, the information contribution vector and matrix $(\mathbf{i}_{i,k}, \mathbf{I}_{i,k})$ are calculated according to (9) and (10). Thereafter, the information metrics are transmitted to the FC. If the camera is not a surprisal camera, then the surprisal update is aborted.

After the surprisal update, each camera c_i in the network receives the global information $(\widehat{\mathbf{y}}_{k|k}, \mathbf{Y}_{k|k})$ from the FC. Hence, each camera in the network has the knowledge of the global state of the object which can also be used as the prior information in the local estimation for the next time step $k + 1$.

In this paper, the FC is assumed to be fixed and not effected by node failures. It is also assumed that the delays in transmitting local information to the FC are all less than the sampling interval of the cameras. Thus, the FC can fuse the arriving information contribution in time. The communication links in the network are assumed to be perfect. Hence, the only cause of a missing information metric from a camera is that the corresponding observations are not informative enough.

Algorithm 1: MOTSC at Fusion Center (FC)

Start: Initialize the filter with $(\widehat{\mathbf{y}}_{0|0}, \mathbf{Y}_{0|0})$.

At each time k, the prior information vector and matrix $(\widehat{\mathbf{y}}_{k-1|k-1}, \mathbf{Y}_{k-1|k-1})$ are known

1. **Surprisal Threshold Calculation**
   if $((|l_k|, |C_k|) \neq (|l_{k-1}|, |C_{k-1}|))$ then

   (a) Calculate the new surprisal threshold β_k
       /* Equation (25) */

   (b) Broadcast β_k in the observation cluster

2. **Local Filtering**

   (a) Compute the predicted information vector and matrix based on the CIF time update
       $[\widehat{\mathbf{y}}_{k|k-1}, \mathbf{Y}_{k|k-1}]$ = Time Update$[\widehat{\mathbf{y}}_{k-1|k-1}, \mathbf{Y}_{k-1|k-1}]$ /* Appendix 1 */

   (b) Upon receiving the measurement $\mathbf{z}_{0,k}$, calculate the information contribution vector and matrix based on the CIF measurement update
       $[\mathbf{i}_{f,k}, \mathbf{I}_{f,k}]$ = Measurement Update $[\widehat{\mathbf{y}}_{k|k-1}, \mathbf{Y}_{k|k-1}, \mathbf{z}_{f,k}]$ /* Appendix 2 */

3. **Information Fusion**

   (a) Receive the information contribution metrics $\{\mathbf{i}_{i,k}, \mathbf{I}_{i,k}\}$, where $i \in C_k$ from the surprisal cameras in the cluster
   (b) Perform information fusion to achieve the global information
       $[\widehat{\mathbf{y}}_{k|k}, \mathbf{Y}_{k|k}]$ = Information Fusion $[\widehat{\mathbf{y}}_{k|k-1}, \mathbf{Y}_{k|k-1}, \{\mathbf{i}_{i,k}, \mathbf{I}_{i,k}\}, (\mathbf{i}_{f,k}, \mathbf{I}_{f,k})]$

4. **Global State Dissemination**
   Broadcast global information $[\widehat{\mathbf{y}}_{k|k}, \mathbf{Y}_{k|k}]$ in the network

---

Algorithm 2: MOTSC at each local camera in the cluster

Start: Initialize the filter at each camera with $(\widehat{\mathbf{y}}_{0|0}, \mathbf{Y}_{0|0})$
At each time k, each camera receives the prior information vector and matrix $(\widehat{\mathbf{y}}_{k-1|k-1}, \mathbf{Y}_{k-1|k-1})$ from the FC.

  1. **Time Update**
     Compute the predicted information vector and matrix based on the CIF time update
     $[\widehat{\mathbf{y}}_{i,k|k-1}, \mathbf{Y}_{i,k|k-1}]$ = Time
     Update$[\widehat{\mathbf{y}}_{k-1|k-1}, \mathbf{Y}_{k-1|k-1}]$ /* Appendix 1 */

  2. **Surprisal Update**

        (a) Upon receiving the measurement $\mathbf{z}_{i,k}$, calculate the measurement residual $\mathbf{e}_{i,k}$
            /* steps 1 to 4 in Appendix 2 */

        (b) Calculate the innovation covariance $\mathbf{P}_{\mathbf{zz},i,k}$

        (c) Compare the information content in the measurement residual $\mathbf{e}_{i,k}$ with surprisal threshold β_k
            if $(\mathbf{e}_{i,k}^T \mathbf{P}_{\mathbf{zz},i,k}^{-1} \mathbf{e}_{i,k} \geq \beta_k)$ then
            I   Calculate the information contribution vector and matrix $(\mathbf{i}_{i,k}, \mathbf{I}_{i,k})$
                        /* steps 5 to 7 in Appendix 2 */

            II   Transmit $(\mathbf{i}_{i,k}, \mathbf{I}_{i,k})$ to the FC

---

6   Results

In this section, the efficiency of the proposed MOTSC method is evaluated based on the simulation and experimental data. In our approach, the efficiency is defined in terms of the sum of the root mean square errors (RMSEs) of the estimated global state and the ground truth in x and y directions. Moreover, the energy and bandwidth efficiency are calculated in terms of the average number of transmissions from the cameras in the observation cluster to the FC.

6.1   Simulation results

The simulation considers a VSN with cameras having overlapping FOVs as shown in Fig. 2. All of the cameras that can observe the xy-plane, where $x \in [-500, 500]$ and $y \in [-500, 500]$ form an observation cluster with a FC. The motion of the object is modeled with Gaussian distributed acceleration as given in (1). The ground truth of the position of the object is simulated by assuming that the process noise covariance $\mathbf{Q}_k$ and measurement noise

covariance $\mathbf{R}_k$ are diag $(5, 5)$ and diag $(1, 1)$, respectively. Each camera c_i in the cluster has its own homography function h_i. Since we assume static cameras, the homography of the cameras do not change with time k and object. The algorithms are evaluated on 1000 different trajectories with different initializations. Figure 3 shows some of the simulated trajectories of the object.

6.1.1   Scenario 1

In this scenario, the accuracy of the CIF- and EIF-based object tracking methods in the VSN are compared. In this comparison, the proposed surprisal selection method is not employed. Hence, all the cameras in the observation cluster participate in the information fusion. In the abovementioned simulation setup, each camera calculates the local information metrics based on the local observations. The information metrics from the local cameras are fused at the FC. Moreover, the process noise covariance $\mathbf{Q}_k$ and measurement noise covariance $\mathbf{R}_k$ are considered to be known to all the cameras in the cluster. The cluster is also assumed to be fully connected with perfect communication links to the FC.

Under the above conditions, Fig. 4 shows the average RMSE (ARMSE) of the multi-camera object tracking methods based on the CIF and EIF for different observation cluster sizes. To achieve statistical reliability, the RMSE is averaged over a thousand simulation runs and 1000 simulated trajectories to yield the ARMSE. From Fig. 4, we can infer that the CIF-based object tracking outperforms the EIF-based method, though the tracking accuracy of the two methods improves with increasing cluster size.

6.1.2   Scenario 2

In this scenario, the accuracy of the proposed MOTSC is analyzed in comparison with multi-camera object

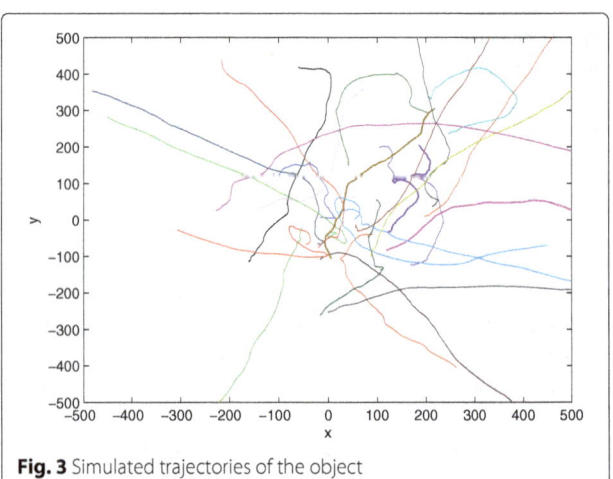

Fig. 3 Simulated trajectories of the object

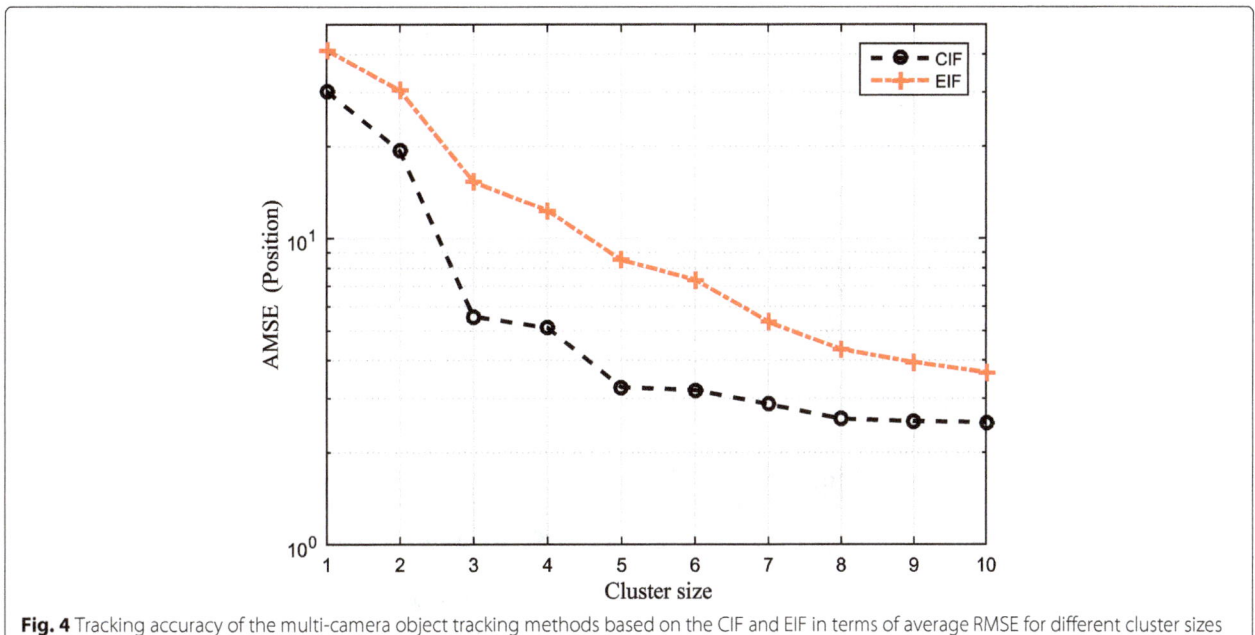

Fig. 4 Tracking accuracy of the multi-camera object tracking methods based on the CIF and EIF in terms of average RMSE for different cluster sizes

tracking with random cameras, fixed cameras, best cameras, and active sensing cameras.

- *Multi-camera object tracking with random cameras (MOTRC):* A random subset of cameras in the observation cluster transmit their local information metrics to the FC independent of the information contained in their measurements.
- *Multi-camera object tracking with fixed cameras (MOTFC):* A fixed subset of cameras in the observation cluster transmit their local information metrics to the FC.
- *Multi-camera object tracking with best cameras (MOTBC):* All the cameras in the observation cluster C_k send their surprisal of the measurement residual to the FC. The FC ranks the cameras in the ascending order of their surprisal score and informs $|l_k|$ best cameras to share their local information metrics. Then, the informed cameras send their local information metrics to the FC. The total number of transmissions to and from the FC involved in this method are $||C_k| + 2|l_k||$. The MOTBC method is an adoption from [36].
- *Multi-camera object tracking method with active sensing cameras (MOTAC):* The FC activates or deactivates the cameras from participating in information exchange by maximizing reward-cost utility function as given in [35]. The reward is expected information gain (EIG). At each time k, the FC evaluates the utility function for all possible activated and deactivated camera combinations

before activating the best cameras to participate in the information fusion. Refer to [35] for complete details.

Figure 5 shows the RMSE of the MOTSC, MOTRC, MOTFC, MOTBC, and MOTAC methods. The x-axis of the figure represents the average number of cameras participated in the information fusion at each time k. The total number of cameras $|C_k|$ in the observation cluster remains 10. From Fig. 5, we can infer that the tracking accuracy of these methods improves with increasing size of the subset that can participate in the information fusion. However, the proposed MOTSC method outperforms both the MOTRC and MOTFC for the same number of cameras $|l_k|$ that can transmit to the FC. The MOTSC, MOTBC, and MOTAC methods approximately achieve the same tracking accuracy. However, in the MOTAC method, at each time k, the FC has to evaluate the reward-cost utility function for all possible activated and deactivated camera combinations (2^{10} in this case) before selecting the best possible cameras to participate in the information fusion. Moreover, the camera selection at time k in the MOTAC method does not depend on the current measurements. In the MOTBC method, in order to select the best possible cameras, the FC has to receive the surprisal scores from all the cameras in the observation cluster. The centralized and complex camera selection restricts the scalability of both the MOTAC and MOTBC methods. On the other hand, in the proposed MOTSC method, the cameras take decision independently whether to participate in information fusion or not.

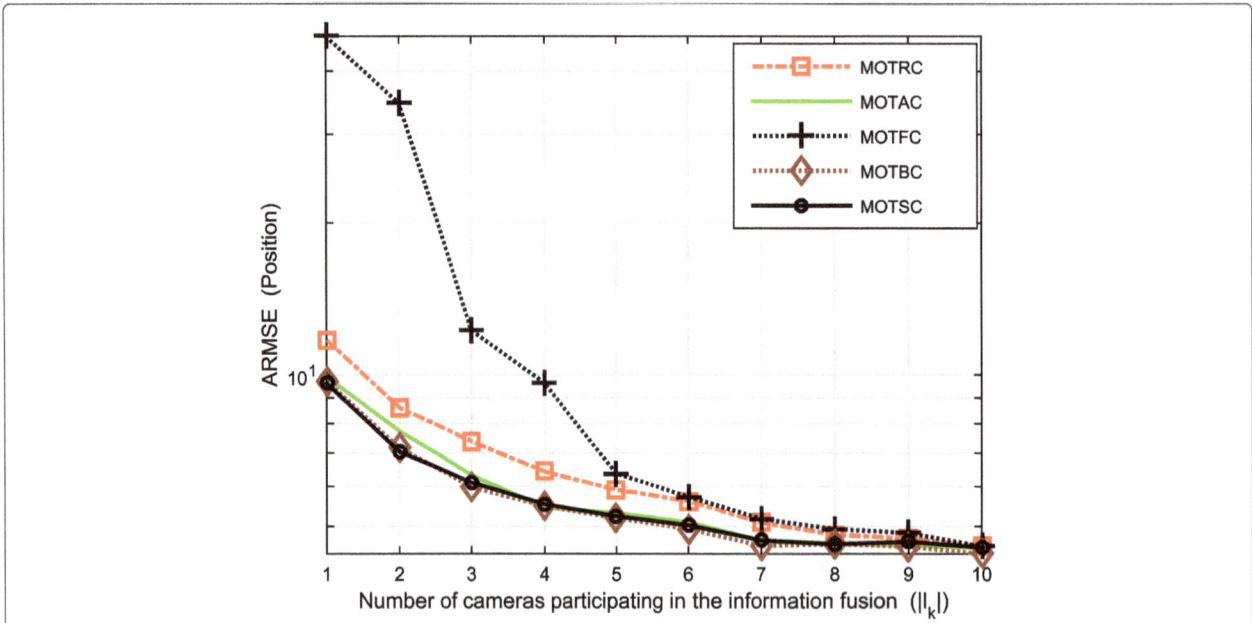

Fig. 5 Tracking accuracy of the MOTSC, MOTRC, MOTFC, MOTBC, and MOTAC methods. The size $|C_k|$ of the observation cluster is 10, and the size of the fixed, random, or surprisal subset varies from 1 to 10

On the other hand, Fig. 6 shows the number of transmissions sent to the FC in the MOTSC and MOTRS methods. The x-axis shows the theoretical number $|l_k|$ of surprisal cameras which is used to calculate the surprisal threshold. The y-axis shows the number of transmissions to the FC from the surprisal and random cameras in the corresponding methods. From the figure, it is illustrated that on average, the number of transmissions to the FC for both methods is approximately equal and matches the theoretical requirements. Even though the MOTBC achieves the same performance as the MOTSC, the number of transmissions in MOTBC is equal to $||C_k| + 2|l_k||$ which can be significantly higher than the average number of transmissions $|l_k|$ in MOTSC.

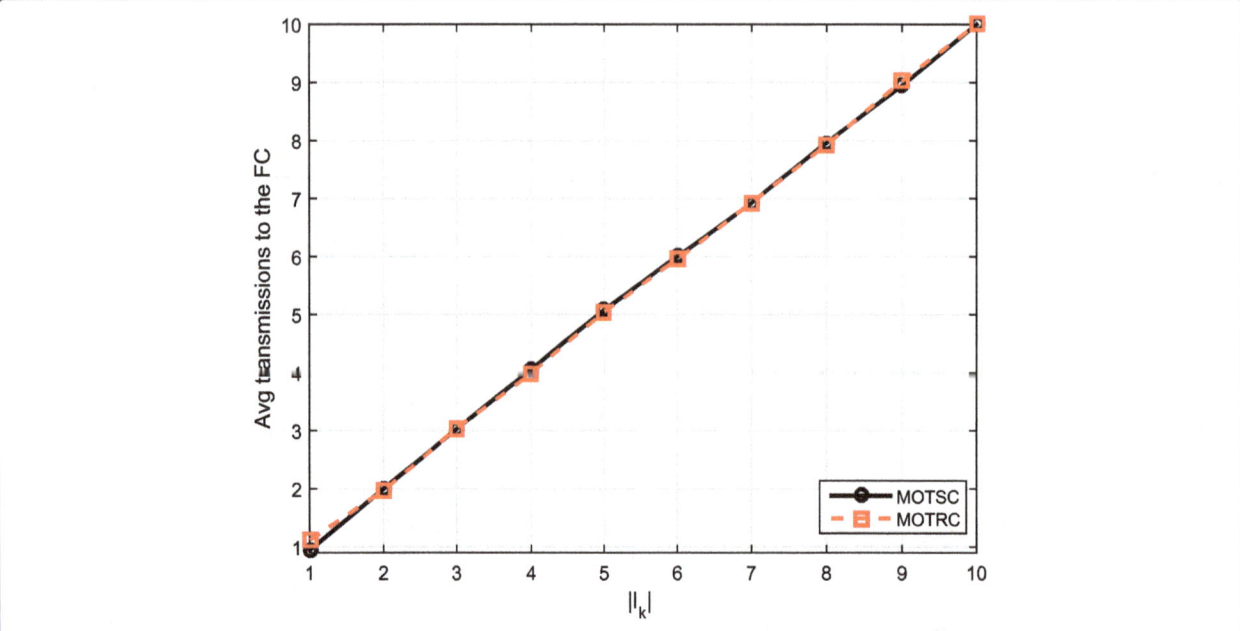

Fig. 6 The average number of transmissions to the FC in the MOTSC and MOTRC methods. The observation cluster size C_k is 10, and the size $|l_k|$ of the random or surprisal subset varies from 1 to 10

6.2 Experimental results

The experimental setup consists of a self-aware multi-camera cluster built in the lab of our institute. The camera cluster consists of four atom-based cameras (1.6 GHz processor, 2 GB RAM, 30 GB internal SSD hard disk) from SLR Engineering and two PandaBoards on which the middle-ware system ELLA [40] is developed. The cameras in the cluster can perform object detection and tracking together with state estimation locally. Moreover, they are connected via Ethernet. In the experimental setup, the four cameras in the network have overlapping FOVs. The motion of the object is modeled by predefined tracks. The experiment considers ten different such predefined tracks within the overlapping FOV of the four cameras. Figure 7 shows some of the object tracks that are used for evaluating the proposed MOTSC method. The x- and y-axes represent the dimensions of the lab where the experimental setup is built. Each track has a duration of 120 s. Each camera c_i in the cluster has its own homography function h_i. Since we assume fixed cameras, the homography of the cameras does not change with time k. The process noise covariance $\mathbf{Q}_k$ and measurement noise covariance $\mathbf{R}_k$ are considered as diag $(10, 10)$ and diag $(2, 2)$, respectively.

Figure 8 shows the average RMSE of the MOTSC, MOTBC, and MOTRC methods. The x-axis of the figure represents the size $|l_k|$ of random and surprisal subset of the cameras that transmit their local information metrics to the FC at each time k. The total number of cameras $|C_k|$ in the observation cluster remains four irrespective of the desired size of the random and surprisal subset. To achieve statistical reliability, the average RMSE is averaged over ten predefined tracks discussed above. From Fig. 8, we can infer that the proposed MOTSC outperforms the MOTRC for the same number of cameras $|l_k|$ that can participate in the information fusion. Even though the MOTBC method achieves approximately the same

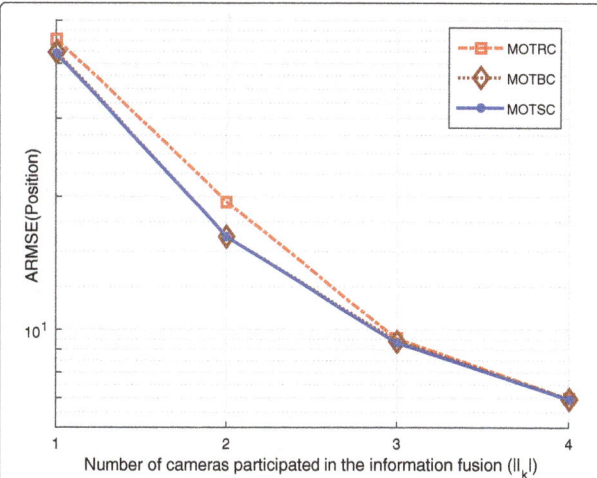

Fig. 8 Tracking accuracy of the MOTSC, MOTRC, and MOTBC methods in the experimental setup defined above. The size of the observation cluster is 4, and the size $|l_k|$ of the random or surprisal subset varies from 1 to 4

tracking accuracy as the MOTSC method, the number of transmissions to the FC is always $||C_k| + 2|l_k||$.

On the other hand, Fig. 9 shows the average number of transmissions sent to the FC in the MOTSC and MOTRC methods. The x-axis shows the theoretical number $|l_k|$ of surprisal cameras which is used to calculate the surprisal threshold. The y-axis shows the average number of transmissions to the FC by the corresponding methods during the experiment. From the figure, it is illustrated that on average, the number of transmissions to the FC for both the methods is approximately equal and matches the theoretical requirements. Hence, the proposed MOTSC shows better accuracy than that of the MOTRC for the same number of average transmissions.

7 Conclusions

In this work, a multi-camera object tracking with surprisal cameras in a VSN is proposed. The cameras in the VSN that can observe an object form an observation cluster with a fixed FC. However, due to bandwidth constraints and energy limitations, it is usually desirable to have only a subset of cameras exchanging their local information to the fusion center. In our approach, each camera runs a local object tracking algorithm based on the on-board CIF. Each camera independently determines whether its observations are informative enough or not by using the surprisal of its measurement residual. Only if a camera's measurements are informative enough (surprisal cameras), it calculates and transmits the local information vector and matrix to the fusion center. The global state of the object is obtained by fusing the local information from surprisal cameras at the fusion center. The proposed scheme also ensures that on average, only a

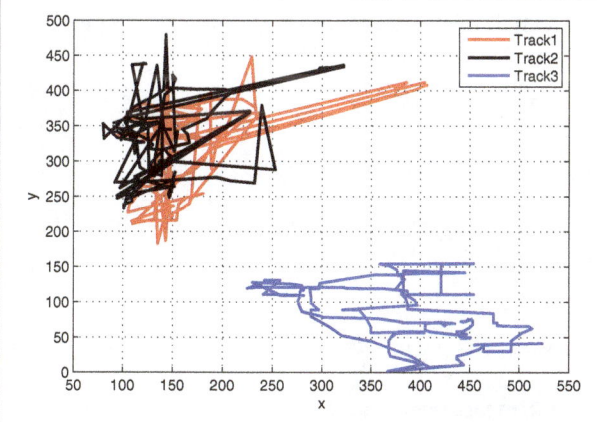

Fig. 7 Examples of the predefined object tracks used for evaluating the MOTSC method

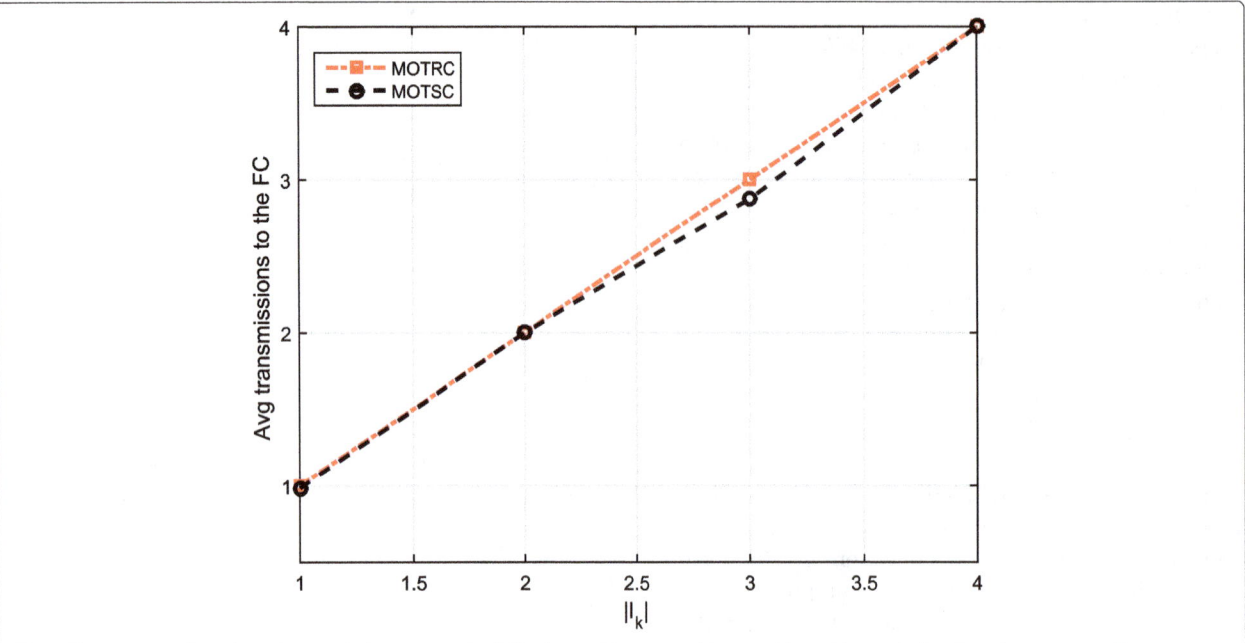

Fig. 9 The average of number of transmissions sent to the FC in the multi-camera object tracking with surprisal and random cameras. The observation cluster size $|C_k|$ is 4, and the size $|I_k|$ of the random or surprisal subset varies from 1 to 4

desired number of cameras participate in the information exchange. The proposed multi-camera object tracking with surprisal cameras shows a considerable improvement in tracking accuracy over the multi-camera object tracking with random and fixed cameras for the same number of transmissions to the fusion center.

Endnote

[1] In general, the surprisal is defined for the discrete random variables (DRV). Hence, we are considering the innovation to be a DRV.

Appendices

The multi-sensor CIF constitutes of three main steps: time update and measurement update at each sensor i and time k.

Appendix 1: time update (TU)

Calculate the predicted information vector and information matrix $\left[\widehat{\mathbf{y}}_{i,k|k-1}, \mathbf{Y}_{i,k|k-1} \right]$ from global prior information $\left[\widehat{\mathbf{y}}_{k-1|k-1}, \mathbf{Y}_{k-1|k-1} \right]$.

1. Calculate the state estimate

$$\widehat{\mathbf{x}}_{k-1|k-1} = \mathbf{Y}_{k-1|k-1} \widehat{\mathbf{y}}_{k-1|k-1}.$$

2. Compute the cubature points $m = (1, 2, \ldots, 2n_\mathbf{x})$

$$\mathbf{cp}_{m,k-1|k-1} = \sqrt{\mathbf{Y}_{k-1|k-1}^{-1}} \xi_m + \widehat{\mathbf{x}}_{k-1|k-1},$$

where $n_\mathbf{x}$ is the length of the state vector. ξ_m represent the mth intersection point of the surface of the n-dimensional unit sphere and its axes.

3. Propagate the cubature points through the motion model

$$\mathbf{x}_{m,k|k-1}^* = \mathbf{f}_{i,k}\left(\mathbf{cp}_{m,i,k-1|k-1} \right).$$

4. Calculate the predicted state as

$$\widehat{\mathbf{x}}_{i,k|k-1} = \frac{1}{2n_\mathbf{x}} \sum_{m=1}^{2n_\mathbf{x}} \mathbf{x}_{m,i,k|k-1}^*.$$

5. Calculate the predicted error covariance as

$$\mathbf{P}_{k|k-1} = \mathbf{M}_{i,k|k-1} \mathbf{M}_{i,k|k-1}^T + \mathbf{Q}_{i,k}^s,$$

where $\mathbf{Q}_{i,k}$ is the process noise covariance. The predicted weighted centered matrix $\mathbf{M}_{i,k|k-1}$ is given as

$$\mathbf{M}_{i,k|k-1} = \frac{1}{\sqrt{2n}} \left[\mathbf{x}_{1,i,k|k-1}^* - \widehat{\mathbf{x}}_{i,k|k-1} \quad \mathbf{x}_{2,i,k|k-1}^* \right.$$
$$\left. - \mathbf{x}_{i,k|k-1} \cdots \mathbf{x}_{2n,i,k|k-1}^* - \mathbf{x}_{i,k|k-1} \right].$$

6. Compute the predicted information matrix and predicted information vector

$$\mathbf{Y}_{i,k|k-1} = \mathbf{P}_{i,k|k-1}^{-1},$$
$$\widehat{\mathbf{y}}_{i,k|k-1} = \mathbf{Y}_{i,k|k-1} \widehat{\mathbf{x}}_{i,k|k-1}.$$

Appendix 2: measurement update (MU)

Each sensor calculates its information contribution vector and matrix $\left[\mathbf{i}_{i,k}, \mathbf{I}_{i,k} \right]$ from the predicted information vector and matrix $\left[\widehat{\mathbf{y}}_{i,k|k-1}, \mathbf{Y}_{i,k|k-1}\right]$ and the measurement $\mathbf{z}_{i,k}$.

1. Calculate the cubature points

$$\mathbf{cp}_{m,i,k|k-1} = \sqrt{\mathbf{P}_{i,k|k-1}}\xi_m + \widehat{\mathbf{x}}_{i,k|k-1}.$$

2. Propagate the cubature points through the observation function

$$\mathbf{z}^*_{m,i,k|k-1} = \mathbf{h}_{i,k}\left(\mathbf{cp}_{m,i,k|k-1}\right).$$

3. Calculate the predicted measurement

$$\widehat{\mathbf{z}}_{i,k|k-1} = \frac{1}{2n_{\mathbf{x}}}\sum_{m=1}^{2n_{\mathbf{x}}}\mathbf{z}^*_{m,i,k|k-1}.$$

4. Calculate the measurement residual

$$\mathbf{e}_{i,k} = \mathbf{z}_{i,k} - \widehat{\mathbf{z}}_{i,k|k-1}.$$

5. Calculate the cross covariance

$$\mathbf{P}_{\mathbf{xz},i,k|k-1} = \frac{1}{2n}\sum_{m=1}^{2n}\mathbf{cp}_{m,i,k|k-1}\mathbf{z}^{*T}_{m,i,k|k-1}$$
$$- \widehat{\mathbf{x}}_{i,k|k-1}\widehat{\mathbf{z}}^T_{i,k|k-1}.$$

6. Calculate the information contribution matrix

$$\mathbf{I}_{i,k} = \mathbf{Y}_{i,k|k-1}\mathbf{P}_{\mathbf{xz},i,k|k-}\mathbf{R}^{-1}_{i,k}\mathbf{P}^T_{\mathbf{xz},i,k|k-1}\mathbf{Y}^T_{i,k|k-1},$$

where $\mathbf{R}_{i,k}$ is the measurement noise covariance matrix.

7. Compute the information contribution vector

$$\mathbf{i}_{i,k} = \mathbf{Y}_{i,k|k-1}\mathbf{P}_{\mathbf{xz},i,k|k-}\mathbf{R}^{-1}_{i,k}$$
$$\left(\mathbf{e}_{i,k} + \mathbf{P}^T_{\mathbf{xz},i,k|k-1}\mathbf{Y}^T_{i,k|k-1}\widehat{\mathbf{x}}_{i,k|k-1}\right).$$

Competing interests

The authors declare that they have no competing interests.

Acknowledgements

This work was supported in part by the EACEA Agency of the European Commission under EMJD ICE FPA no. 2010-0012. The work has also been supported in part by the ERDF, KWF, and BABEG under grant KWF-20214/21530/32602 (ICE Booster). It has been performed in the research cluster Lakeside Labs.

Author details

[1]Institute of Networked and Embedded Systems, Alpen-Adria-Universität, Klagenfurt, Austria. [2]Department of Marine Engineering, Electrical, Electronics, and Telecommunications, University of Genova, Genova, Italy. [3]Institute of Signal Processing, Johannes Kepler University, Linz, Austria.

References

1. B Rinner, M Quaritsch, W Schriebl, T Winkler, W Wolf, *The evolution from single to pervasive smart cameras. Paper presented at the 2nd ACM/IEEE international conference on distributed smart cameras*. (IEEE, Stanford, CA, 2008), pp. 1–10
2. S Soro, W Heinzelman, A survey of visual sensor networks. Adv. Multimed (2009)
3. A Yilmaz, O Javed, M Shah, Object tracking: a survey. ACM Comput. Surv. **38**(4), 1–45 (2006)
4. S-K Weng, C-M Kuo, S-K Tu, Video object tracking using adaptive Kalman filter. J. Visual Commun. Image Represent. **17**(6), 1190–1208 (2006)
5. H Wang, D Suter, K Schindler, C Shen, Adaptive object tracking based on an effective appearance filter. IEEE Trans. Pattern Anal. Mach. Intell. **29**(9), 1661–1667 (2007)
6. W-J Liu, Y-J Zhang, Edge-colour-histogram and Kalman filter-based real-time object tracking. J. Tsinghua Univ. (Sci. Technol). **48**(7) (2008)
7. R Olfati-Saber, NF Sandell, Distributed tracking in sensor networks with limited sensing range. Am. Control Conf, 3157–3162 (2008)
8. C Soto, S Bi, AK Roy-Chowdhury, Distributed multi-target tracking in a self-configuring camera network. IEEE Conf. Comput. Vis. Pattern Recogn, 1486–1493 (2009)
9. H-M Wang, L-L Huo, J Zhang, Target tracking algorithm based on dynamic template and Kalman filter. IEEE Int. Conf. Commun. Softw. Netw, 330–333 (2011)
10. B Song, C Ding, AT Kamal, JA Farrell, AK Roy-Chowdhury, Distributed camera networks. IEEE Signal Process. Mag. **28**(3), 20–31 (2011)
11. SY Chen, Kalman filter for robot vision: a survey. IEEE Trans. Ind. Electron. **59**(11), 4409–4420 (2012)
12. R Rosales, S Sclaroff, Improved tracking of multiple humans with trajectory prediction and occlusion modeling. IEEE CVPR Workshop Int. Vis. Motion (1998)
13. P Li, T Zhang, B Ma, Unscented Kalman filter for visual curve tracking. Image and Vision Comput. **22**(2), 157–164 (2004)
14. M Meuter, U Iurgel, S-B Park, A Kummert, The unscented Kalman filter for pedestrian tracking from a moving host. IEEE Intell. Veh. Symp, 37–42 (2008)
15. VP Bhuvana, M Schranz, M Huemer, B Rinner, Distributed object tracking based on cubature Kalman filter. Asilomar Conf. Signals, Syst. Comput, 423–427 (2013)
16. K Nummiaro, E Koller-Meier, L Van Gool, An adaptive color-based particle filter, image and vision computing. **21**(1), 99–110 (2003)
17. K Okuma, A Taleghani, N Freitas, JJ Little, DG Lowe, A boosted particle filter: multitarget detection and tracking. Eur. Conf. Comput. Vis (2004)
18. Y Rui, Y Chen, Better proposal distributions: object tracking using unscented particle filter. IEEE Comput. Soc. Conf. Comput. Vis Pattern Recognit. **2**, 786–793 (2001)
19. C-C Wang, C Thorpe, S Thrun, M Hebert, H Durrant-Whyte, Simultaneous localization, mapping and moving object tracking. Int. J. Robot. Res. **26**, 889–916 (2007)
20. Y Rathi, N Vaswani, A Tannenbaum, A Yezzi, Tracking deforming objects using particle filtering for geometric active contours. IEEE Transactions on Pattern Analysis and Machine Intelligence. **29**(8), 1470–1475 (2007)
21. Y Li, H Ai, T Yamashita, S Lao, M Kawade, Tracking in low frame rate video: a cascade particle filter with discriminative observers of different life spans. IEEE Trans. Pattern Anal. Mach. Intell. **30**(10), 1728–1740 (2008)
22. MD Breitenstein, F Reichlin, B Leibe, E Koller-Meier, LV Gool, Robust tracking-by-detection using a detector confidence particle filter. IEEE Int. Conf. Comput. Vis, 1515–1522 (2009)
23. AD Bimbo, F Dini, Particle filter-based visual tracking with a first order dynamic model and uncertainty adaptation. Comput. Vis. Image Underst. **115**(6), 771–786 (2011)
24. Z Ni, S Sunderrajan, A Rahimi, BS Manjunath, Distributed particle filter tracking with online multiple instance learning in a camera sensor network. 17th IEEE Int. Conf. Image Process, 37–40 (2010)
25. C Ding, B Song, AA Morye, JA Farrell, AKR Chowdhury, Collaborative sensing in a distributed PTZ camera network. IEEE Trans. Image Process. **21**(7), 3282–3295 (2012)
26. AT Kamal, JA Farrell, AK Roy-Chowdhury, Consensus-based distributed estimation in camera networks. IEEE Int. Conf. Image Process, 1109–1112 (2012)
27. H Medeiros, J Park, AC Kak, Distributed object tracking using a cluster-based Kalman filter in wireless camera networks. IEEE J. Sel. Top. Signal Process. **2**(4), 448–463 (2008)
28. S Dan, *Optimal State Estimation: Kalman, H Infinity, and Nonlinear Approaches*. (John Wiley & Sons, 2006)
29. AT Kamal, C Ding, B Song, JA Farrell, AK Roy-Chowdhury, A generalized Kalman consensus filter for wide-area video networks. IEEE Conf. Decis. Control. Eur. Control, 7863–7869 (2011)

30. AT Kamal, JA Farrell, AK Roy-Chowdhury, Information weighted consensus. IEEE Annu. Conf. Decis. Control (2012)

31. VP Bhuvana, M Huemer, CS Regazzoni, Distributed object tracking based on square root cubature H-infinity information filter. IEEE Int. Conf. Inf. Fusion, 1–6 (2014)

32. K Morioka, K Szilveszter, J-H Lee, P Korondi, H Hashimoto, A cooperative object tracking system with fuzzy-based adaptive camera selection. Int. J. smart Sens. Intell. Syst. **3**, 338–58 (2010)

33. DB Yang, J Shin, AO Ercan, LJ Guibas, Sensor tasking for occupancy reasoning in a network of cameras. Stanf. Netw. Res. Center (2010)

34. L Tessens, M Morbee, H Aghajan, W Philips, Camera selection for tracking in distributed smart camera networks. ACM Trans. Sensor Netw. **10**, 1–33 (2014)

35. A de San Bernabe, JR Martinez-de Dios, A Ollero, Entropy-aware cluster-based object tracking for camera wireless sensor networks. IEEE/RSJ Int. Conf. Intell. Robot. Syst, 3985–3992 (2012)

36. E Shen, R Hornsey, in *Proceedings of the 5th ACM/IEEE International Conference on Distributed Smart Cameras*. Local image quality metric for a distributed smart camera network with overlapping FOVs, (2011), pp. 1–6

37. CE Shannon, A mathematical theory of communications. Bell Syst. Technical J. **27**, 379–423 (1948)

38. I Arasaratnam, S Haykin, Cubature Kalman filters. IEEE Trans. Auto. Control. **54**(6), 1254–1269 (2009)

39. M Schranz, B Rinner, Resource-aware state estimation in visual sensor networks with dynamic clustering. 4th Int. Conf. Sensor Netw, 10 (2015)

40. B Dieber, J Simonjan, L Esterle, B Rinner, G Nebehay, R Pflugfelder, GJ Fernandez, Ella: Middleware for multi-camera surveillance in heterogeneous visual sensor networks. ACM/IEEE Int. Conf. Distrib. Smart Cameras, 1–6 (2013)

4

Multiple descriptions for packetized predictive control

Jan Østergaard[1][*] and Daniel Quevedo[2]

Abstract

In this paper, we propose to use multiple descriptions (MDs) to achieve a high degree of robustness towards random packet delays and erasures in networked control systems. In particular, we consider the scenario, where a data-rate limited channel is located between the controller and the plant input. This forward channel also introduces random delays and dropouts. The feedback channel from the plant output to the controller is assumed noiseless. We show how to design MDs for packetized predicted control (PPC) in order to enhance the robustness. In the proposed scheme, a quantized control vector with future tentative control signals is transmitted to the plant at each discrete time instant. This control vector is then transformed into M redundant descriptions (packets) such that when receiving any $1 \leq J \leq M$ packets, the current control signal as well as $J - 1$ future control signals can be reliably reconstructed at the plant side. For the particular case of LTI plant models and i.i.d. channels, we show that the overall system forms a Markov jump linear system. We provide conditions for mean square stability and derive upper bounds on the operational bit rate of the quantizer to guarantee a desired performance level. Simulations reveal that a significant gain over conventional PPC can be achieved when combining PPC with suitably designed MDs.

Keywords: Quantization, Networked control, Multiple descriptions

1 Introduction

In networked control systems (NCSs), the controller communicates with the plant via a general purpose communication network [1, 2]. When compared to using dedicated hardwired control networks, the use of general purpose and possibly wireless communication technology brings significant benefits in terms of efficiency, interoperability, deployment costs, etc. However, the use of practical communication technology also leads to new challenges, since the network needs to be taken into account in the overall design, see also [1–7].

In this paper, we will focus on the existence of a digital network between the controller and the plant input. This network contains either a single channel that introduces i.i.d. packet delays and erasures or multiple independent channels with i.i.d. packet delays and erasures. The channel between the plant output and the controller is considered ideal, i.e., noiseless and instantaneous. For example, this could be a situation where the controller

and plant communicates over wireless channels. The controller could be battery driven and therefore with limited transmission power. On the other hand, the plant might not have a limitation on the transmission power. In this case, the reverse channel from the plant to the controller has a significantly greater SNR than the forward channel between the controller and the plant. There are many other practical sitations with wireless controller-actuator links but direct sensor-controller connections, e.g., groups of agents/vehicles/robots/drones. Their positions/formation are sensed via a system comprising a camera and attached controller. Activation commands are then sent wirelessly to the agents.

The main contributions of this work is the theoretical analysis and practical design of the quantized control signals. In particular, we propose to combine a recent robust control strategy known as (quantized) packet predictive control (PPC) [8–11] with a joint source-channel coding strategy based on multiple description (MD) coding [12, 13]. We provide computable upper bounds on the operational bit rate required for coding the quantized control signals (descriptions) and provide a practical design based on our theoretical analysis. The simulation study

*Correspondence: jo@es.aau.dk
[1] Department of Electronic Systems, Aalborg University, Fredrik Bajers Vej 7b, Aalborg, Denmark
Full list of author information is available at the end of the article

shows that the combination of MDs and PPC provides a significant improvement over PPC in the case of large packet loss ratios.

In quantized PPC, a control vector with the current and $N-1$ future predicted plant inputs is constructed at the controller side to compensate for random delays and packet dropouts in the channel. Thus, in the case of packet erasures (and if not too many consecutive dropouts occur), the buffer will feed the plant with the appropriate future predicted control value [8]. The key principle of MDs is to encode a source signal into a number of descriptions (packets) that are transmitted over separate channels. Each description is able to approximate the source signal to within a prescribed quality. Moreover, if several descriptions are received, they can be combined to further improve the reconstruction quality. Thus, in the case of packet erasures, it is possible to achieve a graceful degradation of the reconstruction quality [13].

The design of optimal quantized control strategies subject to data rate limitations defines a complicated problem that lies in the intersection of signal processing and controls. In particular, if the quantizers are designed using conventional open-loop source-coding strategies, it cannot be guaranteed that the overall system will be stable, when used in closed-loop control. Indeed, the resulting data rate could exceed the bandwidth of the digital channel, the data rate could be too low to capture the plant uncertainty and thereby not guarantee stability, or the non-linear effects due to quantization could have a negative impact on the overall stability when fed back into the system [8, 9, 14, 15].

The combination of MDs and PPC has to the best of the authors' knowledge not been considered before (except in the conference contributions of the authors [16–18]). In [16], MDs were used for power control in wireless sensor networks. The quantizers were designed under high-resolution assumptions, and no stability assessment was provided. In [17, 18], the preliminary ideas for the current work (without analysis and proofs) were presented. MDs for state-estimation was considered in [19, 20] under high-resolution quantization assumptions. The design of lattice quantizers for PPC without MDs was treated in [16, 21] for the cases of entropy-constrained and resolution-constrained quantization, respectively.

In this work, we will focus on LTI plant models, which are (possibly) open-loop unstable. Thus, it is necessary to provide quantized control signals to the plant in a reliable way to guarantee stability in the presence of data rate limitations, random packet delays and erasures. Our key idea is to design and use MDs in a novel way that differs from how it is traditionally used. Traditionally, when the received descriptions are combined at the decoder, the approximation of a given source signal is improved. On the other hand, in the proposed work, when the received descriptions are combined at the decoder, then rather than improving existing control signals, new future controls signals are instead recovered.

There exists a vast amount on literature on MJLS with delays, cf.,[22–25]. In the present work, we show that the overall system with delays, erasures, quantization effects, and multiple descriptions, can be cast as a Markov jump linear system (MJLS), which makes it possible to use general stability results from the MJLS literature [26, 27].

The paper is organized as follows. Section 2 contains background information on quantized PPC. Section 3 contains the system analysis of a theoretical joint PPC and MD scheme. Section 4 presents the design of the combined practical PPC and MD scheme. Section 5 provides a simulation study of the proposed scheme. Section 6 contains the conclusions. Proofs of lemmas and theorems are deferred to the appendices.

1.1 Notation

Let $S^{\downarrow}$ be the down-shift-by-one matrix operator, which replaces the jth row of an $N \times M$ matrix by its $(j-1)$th row for $j = N, \ldots, 2$. Similarly, define $S^{\uparrow}$ as the up-shift-by-one matrix operator. Let e_i denote the unit-vector aligned with the ith axis of the Cartesian coordinate system, e.g., $e_2 = [0, 1, 0, \cdots, 0]^T$, where the dimension of e_i will be clear from the context. Let $\mathbf{1}_i \in \mathbb{R}^i$ be the all-ones vector of dimension i. Let γ_i be the matrix operator that takes the ith diagonal of an $N \times N$ matrix, where $i = 1$ is the main diagonal and $i > 1$ are diagonals above the main diagonal. Thus, $\gamma_i(A) \in \mathbb{R}^{N-i+1}$ if $A \in \mathbb{R}^{N \times N}$. We will use $\sigma_r(A)$ to denote the spectral radius of the matrix A, and $A \otimes B$ denotes the usual Kronecker product between the matrices A and B. The squared and weighted l_2-norm of a vector, say x, is written as $\|x\|_P^2 = x^T P x$, where $P \succeq 0$, i.e., P is a positive semidefinite matrix.

2 Quantized packetized control over erasure channels

In this section, we provide a summary of existing results on quantized PPC and relate them to the present situation. The system considered is shown in Fig. 1. For a more detailed presentation of quantized PPC, see [11].

2.1 System model

We consider the following discrete-time stochastic linear time invariant (LTI) possibly unstable dynamical plant with state $x_t \in \mathbb{R}^z$, $z \geq 1$ and scalar input $u_t \in \mathbb{R}$:

$$x_{t+1} = Ax_t + B_1 u_t + B_2 w_t, \quad t \in \mathbb{N}. \tag{1}$$

In (1), $w_t \in \mathbb{R}^{z'}$, $z' \geq 1$, is an unmeasured disturbance, modeled as an arbitrarily distributed (and with possibly unbounded support) zero-mean stochastic process with bounded covariance matrix Σ_w, and $B_1 \in \mathbb{R}^z$ and $B_2 \in$

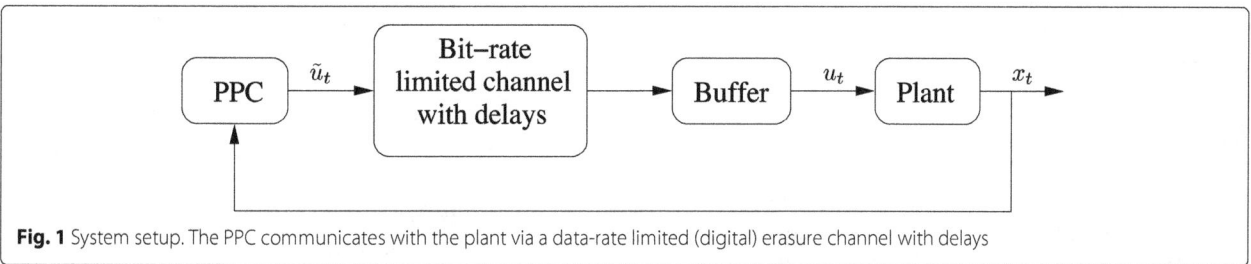

Fig. 1 System setup. The PPC communicates with the plant via a data-rate limited (digital) erasure channel with delays

$\mathbb{R}^{z \times z'}$. We do not assume that $A \in \mathbb{R}^{z \times z}$ is stable; however, we will assume that the pair (A, B_1) is stabilizable. The initial state x_0 is arbitrarily distributed with bounded variance.

2.2 Cost function

In MPC, at each time instant t and for a given plant state x_t, one often uses a linear quadratic cost function on the form [28]:

$$V(\bar{u}', x_t) \triangleq \|x_N'\|_P^2 + \sum_{\ell=0}^{N-1} \left(\|x_\ell'\|_Q^2 + \lambda(u_\ell')^2 \right), \qquad (2)$$

where $N \geq 1$ is the horizon length, and the design variables $P \succeq 0$, $Q \succeq 0$ and $\lambda > 0$ allow one to trade-off control performance versus control effort. The variables x_ℓ' and $\bar{u}_\ell'$ denote tentative variables and are defined below. The final state weighting $\|x_N'\|_P^2$ in (2) aids in stabilizing the feedback loop by approximating the effect of the infinite-horizon behaviour [28]. For example, one may choose P as the unique positive semidefinite solution to the discrete algebraic Riccati equation:

$$P = A^T P A + Q - A^T P B_1 \left(\lambda + B_1^T P B_1 \right)^{-1} B_1^T P A, \qquad (3)$$

which exists if the system (1) is stabilizable [28].

The cost function in (2) examines a prediction of the plant model over a finite horizon of length N. It is common to assume that the predicted state trajectories at time t are independent of the buffer contents at the decoder (i.e., they are independent of what has been received at the plant input side), network effects, and the external disturbances w_t, and are generated by

$$x_{\ell+1}' = A x_\ell' + B_1 u_\ell', \qquad (4)$$

$x_0' = x_t$, while the entries in $\bar{u}' = \left[u_0', \ldots, u_{N-1}' \right]^T$ represent the associated predicted plant inputs. Thus, the current control vector

$$\bar{u}_t = [\, u_t(1), \ldots, u_t(N)\,]^T$$

contains the control signal $u_t(1)$ for the current time instant t as well as $N-1$ future predictive control signals for time up to $t + N - 1$.

One may include the effect of the channel delays in the cost function (2) by, for example, formulating the individual stage costs in terms of their expected stage costs, i.e., weighting by the probabilities of control signals being delayed:

$$\mathbb{E} \sum_{\ell=1}^{N-1} \left(\|x_\ell'\|_Q^2 + \lambda(u_\ell')^2 \right) = \sum_{\ell=1}^{N-1} \left(\|x_\ell'\|_Q^2 + \lambda(u_\ell')^2 \right) p_\ell, \qquad (5)$$

where p_ℓ denotes the probability of using the control signal u_ℓ'. Moreover, in this work, we will also model the effect of the quantizer directly in the design of the control signal u_ℓ', see Section 2.4 for details.

Following the ideas underlying PPCs, see, e.g., [29], at each time instant t, and for current state x_t, the controller sends the *entire* optimizing sequence, $\bar{u}_t$, to the actuator node. Depending upon future packet dropout scenarios, a subsequence of $\bar{u}_t$ will be applied at the plant input, or not. Following the receding horizon paradigm, at the next time instant, x_{t+1} is used to carry out another optimization, yielding $\bar{u}_{t+1}$, etc.

2.3 Network effects

As illustrated in Fig. 1, we shall assume that the backward channel of the network is noiseless and instantaneous, whereas the forward channel is a packet erasure channel, where packets can be delayed and also be received out-of-order. In fact, we allow the delay to be unbounded, which means that packets can be lost. In our setup, if a transmitted packet has not been received within N consecutive time slots, it is considered lost. In MD coding, it is common to assume the availability of either M separate and independent channels or a single (compound) channel where the M packets can be sent simultaneously and yet be subject to independent erasures and delays [13]. Formally, we define $\tau_t^i \in \mathbb{N}_0 \cup \infty$ to be the delay experienced by the ith packet that is constructed at time t. Thus, τ_t^i is a property of the ith channel. We will assume that the delays $\{\tau_t^i\}$ experienced by the different packets are independent and identically distributed (i.i.d.). With this notation, we model transmission effects via the discrete processes $\left\{ d_{t,t'}^i \right\}_{t'=t}^\infty$, where $0 \le t \le t'$ and $i = 1,\ldots,M$, defined via:

$$d_{t,t'}^i \triangleq \begin{cases} 1, & \text{if } \tau_t^i \le t' - t, \\ 0, & \text{else}, \end{cases}$$

where $\tau_t^i \le t' - t$ implies that the ith packet constructed and transmitted at time t has experienced a delay no more than $t' - t$ time instances. We note that even though $\tau_t^i, \forall t$, are mutually independent, the processes $d_{t,t'}^i$ are generally not i.i.d., since if a packet constructed at time t experiences a delay of τ_t^i, then $d_{t,t'}^i = 1$ for all $t' \ge t + \tau_t^i$. However, for $t' = t$, the outcomes $d_{t,t}^i, i = 1,\ldots,M, t \ge 0$, are assumed mutually independent. We will also assume that the packet reception at time t is conditionally independent of the past packet receptions prior to time $t - N$, given the knowledge of the packet reception between time t and $t - N + 1$. Specifically, for $t' \ge t + N$,

$$\text{Prob}\left(d_{t,t'}^i = 1 | d_{t,t'-1}^i, d_{t,t'-2}^i, \ldots, d_{t,t}^i \right)$$
$$= \text{Prob}\left(d_{t,t'}^i = 1 | d_{t,t'-1}^i, d_{t,t'-2}^i, \ldots, d_{t,t'-N+1}^i \right).$$

Finally, we assume that the channel statistics are stationary so that $\text{Prob}\left(d_{t,t'}^i = 1 | d_{t,t'-1}^i, d_{t,t'-2}^i, \ldots, d_{t,t'-N+1}^i \right)$ does not depend upon t. We will make explicit use of the above stationarity and Markov assumptions in Lemma 3.2.

2.4 Quantization constraints

We consider a bit-rate limited digital network between controller output and plant input and all data to be transmitted needs therefore to be quantized. This introduces a quantization constraint into the problem of minimizing $V(\bar{u}', x_t)$.

Let $\overline{Q} \triangleq \text{diag}(Q,\ldots,Q,P) \in \mathbb{R}^{zN \times zN}$ and define:

$$\Phi \triangleq \begin{bmatrix} B_1 & 0 & \ldots & 0 \\ AB_1 & B_1 & \ldots & 0 \\ \vdots & \vdots & \ddots & \vdots \\ A^{N-1}B_1 & A^{N-2}B_1 & \ldots & B_1 \end{bmatrix} \in \mathbb{R}^{zN \times N}, \quad (6)$$

$$\Upsilon \triangleq \begin{bmatrix} A \\ A^2 \\ \vdots \\ A^N \end{bmatrix} \in \mathbb{R}^{zN \times z}, \quad (7)$$

$$F \triangleq \Upsilon^T \overline{Q} \Phi \in \mathbb{R}^{z \times N}, \quad (8)$$

$$\Gamma \triangleq -\Psi^{-T} F^T \in \mathbb{R}^{N \times z}, \quad (9)$$

$$\Psi^T \Psi = \Phi^T \overline{Q} \Phi + \lambda I \in \mathbb{R}^{N \times N},$$

$$\xi_t \triangleq \Gamma x_t \in \mathbb{R}^N. \quad (10)$$

Then using the above and (4), the cost function (2) can be rewritten as

$$V(\bar{u}', x_t) = x_t^T \Upsilon^T \Upsilon x_t + \bar{u}' \Psi^T \Psi \bar{u}' + 2x_t^T F \bar{u}', \quad (11)$$

which has the unique (unquantized) minimizer $\bar{u}^*$ given by

$$\bar{u}^* = (\Psi^T \Psi)^{-1} F^T x_t \quad (12)$$
$$= \Psi^{-1} \Gamma x_t = \Psi^{-1} \xi_t \in \mathbb{R}^N. \quad (13)$$

We note that Ψ is fixed and we may at this point either directly quantize $\bar{u}^*$ or instead quantize ξ_t and then apply the mapping Ψ^{-1} in order to obtain the quantized control vector.[1] Since Ψ is invertible, and we are transmitting the entire quantized control vector, the resulting coding rate is not affected by this operation [30].

When using entropy-constrained (subtractively) dithered (lattice) quantization (ECDQ), a dither vector ζ_t is added to the input prior to quantization and then subtracted again at the decoder to obtain the reconstruction [31].[2] Specifically, let $\mathcal{Q}_\Lambda$ denote an ECDQ with underlying lattice Λ. Then the discrete output ξ_t' of the ECDQ is given by $\xi_t' = \mathcal{Q}_\Lambda(\xi_t + \zeta_t)$. Furthermore, the reconstruction $\hat{\xi}_t$ at the decoder is then obtained by subtracting the dither, i.e., by forming $\hat{\xi}_t = \xi_t' - \zeta_t$. Interestingly, this quantization operation may be exactly modeled by an additive noise channel, i.e., we have $\hat{\xi}_t = \xi_t + n_t$, where the noise n_t is zero-mean with variance σ_n^2 and independent of ξ_t, see [31] for details. With this, the quantized (and reconstructed) control variable $\bar{u}_t$ can be written as

$$\bar{u}_t = \Psi^{-1}(n_t + \xi_t), \quad (14)$$

where n_t and ξ_t are mutually independent and $\xi_t = \Gamma x_t$. We note that $\bar{u}_t$ is the quantized (and reconstructed) control signal, which has been found by using an ECDQ on ξ_t. Thus, $\bar{u}_t$ is a continuous variable whereas $\tilde{u}_t = \Psi^{-1} \xi_t'$ is the corresponding discrete valued variable, which is entropy coded and thereby converted into a bit-stream (to be transmitted over the network), see Fig. 1. Throughout this work, we will use $u_t(i)$ to refer to the ith element of the vector $\bar{u}_t$.

2.5 MD coding for PPC

We design the MDs by explicitly exploiting the *layered* construction of the control signals. In particular, we first generate a quantized control vector based on the principles of PPC. This vector contains the current control signal and $N - 1$ future control signals. Then, we construct M descriptions based on this control vector. The descriptions are constructed so that the current control signal and $J - 1$ future control signals can be obtained by combining any subset of $J \in \{1,\ldots,M\}$ descriptions. Thus, the more packets that are received at the plant, the more future plant predictions become available. Note that on reception of at least one packet out of the M packets, the current quantized control signal can be completely recovered at the plant input side. When receiving and combining more descriptions, the quality of this control signal is not improved. Instead new control signals become available. With this approach, we thus avoid the issue of having to guarantee stability subject to a probabilistic and time-varying accuracy of the control signals. Instead, we can use ideas from quantized PPC, when assessing the stability. A detailed design of the MDs is provided in Section 4.

3 Theoretical analysis of the PPC-MDC scheme

3.1 Markov jump linear system

Let $\bar{x}_t \in \mathbb{R}^{zN \times 1}$ be the $N-1$ past and the present system state vectors, i.e.,

$$\bar{x}_t \triangleq \left[x_t^T, \dots, x_{t+1-N}^T \right]^T, \tag{15}$$

where x_t is given by (1), and let $\bar{n}_t$ be the $N-1$ past and the present quantization noise vectors, i.e.,

$$\bar{n}_t = \left[n_t^T, \dots, n_{t+1-N}^T \right]^T \in \mathbb{R}^{N^2 \times 1}, \tag{16}$$

where n_t is introduced in (14). Moreover, let $\Xi_t \in \mathbb{R}^{N(z+1) \times 1}$ be the augmented state variable given by

$$\Xi_t \triangleq \begin{bmatrix} \bar{x}_t \\ \bar{f}_{t-1} \end{bmatrix}, \tag{17}$$

where $\bar{f}_t = [f_t(1), \dots, f_t(N)]^T \in \mathbb{R}^{N \times 1}$ represents the buffer with the control signals to be applied by the actuator at the plant input side. This buffer holds the present and the $N-1$ tentative future control values. In particular, $f_t(1)$ is the control value to be applied at current time t, and $f_t(i)$ is to be applied at time $t+i-1$. In addition, there is also a buffer $\bar{f}'_t$ at the plant side, which holds all received packets that are no older than $t-N+1$ time instances.

Let $\Delta_t \in \mathbb{R}^{N \times N}$ be an indicator matrix with binary elements $\{0, 1\}$ indicating the complete buffer contents of $\bar{f}'_t$ at time t. In particular, if Δ_t has a "1" at entry (i, j), it shows that at least j packets from time $t-i+1$ have been received and the buffer therefore contains at least $u_{t-i+1}(1), u_{t-i+1}(2), \dots, u_{t-i+1}(j)$. If, in addition, entry $(i, j+1) = 1$, it further means that the buffer also contains $u_{t-i+1}(j+1)$. To better illustrate the relationship between Δ_t and the buffers $\bar{f}'_t$ and $\bar{f}_t$ consider the following example.

Example 3.1. Let $N = 3$ and assume that $\bar{f}'_t$ is empty and that $\bar{f}_t$ is initialized to zero. Moreover, let the three packets constructed at time t be denoted by $s_t(i), i = 1, \dots, 3$. Then at time t, assume that two packets, say $s_t(1)$ and $s_t(3)$, constructed at time t are received, which implies that $u_t(1)$ and $u_t(2)$ can be recovered. At time $t+1$, a single packet, say $s_{t+1}(1)$, from time $t+1$ is received. Finally, at time $t+2$, the third and remaining packet $s_t(2)$ from

time t is received. This leads to the following sequence of variables:

$$\bar{f}'_t = \{s_t(1), s_t(3)\} \Rightarrow \Delta_t = \begin{bmatrix} 1 & 1 & 0 \\ 0 & 0 & 0 \\ 0 & 0 & 0 \end{bmatrix}$$

$$\Rightarrow \bar{f}_t = \begin{bmatrix} u_t(1) \\ u_t(2) \\ 0 \end{bmatrix}$$

$$\bar{f}'_{t+1} = \{s_{t+1}(1), s_t(1), s_t(3)\} \Rightarrow \Delta_{t+1} = \begin{bmatrix} 1 & 0 & 0 \\ 1 & 1 & 0 \\ 0 & 0 & 0 \end{bmatrix}$$

$$\Rightarrow \bar{f}_{t+1} = \begin{bmatrix} u_{t+1}(1) \\ 0 \\ 0 \end{bmatrix}$$

$$\bar{f}'_{t+2} = \{s_{t+1}(1), s_t(1), s_t(2), s_t(3)\}$$

$$\Rightarrow \Delta_{t+2} = \begin{bmatrix} 0 & 0 & 0 \\ 1 & 0 & 0 \\ 1 & 1 & 1 \end{bmatrix} \Rightarrow \bar{f}_{t+2} = \begin{bmatrix} u_t(3) \\ 0 \\ 0 \end{bmatrix}.$$

$\square$

In order to present a formal relationship between Δ_t and the buffer $\bar{f}_t$, we introduce U_t as the upper triangular matrix containing the relevant control signals, that is

$$U_t = \begin{bmatrix} u_t(1) & u_t(2) & u_t(3) & \cdots & u_t(N) \\ 0 & u_{t-1}(2) & u_{t-1}(3) & \cdots & u_{t-1}(N) \\ \vdots & 0 & & \ddots & \vdots \\ 0 & 0 & \cdots & 0 & u_{t-N+1}(N) \end{bmatrix}. \tag{18}$$

The control signal to be applied at time t is given by one of the elements on the main diagonal of U_t, and the control signal to be applied at time $t+j$ is an element on the jth diagonal above the main diagonal (unless the buffer is changed in the mean time). Let

$$\gamma_i(U_t) = [u_t(i), u_{t-1}(i+1), \cdots, u_{t-N+i}(N)]^T$$

and let $\delta_i \triangleq \gamma_i(\Delta_t)$. Moreover, let

$$\tilde{\delta}_i(k) \triangleq \delta_i(k) \prod_{j=1}^{k-1} (1 - \delta_i(j)),$$

where $\delta_i(k)$ is the kth element of the vector δ_i. Thus, for a given i, at most one element of the vector $\tilde{\delta}_i = [\tilde{\delta}_i(1), \dots, \tilde{\delta}_i(N-i+1)]^T$ is 1 and the others are 0. The control signal to be applied at time $t+i-1$ is, thus, given by $\tilde{\delta}_i^T \gamma_i(U_t)$, which could be zero if $\tilde{\delta}_i$ is the all zero vector. With this notation, it follows that

$$\bar{f}_t(i) = \tilde{\delta}_i^T \gamma_i(U_t), \quad i = 1, \dots, N. \tag{19}$$

To avoid updating the buffer $\bar{f}_t$ with information from packets that were already received in previous time instances, it is useful to look only at the changes between Δ_t and Δ_{t-1}. Towards that end, let $\Delta'_t \in \{0, 1\}^{N \times N}$ be the difference indicator matrix that only indicates the packets that are received at current time t, i.e.,

$$\Delta_t' = \Delta_t - S^\downarrow \Delta_{t-1}. \tag{20}$$

In the following, we will show that the number of distinct difference indicator matrices is finite for bounded N, and that the sequence of difference indicator matrices $\{\Delta_t'\}$ is stationary Markov and ergodic. These properties will be helpful in the subsequent analysis.

Lemma 3.1. *The number L of distinct difference indicator matrices is upper bounded by:*

$$L \le (N+1)\left(1 + \frac{1}{2}N(N+1)\right)^{N-1} \tag{21}$$

with equality if $N = M$, i.e., if the number of packets is equal to the horizon length. △

Proof. See Appendix 1. □

Lemma 3.2. *The sequence of difference indicator matrices $\{\Delta_t'\}$ is stationary Markov and ergodic.* △

Proof. See Appendix 2. □

Example 3.2. *Let us briefly consider the special case without delays, i.e., where we do not allow for late packet arrivals but simply discard late packets. Let us assume that $M = N$, i.e., the number of packets equals the horizon length. In this case, the difference indicator matrices Δ_t' take the form of the all-zero matrix except for the first row, which has J_t consecutive ones starting at the beginning of the row. Here J_t denotes the number of packets received at the current time (excluding any late packets). Thus, the number of distinct difference indicator matrices reduces to $L = M + 1$. Let J_{t-1} denote the number of packets received in the previous time slot. Then the transition probability $p_{J_t|J_{t-1}}$, i.e., the probability of receiving J_t packets conditioned upon receiving J_{t-1} packets in the previous time slot does not depend upon J_{t-1}. Indeed, in this particular case:*

$$p_{J_t|J_{t-1}} = \binom{N}{J_t}(1-p)^{J_t}p^{N-J_t}, \quad J_t = 0, \ldots, N. \tag{22}$$

□

We are now in a position to introduce the main technical result of this section, which shows that the sequence of augmented state variables $\{\Xi_t\}$ in (17) and the sequence of difference indicator matrices $\{\Delta_t'\}$ in (20) are *jointly Markovian* and form a Markov jump linear system.

Theorem 3.1. *Let $v_t = \begin{bmatrix} w_t^T, \bar{n}_t^T \end{bmatrix}^T$ be the vector containing the external disturbances and quantization noises. Moreover, let $\delta_i' \triangleq \gamma_i(\Delta_t') \in \mathbb{R}^{N-i+1}$ and let*

$$\tilde{\delta}_i'(k) \triangleq \delta_i'(k) \prod_{j=1}^{k-1}(1 - \delta_i'(j)),$$

where $\delta_i'(k)$ is the kth element of the vector δ_i'. Then, $\{\Xi_t, \Delta_t'\}$ forms a Markov jump linear system with a state recursion that can be written in the following form:

$$\Xi_{t+1} = \mathcal{A}(\Delta_t')\Xi_t + \mathcal{B}(\Delta_t')v_t, \tag{23}$$

where the two switching matrices

$$\mathcal{A}(\Delta_t') \triangleq \begin{bmatrix} \mathcal{A}_1(\Delta_t') & \mathcal{A}_2(\Delta_t') \\ \mathcal{A}_3(\Delta_t') & \mathcal{A}_4(\Delta_t') \end{bmatrix} \in \mathbb{R}^{(zN+N)\times(zN+N)}$$

and

$$\mathcal{B}(\Delta_t') \triangleq \begin{bmatrix} \mathcal{B}_1(\Delta_t') & \mathcal{B}_2(\Delta_t') \\ \mathcal{B}_3(\Delta_t') & \mathcal{B}_4(\Delta_t') \end{bmatrix} \in \mathbb{R}^{(zN+N)\times(z'+N^2)}$$

are given by:

$$\mathcal{A}_1(\Delta_t') = \begin{bmatrix} A & \mathbf{0}_{z\times z(N-1)} \\ \mathbf{0}_{z(N-1)\times z} & \mathbf{0}_{z(N-1)\times z(N-1)} \end{bmatrix} + \begin{bmatrix} B_1\tilde{\delta}_1'^T E_1 \\ \mathbf{0}_{z(N-1)\times zN} \end{bmatrix} \in \mathbb{R}^{zN\times zN} \tag{24}$$

$$\mathcal{A}_2(\Delta_t') = \begin{bmatrix} B_1\left(1 - \mathbf{1}_N^T\tilde{\delta}_1'\right)e_1^T S^\uparrow \\ \mathbf{0}_{z(N-1)\times N} \end{bmatrix} \in \mathbb{R}^{zN\times N} \tag{25}$$

$$\mathcal{A}_3(\Delta_t') = \begin{bmatrix} \tilde{\delta}_1'^T E_1 \\ \tilde{\delta}_2'^T E_2 \quad 0 \\ \vdots \\ \tilde{\delta}_N'^T E_N \quad 0 \cdots 0 \end{bmatrix} \in \mathbb{R}^{N\times zN} \tag{26}$$

$$\mathcal{A}_4(\Delta_t') = \begin{bmatrix} \left(1 - \mathbf{1}_N^T\tilde{\delta}_1'\right)e_1^T \\ \vdots \\ \left(1 - \mathbf{1}_1^T\tilde{\delta}_N'\right)e_N^T \end{bmatrix} S^\uparrow \in \mathbb{R}^{N\times N} \tag{27}$$

and

$$\mathcal{B}_1(\Delta_t') = \begin{bmatrix} B_2 \\ \mathbf{0}_{z(N-1)\times z'} \end{bmatrix} \in \mathbb{R}^{zN\times z'} \tag{28}$$

$$\mathcal{B}_2(\Delta_t') = \begin{bmatrix} B_1\tilde{\delta}_1'^T E_1' \\ \mathbf{0}_{z(N-1)\times N^2} \end{bmatrix} \in \mathbb{R}^{zN\times N^2} \tag{29}$$

$$\mathcal{B}_3(\Delta_t') = \mathbf{0}_{N\times z'} \tag{30}$$

$$\mathcal{B}_4(\Delta_t') = \begin{bmatrix} \tilde{\delta}_1'^T E_1' \\ \tilde{\delta}_2'^T E_2' \quad 0 \\ \vdots \\ \tilde{\delta}_N'^T E_N' \quad 0 \cdots 0 \end{bmatrix} \in \mathbb{R}^{N\times N^2}, \tag{31}$$

where $E_i \in \mathbb{R}^{(N-i+1)\times(N-i+1)z}$ *and* $E'_i \in \mathbb{R}^{(N-i+1)\times(N-i+1)N}$ *are given by*

$$E_i = \begin{bmatrix} e_i^T \Psi^{-1}\Gamma & & & \\ & e_{i+1}^T \Psi^{-1}\Gamma & & \\ & & \ddots & \\ & & & e_N^T \Psi^{-1}\Gamma \end{bmatrix} \tag{32}$$

$$E'_i = \begin{bmatrix} e_i^T \Psi^{-1} & & & \\ & e_{i+1}^T \Psi^{-1} & & \\ & & \ddots & \\ & & & e_N^T \Psi^{-1} \end{bmatrix}. \tag{33}$$

$\triangle$

Proof. See Appendix 3. $\qquad\qquad\square$

3.2 Stability and steady state system analysis

At time step $t+1$, the switching variable jumps from some particular state, say $\Delta'_t = \Delta$ to some state, say $\Delta'_{t+1} = \tilde{\Delta}$, where it is possible that $\Delta = \tilde{\Delta}$. Let the number of distinct states be L, see Lemma 3.1. Thus, without loss of generality, we can enumerate the L (not necessarily distinct) pairs of system matrices that are associated with the L states by $\{(\mathcal{A}(1),\mathcal{B}(1)),(\mathcal{A}(2),\mathcal{B}(2)),\cdots,(\mathcal{A}(L),\mathcal{B}(L))\}$. We note that even though some of the system matrices might be identical, there is a bijection between the state Δ and the index i of the pair of system matrices. Let $p_{i|j} = \text{Prob}(\Delta'_t = i|\Delta'_{t-1} = j)$, i.e., the transition probability due to jumping from state j to state i, where we note that $p_{i|j}$ is independent of t due to stationarity of the switching sequence, see Lemma 3.2.

In order to assess the stability of the MJLS in (23) and find its stationary first- and second-order moments, we will first introduce some new notation and then directly invoke Proposition 3.37 in [27], which we for completeness[3] include as Lemma 3.3 below.

Define $\mathfrak{A}$ and $\mathfrak{B}$ as in (34) and (35), respectively.

$$\mathfrak{A} = \begin{bmatrix} p_{1|1}\mathcal{A}(1)\otimes\mathcal{A}(1) & p_{1|2}\mathcal{A}(2)\otimes\mathcal{A}(2) & \cdots & p_{1|L}\mathcal{A}(L)\otimes\mathcal{A}(L) \\ p_{2|1}\mathcal{A}(1)\otimes\mathcal{A}(1) & p_{2|2}\mathcal{A}(2)\otimes\mathcal{A}(2) & \cdots & p_{2|L}\mathcal{A}(L)\otimes\mathcal{A}(L) \\ \vdots & & \ddots & \vdots \\ p_{L|1}\mathcal{A}(1)\otimes\mathcal{A}(1) & p_{L|2}\mathcal{A}(2)\otimes\mathcal{A}(2) & \cdots & p_{L|L}\mathcal{A}(L)\otimes\mathcal{A}(L) \end{bmatrix}. \tag{34}$$

$$\mathfrak{B} = \begin{bmatrix} p_{1|1}\mathcal{A}(1) & p_{1|2}\mathcal{A}(2) & \cdots & p_{1|L}\mathcal{A}(L) \\ p_{2|1}\mathcal{A}(1) & p_{2|2}\mathcal{A}(2) & \cdots & p_{2|L}\mathcal{A}(L) \\ \vdots & & \ddots & \\ p_{L|1}\mathcal{A}(1) & p_{L|2}\mathcal{A}(2) & \cdots & p_{L|L}\mathcal{A}(L) \end{bmatrix}. \tag{35}$$

Moreover, let $q = [q_1,\ldots,q_L] \triangleq (I - \mathfrak{B})^{-1}\psi$, $\psi \triangleq [\psi_1,\ldots,\psi_L]$, where

$$\psi_j \triangleq \sum_{i=1}^{L} p_{j|i}\mathcal{B}(i)\gamma\pi_i, \tag{36}$$

where $\gamma = \lim_{t\to\infty}\mathbb{E}[v_t]$ and π_i are the state priors. Define the operators ϕ and $\hat{\phi}$ as follows:

$$\phi(V_i) \triangleq \begin{bmatrix} \bar{v}_{i,1} \\ \bar{v}_{i,2} \\ \vdots \\ \bar{v}_{i,L} \end{bmatrix}, \quad \hat{\phi}(V) \triangleq \begin{bmatrix} \phi(V_1) \\ \phi(V_2) \\ \vdots \\ \phi(V_L) \end{bmatrix}, \tag{37}$$

where $\bar{v}_{i,j} \in \mathbb{R}^m$, $V_i = [\bar{v}_{i,1},\ldots,\bar{v}_{i,L}]$ and $V = [V_1,\ldots,V_L]$. Then define

$$Q \triangleq \hat{\phi}^{-1}((I - \mathfrak{A})^{-1}\hat{\phi}(R(q))), \tag{38}$$

where

$$R(q) \triangleq [R_1(q),\ldots,R_L(q)], \tag{39}$$

$$R_j(q) \triangleq \sum_{i=1}^{L} p_{j|i}(\mathcal{B}(i)W\mathcal{B}(i)^*\pi_i$$
$$+ \mathcal{A}(i)q_i\gamma^*\mathcal{B}(i)^* + \mathcal{B}(i)\gamma q_i^*\mathcal{A}(i)^*), \tag{40}$$

where

$$W = \lim_{t\to\infty}\mathbb{E}[v_t v_t^T]$$
$$= \text{diag}\left(\Sigma_w,\sigma_n^2,\ldots,\sigma_n^2\right) \in \mathbb{R}^{(z'+N^2)\times(z'+N^2)}.$$

Definition 3.1 (Definitions 3.8 and 3.32 in [27]). *The MJLS in (23) is mean square stable (MSS) if and only if for any initial condition (Ξ_0,Δ'_0) and ergodic Markov jump sequence $\{\Delta'_t\}$, there exists μ_Ξ and Σ_Ξ such that*

$$\|\mathbb{E}[\Xi_t] - \mu_\Xi\|_2 \to 0 \quad as \quad t \to \infty, \tag{41}$$

$$\|\mathbb{E}[\Xi_t\Xi_t^T] - \Sigma_\Xi\|_2 \to 0 \quad as \quad t \to \infty. \tag{42}$$

Lemma 3.3 (Proposition 3.37 in [27]). *If $\sigma_r(\mathfrak{A}) < 1$, then the system in (23) is MSS.*

Remark 1. *Lemma 3.3 shows that there is an upper limit on the spectral radius of the matrix $\mathfrak{A}$ given by (34) above which the system cannot be stabilized. This matrix $\mathfrak{A}$ depends on the packet loss rates via $p_{i|j}$ and on the delays via the different switching matrices $\mathcal{A}(i), i = 1,\ldots,L$.*

The MJLS in (23) is in general not stationary. However, as can be observed from Definition 3.1, if the system is MSS then asymptotically as $t \to \infty$, its first- and second-order moments do not depend on t. This observation is formalized in ([27] Theorem 3.33), which we include in part below.

Theorem 3.2 ([27] Theorem 3.33). *If the MJLS in (23) is MSS, then it is also asymptotically wide sense stationary (AWSS) and vice versa.*

Lemma 3.4 (Proposition 3.37 in [27]). *If the MJLS is AWSS, then its first- and second-order asymptotically stationary (non-centralized) moments are given by:*

$$\mu_\Xi \triangleq \lim_{t\to\infty} \mathbb{E}[\,\Xi_t] = \sum_{i=1}^{L} q_i, \tag{43}$$

$$\Sigma_\Xi \triangleq \lim_{t\to\infty} \mathbb{E}[\,\Xi_t\Xi_t^T] = \sum_{i=1}^{L} Q_i. \tag{44}$$

$\triangle$

In our case, we note that $\gamma = 0$ since the external disturbance w_t and the quantization noise n_t both are zero mean. This implies that $q_i = 0, \forall i$, in (43).

3.3 Assessing the coding rate of the quantizer

Recall from Section 2.4 that the quantized control vector $\tilde{u}_t$ is obtained by quantizing ξ_t to get the quantized vector ξ_t' and then using $\tilde{u}_t = \Psi^{-1}\xi_t'$. The following result establishes an upper bound on the bit rate required for transmitting ξ_t'.

Theorem 3.3. *Let the system (23) be AWSS. Then, for a given horizon length N, the total coding rate using $M = N$ descriptions of the quantized control vector $\tilde{u}_t$, can be upper bounded by R_u:*

$$R_u \triangleq \frac{N}{2} \log_2 \left(\prod_{i=1}^{N} \left(1 + \frac{\sigma^2_{\bar{\xi}(i)|\bar{\xi}(1),\cdots,\bar{\xi}(i-1)}}{\sigma_n^2} \right)^{\frac{1}{i}} \right)$$
$$+ \frac{N}{2} \log_2 \left(\frac{\pi e}{6} \right) + 1, \tag{45}$$

where $\sigma^2_{\bar{\xi}(i)|\bar{\xi}(1),\cdots,\bar{\xi}(i-1)}$ denotes the conditional variance of $\bar{\xi}(i)$ given $(\bar{\xi}(1),\cdots,\bar{\xi}(i-1))$, and where $\bar{\xi}$ denotes Gaussian random variables with the same first- and second-order moments as the asymptotically stationary moments of ξ_t.

Proof. See Appendix 4. □

Remark 2. *It is straight-forward to extend Theorem 3.3 to the case of $M \leq N$ descriptions by considering M (instead of N) subsets of the vector ξ_t. For example, if $N = 4$ and $M = 3$, one could make the split $\{\xi_t'(1), \xi_t'(2), (\xi_t'(3), \xi_t'(4))\}$, where upon receiving a single description only $\xi_t'(1)$ is recovered, receiving any two descriptions makes it possible to recover $\xi_t'(1)$ and $\xi_t'(2)$, and receiving all $M = 3$ descriptions, the entire vector $\xi_t'(1),\ldots,\xi_t'(4)$ is recovered.* □

Remark 3. *In (45), the conditional variances can easily be obtained using Schur's complement on the covariance matrix Σ_ξ of ξ, which is implicitly given via Σ_Ξ in (44) using (10), that is*

$$\Sigma_\xi = [\Gamma \quad 0]\, \Sigma_\Xi\, [\Gamma \quad 0]^T. \tag{46}$$

This makes the upper bound on the bit rate in (45) computable and thereby relevant from a practical perspective. Indeed, we show in the simulation study in Section 5, that the bound in (45) is very close to (only 1 bit above) the resulting operational bit rate.[4] □

4 Practical design of the PPC-MDC scheme

In this section, we design a scheme that satisfies the theoretical analysis provided in the previous section. We first present the idea behind our design of MDs and then show the connection to PPC that was sketched in Section 2.5. The proposed scheme is illustrated in Fig. 2.

There are many ways to design MD coding schemes, for example, by use of lattice quantization and index assignment techniques [32, 33], frame expansions followed by quantization [34], oversampling and delta-sigma quantization [35], or layered source coding followed by unequal error protection [36, 37]. In this work, we will be using the latter technique, where the source is decomposed into a number of layers and encoded in such a way that upon reception of say k descriptions, all layers up till the kth layer are revealed [36]. In particular, we rely on a common practical implementation of this strategy, which is based on conventional forward error correction (FEC)

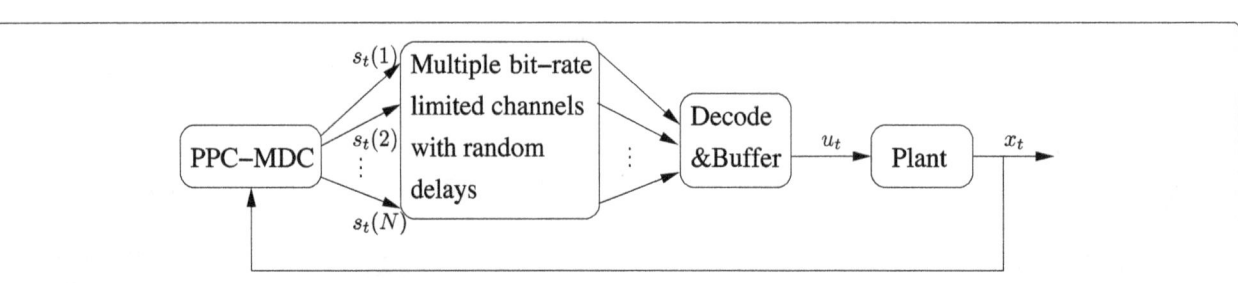

Fig. 2 The proposed combined PPC and MD scheme. The PPC-MDC (controller-encoder) communicates with the plant via multiple independent data-rate limited (digital) erasure channels with delays. The received descriptions are decoded and combined in the buffer

codes that are applied on the individual source layers [37]. It will be shown that there exists a natural connection between PPC and MD based on FEC codes, in the sense that a quantized control vector $\tilde{u}_t$ with $N-1$ future predictions, can be split into $M \leq N$ "layers", where each layer contains at least one control value. Then, based on these M "layers", we construct M packets $s_t(i), i = 1, \ldots, M$, so that upon reception of any $k \leq M$ packets, the control signals $\tilde{u}_t(1), \ldots, \tilde{u}_t(k)$ can be exactly obtained at the decoder. Thus, as more packets are received, more information about future predicted control signals will become available at the plant input side.

4.1 Forward error correction codes

Consider an (n, k)-erasure code, which as input takes k symbols $y_t^k = (y_t(1), \ldots, y_t(k))$ and outputs n symbols $\tilde{y}_t^n = (\tilde{y}_t(1), \ldots, \tilde{y}_t(n))$, where $n \geq k$, and where $y_t, \tilde{y}_t$ belong to some (yet to be specified) discrete alphabets. With an (n, k)-erasure code, the original k input symbols can be completely recovered using any subset of at least k output symbols. For example, a $(3, 2)$-erasure code may be constructed by letting $\tilde{y}_t(1) = y_t(1), \tilde{y}_t(2) = y_t(2)$, and $\tilde{y}_t(3) = y_t(1)\text{XOR}y_t(2)$, where the XOR operation is performed on, e.g., the binary expansions of $y_t(1)$ and $y_t(2)$. Thus, using any two $\tilde{y}_t(i), \tilde{y}_t(j), i \neq j$ both $y_t(1)$ and $y_t(2)$ may be perfectly recovered. This principle extends to any $n > k$ by using, e.g., erasure codes that are maximum distance separable cf. [38].

4.2 Combining PPC- and FEC-based MDs

For the NCS studied, we apply a sequence of erasure codes on the quantized control vector $\tilde{u}_t = (\tilde{u}_t(1), \ldots, \tilde{u}_t(N))$ in order to obtain M packets. This process is illustrated in Fig. 3 and described in detail below. We first split $\tilde{u}_t$ into M subsets. For example, if $M = N$, the kth set consists of the kth control signal (i.e., $\tilde{u}_t(k)$). In general, we allow several control signals within the same set so that $M < N$. To simplify the exposition and without loss of generality, we will in the following assume that $M = N$. Due to quantization, each distinct $\tilde{u}_t(k)$ can be mapped (entropy coded) to a unique bit stream (codeword), say $b_t(k)$. The bitstream is then split into k non-overlapping sub-bitstreams $b_t^{(i)}(k), i = 1, \ldots, k$ of equal length.[5] These k bitstreams (whose union yields $b_t(k)$) are now considered as input to an (M, k)-erasure code, whose M outputs are denoted by $\phi_t^{(i)}(k), i = 1, \ldots, M$. To summarize, $\tilde{u}_t(1)$ is first mapped to bits $b_t(1)$ and then an $(M, 1)$-erasure code is applied,

which outputs M symbols $\phi_t^{(i)}(1), i = 1, \ldots, M$. Then, the second control signal $\tilde{u}_t(2)$ is mapped to $b_t(2)$. Hereafter, $b_t(2)$ is split into two bitstreams $b_t^{(1)}(2)$ and $b_t^{(2)}(1)$ and an $(M, 2)$-erasure code is applied, which outputs $\phi_t^{(i)}(2), i = 1, \ldots, M$. This process is repeated for all the M control signals.

The M packets $s_t(i), i = 1, \ldots, M$, to be sent over the network at time t are then finally constructed as:

$$s_t(i) = (\phi_t^{(i)}(1), \phi_t^{(i)}(2), \ldots, \phi_t^{(i)}(M)), i = 1, \ldots, M.$$

To further illustrate the usefulness of the above approach, consider the case where $M = 5$ and where the decoder receives three packets say $s_t(2), s_t(3)$, and $s_t(5)$. Then from say $s_t(2)$, we first recover $\phi_t^{(2)}(1)$, which is in fact identical to $\tilde{u}_t(1)$. Then, from say $s_t(2)$ and $s_t(3)$, we then recover $\phi_t^{(2)}(2)$ and $\phi_t^{(3)}(2)$ from which we can decode $\tilde{u}_t(2)$. Finally, using all three received packets, we recover $\phi_t^{(2)}(3), \phi_t^{(3)}(3)$, and $\phi_t^{(5)}(3)$, which can be uniquely decoded to obtain $\tilde{u}_t(3)$.

The foregoing discussion shows that the presence of packet dropouts together with the use of MDs makes the length of the received control packets stochastic and time-varying, while the prediction horizon N is fixed. This aspect makes the analysis of the resultant NCS significantly more involved than that of earlier PPC schemes, as presented in [11]. For example, the number of switching states L, as given by Lemma 3.1, grows exponentially in the horizon length N, whereas in [11] it was enough to consider only two states irrespective of the horizon length.

4.3 Buffering and reconstruction of control signals

At time t, the buffer at the plant input side contains all received packets, which are not older than $t - N + 1$. These will be used for obtaining the current control signal $\hat{u}_t$ giving preference to newer data. For example, assume the buffer is initially empty. Then, for the case of $M = N = 3$, if we at time t receive $s_t(2)$, then clearly we obtain $\hat{u}_t = u_t(1)$. If we then at time $t + 1$ receive $s_{t+1}(1)$ and the delayed packet $s_t(3)$ then we should form $\hat{u}_{t+1} = u_{t+1}(1)$ from $s_{t+1}(1)$ and, thus, simply ignore $s_t(3)$. However, if we now at time $t + 2$, only receive the very late $s_t(1)$, then we recover $\hat{u}_{t+2} = u_t(3)$. Thus, we use the older packets to obtain the control signal. This process is clarified in Table 1 for $M = N = 3$.

$$\tilde{u}_t(k) \longleftrightarrow b_t(k) \longleftrightarrow \underbrace{\left(b_t^{(1)}(k), b_t^{(2)}(k), \ldots, b_t^{(k)}(k) \right)}_{k \text{ input symbols}} \xleftrightarrow{\text{FEC}} \underbrace{\left(\phi_t^{(1)}(k), \phi_t^{(2)}(k), \ldots, \phi_t^{(M)}(k) \right)}_{M \text{ output symbols}}$$

Fig. 3 MPC control vector conversion into MDs. The kth control signal at time t is mapped into M output symbols

Table 1 Control value $\hat{u}_t$ at time t from available buffer contents

$\hat{u}_t$	$s_t(1)$	$s_t(2)$	$s_t(3)$	$s_{t-1}(1)$	$s_{t-1}(2)$	$s_{t-1}(3)$	$s_{t-2}(1)$	$s_{t-2}(2)$	$s_{t-2}(3)$
$u_t(1)$	1	x	x	x	x	x	x	x	x
$u_t(1)$	x	1	x	x	x	x	x	x	x
$u_t(1)$	x	x	1	x	x	x	x	x	x
$u_{t-1}(2)$	0	0	0	1	1	x	x	x	x
$u_{t-1}(2)$	0	0	0	1	x	1	x	x	x
$u_{t-1}(2)$	0	0	0	x	1	1	x	x	x
$u_{t-2}(3)$	0	0	0	x	0	0	1	1	1
$u_{t-2}(3)$	0	0	0	0	x	0	1	1	1
$u_{t-2}(3)$	0	0	0	0	0	x	1	1	1

"1" indicates that the packet is in the buffer and "0" indicates that it is not. "x" indicates that the control value does not depend on the given packet. In all other cases, we set $\hat{u}_t = 0$

4.4 Quantization and coding rates

In order to construct the MDs, we need to split the quantized control vector into individual components. It is therefore not possible to directly quantize the vector ξ_t by use of vector quantization as we have done in our previous work on NCS [11], which did not include the use of MDs. Instead, we will in this work use a scalar quantizer separately along each dimension of the vector ξ_t. Of course, a scalar quantizer is not as efficient as a vector quantizer, but the gap from optimality, which is given by $N/2 \log_2(\pi e/6)$, is included in the upper bound in (52). Interestingly enough, we can still do vector entropy coding by making use of conditional entropy coding. In particular, we first entropy code the first element of the quantized control vector, i.e., $\tilde{u}_t(1)$. This results in an average discrete entropy of $H(\tilde{u}_t(1)|\zeta_t)$. Next, we conditional entropy code the second element $\tilde{u}_t(2)$, which results in an average entropy of $H(\tilde{u}_t(2)|\tilde{u}_t(1), \zeta_t)$. This procedure is repeated for the entire vector $\tilde{u}_t$. The FEC code is now applied on outputs of the conditional entropy coders following the approach described in Section 4.2.

As pointed out in Section 2.4, we transmit the elements of $\tilde{u}_t$ and not those of ξ'_t. The reason for this is that if we receive $\xi'_t(1)$ for the case of $N > 1$, then we are actually not able to reconstruct $\tilde{u}_t(1)$, since $\tilde{u}_t = \Psi^{-1}\xi'_t$. Thus, $\tilde{u}_t(1)$ depends upon the whole vector ξ'_t and not just the first element. Since Ψ^{-1} is fixed and full rank, it simply maps elements one from discrete set into another discrete set. Thus, the coding rate is not affected by sending $\tilde{u}_t(i)$ instead of $\xi'_t(i)$.

The size R (in bits) of a single packet is then on average given by:

$$R = H(\tilde{u}_t(1)|\zeta_t) + \frac{1}{2}H(\tilde{u}_t(2)|\tilde{u}_t(1), \zeta_t) + \cdots$$

$$+ \frac{1}{M}H(\tilde{u}_t(N)|\tilde{u}_t(1), \ldots, \tilde{u}_t(N-1), \zeta_t). \quad (47)$$

Since we have M of these packets, i.e., we have M descriptions, the resulting coding rate is RM.

5 Simulation study

We will now use the analysis and design presented in Sections 3 and 4 in a simulation study in MATLAB.[6]

5.1 System setup

In the state recursion given in (1), we let $z = 5$ and randomly select the system matrix $A \in \mathbb{R}^{z \times z}$ to be

$$A = \begin{bmatrix} -0.1065 & -0.4330 & -0.0006 & -0.8232 & -0.9397 \\ -1.0164 & -1.0668 & -0.1995 & 0.1945 & -0.8169 \\ -1.3309 & 0.8582 & 0.3173 & -1.0053 & -0.3214 \\ -0.5629 & -0.5697 & -0.2112 & -0.2778 & 0.1390 \\ 0.2247 & -0.0090 & -1.3312 & -0.7531 & -0.0929 \end{bmatrix},$$

where the absolute values of the eigenvalues of A are $\{1.9829, 1.2265, 1.2265, 0.9455, 0.9455\}$. Thus, the system is open-loop unstable. We let the external disturbance $w_t \in \mathbb{R}^2$ in (1) be Gaussian distributed with zero mean and covariance matrix $\Sigma_w = I_2$, where I_2 denotes the 2×2 identity matrix. The remaining constants in (1) are set to $B_1 = \mathbf{1}_z$ and $B_2 = [B_1, B_1]$. In these simulations, we have used $T = 4 \times 10^6$ vectors each of dimension $z = 5$ in the sequence $\{x_t\}_{t=0}^{T}$ in (1). x_0 is initialized to the zero vector.

5.2 Cost function

For the cost function in (2), we let $Q = I_5, \lambda = 1/20$, and P is found by (3) and given by:

$$P = $$

$$\begin{bmatrix} 259.5872 & -100.8986 & -76.8526 & -63.0725 & -59.5344 \\ -100.8986 & 46.9687 & 40.4038 & 15.9182 & 15.0465 \\ -76.8526 & 40.4038 & 73.9883 & -10.3694 & -32.3071 \\ -63.0725 & 15.9182 & -10.3694 & 34.9787 & 42.6824 \\ -59.5344 & 15.0465 & -32.3071 & 42.6824 & 68.9741 \end{bmatrix}.$$

5.3 Horizon length and number of packets

We consider the cases where $N = 1, 2, 3$ and compare the proposed scheme that includes multiple descriptions, with the same scheme without multiple descriptions, i.e., that of our earlier work [11]. The two schemes are hereafter referred to as PPC-MDC and PPC, respectively. For the case of PPC-MDC, we let the number of packets M be equal to the horizon length N. For the case of PPC, the entire N-horizon vector is encoded into a single packet. For the case of $N = 1$, the two schemes are identical.

5.4 Network

To simplify the simulations and to be able to compare to existing works on PPC, we will not consider delayed or out-of-order packets. Specifically, if at time t, packet $s_{t-\ell}, \ell > 0$ is received, it is discarded. This means that for the case of $N = M = 3$, the number of jump states reduces to $L = 4$ instead of $L = 196$ as given by Lemma 3.1. Note that even though we do not consider late packet arrivals, control signals can still be applied out of order. To see this, assume that $M = N = 3$, and that all three packets $\{s_t(1), s_t(2), s_t(3)\}$ are received at time t. Then, at time $t+1$, a single packet is received, say $s_{t+1}(1)$, and at time $t + 2$ no packets are received. Then, the control signal $u_{t+1}(1)$ applied at time $t + 1$ is constructed later than the control signal $u_t(3)$ to be applied at time $t + 2$.

We let the packet losses be mutually independent and identically distributed with probability p that a packet is lost (erased). For this case, the state transition probabilities are given by (22).

5.5 Stability

To assess the stability of the system, we need to compute the spectral radius $\sigma_r(\mathfrak{A})$ of $\mathfrak{A}$ in (34). In order to compute $\mathfrak{A}$ we simply insert the above presented system and network parameters into (24) – (27) and (34). We then obtain the spectral radius by using MATLAB to find the eigenvalue of $\mathfrak{A}$ with the largest absolute value. For the case of $N = 1, 2, 3$, we have in Fig. 4 shown the spectral radius $\sigma_r(\mathfrak{A})$ as a function of the packet loss probability $p \in [0, 0.5]$. According to Lemma 3.3, the MJLS is MSS and AWSS if $\sigma_r(\mathfrak{A}) < 1$. As can be observed from Fig. 4, the MJLS is guaranteed to be MSS for $p < 0.06, p < 0.3$, and $p < 0.5$ for the cases of $N = 1, N = 2$, and $N = 3$, respectively. Thus, choosing a larger horizon brings stability benefits.

5.6 Quantization

Each scalar control value in the control vector $\bar{u}_t$ is quantized using a uniform scalar quantizer with some step size δ. Specifically, for the case of PPC, we simply keep the step size fixed at $\delta = 10$. On the other hand, for the case of PPC-MDC, we need to use a larger step than what is used for PPC, since PPC-MDC introduces redundancy across

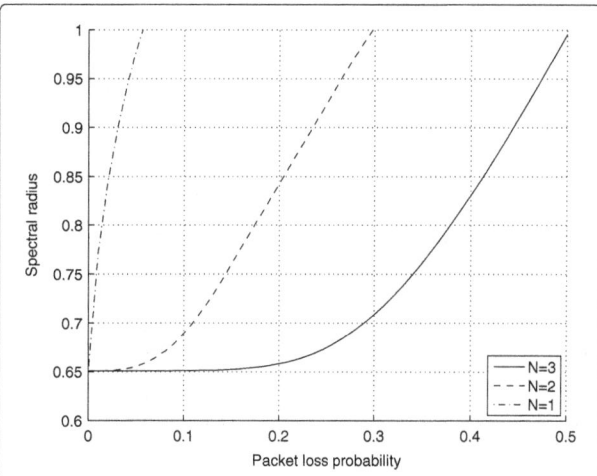

Fig. 4 Spectral radius. Spectral radius $\sigma_r(\mathfrak{A})$ of $\mathfrak{A}$ in (34) for $N = 1, 2, 3$ and as a function of the packet loss probability $p \in [0, 0.5]$

the $M = N$ descriptions. Thus, to keep the bit rate from growing too much as a function of N, we have experimentally found that $\delta = 25N^2$ to be a suitable choice, i.e., $\delta = 25, 100, 225$, for $N = 1, 2, 3$, respectively.

5.7 Bit-rates

In order to compute the upper bound (45) on the bit-rate, we need to estimate the conditional variances $\sigma^2_{\bar{\xi}(i)|\bar{\xi}(1),\cdots,\bar{\xi}(i-1)}$ for $i = 1, \ldots, N$, and the quantization noise variance σ^2_n. To find $\sigma^2_{\bar{\xi}(i)|\bar{\xi}(1),\cdots,\bar{\xi}(i-1)}$, we first find Σ_Ξ in (44) by use of (34) – (40). Then, we use (46) to obtain Σ_ξ from Σ_Ξ, where Σ_ξ is the steady state covariance matrix of ξ_t. Finally, we simply use the Schur complement [39] of Σ_ξ to obtain the desired conditional variances. To estimate the quantization noise variance σ^2_n, we use the relationship $\sigma^2_n \approx \delta^2/12$, which is exact for a dithered uniform quantizer and a good approximation for a non-dithered scalar uniform quantizer. We have plotted the theoretical upper bound (45) in Fig. 5 as a function of the packet loss probability and for $N = 1, 2, 3$.

To estimate the bit-rate of the quantized control signals, we use (47), which require the computations of discrete conditional entropies. To estimate these conditional entropies, we use a histogram-based entropy estimation on the sequence of discrete (quantized) control signals $\{\tilde{u}_t\}_{t=0}^T$. Specifically, we first estimate $H(\tilde{u}_t(1))$ directly from $\{\tilde{u}_t(1)\}_{t=0}^T$. Then, we estimate $H(\tilde{u}_t(1), \tilde{u}_t(2))$ from $\{\tilde{u}_t(1), \tilde{u}_t(2)\}_{t=0}^T$ and use that $H(\tilde{u}_t(2)|\tilde{u}_t(1)) = H(\tilde{u}_t(1), \tilde{u}_t(2)) - H(\tilde{u}_t(1))$. We obtain $H(\tilde{u}_t(3)|\tilde{u}_t(1), \tilde{u}_t(2)) = H(\tilde{u}_t) - H(\tilde{u}_t(1), \tilde{u}_t(2))$ in a similar way. Finally, these estimates of the conditional entropies are inserted into (47) in order to approximate the resulting operational bit-rate R. The resulting total discrete entropy $R_T = RM$ obtained by adding the entropies

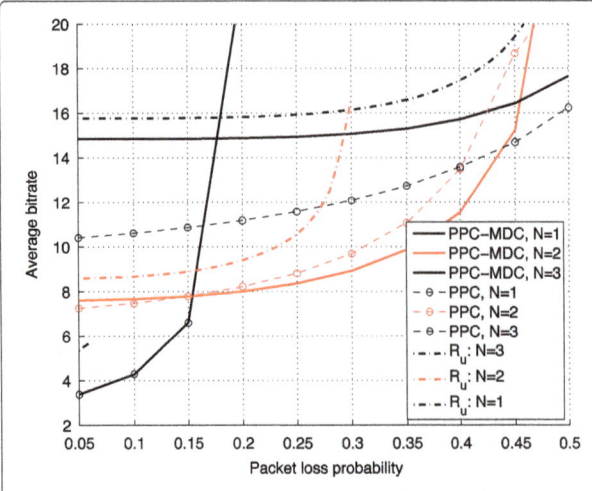

Fig. 5 Average entropy. Average entropy as a function of the packet loss probability

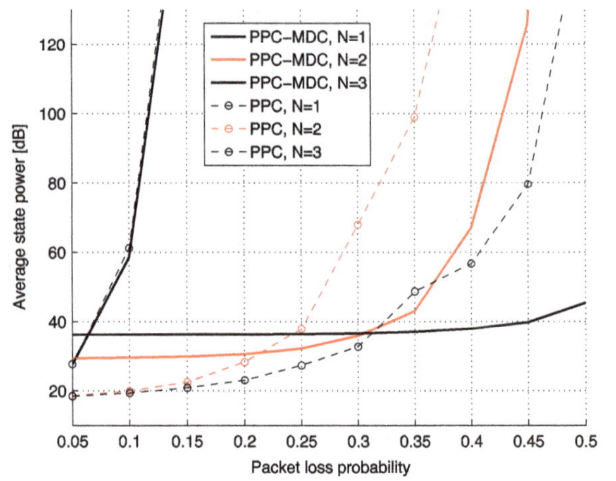

Fig. 6 Average state power. Average state power as a function of the packet loss probability

of the M descriptions is shown in Fig. 5 as a function of the packet loss rate and $M = N$.

It may be noticed in Fig. 5 that the upper bound (45) is approximately 1 bit above the estimate R_T of the operational bit-rates except in the region, where the packet loss rates approach and exceed the critical point, where the system becomes unstable. This excess 1 bit accounts for the theoretical loss of an entropy coder. While we have not applied actual entropy coding, it is well known that the loss of the entropy coder diminishes at moderate to large bit rates.

Note that the multiple descriptions of PPC-MDC have a certain amount of controlled redundancy, and one might therefore expect that the total coding rate for all $M = N$ descriptions would be much greater than what is used for the single description in PPC. However, due to being a closed-loop system, packet losses affect the variance of the input to the quantizer. Consequently, the resulting coding rate for PPC as well as for PPC-MDC also depend upon the packet loss rate.

5.8   Performance

We have measured the performance of the system in terms of the average state power $\frac{1}{T} \sum_{t=1}^{T} \|x_t\|_2^2$. This is shown in Fig. 6. For smaller packet loss rates, the performance of PPC is better than that of PPC-MDC for $N > 1$. This is because the negative impact on the performance due to quantization in PPC-MDC out-weights the impact due to using future predicted control values in PPC in case of packet losses. Recall that the quantizer in PPC-MDC is coarser than that used in PPC. When the packet loss rate is increased, PPC-MDC is often able to apply the most recent control value $\tilde{u}_t(1)$ due to the construction of the MDs. On the other hand, PPC will frequently be applying the future predicted control values $\tilde{u}_t(2)$ and $\tilde{u}_t(3)$ due the

packet dropouts. This leads to a significant performance gain of PPC-MDC at higher packet loss rates.

5.9   Complexity

From the analysis of the MJLS in Section 3, it is not easy to assess the computational burden required, when using the proposed system in practice. In this section, we provide a brief overview of the complexity of the encoder and decoder. The encoder includes the controller, quantizer, entropy coder, and channel (FEC) coder. The decoder includes channel decoder, entropy decoder, buffering, and selection of the control values:

Encoder

1. At any given time, say t, the control vector $\bar{u}_t$ is constructed as in (10) and (13), which amount to a few matrix vector multiplications. The matrices in question are $\Gamma \in \mathbb{R}^{N \times z}$ and $\Psi \in \mathbb{R}^{N \times N}$, where N is the horizon length and z is the state dimension. For many applications, both the horizon length and the state dimension are moderately small.

2. Each scalar element in either the control vector $\bar{u}_t \in \mathbb{R}^N$ or in $\xi_t \in \mathbb{R}^N$ is quantized using a scalar quantizer as described in Section 5.6. This amounts to N simple rounding operations, which can be done efficiently in hardware.

3. The quantized elements are entropy encoded either independently, conditionally, or jointly. In either case, it is done in practice by look-up tables and is therefore of low complexity, i.e., $\mathcal{O}(N)$.

4. The resulting bitstream after entropy coding is converted into M packets by applying M FEC codes, which amounts to matrix-vector multiplications over finite fields [40]. If we use $M = N$ packets, and

thereby split the control vector into N "layers", then the ith layer uses $i \times N$ multiplications due to the (N, i) FEC code. Thus, the total number of multiplications is $N \times (1 + 2 + \cdots + N) = \mathcal{O}(N^3)$.

Decoder

1. At the decoder at time t, all received packets that are no older than time $t - N + 1$, are stored in a buffer. Moreover, all decoded control values that are no older than time $t - N + 1$ time delays are stored in another buffer. Thus, since there can be $M \le N$ packets in each time slot, the storage complexity is $\mathcal{O}(MN)$.

2. Decoding of received packets involves decoding the FEC code and decoding the entropy code. Decoding the FEC code can be done by, e.g., Gaussian elimination, which has complexity $\mathcal{O}(N^3)$ per layer, and therefore at most $\mathcal{O}(N^4)$ for decoding the entire control vector. Decoding of the entropy code is done by a look-up table and has, thus, complexity $\mathcal{O}(N)$, since the control vector contains N elements.

3. If the decoded control signals are stored in U_t (18), then the selection of the control signal from the buffer can be done as suggested in (19). This includes construction of the vector $\tilde{\delta}_i$ in addition to forming the inner product of $\tilde{\delta}_i$ and the diagonal of U_t indexed by γ_i. The inner product has complexity $\mathcal{O}(N)$.

6 Conclusions

We have shown how to combine multiple description coding with quantized packetized predictive control, in order to get a high degree of robustness towards packet delays and erasures in network control systems. We focused on a digital network located between the controller and the plant input. In our scheme, when any single packet is received, the most recent control value becomes available at the plant input. Moreover, when any J out of M packets are received, the most recent control value and $J - 1$ future predicted control values become available at the plant input. These future-predicted control values can then be applied at time instances, where no packets are received. The key motivation for this design was twofold. From a practical point of view, it was shown that a significant gain over existing packetized predictive control was possible in the range of large packet loss rates. Moreover, from a theoretical point of view, computable guarantees for stability and upper bounds on the operational bit rate could be established. Indeed, a simulation study revealed that the upper bounds on the bit rate was a good indicator for the operational bit rate of the system in the range of packet loss probabilities that were not too close to the region of system instability.

Future works could include source coding in the feedback channel as well as the forward channel, which is a non-trivial extension. Indeed, the design and analysis of optimal joint controller, encoders, and decoders in both forward and backward channels is an open problem even in the absence of erasures and delays. The main difficulty is that the design of the source coder in the forward channel hinges heavily on the design of the source coder in the backward channel as well as on the controller. Another interesting open research direction is to establish lower bounds on the bit rates, which will then make it possible to assess the optimality of the overall system architecture from an information theoretic point of view.

Endnotes

[1] For the case of quantized MPC with fixed-rate quantization and without dithering, it was shown in [41], that the optimal quantized control vector is given by nearest neighbour quantization of ξ_t in (10).

[2] It follows that we require the dither sequence to be known both at the encoder and at the decoder.

[3] We will explicitly make use of (34) – (40) and Lemma 3.3, when assessing the stability of the system in the simulation study in Section 5.

[4] The excess 1 bit is due to the conservative estimate of the loss of the entropy coder, which is characterized by 1 bit.

[5] If they are not of equal length, it is always possible to augment one of the sub-bitstreams with a fixed (known) bit pattern to make them of equal length.

[6] Matlab code to reproduce all results (figures and tables) will be made available online on the authors webpage.

[7] Of course, information about what time instances the packets were received can be learned from past Δ's. However, we are not exploiting this knowledge here.

Appendix 1: Proof of Lemma 3.1

Let us first consider the case $M = N$. In this case, each row of Δ_t can take on $N + 1$ distinct patterns, i.e.,

$$[\overbrace{1 \cdots 1}^{m} \overbrace{0 \cdots 0}^{N-m}], \quad m = 0, \ldots, N,$$

where m describes the number of packets received for the time slot corresponding to that particular row. The first row of Δ'_t is equivalent to the first row of Δ_t. The remaining rows of Δ'_t can each either be the zero vector or any one of the following:

$$[\overbrace{0 \cdots 0}^{m-k} \overbrace{1 \cdots 1}^{k} \overbrace{0 \cdots 0}^{N-m}], m = 0, \ldots, N, k = 1, \ldots, m,$$

where k describes the number of packets received at time t and which contain control signals for that particular row in the buffer. Thus, the number of distinct patterns for each of these rows are $1 + \sum_{m=1}^{N} m$. Since there is a total of $N - 1$ of such rows, the total number of distinct difference matrices is

$$(N+1)\left(1 + \sum_{m=1}^{N} m\right)^{N-1} = (N+1)\left(1 + \frac{1}{2}N(N+1)\right)^{N-1}.$$

The case of $M < N$ follows easily from the above analysis. In this case, each row of Δ_t can only take on $M + 1$ distinct patterns, i.e., the zero vector, or a vector containing the number of consecutive ones corresponding to the number of control values that are recovered, when receiving J out of the M packets, where $J = 1, \ldots, M$. It follows immediately that the number of possible difference indicator matrices is less for $M < N$ compared to $M = N$.

$\square$

Appendix 2: Proof of Lemma 3.2

We first prove ergodicity. Clearly, from the all zero difference indicator matrix, it is possible to get to any other difference indicator matrix in a finite number of steps. Moreover, the probability of not receiving any packets in N consecutive time steps is positively bounded away from zero for any finite N. The all zero difference indicator matrix can therefore be reached in a finite number of steps (from any other difference indicator matrix). Thus, it is possible to jump between any two difference indicator matrices in a finite number of steps. We may therefore view the difference indicator matrices as being the different nodes in a fully connected graph. In this graph, any node can be reached at irregular times. Thus, the nodes are recurrent and aperiodic, which implies that they are are ergodic and the sequence $\{\Delta_t'\}$ of difference indicator matrices is therefore also ergodic.

We now prove the Markovian property. Observe that the matrices in the sequence $\{\Delta_t\}$ are not mutually independent. However, the sequence does satisfy a first-order Markov condition due to the Markov assumption on the data reception, see Section 2.3, i.e.,

$$\Delta_0^{t-1} \leftrightarrow \Delta_t \leftrightarrow \Delta_{t+1}, \quad \forall t, \tag{48}$$

which implies that knowledge of the buffer $\bar{f}_{t-1}'$ does not bring more *useful*[7] information about the buffer $\bar{f}_{t+1}'$ if the buffer $\bar{f}_t'$ is already known. Similarly, it is easy to see that the sequence $\{\Delta_t'\}$ of difference matrices form a Markov chain similar to (48), i.e.,

$$\{\Delta_i'\}_{i=0}^{t-1} \leftrightarrow \Delta_t' \leftrightarrow \Delta_{t+1}', \quad \forall t. \tag{49}$$

Finally, the stationarity of the channel, see Section 2.3, implies that the sequence of difference matrices $\{\Delta_t'\}$ is stationary. This proves the lemma.

$\square$

Appendix 3: Proof of Theorem 3.1

In Lemma 3.2, we have established ergodicity and Markov properties of the switching sequence $\{\Delta_t'\}$ as is required by Lemma 3.3. We then need to derive the recursive form for the system evolution, which guarantees that the combined system $\{\Xi_t, \Delta_t'\}$ will be Markovian.

Recall from (19) that $\bar{f}_t(i) = \tilde{\delta}_i^T \gamma_i(U_t)$ for $i = 1, \ldots, N$. However, to avoid updating the buffer with information about packets that was already received in previous time instances, we need to look only at the changes between Δ_t and Δ_{t-1}. Towards that end, let $\delta_i' \triangleq \gamma_i(\Delta_t - S^{\downarrow}\Delta_{t-1})$. Define $\tilde{\delta}_i'$ in a similar manner as $\tilde{\delta}_i$. If δ_i' is the all zero vector for some i, it means that no new control signals to be used at time $t + i - 1$ has been received yet. Thus, the ith element of the buffer is then simply obtained by using older control signals, i.e., $\bar{f}_t(i) = \bar{f}_{t-1}(i + 1)$. With this, we obtain the following recursion:

$$\bar{f}_t(i) = \tilde{\delta}_i'^T \gamma_i(U_t) + \left(1 - \mathbf{1}_{N-i+1}^T \tilde{\delta}_i'\right)\bar{f}_{t-1}(i+1). \tag{50}$$

Using (10) and (14), it is possible to write $\gamma_i(U_t)$ as a function of $\bar{x}_t$ and $\bar{n}_t$, that is:

$$\gamma_i(U_t) = [u_t(i), u_{t-1}(i+1), \cdots, u_{t-N+i}(N)]^T$$

$$= \begin{bmatrix} e_i^T \Psi^{-1} \Gamma x_t \\ e_{i+1}^T \Psi^{-1} \Gamma x_{t-1} \\ \ddots \\ e_N^T \Psi^{-1} \Gamma x_{t-N+i} \end{bmatrix} + \begin{bmatrix} e_i^T \Psi^{-1} n_t \\ e_{i+1}^T \Psi^{-1} n_{t-1} \\ \vdots \\ e_N^T \Psi^{-1} n_{t-N+i} \end{bmatrix}.$$

With the above notation, the system state vector recursions can be written as:

$$x_{t+1} = Ax_t + B_1(\tilde{\delta}_1'^T \gamma_1(U_t) + (1 - \mathbf{1}_N^T \tilde{\delta}_1')e_1^T S^{\uparrow} \bar{f}_{t-1}) + B_2 w_t. \tag{51}$$

Using that $\Xi_t = \begin{bmatrix} \bar{x}_t \\ \bar{f}_t \end{bmatrix}$ and combining (51) and (50) and using the matrix definitions in (24) – (33) yields (23). This proves the theorem.

$\square$

Appendix 4: Proof of Theorem 3.3

In order to provide an upper bound on the required bit rate for transmitting the quantized control vector ξ_t', we assume that the system is designed such that the loop is AWSS. For such a system, the bit rate R of the ECDQ is related to the discrete entropy $H(\xi_t'|\zeta_t)$ of the quantized signal ξ_t', conditioned upon the dither signal ζ_t [31]. That is,

$$H(\xi_t'|\zeta_t) \leq R \leq H(\xi_t'|\zeta_t) + 1/N, \tag{52}$$

where the term $1/N$ is the loss due to using entropy coding on finite dimensional vectors [42]. At this point, we could continue upper bounding $H(\xi_t'|\zeta_t)$, which would then provide an upper bound on the bit rate required for coding the entire control vector. However, recall that we need to send ξ_t' using M descriptions such that upon receiving any $0 < J \le M$ descriptions, the J control signals $\tilde{u}_t(1), \ldots, \tilde{u}_t'$ can be reliably recovered. This is clearly not possible if ξ_t' is arbitrarily split into M sub-streams having a total bit rate of $H(\xi_t'|\zeta_t)$. In Section 4.2, we introduced a practical scheme for MDs based on forward error correction codes. With this scheme, a description is constructed by concatenating the entire bitstream used for representing the encoded version of $\tilde{u}_t(1)$ with half the bitstream used for $\tilde{u}_t(2)$, one third of the bits allocated for $\tilde{u}_t(3)$, and so on. With such a scheme in mind, we first invoke the chain rule of entropies, in order to expand $H(\xi_t'|\zeta_t)$ in (52) as:

$$H(\xi_t'|\zeta_t) = H(\xi_t'(1)|\zeta_t) + H(\xi_t'(2)|\xi_t'(1), \zeta_t) + \cdots \\ + H(\xi_t'(N)|\xi_t'(1), \ldots, \xi_t'(N-1), \zeta_t). \quad (53)$$

The ith term on the r.h.s. of (53), describes the minimum bit rate required for conditionally encoding $\xi_t'(i)$. With this, and using the MD construction sketched above, the total rate R_T required for all $M = N$ descriptions is given by:

$$R_T \le M\left(H(\xi_t'(1)|\zeta_t) + \frac{1}{2}H(\xi_t'(2)|\xi_t'(1), \zeta_t) + \cdots \\ + \frac{1}{M}H(\xi_t'(N)|\xi_t'(1), \ldots, \xi_t'(N-1), \zeta_t) \right) + 1. \quad (54)$$

Since we are using an ECDQ, the discrete entropy of the quantized variables satisfies:

$$H(\xi_t'|\zeta_t) = I(\xi_t; \hat{\xi}_t) \quad (55)$$
$$= I(\xi_t; \xi_t + n_t) \quad (56)$$
$$\le I(\bar{\xi}_t; \bar{\xi}_t + \bar{n}_t) + \mathcal{D}(n_t \| \bar{n}_t), \quad (57)$$

where equality in (55) follows from [31] and where $I(\xi_t; \hat{\xi}_t)$ denotes the mutual information [30] between the input ξ_t and the output $\hat{\xi}_t$ of the ECDQ [31]. In (56), the equality follows by replacing the quantization operation by its additive noise model, which is exact from a statistical point of view [31]. The upper bound in (57) follows from ([43] Lemma 2) by replacing the variables in play by their Gaussian counterparts, i.e., $\bar{\xi}_t$ and $\bar{n}_t$ are Gaussian distributed with the same first- and second moments as ξ_t and n_t, respectively. The Divergence operator $\mathcal{D}(n_t \| \bar{n}_t)$ describes the Kullback-Leibler distance (in bits) between the distribution of the quantization noise n_t to that of a Gaussian distribution [30] and is in our case upper bounded by $\mathcal{D}(n_t \| \bar{n}_t) \le N/2 \log_2(\pi e/6)$. This upper bound is achieved if n_t is uniformly distributed over an N-dimensional cube [44]. We may now proceed by expressing the mutual information in terms of differential entropies provided the latter exists [30]. Thus, we obtain that $I(\bar{\xi}_t; \bar{\xi}_t + \bar{n}_t) = h(\bar{\xi}_t + \bar{n}_t) - h(\bar{n}_t)$. Using the same idea on the conditional estimates $\xi_t(i)|\xi_t(1), \cdots, \xi_t(i-1)$ instead of the entire vector ξ_t leads to:

$$H(\xi_t'(i)|\xi_t'(1), \ldots, \xi_t'(i-1), \zeta_t) \\ \le h(\bar{\xi}_t(i)|\bar{\xi}_t(1), \cdots, \bar{\xi}_t(i-1) + \bar{n}_t(i)) - h(\bar{n}_t(i)) \\ + 1/2 \log_2(\pi e/6) \\ = \frac{1}{2} \log_2 \left(1 + \frac{\sigma^2_{\xi'(i)|\xi'(1), \cdots, \xi'(i-1)}}{\sigma_n^2} \right) + \frac{1}{2} \log_2 \left(\frac{\pi e}{6} \right), \quad (58)$$

where the last equality follows from the definition of differential entropy of a Gaussian variable [30]. Inserting (58) into (54) and (52) yields (45). This completes the proof. □

Competing interests
The authors declare that they have no competing interests.

Acknowledgements
This research was partially supported by VILLUM FONDEN Young Investigator Programme, Project No. 10095. The authors would like to thank the reviewers for pointing us to references [23–25] and to [6, 7].

Author details
[1]Department of Electronic Systems, Aalborg University, Fredrik Bajers Vej 7b, Aalborg, Denmark. [2]Department of Electrical Engineering (EIM-E), Paderborn University, Paderborn, Germany.

References
1. JP Hespanha, P Naghshtabrizi, Y Xu, A survey of recent results in networked control systems. Proc. IEEE. **1**(95), 138–162 (2007)
2. GN Nair, F Fagnani, S Zampieri, RJ Evans, Feedback control under data rate constraints: an overview. Proc. IEEE. **95**(1), 108–137 (2007)
3. S Wong, RW Brockett, Systems with finite communication band- width constraints ii: stabilization with limited information feedback. IEEE Trans. Autom. Control. **44**(5), 1049–1053 (1999)
4. M Trivellato, N Benvenuto, State control in networked control systems under packet drops and limited transmission bandwidth. IEEE Trans. Commun. **58**(2), 611–622 (2010)
5. DE Quevedo, A Ahlén, J Østergaard, Energy efficient state estimation with wireless sensors through the use of predictive power control and coding. IEEE Trans. Signal Process. **58**(9), 4811–4823 (2010)
6. T Wang, Y Zhang, J Qiu, H Gao, Adaptive fuzzy backstepping control for a class of nonlinear systems with sampled and delayed measurements. IEEE Trans. Fuzzy Syst. **23**(2), 302–312 (2015)
7. T Wang, H Gao, J Qiu, A combined adaptive neural network and nonlinear model predictive control for multirate networked industrial process control. IEEE Trans. Neural Netw. Learn. Syst. **27**(2), 416–425 (2016)
8. PL Tang, CW de Silva, Compensation for transmission delays in an Ethernet-based control network using variable-horizon predictive control. IEEE Trans. Contr. Syst. Technol. **14**(4), 707–718 (2006)
9. G Liu, Y Xia, J Chen, D Rees, W Hu, Networked predictive control of systems with random network delays in both forward and feedback channels. IEEE Trans. Ind. Electron. **54**(3), 1282–1297 (2007)
10. DE Quevedo, D Nešić, Input-to-state stability of packetized predictive control over unreliable networks affected by packet-dropouts. IEEE Trans. Autom. Control. **56**(2), 370–375 (2011)
11. DE Quevedo, J Østergaard, Nešić, Packetized predictive control of stochastic systems over digital channels with random packet loss. IEEE Trans. Autom. Control. **56**(12), 2854–2868 (2011)
12. AAE Gamal, TM Cover, Achievable rates for multiple descriptions. IEEE Trans. Inf. Theory. **IT-28**(6), 851–857 (1982)
13. VK Goyal, Multiple description coding: compression meets the network. IEEE Signal Proc. Mag. **18**(5), 74–93 (2001)
14. EI Silva, MS Derpich, J Østergaard, An achievable data-rate region subject to a stationary performance constraint for LTI plants. IEEE Trans. Autom. Control. **56**, 1968–1973 (2011)

15. AS Leong, S Dey, GN Nair, Quantized filtering schemes for multi-sensor linear state estimation: stability and performance under high rate quantization. IEEE Trans. Signal Process. **61**(15), 3852–3865 (2013)

16. J Østergaard, DE Quevedo, A Ahlen, in *IEEE Int. Conference on Audio, Speech and Signal Processing*. Predictive power control and multiple-description coding for wireless sensor networks (IEEE, 2009), pp. 2785–2788

17. J Østergaard, DE Quevedo, in *9th IEEE International Conference on Control & Automation*. Multiple descriptions for packetized predictive control over erasure channel (IEEE, Santiago, Chile, 2011)

18. J Østergaard, DE Quevedo, in *IEEE Data Compression Conference*. Multiple description coding for closed loop systems over erasure channels (IEEE, 2013)

19. Z Jin, V Gupta, R Murray, State estimation over packet dropping networks using multiple description coding. IEEE Trans. Autom. Control. **42**(9), 1441–1452 (2006)

20. DE Quevedo, J Østergaard, A Ahlen, A power control and coding formulation for state estimation with wireless sensors. IEEE Trans. Control Syst. Technol. **22**(2), 413–427 (2014)

21. E Peters, DE Quevedo, J Østergaard, Shaped Gaussian dictionaries for quantized networked control systems with correlated dropouts. IEEE Trans. Signal Process. **64**(1), 203–213 (2016)

22. Y Ge, Q Chen, M Jiang, Y Huang, J. Control Sci. Eng. **2013** (2013)

23. Y Wei, J Qiu, H Karimi, M Wang, A new design of H_∞ filtering for continuous-time Markovian jump systems with time-varying delay and partially accessible mode information. Signal Process. **93**(9), 2392–2407 (2013)

24. Y Wei, J Qiu, H Karimi, M Wang, Filtering design for two-dimensional Markovian jump systems with state-delays and deficient mode information. Inf. Sci. **269**(10), 316–331 (2014)

25. J Qiu, Y Wei, H Karimi, New approach to delay-dependent H_∞ control for continuous-time Markovian jump systems with time-varying delay and deficient transition descriptions. J. Frankl. Inst. **352**, 189–215 (2015)

26. Y Fang, KA Loparo, Stochastic stability of jump linear systems. IEEE Trans. Autom. Control. **7**, 1204–1208 (2002)

27. OLV Costa, MD Fragoso, RP Marques, *Discrete-time Markov Jump Linear Systems*. (Springer, 2005)

28. JB Rawlings, DQ Mayne, *Model Predictive Control: Theory And Design*. (Nob Hill Publishing, 2009)

29. DE Quevedo, E Silva, GC Goodwin, Packetized predictive control over erasure channels. Proc. Amer. Contr. Conf, 1003–1008 (2007)

30. TM Cover, JA Thomas, *Elements of Information Theory*, 2nd edn. (Wiley-Interscience, 2006)

31. R Zamir, M Feder, On universal quantization by randomized uniform/lattice quantizers. IEEE Trans. Inform. Theory. **38**(2), 428–436 (1992)

32. VA Vaishampayan, NJA Sloane, SD Servetto, Multiple-description vector quantization with lattice codebooks: design and analysis. IEEE Trans. Inf. Theory. **47**(5), 1718–1734 (2001)

33. J Østergaard, J Jensen, R Heusdens, n-channel entropy-constrained multiple-description lattice vector quantization. IEEE Trans. Inf. Theory. **52**(5), 1956–1973 (2006)

34. J Kovačević, PL Dragotti, VK Goyal, Filter bank frame expansions with erasures. IEEE Trans. Inf. Theory. **48**(6), 1439–1450 (2002)

35. J Østergaard, R Zamir, Multiple description coding by dithered delta-sigma quantization. IEEE Trans. Inf. Theory. **55**(10), 4661–4675 (2009)

36. R Yeung, R Zamir, in *Proceedings of IEEE International Symposium on Information Theory*. Multilevel diversity coding via successive refinement (IEEE, Ulm, Germany, 1996)

37. R Puri, K Ramchandran, Conf. Record Thirty-Third Asilomar Conf Signals Syst. Comput. **1**, 342–346 (1999)

38. RC Singleton, Maximum distance q-nary codes. IEEE Trans. Inf. Theory. **10**(2), 116–118 (1964)

39. S Boyd, L Vandenberghe, *Convex Optimization*. (Cambridge University Press, 2004). Appendix A.5.5

40. J Sundararajan, D Shah, M Medard, S Jakubczak, M Mitzenmacher, J Barros, Network coding meets tcp: theory and implementation. Proc. IEEE. **99**(3), 490–512 (2011)

41. DE Quevedo, GC Goodwin, JA De Doná, Finite constraint set receding horizon quadratic control. Int. J. Robust Nonlin. Contr. **14**(4), 355–377 (2004)

42. CE Shannon, A mathematical theory of communications. Bell Syst. Tech. J. **27** (1948)

43. MS Derpich, J Østergaard, DE Quevedo. Achieving the quadratic Gaussian rate-distortion function for source uncorrelated distortions, (2008). Electronically available on arXiv.org: http://arxiv.org/pdf/0801.1718.pdf. Accessed Oct 2015

44. R Zamir, M Feder, On lattice quantization noise. IEEE Trans. Inf. Theory. **42**(4), 1152–1159 (1996)

5

Exploiting sensor mobility and covariance sparsity for distributed tracking of multiple sparse targets

Guohua Ren[1], Vasileios Maroulas[2] and Ioannis D. Schizas[1*]

Abstract

The problem of distributed tracking of multiple targets is tackled by exploiting sensor mobility and the presence of sparsity in the sensor data covariance matrix. Sparse matrix decomposition relying on norm-one/two regularization is integrated with a kinematic framework to identify informative sensors, associate them with the targets, and enable them to follow closely the moving targets. Coordinate descent techniques are employed to determine in a distributed way the target-informative sensors, while the modified barrier method is employed to minimize proper error covariance matrices acquired by extended Kalman filtering. Different from existing approaches which force all sensors to move, here, local updating recursive rules are obtained only for the target-informative sensors that can update their location and follow closely the corresponding targets while staying connected. Simulations advocate that the proposed scheme outperforms alternative tracking schemes while accurately tracks multiple targets by imposing mobility only on the target-informative sensors.

Keywords: Mobile sensor networks, Distributed processing, Sparse decomposition, Multi-target tracking

1 Introduction

In recent years, potential applications of sensor networks (SN) have expanded due to the low cost of the sensing units, their ability to cover large areas, and the robustness distributed processing offers. One characteristic exploited more and more in sensor networks is sensor mobility and the design of kinematic rules that control sensor movement. Sensor mobility adds extra flexibility to a sensor network making it capable of covering larger areas, as well as being more energy efficient and robust [1]. Mobile sensors have been extensively utilized in target tracking applications to enable sensors to closely follow the moving target(s) and provide accurate target location estimates [2–4]. The aforementioned approaches require all the sensors to keep active [2–4], which may lead to excessive resource consumption despite the targets' locality and the fact that in practice a small portion of sensors may possess useful information about the present targets. The aim here is to design an adaptive scheme that exploits mobility

and covariance sparsity to associate targets with sensors and then properly determine kinematic strategies only for the informative sensors which will closely follow the field targets.

In the absence of sensor mobility, there has been a plethora of approaches for tracking multiple targets while associating targets with sensor measurements. Existing works [5–8] associate measurements acquired at static sensors with targets across time and rely heavily on probability models. A distributed Kalman filtering scheme is proposed in [9] relying on information diffusion strategies. In [9], only neighboring sensors collaborate, though all sensors in the network are utilized to track a single source while sensors have fixed locations. A different approach is followed in [10], where consensus-averaging is employed across the whole sensor network and all the sensors are forced to be active irrespective of the quality of their measurements. In [11], a related single-target distributed tracking approach is proposed, in which extended Kalman filtering is employed for tracking. A probability model is assumed to determine informative sensors which may lead to instability due to its dependence on the tracking estimates. Different in this paper,

*Correspondence: schizas@uta.edu
[1] Department of Electrical Engineering, University of Texas at Arlington, Arlington, TX, USA
Full list of author information is available at the end of the article

distributed tracking of multiple targets will be considered, while sensor mobility will be exploited and combined with a sensor-to-target association scheme for selecting target-informative sensors without the need of relying on model parameters and state estimators that maybe inaccurate and result divergence. It should be pointed out that the distributed characterization here is referring to the fact that (i) only neighboring sensors need to communicate with each other and collaborate for multi-target tracking, while (ii) processing will take place in a few head sensors and will not involve all sensors in the network but only those sensors that bear information about the moving targets.

Single-target tracking using mobile sensor networks has been studied for a variety of different scenarios [12–14]. Most of these approaches control the movement of *all* sensors by minimizing the estimation error covariance, [4, 13], while the approach in [12] manages sensor mobility based on a Bayesian estimation model and restricting sensors to move only on a grid of locations. A path planning strategy for a setting involving a fixed-location target and a single moving sensor is designed in [15] by maximizing the determinant of the Fisher information matrix corresponding to the configuration. In [16], an approach is proposed for controlling the trajectories of multiple UAVs by minimizing the localization uncertainty for a fixed-location target setting where the target is emitting a radio signal. The work in [17] rigorously presents how sensor mobility can increase spatial resolution when tracking a target with mobile sensors.

When tracking multiple targets with mobile sensors, the approach in [3] proposed an active sensing model, whereas the target-sensor association is based on a nearest neighbor rule which heavily relies on the accuracy of the state estimator while a central processing center is required. The scheme in [18] tackled the problem of moving sensors using a flock control law where all sensors are utilized, while the targets are some of the moving sensors whose position is known. The approach in [19] is utilizing clustering and neural networks to move sensors under the assumption that target locations are available. The scheme in [20] designs a Kalman filtering approach with gradient descent-based kinematic rules under the assumption that it is known which targets every sensor observes bypassing in that way the essential sensor-to-target association step. These schemes involve movement of all sensors at every time instant leading to resource-demanding algorithms that do not exploit spatial locality of the field targets.

Measurements corresponding to sensors which are close to the same target tend to be statistically correlated. Given that targets are spatially localized and affect small portions of the sensor network, an approximately sparse sensor data covariance matrix is emerging. Sparsity (presence of a many zero entries in a vector or matrix) has

been exploited in a wide range of applications including sparse regression and statistical inference, e.g., see [21, 22]. The problem of associating targets to sensors, as well as determining the sensors with target-informative measurements, is formulated here as the task of decomposing a matrix into sparse factors. The sparse matrix factorization techniques in [23, 24] are integrated here with proper sensor kinematic strategies and tracking techniques to exploit sensor mobility. Note that in [23, 24], a stationary (immobile) sensor network is considered where sensors have fixed locations. Tracking in [23, 24] is performed by immobile sensors, whereas here tracking is generalized to a mobile network with the more challenging task of designing and integrating with multi-target tracking, sensor kinematic strategies that improve tracking accuracy while preserving local sensor network connectivity. Norm-one and norm-two regularization mechanisms are employed to formulate a pertinent minimization framework that recovers sparse covariance factors, while estimates the number of targets on the field. Coordinate descent techniques [25, 26] are employed to derive local updating recursions that allow sensors to associate with targets.

Different from the aforementioned tracking schemes using sensor mobility, here, only the target-informative sensors will be enabled to move at every time instant and track closely the moving targets. Thus, only target-affected portions of the sensor network will be used for tracking the moving targets, potentially resulting better resource consumption and prolongation of the network lifetime. Kinematic rules will be designed by minimizing proper error covariance matrices obtained by extended Kalman filtering recursions [27] used to track each of the targets. The minimization will be performed under connectivity constraints that ensure the moving sensors stay connected and are able to communicate. The modified barrier method ([26], pg. 423) is employed to solve a pertinent constrained minimization problem and obtain distributed kinematic rules that the mobile sensors can apply locally without the need of a central controller. In contrast to existing approaches, the novel framework identifies and controls the movement *only* of target-informative sensors allowing for accurate tracking.

The novelties of the proposed framework with respect to existing work involves the following: (i) utilization of covariance sparsity and properly designed kinematic rules to perform dynamic spatio-temporal sensor-to-target association and tracking using mobile sensors not present in [11, 23, 24]; (ii) different from existing approaches [12–14], it enables only the sensors acquiring informative measurements about the targets to move instead of moving all sensors to track a single target; (iii) different from [11], it does not rely on probabilistic models or state estimates to associate sensors with targets; (iv) it takes local sensor network connectivity into account to ensure that the tracking process carries out continuously; and (v) different from [15, 16] that consider fixed-location target(s), here, kinematic strategies are developed for tracking *multiple moving* targets. The proposed approach here achieves this task only by using the sensor data and sparsity-imposing mechanisms.

The paper is structured as follows. The problem formulation and setting are given in Section 2. The sensor-to-target association scheme which provides the target-informative sensors is delineated in Section 3.1 and is combined with extended Kalman filtering techniques in Section 3.2. Novel kinematic rules are developed in Section 3.3 after minimizing a pertinent error covariance matrix under connectivity constraints and employing the modified barrier method to derive local kinematic rules that enable the target-informative sensors to update their location and follow closely to moving targets. The proposed algorithmic framework is detailed in Section 4.1, while the communication and computational costs are discussed in Section 4.2. Extensive numerical tests are carried out in Section 5 that corroborate the advantages of the novel method over existing alternatives.

2   Problem setting

An ad hoc sensor network conformed by p mobile sensors is considered here. The sensors monitor a field where an unknown and possibly time-varying number of moving targets is present. Each sensor communicates only with its neighboring sensors which are within its communication range and are able to exchange information with a single-hop of communication. The single-hop neighborhood of sensor j is denoted as $\mathcal{N}_j(t)$, where t denotes the time index.

In general, all targets are assumed to be moving in a K-dimensional space. Then, every target, say the ρth, is characterized by a $2K \times 1$ state vector which contains both its position coordinates and velocity information for each coordinate. The position coordinates for the ρth target at time t are stacked in vector $\mathbf{p}_\rho(t) = \left[p_{\rho,x_1}(t),\dots,p_{\rho,x_K}(t)\right]^T$, while the velocity per coordinate is in vector $\mathbf{v}_\rho(t) = \left[v_{\rho,x_1}(t),\dots,v_{\rho,x_K}(t)\right]^T$. So at time t, the state vector can be written as $\mathbf{s}_\rho(t) = \left[\mathbf{p}_\rho^T(t),\mathbf{v}_\rho^T(t)\right]^T$, while it evolves according to a near constant velocity model [28]. Specifically, the ρth target's state vector evolves according to the following constant velocity model, see e.g., [28]:

$$\mathbf{s}_\rho(t+1) = \mathbf{F}\mathbf{s}_\rho(t) + \mathbf{u}_\rho(t), \tag{1}$$

where $\mathbf{F}$ is the $2K \times 2K$ state transition matrix and $\mathbf{u}_\rho(t)$ the zero-mean Gaussian state noise with variance $\mathbf{\Sigma}_u$. Matrices $\mathbf{F}$ and $\mathbf{\Sigma}_u$ are given as follows:

$$\mathbf{F} = \begin{bmatrix} 1 & 0 & \dots & \Delta T & \dots & 0 \\ \vdots & \vdots & \vdots & \vdots & \vdots & \vdots \\ 0 & 1 & 0 & 0 & \dots & \Delta T \\ 0 & 0 & 1 & \dots & 1 & 0 \\ 0 & 0 & 0 & \dots & 0 & 1 \end{bmatrix}, \tag{2}$$

$$\mathbf{\Sigma}_u = \sigma_u^2 \begin{bmatrix} (\Delta T)^3/3 \cdot \mathbf{I}_K & (\Delta T)^2/2 \cdot \mathbf{I}_K \\ (\Delta T)^2/2 \cdot \mathbf{I}_K & \Delta T \cdot \mathbf{I}_K \end{bmatrix},$$

where σ_u^2 is the noise variance and $\mathbf{I}_K$ denotes the identity matrix of size $K \times K$, while ΔT denotes the sampling period.

Sensor j senses the moving targets, by acquiring at time t a scalar measurement depending on the target location according to the following nonlinear model:

$$x_j(t) = \sum_{\rho=1}^{R} a_\rho(t) d_{j,\rho}^{-2}(t) + w_j(t), \ j = 1,\dots,p, \tag{3}$$

where $a_\rho(t)$ denotes the intensity of a signal emitted from the ρth target and $d_{j,\rho}(t) = \|\mathbf{p}_j(t) - \mathbf{p}_\rho(t)\|$ is the distance between sensor j and the ρth target at time t. The total number of targets which move in the field through the lifespan of the SN is indicated as R, while $w_j(t)$ denotes the white sensing noise with variance σ_w^2 and zero-mean. The following assumptions are introduced in the considered setting:

- **A1:** In the measurement model in (3), it is assumed that the targets act as transmitters and each sensor will receive one reflection of the signal emitted from the targets. Signals emitted from the targets propagate via free space, explaining the $d_{j,\rho}^{-2}(t)$ attenuation coefficients, and are superimposed as shown in (3), see e.g., [29].
- **A2:** The signal amplitudes $a_\rho(t)$ are considered to be uncorrelated across the different targets.
- **A3:** Among the summands $a_\rho(t) d_{j,\rho}^{-2}(t)$ in (3), only one has a large amplitude when sensor j is close to the ρth target, whereas others are negligible due to the square-law attenuation $d_{j,\rho}^{-2}(t)$ caused by the free space propagation.

Note that assumption A3 corresponds to a setting where at most, one target is present within the sensing range of a sensor. Note that this is a more relaxed version of the common assumption that one sensor just contains the measurement of a specific target [6–8]. The signal amplitudes $a_\rho(t)$ will be nonzero for the interval in which the corresponding target is active and moving while is kept at zero when the target is inactive and disappears.

The emitted, from the targets, signals $a_\rho(t)$ could correspond to communication radio signals that possibly the targets are transmitting, e.g., targets could correspond to cell phone users moving in an area or unnamed aerial/ground vehicles or military vehicles that move within the monitored area and need to be tracked, see e.g.,

[16]. The deployed sensors are listening for these signals to track the moving entities. The targets could correspond to independent entities; thus, it is expected that the information bits they transmit are uncorrelated, giving rise to uncorrelated transmission signals [30]. Thus, the communication radio signals that the targets may be emitting are utilized to perform tracking and move the sensors appropriately. Applications include localization and tracking of mobile users in wireless networks, as well as tracking of vehicles in tactical environments [16].

Stacking all the sensor measurements in (3) on a $p \times 1$ vector, it follows:

$$\mathbf{x}_t = \mathbf{D}_t \mathbf{a}_t + \mathbf{w}_t, \text{ where } \mathbf{a}_t := [a_1(t) \ a_2(t) \ldots a_R(t)]^T, \tag{4}$$

where $\mathbf{D}_t$ is a $p \times R$ matrix with entries $\mathbf{D}_t(j, \rho) = d_{j,\rho}^{-2}(t)$ with $j = 1, \ldots, p$ and $\rho = 1, \ldots, R$. The noise $\mathbf{w}_t$ has covariance $\Sigma_w = \sigma_w^2 \mathbf{I}_p$. Given that the entries of $\mathbf{a}_t$ are uncorrelated, it follows readily that the data covariance matrix is

$$\Sigma_{x,t} = \mathbf{D}_t \Sigma_a \mathbf{D}_t^T + \sigma_w^2 \mathbf{I}_p = \bar{\mathbf{D}}_t \bar{\mathbf{D}}_t^T + \sigma_w^2 \mathbf{I}_p, \tag{5}$$

where Σ_a is the diagonal covariance matrix of $\mathbf{a}_t$, while $\bar{\mathbf{D}}_t := \mathbf{D}_t \Sigma_a^{1/2}$. Among the R entries in $\mathbf{a}_t$, there will be $r(t)$ nonzero entries corresponding to the active targets moving at the sensed field at t. In the setting here, once a target becomes inactive (i.e., $a_\rho(t) = 0$), it remains inactive for the rest of time.

The goal is to enable the mobile sensors to track an unknown number of targets present in the monitored field. Novel target association and sensor mobility strategies will be combined with tracking techniques to enable sensors to accurately track the different target trajectories. Proper kinematic strategies will be developed to allow only a small percentage of target-informative sensors to move, different from existing approaches [12, 14] where *all* sensors are moving at every time instant that may be more resource demanding. Judiciously selecting and moving sensors will enable target tracking even when the targets move outside the area originally monitored by the sensors.

3 Distributed association, tracking, and sensor kinematic strategies

3.1 Target-informative sensor selection

Due to the presence of multiple target in the monitored field, the first goal is to determine sets of sensors, namely $\mathcal{S}_{\rho,t}$, that acquire information bearing measurements about the ρth target. From the observation model in (4), note that the strong-amplitude entries of the ρ column in $\mathbf{D}_t$, namely $\{\mathbf{D}_{t,:\rho}\}_{\rho=1}^R$, can reveal the sensors within subset $\mathcal{S}_{\rho,t}$. Specifically, recall that $\mathbf{D}_t(j, \rho) = $

$d_{j,\rho}^{-2}(t)$; thus, when sensor j and target ρ are close in distance then the corresponding entry is expected to have large amplitude, while the further away they get from each other the closer to zero the corresponding entry gets. The matrix $\mathbf{D}_t$ can be assumed approximately sparse. Thus, the strong-amplitude entries (away from zero) in $\mathbf{D}_{t,:\rho}$ can be used to determine the informative sensor members of $\mathcal{S}_{\rho,t}$ at time instant t. Thus, determining $\mathcal{S}_{\rho,t}$ boils down to the problem of recovering the support of the columns of $\mathbf{D}_t$.

To recover the sparse matrix $\bar{\mathbf{D}}_t$ in (4), the data covariance matrix will be decomposed into sparse factors. Due to the fact that the targets and sensors may be moving while the number of targets is changing, the sparse sensor data covariance $\Sigma_{x,t}$ is also time-varying. In practice, the ensemble covariance is not available and needs to be estimated. To this end, exponential weighing is employed to estimate the time-varying covariance entries. The notion of exponential weighing in recursive least-squares used in processing non-stationary signals, see e.g., ([31], Ch. 9), is estimating the time-varying covariance matrix here as follows:

$$\hat{\Sigma}_{x,t} = \frac{1-\omega}{1-\omega^{t+1}} \sum_{\tau=0}^{t} \omega^{t-\tau} (\mathbf{x}_\tau - \bar{\mathbf{x}}_t)(\mathbf{x}_\tau - \bar{\mathbf{x}}_t)^T, \tag{6}$$

where $\omega \in (0,1)$ denotes a forgetting factor and

$$\bar{\mathbf{x}}_t = \frac{1-\omega}{1-\omega^{t+1}} \sum_{\tau=0}^{t} \omega^{t-\tau} \mathbf{x}_\tau, \tag{7}$$

corresponds to a real-time estimate for the ensemble mean at time instant t. Note that ω in Eqs. (6) and (7) is used in a way that puts more emphasis to the recent data while it gradually forgets the past data, which is exactly what an up-to-date estimator needs to do for the time-varying setting considered here. The scaling $(1 - \omega)(1 - \omega^{t+1})^{-1}$ in (6) and (7) is to ensure that the two estimates $\hat{\Sigma}_{x,t}$ and $\bar{\mathbf{x}}_t$ for the ensemble quantities $\Sigma_{x,t}$ and $\mathbb{E}[\mathbf{x}_t]$ will be unbiased.

To account for the nearly sparse structure of $\bar{\mathbf{D}}_t$, the unknown number of targets and single-hop connectivity of the sensor nodes the following formulation relying on norm-one/norm-two regularization is utilized:

$$\left(\hat{\mathbf{M}}_t, \{\hat{\sigma}_j\}_{j=1}^m\right) := \arg \min_{\mathbf{M}_t, \{\sigma_j\}_{j=1}^m} \left\| \mathbf{E} \odot \left(\hat{\Sigma}_{x,t} - \mathbf{M}_t \mathbf{M}_t^T\right.\right.$$

$$\left.\left. -\text{diag}\left(\sigma_{1,t}^2, \ldots, \sigma_{p,t}^2\right)\right) \right\|_F^2$$

$$+ \sum_{\ell=1}^{L} \left(\lambda_\ell \|\mathbf{M}_{t,:\ell}\|_1 + \phi \|\mathbf{M}_{t,:\ell}\|_2\right), \tag{8}$$

where $\odot$ denotes the Hadamard operator (entry-wise matrix product), while σ_j^2 is the noise variance estimate at sensor j, and L is an upper bound for the number of active sensed targets $r(t)$ $(L \geq r(t))$ and $\mathbf{M}_t \in \mathbb{R}^{p \times L}$ contains L columns that estimate the sparse columns of $\bar{\mathbf{D}}_t$. $\mathbf{M}_{t,:\ell}$ denotes the ℓth column of $\mathbf{M}_t$. The formulation was first proposed in [23, 24] to perform target-sensor association in a network of stationary sensors that do not have moving capabilities. Here, this formulation will be utilized to determine the different sets of informative sensors observing different targets before being integrated with kinematic control rules.

The Hadamand operator $\odot$ along with the adjacency matrix $\mathbf{E}$ in (8) allows only the single-hop covariance entries to be used since they can be calculated by direct communication of the corresponding neighboring sensors. The first term in (8) accounts for the structure in (5). Sparsity is induced in the columns of $\mathbf{M}_t$ using the norm-one term in (8), (see e.g., [22]), while λ_ℓ denotes the nonnegative sparsity-controlling coefficient used to adjust the number of zeros in $\hat{\mathbf{M}}_{t,:\ell}$. The coefficient $\phi \geq 0$ in the last term of (8) promotes group sparsity among rows, [32], thus is introduced to adjust the number of nonzero columns of $\hat{\mathbf{M}}_t$ needed to approximate $\hat{\boldsymbol{\Sigma}}_{x,t}$. This is done to zero-out unnecessary columns in $\hat{\mathbf{M}}$ when the number of active targets in the field is smaller than R. The number of nonzero columns in $\mathbf{M}_t$ indirectly estimates the number of targets at time instant t, namely $\hat{r}(t)$.

The cost in (8) is minimized by an iterative minimization scheme based on coordinate descent [25, 26], where sensor j is responsible for updating the jth row of $\mathbf{M}_t$, namely $\mathbf{M}_{t,j:}$. Specifically, the cost is minimized wrt one entry of $\mathbf{M}_t$ or $\text{diag}(\sigma_1^2, \ldots, \sigma_p^2)$, while keeping the rest fixed to their most up-to-date values. Sensor j updates the entries $\{\mathbf{M}_t(j, \ell)\}_{\ell=1}^L$ and variance $\sigma_{j,t}^2$. During one coordinate cycle, all the entries of $\mathbf{M}_t$ and $\text{diag}(\sigma_{1,t}^2, \ldots, \sigma_{p,t}^2)$ will be updated.

The updates for entries $\hat{\mathbf{M}}_t^k(j, \ell)$ will be formed by differentiating (8) wrt $\mathbf{M}_t(j, \ell)$ and setting the derivative equal to zero, while fixing the rest of the entries of $\mathbf{M}_t$ and $\sigma_{j,t}$ to their most recent updates in $\hat{\mathbf{M}}_t^{k-1}$ and $\{\sigma_{j,t,k-1}^2\}$ evaluated at cycle $k - 1$. It turns out (details in [23, 24]) that during coordinate cycle k, the update $\hat{\mathbf{M}}_t^k(j, \ell)$ can be obtained as the value that gets the minimum possible cost in (8) (while fixing the rest of the variables) among the candidate values: (i) $z = 0$; (ii) the real positive roots of the third-degree polynomial

$$z^3 + \left[\sum_{i \in \mathcal{N}_j} [\hat{\mathbf{M}}_t^{k-1}(i, \ell)]^2 - \psi_{t,\Sigma}^k(j, j, \ell) + \frac{\phi}{2} \right] \tag{9}$$
$$z - \left[\sum_{i \in \mathcal{N}_j} \psi_{t,\Sigma}^k(j, \mu, \ell) \hat{\mathbf{M}}_t^{k-1}(i, \ell) \right] + \frac{\lambda_\ell}{4} = 0$$

and (iii) the real negative roots of the third-degree polynomial

$$z^3 + \left[\sum_{i \in \mathcal{N}_j} [\hat{\mathbf{M}}_t^{k-1}(\mu, \ell)]^2 - \psi_{t,\Sigma}^k(j, j, \ell) + \frac{\phi}{2} \right] \tag{10}$$
$$z - \left[\sum_{i \in \mathcal{N}_j} \psi_{t,\Sigma}^k(j, i, \ell) \hat{\mathbf{M}}_t^{k-1}(i, \ell) \right] - \frac{\lambda_\ell}{4} = 0$$

where

$$\psi_{t,\Sigma}^k(j, i, \ell) := \hat{\boldsymbol{\Sigma}}_{x,t}(j, i) - \delta_{j,i} \hat{\sigma}_{j,t,k-1}^2 \tag{11}$$
$$- \sum_{\ell'=1, \ell' \neq \ell}^L \hat{\mathbf{M}}_t^{k-1}(j, \ell') \hat{\mathbf{M}}_t^{k-1}(i, \ell'),$$

while $\delta_{j,i}$ denotes the Kronecker delta function, i.e., $\delta_{j,i} = 1$ if $j = i$, and $\delta_{j,i} = 0$ if $j \neq i$. The roots of the two aforementioned polynomials can be calculated using companion matrices, see e.g., [33].

Furthermore, during cycle k at time instant t, the noise variance estimates across sensors can be updated as

$$\hat{\sigma}_{j,t,k}^2 = \hat{\boldsymbol{\Sigma}}_{x,t}(j, j) - \hat{\mathbf{M}}_{t,j:}^k \left(\hat{\mathbf{M}}_{t,j:}^k \right)^T, \quad j = 1, \ldots, p. \tag{12}$$

Sensor j needs to communicate only with its single-hop neighbors in $\mathcal{N}_j$, in order to evaluate the coefficients of the polynomials in (9) and (10) and to update the noise variance estimates in (12). It can be shown that as $k \to \infty$, the updates $\hat{\mathbf{M}}_t^{k-1}$ converge at least to a stationary point of (8). Further, the sparsity-controlling coefficients $\{\lambda_\ell\}_{\ell=1}^L$ can be set using the strategy proposed in ([23], Sec. V.A). Once the sparse columns $\{\hat{\mathbf{M}}_{t,:\ell}\}$ are estimated, their support (the indices of relatively strong-amplitude entries) is used to determine which sensors sense a specific target at time t.

3.2 Tracking via extended Kalman filtering

The target-informative sensor subsets $\mathcal{S}_{\rho_\ell,t}$ for $\ell = 1, \ldots, \hat{r}(t)$, where $\hat{r}(t)$ corresponds to an estimate of the number of targets at time instant t obtained from the number of nonzero columns of $\hat{\mathbf{M}}_t := \hat{\mathbf{M}}_t^{\bar{K}}$, after applying $\bar{K}$ coordinate cycles. Extended Kalman filtering is employed to process the nonlinear observations and track each target's location using the observations of the corresponding set $\mathcal{S}_{\rho_\ell,t}$. For simplicity in exposition, the specifics of extended Kalman filtering (EKF) will be delineated here for $K = 2$ dimensions, but it can be readily generalized to more dimensions. The target state estimator and corresponding error covariance matrix, obtained by the extended Kalman filter using the observations in $\mathcal{S}_{\rho_\ell,t}$ for target ρ_ℓ are denoted as $\hat{\mathbf{s}}_{\rho_\ell}(t|t)$ and $\mathbf{M}_{\rho_\ell}(t|t)$, respectively. The prediction step, see e.g., [27], involves the following updating recursions for the state estimator and covariance at time instant t

$$\hat{\mathbf{s}}_{\rho_\ell}(t+1|t) = \mathbf{F}\hat{\mathbf{s}}_{\rho_\ell}(t|t), \quad \hat{\mathbf{M}}_{\rho_\ell}(t+1|t) = \mathbf{F}\hat{\mathbf{M}}_{\rho_\ell}(t|t)\mathbf{F}^T + \mathbf{\Sigma}_u. \tag{13}$$

The measurements of the sensors within set $\mathcal{S}_{\rho_\ell,t}$ will then be used to carry out the correction step of the extended Kalman filter which involves the following update recursions:

$$\hat{\mathbf{s}}_{\rho_\ell}(t+1|t+1) = \hat{\mathbf{s}}_{\rho_\ell}(t+1|t) + \mathbf{K}(t+1) \tag{14}$$
$$\cdot \left[\mathbf{x}_{t+1} - a_\rho(t)\hat{\mathbf{D}}_{\mathcal{S}_{\rho_\ell,t}} \right]$$

$$\mathbf{M}_{\rho_\ell}(t+1|t+1) = \mathbf{M}_{\rho_\ell}(t+1|t) + \mathbf{D}_{\nabla,\rho_\ell}^T(t+1|t)$$
$$\cdot \sigma_w^2 \mathbf{I}_{|\mathcal{S}_{\rho_\ell,t}|} \cdot \mathbf{D}_{\nabla,\rho_\ell}(t+1|t), \tag{15}$$

for $\ell = 1,\ldots,\hat{r}(t)$, while the matrix $\mathbf{K}_{\rho_\ell}(t+1)$ corresponds to the Kalman gain given as

$$\mathbf{K}_{\rho_\ell}(t+1) = \mathbf{M}_{\rho_\ell}(t+1|t+1) \cdot \mathbf{D}_{\nabla,\rho_\ell}^T(t+1|t) \cdot \sigma_w^2 \mathbf{I}_{|\mathcal{S}_{\rho_\ell,t}|}, \tag{16}$$

where $\hat{\mathbf{D}}_{\mathcal{S}_{\rho_\ell,t}}$ is a $|\mathcal{S}_{\rho_\ell,t}| \times 1$ vector whose entries are given by $\{\|\mathbf{p}_j(t) - \hat{\mathbf{p}}_{\rho_\ell}(t+1|t)\|^{-2}\}_{j\in\mathcal{S}_{\rho_\ell,t}}$, in which $\hat{\mathbf{p}}_{\rho_\ell}(t+1|t)$ is the ρ_ℓth target position extracted from the state prediction $\hat{\mathbf{s}}_{\rho_\ell}(t+1|t)$. Further, $\mathbf{D}_{\nabla,\rho_\ell}(t+1|t)$ is the $|\mathcal{S}_{\rho_\ell,t}| \times 4$ matrix whose rows constitute of gradient $\nabla \mathbf{D}_t(j,\rho_\ell)$ with respect to the state vector $\mathbf{s}_{\rho_\ell}$ and evaluated at $\hat{\mathbf{s}}_{\rho_\ell}(t+1|t)$ for $j \in \mathcal{S}_{\rho_\ell,t}$, i.e.,

$$\nabla_{\mathbf{s}_\rho}\mathbf{D}_t(j,\rho_\ell)\Big|_{\mathbf{s}_{\rho_\ell}=\hat{\mathbf{s}}_{\rho_\ell}(t+1|t)}$$
$$= \frac{2 \cdot \left[\mathbf{p}_{j,x}(t) - \hat{\mathbf{p}}_{\rho_\ell,x}(t+1|t), \mathbf{p}_{j,y}(t) - \hat{\mathbf{p}}_{\rho_\ell,y}(t+1|t), 0, 0 \right]^T}{\left[(\mathbf{p}_{j,x}(t) - \hat{\mathbf{p}}_{\rho_\ell,x}(t+1|t))^2 + (\mathbf{p}_{j,y}(t) - \hat{\mathbf{p}}_{\rho_\ell,y}(t+1|t))^2 \right]^2}. \tag{17}$$

Within each informative subset of sensors $\mathcal{S}_{\rho_\ell,t}$, the sensor closest in distance to the predicted position of the ρ_ℓth target, namely $\hat{\mathbf{s}}_{\rho_\ell}(t+1|t)$, is set as a the subset head sensor that will gather the measurements of all other sensors in $\mathcal{S}_{\rho_\ell,t}$ and perform the EKF tracking recursions.

3.3 Sensor kinematics

The focus in this section is to derive kinematic rules for the target-informative sensors, which are selected according to the scheme in Section 3.1, such that they follow closely the moving targets and give accurate position estimates. The benefit from having a few sensors moving is that targets can be tracked even when they move away from the original field monitored by the sensors. Having sensors following closely, the moving targets can provide more reliable measurements about the targets than just using static sensors. Note that only informative sensors

close to the targets will be responsible for carrying out the tracking procedure leading to resource savings. Toward this end, the informative sensors in each subset $\mathcal{S}_{\rho_\ell}$ will be placed/move in locations that minimize the trace of the error covariance associated with the estimator $\hat{\mathbf{s}}_{\rho_\ell}(t|t)$. This will ensure that the informative sensors associated with each target move to a location that will provide measurements that result good tracking accuracy. The idea of minimizing a scalar function of the predicted error covariance was also applied in moving all sensors in a network for tracking *a single* target [2, 3, 12]. Here kinematic strategies are derived in the presence of multiple targets, while a judiciously selected small portion of target-informative sensors will be moving instead of all sensors moving.

Among the two terms in the covariance matrix in Eq. (15), only the second term is affected by the sensors' location. The latter term, after using (17), can be written as:

$$\sum_{j\in\mathcal{S}_{\rho_\ell,t}} \frac{4}{\left[(\mathbf{p}_{j,x}(t+1) - \hat{\mathbf{p}}_{\rho_\ell,x}(t+1|t))^2 + (\mathbf{p}_{j,y}(t+1) - \hat{\mathbf{p}}_{\rho_\ell,y}(t+1|t))^2 \right]^3}. \tag{18}$$

Clearly, (18) depends on the position of the sensors associated with target ρ_ℓ, namely the sensors in subset $\mathcal{S}_{\rho_\ell,t}$, at time instant t. Letting $\mathbf{p}_j(t+1) := [\mathbf{p}_{j,x}(t+1) \quad \mathbf{p}_{j,y}(t+1)]^T$ notice that the trace cost in (18) is separable with respect to the position of each sensor j within subset $\mathcal{S}_{\rho_\ell,t}$. Thus, the position of sensor $j \in \mathcal{S}_{\rho_\ell,t}$ at time instant $t+1$ is determined by minimizing the corresponding summand in (18), i.e., the updated location for sensors $j \in \mathcal{S}_{\rho_\ell,t}$ can be found as

$$\mathbf{p}_j(t+1) = \arg\min_{\mathbf{p}_{j,x},\mathbf{p}_{j,y}} \frac{4}{\left([\mathbf{p}_{j,x} - \hat{\mathbf{p}}_{\rho_\ell,x}(t+1|t)]^2 + [\mathbf{p}_{j,y} - \hat{\mathbf{p}}_{\rho,y}(t+1|t)]^2 \right)^3}$$
$$\text{s. to } [\mathbf{p}_{j,x} - \hat{\mathbf{p}}_{\rho_\ell,x}(t+1|t)]^2 + [\mathbf{p}_{j,y} - \hat{\mathbf{p}}_{\rho_\ell,y}(t+1|t)]^2 < R^2 \tag{19}$$

Note that the inequality constraint in (19) ensures that the new location of the moving sensors $j \in \mathcal{S}_{\rho_\ell}$ will be within distance R_j from the latest target location estimate $\hat{\mathbf{p}}_{\rho_\ell}(t+1|t)$. This inequality further ensures that all sensors in $\mathcal{S}_{\rho_\ell}$ will move to new locations which are "close" to the target. After applying the triangle inequality for the new locations of two sensors j and j' within $\mathcal{S}_{\rho_\ell}$ and using the constraint in (19), it turns out that the new location should satisfy

$$\|\mathbf{p}_j(t+1) - \mathbf{p}_{j'}(t+1)\|_2 \leq \sqrt{2}R, \tag{20}$$

which ensures that each subset $\mathcal{S}_{\rho_\ell}$ of moving sensors will stay connected, as long as the communication range of the sensing units is at least $\sqrt{2}R$. Thus, R can be set such that the moving sensors stay connected. Connectivity is necessary to elect a head sensor for each moving subset of sensors that will acquire the measurements of all other sensors and perform clustering. Details of the algorithm are given in Section 4.1. Note that existing approaches do not entail mechanisms as the one introduced here to ensure that sensors will be connected.

Next, the modified barrier method (MBM) is utilized ([26], pg. 423) to allow every sensor $j \in \mathcal{S}_{\rho_\ell}$ to solve (19) and determine its next location. To this end, let $f(\mathbf{p}_j)$ denote the cost in (19) and $g(\mathbf{p}_j)$ denote the left hand side function of the inequality constraint in (19). MBM involves an iterative application of the following unconstrained minimization problem (where κ denotes the iteration index within time instant $t + 1$):

$$\mathbf{p}_j^\kappa(t+1) \in \arg\min_{\mathbf{p}_{j,x}, \mathbf{p}_{j,y}} \left\{ f(\mathbf{p}_j) + \frac{\mu^\kappa}{c^\kappa} \phi \left[c^\kappa \cdot g(\mathbf{p}_j) \right] \right\},$$
(21)

where the Lagrange multiplier-like scalar μ^κ is updated as

$$\mu^{\kappa+1} = \mu^\kappa \cdot \nabla \phi \left[c^\kappa \cdot g(\mathbf{p}_j^\kappa(t+1)) \right],$$
(22)

while the barrier function $\phi[\tau]$ is chosen as a logarithmic function having the form

$$\phi(\tau) = -ln(1 - \tau)$$
(23)

and c^κ is a penalty parameter associated with the inequality constraint in (19) that is updated according to the recursion

$$c^\kappa = \frac{\gamma^\kappa}{\mu^\kappa}$$
(24)

where $\{\gamma^k\}$ is a positive monotonically increasing scalar sequence [26].

For the logarithmic barrier function in (23) the updating recursion of the multipliers in (22) takes the following form

$$\mu^{k+1} = \frac{\mu^\kappa}{1 - c^\kappa g(\mathbf{p}_j^\kappa(t+1))}.$$
(25)

Further, letting $F(\mathbf{p}_j) := f(\mathbf{p}_j) + \frac{\mu^\kappa}{c^\kappa} \phi \left[c^\kappa \cdot g(\mathbf{p}_j) \right]$, the coordinates of the new sensor location $\mathbf{p}_j(t+1)$ are updated during iteration κ according to the following gradient descent recursions

$$\mathbf{p}_{j,x}^{\kappa+1}(t+1) = \mathbf{p}_{j,x}^\kappa(t+1) - \Gamma \cdot \frac{dF(\mathbf{p}_j)}{d\mathbf{p}_{j,x}} \bigg|_{\mathbf{p}_{j,x} = \mathbf{p}_{j,x}^\kappa(t+1)}$$
(26)

$$\mathbf{p}_{j,y}^{\kappa+1}(t+1) = \mathbf{p}_{j,y}^\kappa(t+1) - \Gamma \cdot \frac{dF(\mathbf{p}_j)}{d\mathbf{p}_{j,y}} \bigg|_{\mathbf{p}_{j,y} = \mathbf{p}_{j,y}^\kappa(t+1)}$$
(27)

where Γ is the step size for the gradient descent method, while the derivatives in (26) are given as

$$\frac{dF(\mathbf{p}_j)}{d\mathbf{p}_{j,x}} \bigg|_{\mathbf{p}_{j,x} = \mathbf{p}_{j,x}^\kappa(t+1)}$$

$$= \frac{24 \cdot \left[-\mathbf{p}_{j,x}^\kappa(t+1) + \hat{\mathbf{p}}_{\rho,x}(t+1|t) \right]}{\left(\left[\mathbf{p}_{j,x}^\kappa(t+1) - \hat{\mathbf{p}}_{\rho,x}(t+1|t) \right]^2 + \left[\mathbf{p}_{j,y}^\kappa - \hat{\mathbf{p}}_{\rho,y}(t+1|t) \right]^2 \right)^4}$$

$$+ \frac{\mu^\kappa}{1 - c^\kappa g(\mathbf{p}_j^\kappa(t+1))} \cdot 2 \cdot (\mathbf{p}_{j,x}^\kappa(t+1) - \hat{\mathbf{p}}_{\rho,x}(t+1|t)),$$

$$\frac{dF(\mathbf{p}_j)}{d\mathbf{p}_{j,y}} \bigg|_{\mathbf{p}_{j,y} = \mathbf{p}_{j,y}^\kappa(t+1)}$$

$$= \frac{24 \cdot \left[-\mathbf{p}_{j,y}^\kappa(t+1) + \hat{\mathbf{p}}_{\rho,y}(t+1|t) \right]}{\left(\left[\mathbf{p}_{j,x}^\kappa(t+1) - \hat{\mathbf{p}}_{\rho,x}(t+1|t) \right]^2 + \left[\mathbf{p}_{j,y}^\kappa(t+1) - \hat{\mathbf{p}}_{\rho,y}(t+1|t) \right]^2 \right)^4}$$

$$+ \frac{\mu^\kappa}{1 - c^\kappa g \left(\mathbf{p}_j^\kappa(t+1) \right)} \cdot 2 \cdot \left(\mathbf{p}_{j,y}^\kappa(t+1) - \hat{\mathbf{p}}_{\rho,y}(t+1|t) \right).$$
(28)

During time instant $t + 1$, each sensor j within the subset $\mathcal{S}_{\rho_\ell,t}$ will keep updating their location until the cost function in (19) is not reduced more than a predefined threshold ϵ within two consecutive updating steps $\kappa, \kappa + 1$. The location $\mathbf{p}_j(t+1)$ will be set to the last update $\mathbf{p}_j^{K'}(t+1)$ obtained after K' MBM iterations during time instant $t + 1$. The following steps are carried out during the determination of the sensor's new location:

S1) The head sensor in each subset $\mathcal{S}_{\rho_\ell,t}$ sends the predicted position estimate of target ρ_ℓ, namely $\hat{\mathbf{p}}_{\rho_\ell}(t+1|t)$, to all sensors in $\mathcal{S}_{\rho_\ell,t}$.

S2) Each sensor in $j \in \mathcal{S}_{\rho_\ell,t}$, determines its new location using the MBM scheme. Sensors in $\mathcal{S}_{\rho_\ell,t}$ check their distances to other neighboring sensors, and if their future location is too close, they adjust their coordinates to avoid collision when moving. Similarly, each moving sensor checks the distance between its updated location and the position estimate of target ρ_ℓ, and if too close, adjustments will be made to the sensor's location such that a minimum distance will be kept from the target. Specifically, if sensor j has an updated location $\mathbf{p}_j^{K'}(t+1)$ which is too close to the already updated location of sensor j', namely $\mathbf{p}_{j'}(t+1)$, i.e., $\|\mathbf{p}_j^{K'}(t+1) - \mathbf{p}_{j'}(t+1)\|_2 \leq R_{\min,a}$, where $R_{\min,a}$ is the smallest distance allowed that two sensors can be separated from each other, then $\mathbf{p}_j^{K'}(t+1)$ is updated as follows:

$$\mathbf{p}_j(t+1) = \mathbf{p}_{j'}(t+1) + R_{\min,a} \frac{\mathbf{p}_j^{K'}(t+1) - \mathbf{p}_{j'}(t+1)}{\|\mathbf{p}_j^{K'}(t+1) - \mathbf{p}_{j'}(t+1)\|_2}.$$
(29)

Similarly, if sensor j has an updated location $\mathbf{p}_j^{K'}(t+1)$ which is too close to the ρ_ℓ target location estimate, i.e.,

$\|\mathbf{p}_j^{K'}(t+1) - \hat{\mathbf{p}}_{\rho_\ell}(t+1|t)\|_2 \leq R_{\min,b}$ where $R_{\min,b}$ is the smallest distance that a sensor can be placed from a target, then location $\mathbf{p}_j^{K'}(t+1)$ is updated as follows:

$$\mathbf{p}_j(t+1) = \hat{\mathbf{p}}_{\rho_\ell}(t+1|t) + R_{\min,b}\frac{\mathbf{p}_j^{K'}(t+1) - \hat{\mathbf{p}}_{\rho_\ell}(t+1|t)}{\|\mathbf{p}_j^{K'}(t+1) - \hat{\mathbf{p}}_{\rho_\ell}(t+1|t)\|_2}.$$

(30)

The collision-avoidance position modifications in (30) were proposed in [34] to prevent collision of unmanned aerial vehicles with a stationary target. The position updates in (29) and (30) ensure that the updated locations are at distance $R_{\min,a}$ and $R_{\min,b}$ from another moving sensor or moving target, respectively, satisfying the minimum distance required to prevent collision.

The actual movement can be achieved using for example robotic sensors, see e.g., [35, 36]. Each sensor $j \in \mathcal{S}_{\rho_\ell,t}$ updates its location $\mathbf{p}_j(t+1)$, in a coordinate fashion while the remaining sensors in $\mathcal{S}_{\rho_\ell,t}$ are kept stationary waiting for their turn to update their location.

4 Algorithmic summary

4.1 Implementation

At the start-up stage, fast sampling is used to acquire Q measurements fast enough that the initial number of targets $r(0)$ can be assumed stationary. By utilizing the Q acquired data, the subsets of target-informative sensors $\{\mathcal{S}_{\rho_\ell,0}\}$ are initialized, where $\ell = 1, \ldots, \hat{r}(0)$ and $\hat{r}(0)$ is the estimated number of $r(0)$ sensed targets at time $t = 0$ (number of nonzero columns in the sparse matrix $\hat{\mathcal{M}}_0$). One sensor within each $\mathcal{S}_{\rho_\ell,0}$ will be randomly selected as the head sensor, which will collect the measurements $x_j(0)$ and their positions $\mathbf{p}_j(0)$ from all the other sensors $j \in \mathcal{S}_{\rho_\ell,0}$. Each head sensor $C_{\rho_\ell,0}$ averages the positions of the informative sensors in subset $\mathcal{S}_{\rho_\ell,0}$ to be the initial estimate of the corresponding target ρ_ℓ. The latter target location estimate along with the informative measurements $x_j(0)$, for $j \in \mathcal{S}_{\rho_\ell,0}$ are utilized to initialize the recursions of the extended Kalman filtering carrying out the target tracking in Section 3.2.

At time instant t, every head sensor $C_{\rho_\ell,t}$ has available the state estimates for active target ρ_ℓ, namely $\hat{\int}_{\rho_\ell}(t|t)$, obtained from EKF in Section 3.2. The target's estimated position $\hat{\mathbf{p}}_{\rho_\ell}(t|t)$ is then used to select a group of "candidate informative" sensors, which are denoted as $\mathcal{J}_{\rho_\ell,t}$ for target ρ_ℓ at time instant t. This set is formed by having the head sensor transmit the estimated state $\hat{\int}_{\rho_\ell}(t|t)$ to its single-hop neighboring sensors which then transmit the same information to their own neighbors. Every sensor j who receives $\hat{\int}_{\rho_\ell}(t|t)$, from a neighboring sensor, subsequently forwards this estimate only to those sensors $j' \in \mathcal{N}_j$ whose present position is within radius R_s from the estimated target location, i.e., $\|\mathbf{p}_{j'}(t) - \hat{\mathbf{p}}_{\rho_\ell}(t|t)\|_2 \leq R_s$. The parameter R_s can be set to be sufficiently large in

order for all ρ_ℓ-target-informative sensors to be incorporated in subset $\mathcal{J}_{\rho_\ell,t}$. The sensor subset $\mathcal{J}_{\rho_\ell,t}$ by construction is connected.

Since not all sensors within the candidate subsets $\mathcal{J}_{\rho_\ell,t}$ maybe informative, the scheme in Section 3.1 is employed among the sensors in $\mathcal{J}_{\rho_\ell,t}$ to find out the target-informative sensor subset $\mathcal{S}_{\rho_\ell,t+1} \subseteq \mathcal{J}_{\rho_\ell,t}$ for all the active targets. Rather than running the target-sensor association scheme in Section 3.1 in the whole sensor network, it is performed independently in the different sensor subsets $\mathcal{J}_{\rho_\ell,t}$ associated with each target.

Once subsets $\mathcal{S}_{\rho_\ell,t+1}$ are found, the head sensor in each of these subsets is chosen to be the sensor whose distance is the closest to the estimated position of the corresponding target ρ_ℓ, i.e.,

$$C_{\rho_\ell,t+1} = \arg\min_{j\in\mathcal{S}_{\rho_\ell,t+1}} \|\mathbf{p}_j(t) - \hat{\mathbf{p}}_{\rho_\ell}(t|t)\|_2.$$

The head sensor $C_{\rho_\ell,t+1}$ gathers the sensor measurements $x_j(t+1)$ from the informative sensors $j \in \mathcal{S}_{\rho_\ell,t+1}$ to carry out the extended Kalman filtering recursions at time instant $t+1$ as outlined in Section 3.2. Then, steps S1 and S2 in Section 3.3 are employed to allow all sensors in $\mathcal{S}_{\rho_\ell,t+1}$ to determine and move to their new positions $\mathbf{p}_j(t+1)$. Note that connectivity of the sensors in $\mathcal{S}_{\rho_\ell,t+1}$ is preserved as explained in Section 3.3. The head sensor $C_{\rho_\ell,t+1}$ broadcasts the latest state estimate $\hat{\mathbf{s}}_{\rho_\ell}(t+1|t+1)$ to its single-hop neighbors and repeats the process described earlier to update the candidate informative sensor subsets $\mathcal{J}_{\rho_\ell,t+1}$.

It is worth mentioning that the kinematic rules implemented in Section 3.3 at each sensor are fully distributed since each sensor requires knowledge only of its location and the estimated target position obtained from the head sensor in $\mathcal{S}_{\rho_\ell,t+1}$. Connectivity of the candidate informative subsets $\mathcal{J}_{\rho_\ell,t+1}$ is ensured by construction irrespective of the sensor movement. This way the sensor-to-target association scheme in Section 3.1 can still be applied in $\mathcal{J}_{\rho_\ell,t+1}$ and determine the informative sensors.

The target-informative sensor selection scheme in Section 3.1 may also need to be reapplied across the whole sensor network since moving targets may disappear and not being sensed anymore, while new targets may appear at different regions of the sensor network. The following conditions are checked to determine such events: (i) If any of the sensor subsets $\mathcal{S}_{\rho_\ell,t+1}$ becomes empty, this implies that some of the targets previously sensed are not present anymore; (ii) If at $t+1$, the energy of a sensor, not previously selected, exceeds a certain threshold, this indicates that most likely a new target enters the sensed field. The two aforementioned conditions signify that the target configuration has changed and the sensor selection scheme in Section 3.1 needs to be reapplied in the sensor network to update the sensor-informative subsets. The novel tracking scheme is summarized as Algorithm 1.

Algorithm 1 Multi-target tracking using sensor mobility and informative sensor selection

---

1: **Start-up stage ($t = 0$)/Reconfiguration ($t \neq 0$):** Every sensor j collects Q measurements $x_j(t)$ and the sensors-targets association scheme in Section 3.1 is applied in the network to determine the subsets $\mathcal{S}_{\rho_\ell, t}$ with $\ell = 1, \ldots, \hat{r}(t)$, and $\hat{r}(t)$ is the estimated number of targets.

2: **for** $\tau = t, \ldots,$ **do**

3:     Determine the head sensor $C_{\rho_\ell, \tau}$ in each $\mathcal{S}_{\rho_\ell, \tau}$ for $\ell = 1, \ldots, \hat{r}(t)$.

4:     Each head sensor $C_{\rho_\ell, \tau}$ receives measurements $x_j(\tau)$ from $j \in \mathcal{S}_{\rho_\ell, \tau}$ to perform tracking for targets $\rho_\ell = 1, \ldots, \hat{r}(t)$ via the EKF recursions in Section 3.2.

5:     Informative sensors $j \in \mathcal{S}_{\rho_\ell, \tau}$ relocate themselves according to the sensor kinematics introduced in Section 3.3.

6:     Each head sensor $C_{\rho_\ell, \tau}$ propagates the state estimates $\hat{\mathbf{s}}_{\rho_\ell}(\tau)$ to every sensor j that can be reached from $C_{\rho_\ell, \tau}$ by a multi-hop path and satisfies $\|\mathbf{p}_j(\tau) - \hat{\mathbf{p}}_{\rho_\ell}(\tau|\tau)\|_2 < R_s$. Then, the candidate informative sets $\{\mathcal{J}_{\rho_\ell, \tau+1}\}_{\ell=1}^{\hat{r}(t)}$ are formed.

7:     The sensor selection scheme in Section 3.1 is carried out in each subset $\mathcal{J}_{\rho_\ell, \tau+1}$ to identify the target-informative sets $\mathcal{S}_{\rho_\ell, \tau+1}$.

8:     If target configuration has changed then go to step 1, otherwise go to step 2.

9: **end for**

4.2 Communication and computational expenses

The communication cost of the proposed algorithm is studied next. Note that inter-sensor communication takes place during (i) the sensor-to-target association scheme in Section 3.1; (ii) carrying out the EKF tracking steps in Section 3.2; and (iii) when applying the kinematic strategy in Section 3.3 to move the informative sensors. In detail, at time instant t, sensor j has to receive $|\mathcal{N}_j|$ scalar measurements from its neighbors, namely $\{x_{j'}(t+1)\}_{j' \in \mathcal{N}_j}$, to update $\hat{\mathbf{\Sigma}}_{x,t+1}(j,j')$. Furthermore, to implement the association scheme in Section 3.1, each sensor j receives the updates $\{\hat{\mathbf{M}}_{t+1}^{k-1}(j',\ell)\}_{\ell=1}^L$ from neighborhood $\mathcal{N}_j$, corresponding to $L|\mathcal{N}_j|$ scalars in total, to form its local updates $\{\hat{\mathbf{M}}_{t+1}^k(j,\ell)\}_{\ell=1}^L$. Thus, sensor j receives $(L+1)|\mathcal{N}_j|$ scalars in total. Similarly, sensor j has to transmit $x_j(t+1)$ and $\{\hat{\mathbf{M}}_{t+1}^{k-1}(j,\ell)\}_{\ell=1}^L$, a total of $L+1$ scalars to its neighbors, per iteration k.

After the target-informative sensors are determined, each head sensor has to carry out the estimation process about the corresponding target's states. Thus, the head sensor $\{C_{\rho_\ell, t}\}$ will collect the measurements $x_j(t)$ from the sensors within $\mathcal{S}_{\rho_\ell, t+1}$. This involves $|\mathcal{S}_{\rho_\ell, t+1}|$ scalar exchanges. Further, all sensors in the subset $\mathcal{S}_{\rho_\ell, t+1}$ will

receive four scalars corresponding to the current state estimate. Once the state estimation process (Section 3.2) is carried out by the head sensors, sensor communication also occurs among the informative sensors when adjusting their new location to avoid collision with closely located sensors (Section 3.3). Specifically, sensor j receives $2|\mathcal{N}_j|$ scalars from its neighbors, corresponding to their two location coordinates, while it sends out its own location. It is worth mentioning that the communication complexity for each sensor is linear with respect to its neighborhood size $|\mathcal{N}_j|$, and the upper bound number of present targets L. The latter linear cost advocates that the proposed framework is a communication-affordable distributed approach.

When applying the scheme in Section 3.1 during each coordinate cycle k and time instant t, each sensor j has to form the coefficients in (10), (9) with a computational complexity of the order $\mathcal{O}(|\mathcal{N}_j|)$, while determining the roots of the two third-order polynomial in (10), (9) involves determining the eigenvalues of two 3×3 companion matrices whose complexity is fixed and non-dependent on any algorithmic parameters. The EKF in Section 3.2 can be carried out at a complexity of the order $\mathcal{O}(K^2 + |\mathcal{S}_\rho|)$, where $K = 4$ here and $|\mathcal{S}_\rho|$ corresponds to the size of the target-informative subsets. The kinematic rules implemented at the target-informative sensors in Section 3.3 have a computational complexity $\mathcal{O}(K)$.

5 Simulations

The tracking performance of the novel scheme is tested in a network with $p = 80$ sensors, which are deployed randomly in the region of $[0, 15] \times [0, 15] \, m^2$. The tracking root mean-square error (RMSE) is studied and compared with the RMSE attained by the tracking schemes in [2, 12]. The comparison is done using one target since the aforementioned existing approaches can handle one target. Target $\rho = 1$ is initialized at location $[1.5, 11.5]$ and moves with velocities of $[0.15, 0.1]$m/s respectively along the x-axis and y-axis. The tracking process is carried out for a total of 30 s, with the state noise and observation noise variances set to be $\sigma_u^2 = 0.08$ and $\sigma_w^2 = 0.08$ (corresponding to a sensing SNR of 11 dB). Figure 1 depicts in logarithmic scale the tracking RMSE (for better display) of (i) the novel approach proposed here; (ii) the tracking scheme in [12]; and (iii) the tracking approach in [2]. Note that for all the three tracking schemes, the initial position of the target is found by applying the sparsity matrix decomposition scheme in Section 3.1, ensuring the same initial error for all the three different tracking approaches. As corroborated by Fig. 1, our tracking scheme exhibits the lowest tracking RMSE. The approach in [12] attains the worst performance since the sensors can only move on a grid which reduces accuracy. The scheme in [2] performs worse than our approach since it does

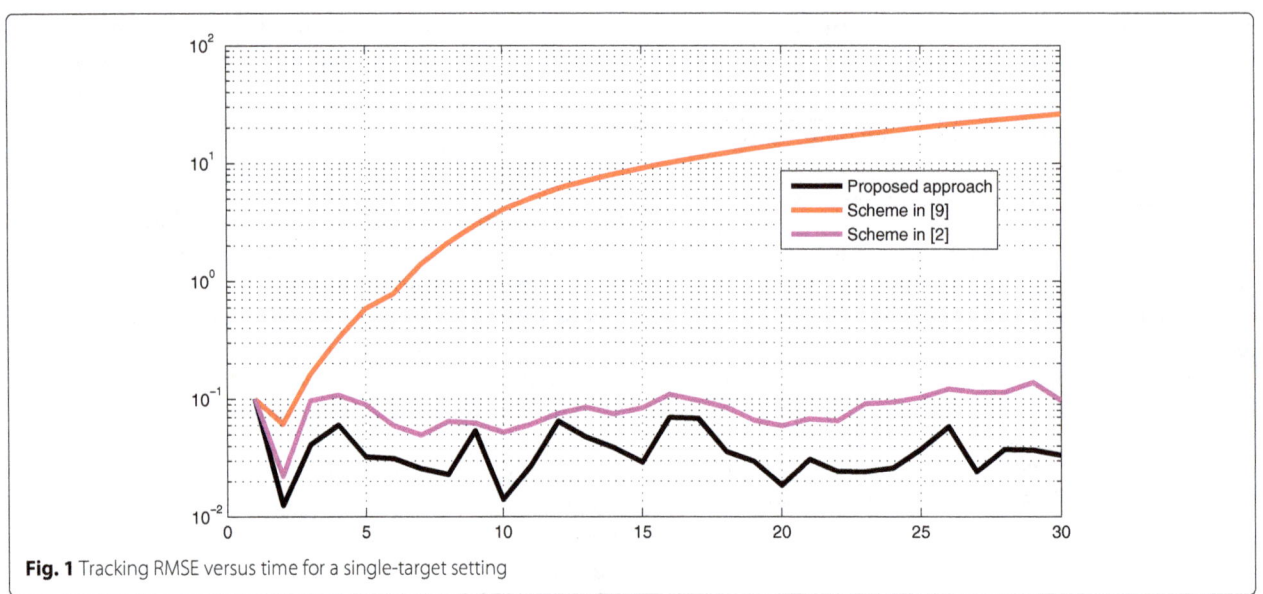

Fig. 1 Tracking RMSE versus time for a single-target setting

not have an informative sensor selection scheme, which results all sensors to move and participate in the tracking process which may reduce accuracy when noisy sensors are utilized.

Figure 2 depicts the distance (in meters) between two moving sensors and the corresponding moving target during the 30-s tracking period. It can be seen that the distance is decreasing with time which further implies that the proposed approach allows the informative sensors to closely follow the target.

Next, the performance of our novel tracking scheme is tested in a setting where the number of targets is changing. Specifically, targets $\rho = 1, 2$ start moving at positions [1.5, 11.5], [5, 7] and follow the dynamics in (1), with velocities of [0.15, 0.1] and [0.4, 0.13] m/s along the x-axis and y-axis, respectively. Targets $\rho = 1, 2$ move in the field for the time interval [1, 30] s and then are not sensed anymore. In the interval [15, 17] s, no targets are present in the field. Then, targets $\rho = 3, 4$ start at positions [6.1, 4.8], [9.0, 4.0] and move according to same state model

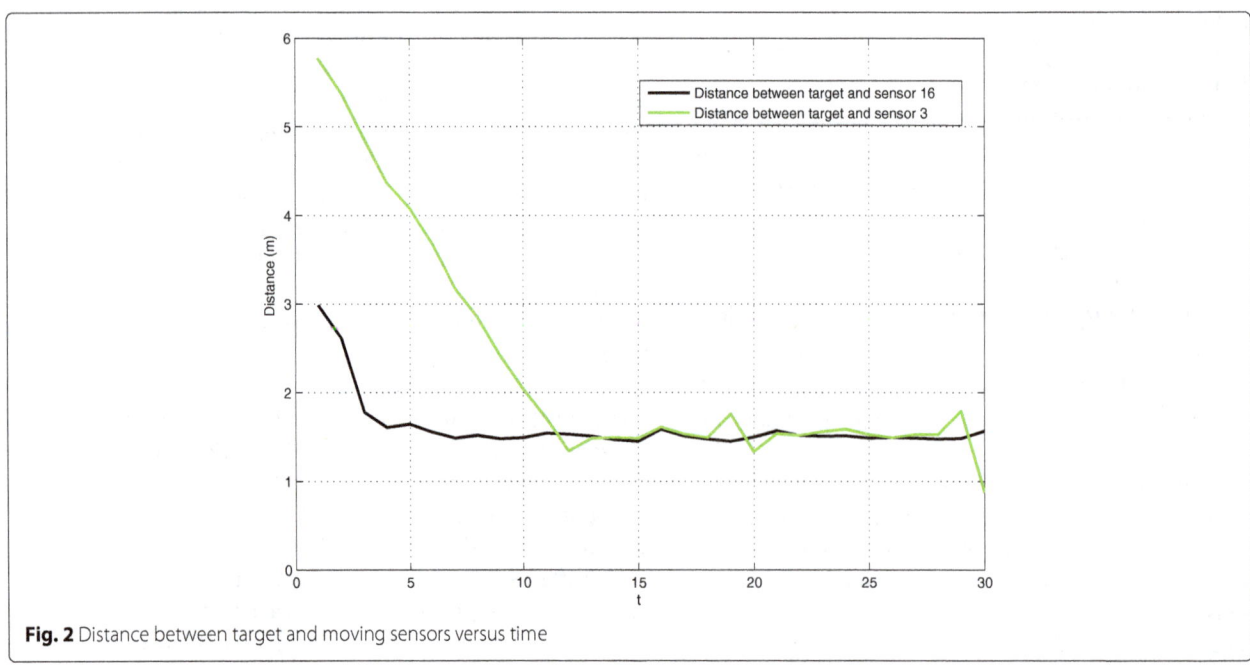

Fig. 2 Distance between target and moving sensors versus time

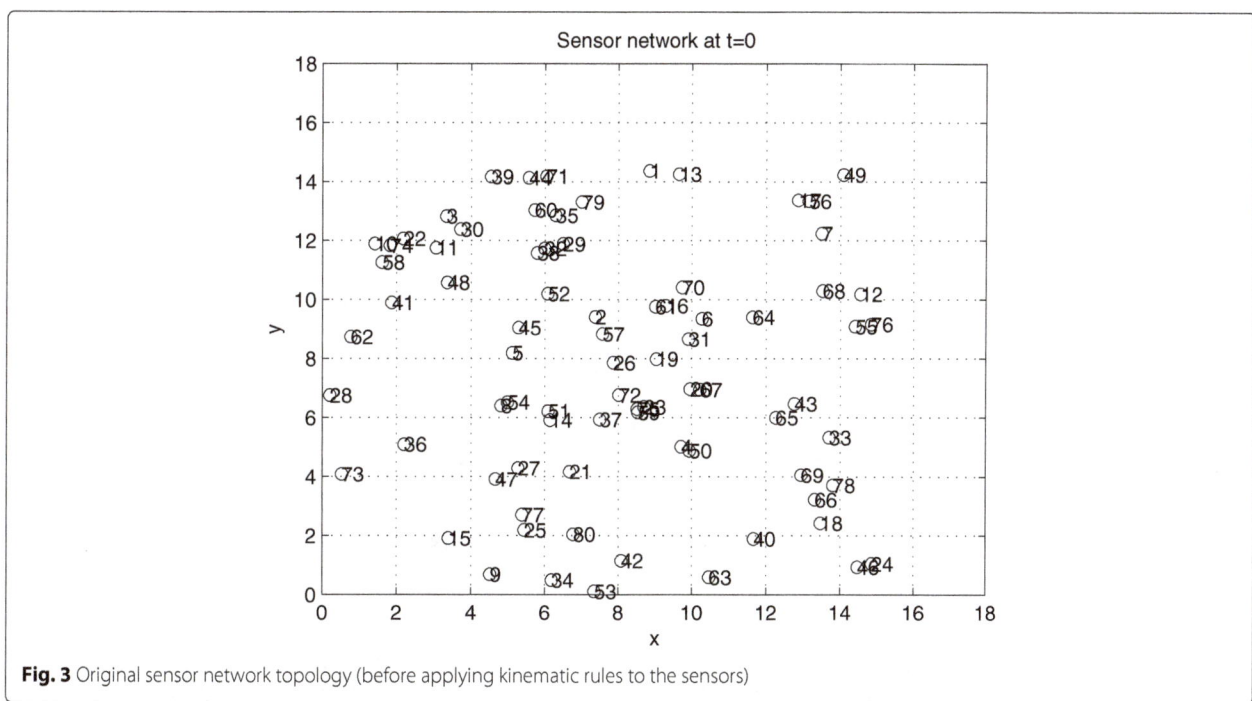

Fig. 3 Original sensor network topology (before applying kinematic rules to the sensors)

followed by the first two for the time interval [32, 50] s but with velocities [0.03, 0.35] and [0.25, −0.25] m/s. Again, no targets are present during the interval [50, 52] s. Then, three targets $\rho = 5, 6, 7$ start moving at positions [15, 1.1], [13, 13.5], and [12, 6], according to (1), for the time interval [52, 70] s and velocities [−0.1, 0] m/s for target $\rho = 5$, and [0.12, −0.03] for both $\rho = 6, 7$ along the x-axis and y-axis, respectively. Figure 3 depicts the original positions of the sensors represented by blue circles.

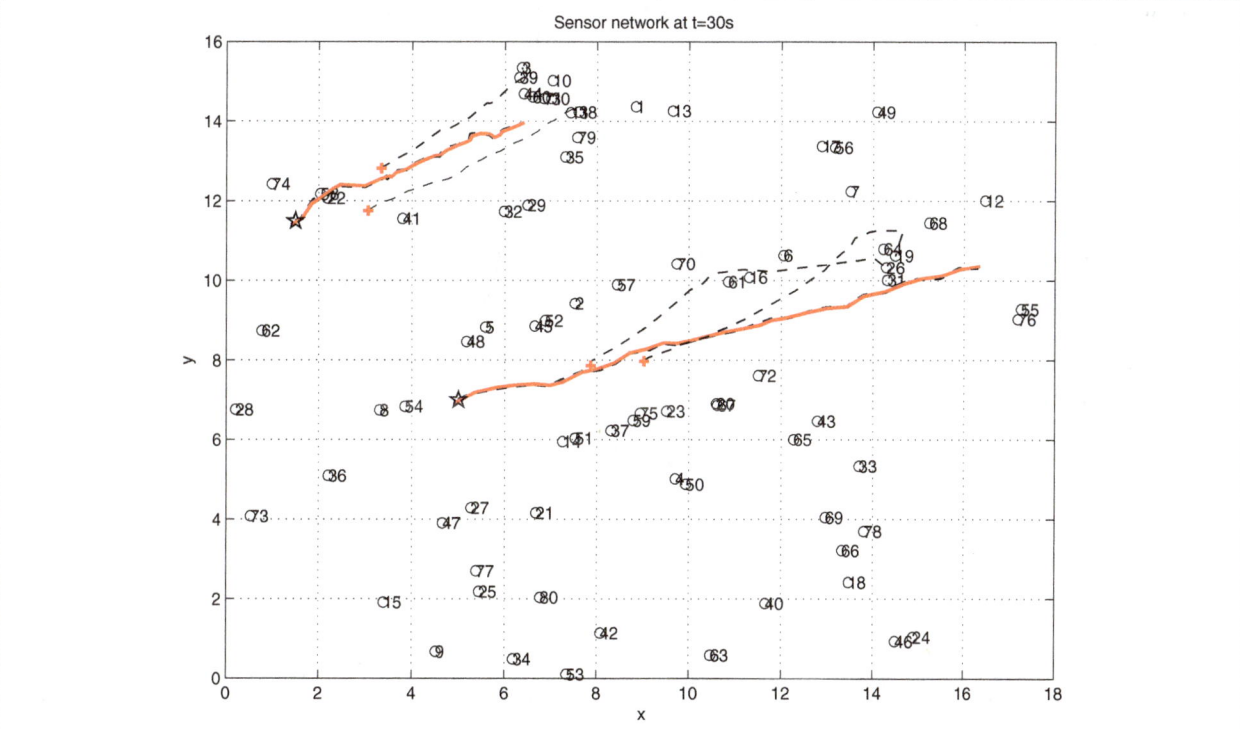

Fig. 4 Snapshot of the trajectories of targets and moving sensors at time instant $t = 30$ s. *Red solid curves* correspond to the true target trajectories; *blue dashed curves* represent the estimated track. *Blue stars* indicate the starting locations of the targets, while the *red crosses* denote the initial locations of some moving sensors. The *black dashed curves* show the trajectories of some of the moving sensors during the tracking process

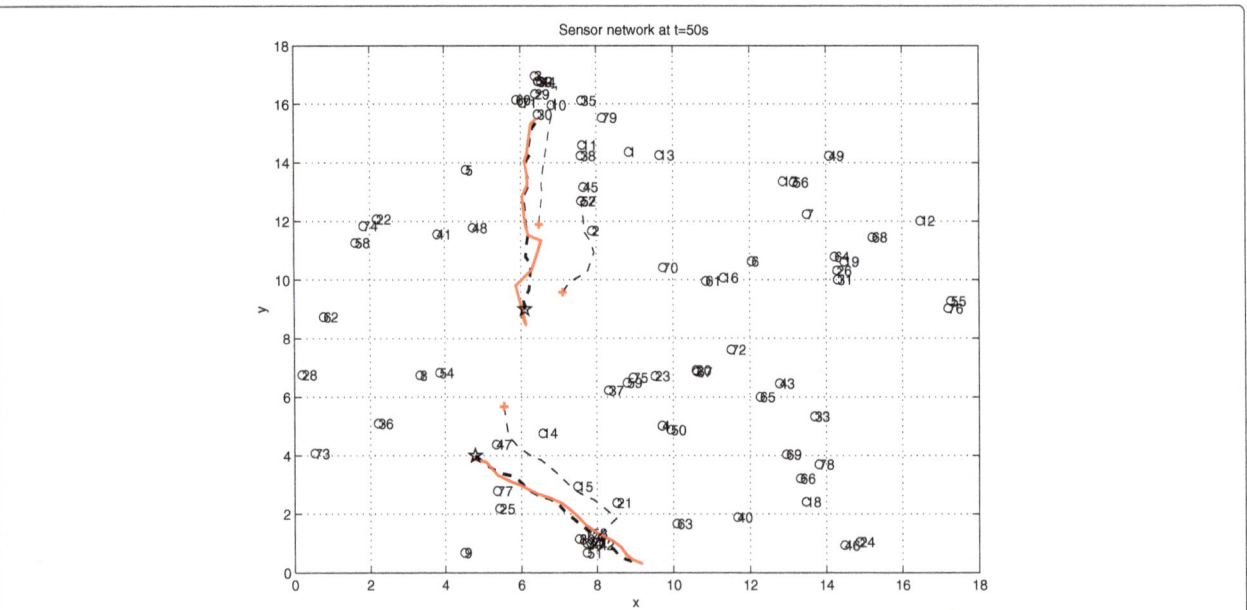

Fig. 5 Snapshot of the trajectories of targets and moving sensors at time instant $t = 50$ s. *Red solid curves* correspond to the true target trajectories; *blue dashed curves* represent the estimated track. *Blue stars* indicate the starting locations of the targets, while the *red crosses* denote the initial locations of some moving sensors. The *black dashed curves* show the trajectories of some of the moving sensors during the tracking process

Figures 4, 5 and 6 show snapshots of the configuration of the targets and the moving sensors at different time instances. Details for the different curves and coloring on those figures is given in the caption below the figures. From Fig. 4, it is clear that for the first two targets, even though the targets move out of the original

$[0, 15] \times [0, 15]$ region, both targets are still tracked well as some of the sensors follow them closely. Note that only informative sensors, on average around 10 % of the total number of sensors, move according to the proposed kinematic rules in Section 3.3, while the majority of other sensors which are not close to the moving targets are

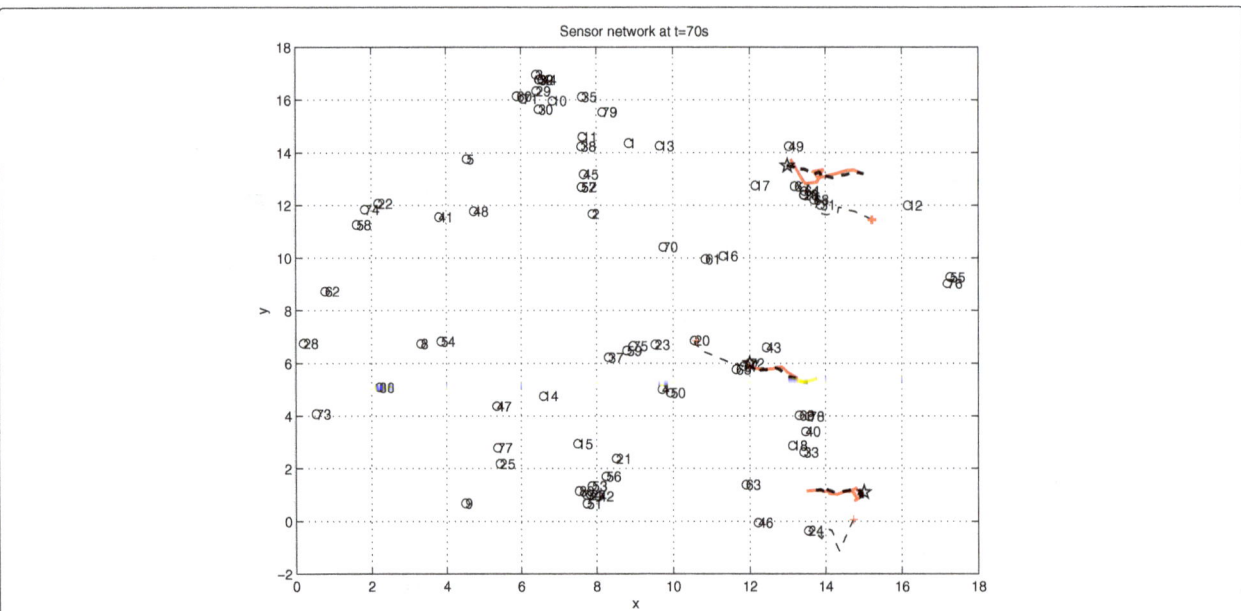

Fig. 6 Snapshot of the trajectories of targets and moving sensors at time instant $t = 70$ s. *Red solid curves* correspond to the true target trajectories; *blue dashed lines* represent the estimated track. *Blue stars* indicate the starting locations of the targets, while the *red crosses* denote the initial locations of some moving sensors. The *black dashed curves* show the trajectories of some of the moving sensors during the tracking process

not moving and maintain their original positions. Similar conclusions can be extracted from Figs. 5 and 6, where different sets of targets are moving on the field. It should be emphasized that it is not known that at $t = 30$ and $t = 50$, the target configuration is changing. As discussed in Section 4, such changes can be determined by having all sensors self-checking the energy level of their measurements, while the head sensors monitor the informative sensor subsets whether they are empty or not.

The tracking RMSE for the above tracking setting is plotted in Fig. 7 in logarithmic scale. The proposed tracking framework exploiting sensor mobility is compared with a tracking scheme where sensors are stationary and not moving. In the immobile sensor network, when the targets are moving away from the sensed field, the sensors will acquire less and less reliable measurements leading to the dramatic increase of the tracking RMSE (blue dashed curves). In contrast, the proposed framework here enables sensors to follows closely the targets and achieve a much lower tracking RMSE. So even though the targets move out of the original sensed field, there is always a group of sensors keeping adjacent to it and can be selected again to provide accurate measurements. Note that in Fig. 7, there are three discontinued curves which corresponds to the error of tracking the three different groups of targets that appear and cease to exist in the monitored field at different time periods. This leads to the RMSE discontinuity at time $t = 31$ s and $t = 51$ s since there are no targets moving during that time interval and no need for tracking.

Next, a tracking setting is considered with two targets where one of targets splits into two targets at a certain time. Similarly to the previous tracking scenario,

two targets $\rho = 1, 2$ initialized at positions $[1.6, 11.5]$, $[5.4, 7]$ (indicated by the blue stars) start moving according to the dynamics in (1), with velocities of $\mathbf{v}_1 = [v_{1,x}, v_{1,y}] = [0.15, 0.1]$ m/s and $\mathbf{v}_2 = [v_{2,x}, v_{2,y}] = [0.4, 0.13]$ m/s, respectively. As $t = 30$, the second target stops moving while the first one splits into two targets. Target $\rho = 3$ continues to move according to the dynamics of target $\rho = 1$, while target $\rho = 4$ moves with velocities $v_x = 0.4$ m/s and $v_y = -0.5$ m/s along the x-axis and y-axis. The two new targets move for the time interval $[31, 42]$ s. The splitting point is indicated by the green star in Fig. 8. Figure 8 shows the trajectories of the targets and some moving sensors; details on the coloring and curve types used can be found in the caption of Fig. 8 . The target trajectories in Fig. 8 are depicted by blue dashed lines for the time interval $[1, 30]$ s and by blue crossed lines after $t = 30$ s. When sensors do not move, the violet estimated trajectories indicate that the split of targets cannot be followed, while target $\rho = 1$ cannot be tracked after a while since is moving away from the immobile sensors. When the kinematic rules in Section 3.3 are employed, informative sensors follow closely the targets as depicted by the black dashed sensor trajectories. Note that the corresponding estimated red trajectories accurately follow the multiple targets present in the field. As before, the tracking RMSE (logarithm) is compared for the cases where sensors cannot move with the case where the approach in Section 3.3 is applied. As Fig. 9 shows, our active tracking scheme outperforms in terms of tracking accuracy the utilization of stationary sensors. Notice that in Fig. 9, after the target splits, tracking using stationary sensors performs much better than before splitting. The reason is that when using stationary sensors, for the second tracking

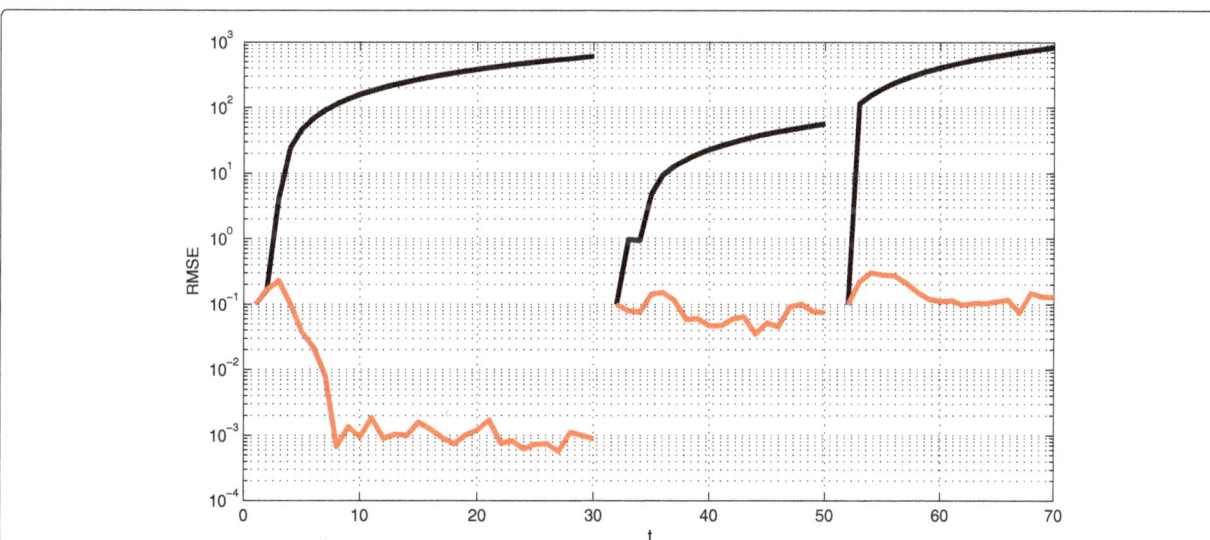

Fig. 7 Tracking RMSE versus time for tracking the multiple objects in the setting depicted in Figs. 4, 5, and 6. Two different cases are studied where sensor mobility is utilized (*red curves*) and when sensors are immobile (*blue curves*)

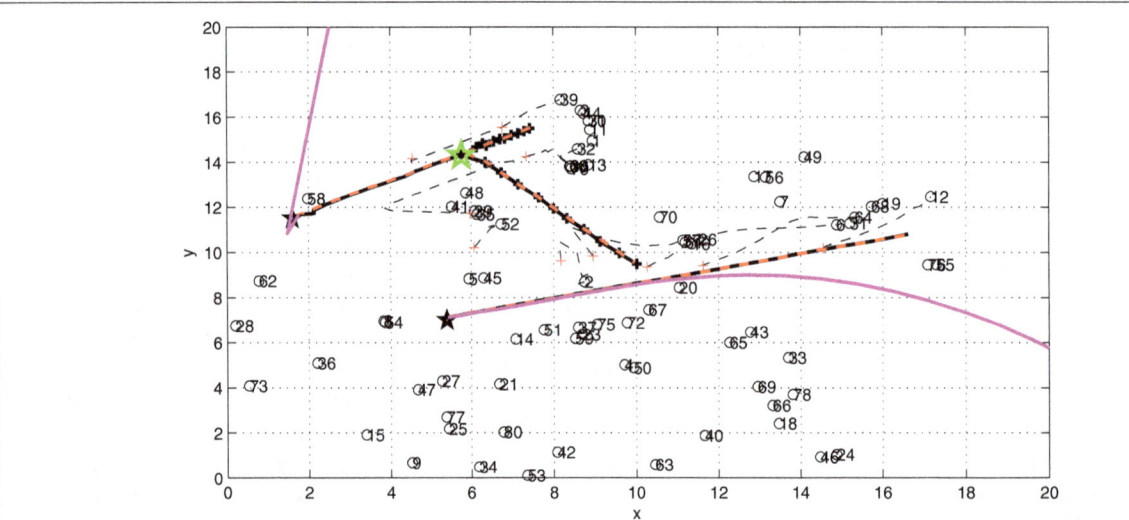

Fig. 8 Tracking multiple objects; target trajectories and mobile sensor trajectories. *Blue dashed curves* and *blue crossed curves* correspond to the true target trajectories; *red dashed curves* represent the estimated track using our novel framework. *Blue stars* indicate the starting locations of the targets, while the *red crosses* represent the initial locations of some moving sensors. The *black dashed curves* show the trajectories of some of the moving sensors during the tracking process using the kinematic rules in Section 3.3. *Violet colored curves* correspond to estimated trajectories using immobile sensors, while the *green star* points to the position where the splitting of the targets takes place

phase ($t = [31, 42]$ s), there are more sensors originally located close to the trajectories of the targets, compared to the first tracking phase ($t = [1, 30]$ s) which makes the tracking error much smaller than the first 30 s, though the performance when using stationary sensors is still worse than tracking using our sensor mobility-based tracking scheme.

6   Conclusions

A novel framework combining sparse matrix factorization with proper kinematic rules enables multiple mobile

sensors to track multiple targets. A norm-one/norm-two regularized matrix decomposition formulation is utilized to perform sensor-to-target association and select the target-informative sensors. Optimal kinematic rules are obtained by minimizing the covariances of parallel extended Kalman filters that track multiple targets using only target-informative sensors. The modified barrier method is utilized to obtain the sensors' location updates while ensuring that the moving sensors remain connected. Numerical tests in multi-sensor networks corroborate that our novel scheme outperforms related approaches

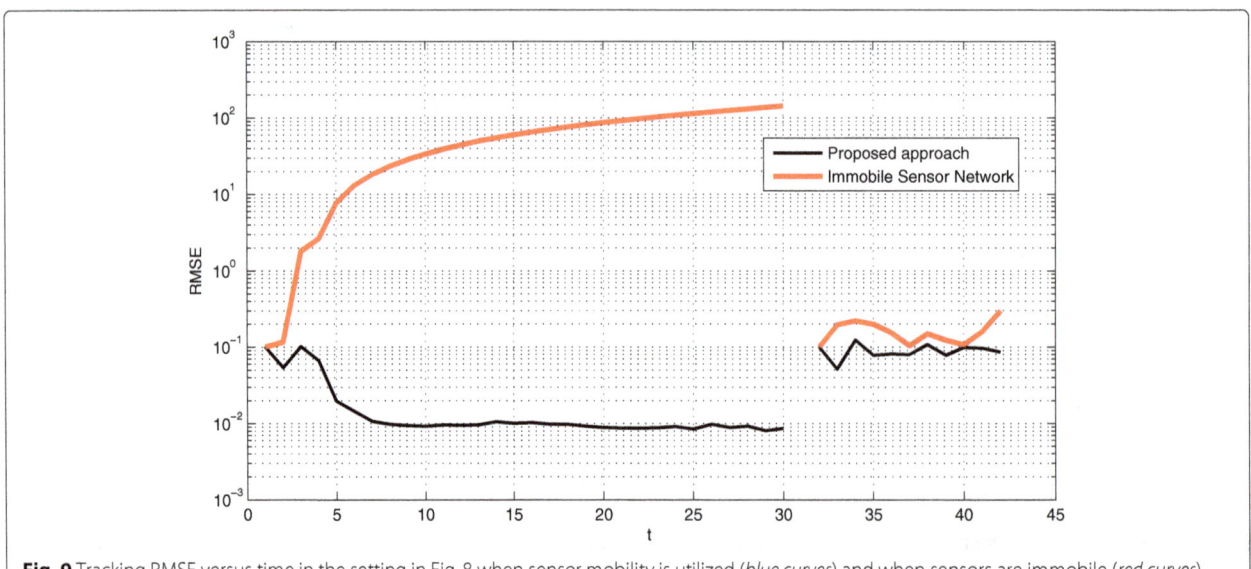

Fig. 9 Tracking RMSE versus time in the setting in Fig. 8 when sensor mobility is utilized (*blue curves*) and when sensors are immobile (*red curves*)

and accurately tracks multiple targets utilizing only a small percentage of moving sensors that closely follow the targets.

Competing interests
The authors declare that they have no competing interests.

Acknowledgements
Work in this paper is supported by the NSF grant CCF 1218079 and the AirForce grant #FA9550-15-1-0103, Simons Foundation Award # 279870 and UTA.

Author details
[1] Department of Electrical Engineering, University of Texas at Arlington, Arlington, TX, USA. [2] Department of Math, University of Tennessee at Knoxville, Knoxville, TN, USA.

References

1. R Silva, J Silva, F Boavida, Mobility in wireless sensor networks—survey and proposal. Elsevier Comput. Commun. **52**, 1–20 (2014)
2. P Yang, R Freeman, K Lynch, in *Proc. of the 2007 IEEE Intl. Conf. on Robotics and Automation*. Distributed Cooperative Active Sensing Using Consensus Filters, (Rome, Italy, 2007), pp. 405–410
3. T Mukai, M Ishikawa, An active sensing method using estimated errors for multisensor fusion systems. IEEE Trans. Ind. Electron. **43**(3), 380–386 (1996)
4. T Chung, V Gupta, J Burdick, R Murray, in *Proc. of 43rd IEEE Conference on Decision and Control*. On a Decentralized Active Sensing Strategy Using Mobile Sensor Platforms in a Network, vol. 2, (Nassau, The Bahamas, 2004), pp. 1914–1919
5. A Gorji, M Menhaj, Multiple target tracking for mobile robots using the jpdaf algorithm. IEEE Int. Conf. Tools Artif. Intell. **1**, 137–145 (2007)
6. S Oh, S Russell, S Sastry, Markov chain monte carlo data association for multi-target tracking. IEEE Trans. Autom. Control. **54**, 481–497 (2009)
7. J Vermaak, S Godsill, P Perez, Monte Carlo filtering for multi target tracking and data association. IEEE Trans. Aerosp. Electron. Syst. **41**(1), 309–332 (2005)
8. JLC C Hue, P Perez, Tracking multiple objects with particle filtering. IEEE Trans. Aerosp. Electron. Syst. **38**(3), 1457–1469 (2002)
9. FS Cattivelli, AH Sayed, Diffusion strategies for distributed Kalman filtering and smoothing. IEEE Trans. Autom. Control. **55**(9), 2069–2084 (2010)
10. NF Sandell, R Olfati-Saber, in *Proc. of IEEE Conf. on Decision and Control*. Distributed Data Association for Multi-target Tracking in Sensor Networks (IEEE, Cancun, Mexico, 2008), pp. 1085–1090
11. J Lin, W Xiao, FL Lewis, L Xie, Energy-efficient distributed adaptive multisensor scheduling for target tracking in wireless sensor networks. IEEE Trans. Instrum. Meas. **58**(6), 1886–1896 (2009)
12. Y Zou, K Chakrabarty, Distributed mobility management for target tracking in mobile sensor networks. IEEE Trans. Mob. Comput. **6**(8), 872–887 (2007)
13. K Zhou, S Roumeliotis, Multirobot active target tracking with combinations of relative observations. IEEE Trans. Robot. **27**(4), 678–695 (2011)
14. K Lynch, I Schwartz, P Yang, R Freeman, Decentralized environmental modeling by mobile sensor networks. IEEE Trans. Robot. **24**(3), 710–724 (2008)
15. Y Oshman, P Davidson, Optimization of observer trajectories for bearings-only target localization. IEEE Trans. Aerosp. Electron. Syst. **35**(3), 892–902 (1999)
16. Doğançay, UAV path planning for passive emitter localization. IEEE Trans. Aerosp. Electron. Syst. **48**(2), 1150–1166 (2012)
17. G Keung, B Li, Q Zhang, H Yang, in *Proc. of 2011 Global Telecommunications Conference*. The Target Tracking in Mobile Sensor Networks, (Houston, TX, 2011), pp. 1–5
18. H La, W Sheng, Dynamic target tracking and observing in a mobile sensor network. Elsevier Robot. Auton. Syst. **60**(7), 996–1009 (2012)
19. A Elmogy, F Karray, Cooperative multi-target tracking using multi-sensor network. Int. J. Smart Sens. Intell. Syst. **1**(3), 716–734 (2008)
20. Y Fu, L Yang, Sensor mobility control for multitarget tracking in mobile sensor networks. International Journal of Distributed Sensor Networks. **2014**(3), 1–15 (2014)
21. R Tibshirani, Regression shrinkage and selection via the lasso. J. R. Stat. Soc. Ser. B. **58**(1), 276–288 (1996)
22. H Zou, T Hastie, R Tibshirani, Sparse principal component analysis. J. Comput. Graph. Stat. **15**(2), 265–286 (2006)
23. I Schizas, Distributed informative-sensor identification via sparsity-aware matrix decomposition. IEEE Trans. Sig. Proc. **61**(18), 4610–4624 (2013)
24. G Ren, V Maroulas, I Schizas, Distributed spatio-temporal association and tracking of multiple targets using multiple sensors. IEEE Trans. Aerosp. Electron. Syst. **51**(4), 2570–2589 (2015)
25. P Tseng, Convergence of a block coordinate descent method for nondifferentiable minimization. J. Opt. Theory Appl. **109**(3), 475–494 (2001)
26. D Bertsekas, *Nonlinear Programming*, Second edition. (Athena Scientific, Nashua, 2003)
27. S Kay, *Fundamental of Statistical Signal Processing: Estimation Theory*. (Prentice Hall, New Jersey, 1993)
28. Y Bar-Shalom, X Li, T Kirubarajan, *Estimation With Applications to Tracking and Navigation*. (Wiley, New York, 2001)
29. A Goldsmith, *Wireless Communications*. (Cambridge University Press, New York, 2005)
30. JG Proakis, M Salehi, *Digital Communications*, 5th edition. (McGrawHill, New York, 2008)
31. SS Haykin, *Adaptive Filter Theory*. (Pearson Education India, India, 2008)
32. M Yuan, Y Lin, Model selection and estimation in regression with grouped variables. J. R. Stat. Soc. Ser. B Stat Methodol. **68**(1), 49–67 (2006)
33. R Horn, C Johnson, *Matrix Analysis*. (Cambridge Univ. Press, Cambridge, 1985)
34. K Dogancay, Online optimization of receiver trajectories for scan-based emitter localization. IEEE Trans. Aerosp. Electron. Syst. **43**(3), 1117–1125 (2007)
35. J Borenstein, L Feng, H Everett, *Navigating Mobile Robots: Systems and Techniques*. (AK Peters, Ltd, Wellesley, 1996)
36. B Barshan, H Durrant-Whyte, Inertial navigation systems for mobile robots. IEEE Trans. Robot. Autom. **11**(3), 328–342 (1995)

6

Robust detection in ultra-wideband impulse radar using DPSS-MMSE estimator

S. M. Ali Tayaranian Hosseini[1], Hamidreza Amindavar[1*] and James A. Ritcey[2]

Abstract

In this paper, it has been shown that non-stationary received signals at ultra-wideband impulse radars can be addressed by Fourier series model with time-varying coefficients. Next, simply by computing statistical features of the coefficients, we show that this model can be considered as a sum of band-limited sources. Based on the unconditional orthonormal representation of band-limited signals, a MMSE estimator is introduced to determine the Fourier coefficients. Getting the most out of our novel estimator, we suggest a new method for blind and robust detection that enables us to determine the range and velocity of moving targets, accurately without utilizing any matched filters. In this new approach, no prior information is required for detection, except pulse repetition interval. Since the novel method is based on the non-stationary analysis, the signal is analyzed in a long period of time to estimate the velocity in high resolution. Furthermore, since there is no assumption on the noise distribution, the signal of interest can be simply detected in the presence of correlated and non-Gaussian noise, i.e., encompassing the conglomerate effects of clutter and interference. To verify our result, an experimental test and simulations are also presented comparing the new detector with conventional ultra-wideband impulse radar detectors referred to as interleaved periodic correlation processing (IPCP) in the literature.

Keywords: Robust detection, Ultra-wideband impulse, MMSE estimator

1 Introduction

The vast majority of traditional radars use harmonic pulse signals to detect targets. The bandwidth of signals used in such radars is much less than the carrier frequency. Consequently, these radars can provide only low resolution detection. However, in today's applications, high-resolution radars are sought for fine and sensitive surveillance of the environment.

Improving detected target range measurement accuracy, identifying target classes and types, low-altitude detection, ground penetrating, immunity to passive interference, and some other advantages have been provided by reducing the pulse width of signals in ultra-short pulse or ultra-wideband (UWB) impulse radars [1]. This type of the UWB radars, as a high-resolution radar, is widely used for remote sensing of the objects in medicine, psychophysiology [2], human being detection [3], through the wall imaging [4], and stealth target detection [1].

In some applications of ranging and velocity detection traditional correlation is used to increase signal-to-noise ratio [3]. Some factors such as multiple scattering points and dispersion make echo waveforms seriously distorted in ultra-short pulse radars; hence, the backscattered signal from a target upon impingement of an ultra-short pulse signal has a complex shape in time domain and is completely different from the transmitted waveform [5]. In addition, the radar cross section (RCS) in UWB target detection becomes time-dependent, so the concept of instantaneous target RCS has been introduced [6]. In such cases, the target scattered signal is non-stationary in time. These phenomena cause the traditional correlation detection method, which selects the transmitted signal as the reference waveform in the correlator, to degrade in detecting targets. The parameters such as duration, location, and amplitude of the scattered signal are strongly dependent on the target geometry. Since the scattered signal from a target is represented as a stochastic time series, hence, generally speaking, its mathematical description is unknown. Therefore, blind detection has attracted the attention of advanced ultra-short pulse radar designers.

*Correspondence: hamidami@aut.ac.ir
[1] Amirkabir University of Technology, Tehran, Iran
Full list of author information is available at the end of the article

A blind detector based on the correlation of the received signals in two adjacent periods, has been introduced as the interleaved periodic correlation processing (IPCP) detector [7]. In the IPCP, an invariable channel in at least two adjacent periods is assumed. The first period is considered as the reference signal and matched filter for the second one. Because the reference signal may be completely different from a real reference (transmitted signal), this approach causes noticeable degradation in detection compared to the optimum matched filter in complicated scenarios such as moving target scenarios. In the aforementioned detector, the integration period of the correlation processing is determined by the observation interval or the scattered signal duration which depends on the target size. Hence, the detection threshold is unknown and depends on the target size [8]. Another prominent issue, which we will show in this article, is that the mentioned correlation processing, IPCP, causes removal of the Doppler information, a major drawback in determining the velocity of a moving target accurately. Another approach is introduced in [9], used in human being detection; there is no correlation processing and detection is based on the frequency characteristics of the wideband backscattered signals. In this application, Doppler shift is negligible, and generally speaking, velocity estimation is not important. Although the Doppler shift is really small, extraction of Doppler shift is the base of the some nowadays strategies in this application [10, 11]. The velocity measurements in ultra-short pulse radar is considered in [6, 12] by taking the advantage of the conventional correlation detectors. However, these measurements are also limited in the resolution of velocity estimation. Due to these limitations, velocity, which can be used as an important parameter in the tracking filters in applications such as UWB radar sensor networks, has often been ignored [13]. In another procedure, a UWB detection radar is introduced based on cell averaging constant false alarm rate (CA-CFAR)[14]. CA-CFAR is a powerful technique that reduces the false alarm rate in a given probability of detection, especially in the presence of clutter and jamming signals. However, it is not consider the non-stationarity issues in the interest signal and clutter. Therefore, extraction of velocity is not possible by only CA-CFAR technique.

There are some limitations to produce short time pulses [6]. Since the Fourier series method for waveform generation overcomes these limitations [15], UWB radar signals are generated using the Fourier series-based waveform paradigm [15]. The present article is based on the Fourier series signal generation. Here, it is shown that the UWB received signal is an almost periodic signal which can be represented as a Fourier series expansion with time-varying coefficients. Cyclostationary characteristics which appear in this type of waveform can be exploited to determine a blind detector which is based on detecting the presence of cyclostationarity. Cyclostationary detection is examined in [16]. Although cyclostationary features appear in these kind of signals, actually, there is no pure cyclostationarity, especially, when the observations have to be made in large intervals and high resolution, velocity estimation is intended. In this article, in Section 2, we first describe the model of received ultra-short pulse signal scattered from a target, based on Fourier series model for transmitted signal. Then, in Section 3, we will do an analysis on the IPCP detector and demonstrate its inability in velocity detection. Next, we aim to show that the non-stationary received signal can be modeled by a Fourier series with time-varying coefficients. In Section 4, we have used here time-varying orthonormal sequences as the optimal weights, that is an index-limited sequence with maximum energy concentration in a finite sample interval to introduce a linear MMSE estimation for the time-varying coefficients. The aforementioned optimal window is related to the discrete prolate spheroidal sequences (DPSS) [17]. In Sections 5 and 6, based on the estimation of the coefficients, we extract range and velocity of targets. We fully describe the analysis of resolution of range and velocity and the error estimation. Because of utilization of time-varying weights in the estimation of range and doppler, which leads to the estimation of non-stationary signals, the novel estimator is robust for a wide variety of lifelike scenarios including slow-moving or hovering platforms. Since there is no assumption on noise distribution, all analyses are done based on general correlated non-Gaussian noise. Therefore, the noise considered herein accounts for all interferences, clutters, and thermal noise. In Section 7, in simulations, using this signal, we have sketched the range-velocity plot to visualize the results of detection and ROC plot to verify the advantages of our work compared to IPCP. This will be done for both cases of Gaussian and non-Gaussian noise, as well as colored and white noise.

2 System model and problem formulation

In UWB radars, anomalies caused by the array antenna, the propagation path and scattering from the target, will change shape and bandwidth of the transmitted signal. In this section, the model of transmitted signal and the factors that may cause variations in the shape of a UWB signal during the radar observation of a target is discussed.

2.1 Fourier series model in ultra-short pulse

Most of the UWB radar waveforms considered in the literature are of the form of impulse signals which could be implemented by Marx-Bank or similar techniques [18]. In pulse generation, the energy stored in a long period of time have to be released in a short while. In Marx-Bank, as a typical way of energy storage, the capacitive stored energy is released by switches such as spark gap, diode, and laser-actuated semiconductors which cause the pulse shape, and PRI is not precisely controllable [1]. The Fourier method of waveform generation [15] (combination of conventional heterodyning method in [1]) is an appropriate method to overcome these problems [1, 15]. In this method, the transmitted signal $s_1(t)$ is produced by the summation of a truncated Fourier series components.

The number of transmitting sources required to generate a short pulse train is a function of pulse width and duration. In this case, the pulse train $s_1(t)$ can be represented by a series of $N+1$ Fourier components as

$$s_1(t) = \sum_{n=-N/2}^{N/2} c_n e^{j\omega_n t}, \tag{1}$$

where $\omega_n = \frac{2\pi n}{T}$, T is the total pulse duration of the signal and c_n are the complex harmonic coefficients. $s_1(t)$ assumed the transmitted baseband signal to be serve as an ultra-short pulse signal. For example the complex coefficients for the 11-chip Barker sequence with chip width of τ and pulse duration $T = 11\tau$, can be written as

$$c_n = \frac{1}{\pi n}\frac{1}{2j}\left(1 - 2e^{-j\omega_n 3\tau} + 2e^{-j\omega_n 6\tau} - 2e^{-j\omega_n 7\tau} + 2e^{-j\omega_n 9\tau} - 2e^{-j\omega_n 10\tau} + e^{-j\omega_n 11\tau}\right). \tag{2}$$

It should be noted that after transmitting $s_1(t)$, there will be no other transmissions during the pulse repetition interval (PRI). So, the duration of the transmitted signal, comprising both the transmitted pulse $s_1(t)$ and zero, is PRI . As an example, the 11-chip Barker code is depicted in Fig. 1.

2.2 Effect of array antenna

We assume the UWB signal $s_1^c(t) = \text{Re}\left\{s_1(t) e^{j2\pi f_c t}\right\}$ to be transmitted with the carrier frequency f_c in a form of current pulse by an array antenna with P radiators shown in Fig. 2. The effects of the array antenna change $s_1^c(t)$ to another pulse train $s_2^c(t)$. These effects, discussed in [7], are briefly explained in the following.

If the antenna radiators have a length of $L_r \gg c\tau$ (c is the velocity of light and τ is the pulse width), the antennae radiate several pulses of the electromagnetic wave serially. As a result, a single pulse transforms into a sequence of K pulses each of which radiates in a time interval ε_k. Another phenomenon occurs due to the spatial delay of radiated signals which is $(d_p/c)\cos(\phi)$, for adjacent radiators. Here d_p is the spacing between the radiators and ϕ is the observation angle of the radar. Therefore, the baseband signal $s_2(t)$ in the radiated waveform $s_2^c(t) = \text{Re}\left\{s_2(t) e^{j2\pi f_c t}\right\}$ can be written as:

$$s_2(t) = \sum_{p=1}^{P}\sum_{k=1}^{K} e^{j\phi_{kp}} s_1\left(t + \varepsilon_k + \frac{d_p}{c}\cos\phi\right) \tag{3}$$

$$\phi_{kp} = 2\pi f_c\left(\varepsilon_k + \frac{d_p}{c}\cos\phi\right).$$

If we assume uniform spacing between the radiators of distance d, Eq. (3) can be written as:

$$s_2(t) = P\sum_{k=1}^{K} e^{j\phi_k} s_1\left(t + \varepsilon_k + \frac{d}{c}\cos\phi\right) \tag{4}$$

$$\phi_k = 2\pi f_c\left(\varepsilon_k + \frac{d}{c}\cos\phi\right).$$

2.3 Model of the target in UWB radar

Assume the maximum dimension of the target be L_t satisfying the condition $c\tau \ll L_t$. Then, the target can be modeled as a combination of M local scattering elements (bright point [8]). The delay of radiated signal $s_2^c(t)$ in arriving at the mth bright point is:

$$\tau_d^m = R(t)/c + \delta_m + \kappa_m(t), \tag{5}$$

where $R(t) = R - vt$, R is the distance from the receiver to the center of the target, v is the radial velocity of the target's combination, δ_m models the distance from the center of the target to the mth scatterer; and $\kappa_m(t)$ models time-varying aspect angle in the target's plate [19–22]. The signal $s_2^c(t)$ is reflected by the discrete target elements with various time delays τ_d^m and various attenuations α_m. The reflected signal will come back to the radar receiver

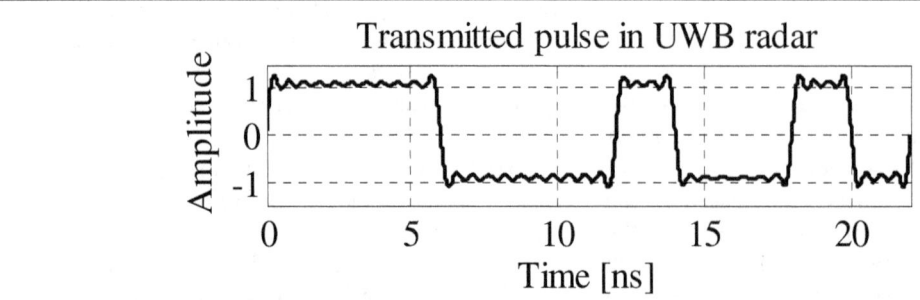

Fig. 1 A UWB pulse signal. This signal is generated by Fourier series model using 40 harmonies. The pulse is produced by the Barker code with the length of 11, chip width of 2 *ns* and the total pulse duration 22 of *ns*

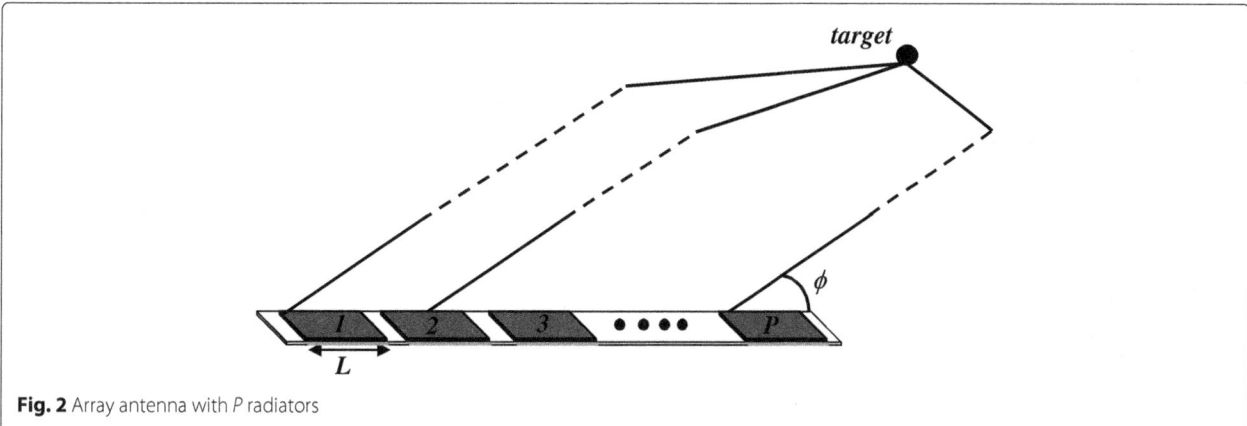

Fig. 2 Array antenna with P radiators

after τ_d^m. Hence, $s_3(t)$, the baseband received signal during one pulse repetition interval (T_{PRI}), is expressed as

$$s_3(t) = \sum_{m=1}^{M} \alpha_m e^{-j2\pi f_c(2\kappa_m(t))} e^{-j2\pi f_c(2\delta_m + 2R/c)} e^{j2\pi F_d t}$$
$$s_2(t - 2\delta_m - 2R/c). \tag{6}$$

In Eq. (6), because $v << c$, the vt/c and $\kappa_m(t)$, due to far field assumption, are very small and can be neglected in the argument of $s_2(t)$. α_m, the intensity of reflection of each bright point, has an statistical model based on the RCS models which is discussed in several literatures [8]. An ultra-short pulse signal possesses a large bandwidth that causes the radar cross section (RCS) to vary significantly. Therefore, in practical applications, the target RCS depends on the frequency [7]. The frequency-dependent RCS implies that α_m, previously assumed to be a single coefficient, should act as a FIR filter h_m with complex coefficients for each bright point. Hence, $s_3(t)$ is expressed as:

$$s_3(t) = \sum_{m=1}^{M} \int_0^{T_{PRI}} h_m(\xi - \delta_m - R/c)$$
$$e^{j2\pi F_d \xi} s_2(\xi - t + 2\delta_m + 2R/c) \, d\xi. \tag{7}$$

Herein, instead of using $h_m(t)$ as a FIR filter, we utilize α_{mn} coefficients which depend on nth harmonic frequency. By substitution Eq. (1) in (4) and (4) in (6) and considering α_{mn} instead of α_m, the $s_3(t)$ during a pulse repetition interval can be represent as:

$$s_3(t) = \sum_{n=-N/2}^{N/2} a_n(t) e^{j\omega_n t} \tag{8}$$

$$a_n(t) = \sum_{m=1}^{M} \sum_{k=1}^{K} Pc_n \alpha_{mn} e^{-j2\pi f_c(2\kappa_m(t))} e^{j(\varphi_k + \phi_m + \theta_{nkm})} e^{j2\pi F_d t}$$

$$\varphi_m = 2\pi f_c(2\delta_m + 2R/c)$$

$$\theta_{nkm} = \omega_n \left(\varepsilon_k - 2\delta_m + \frac{d \cos(\phi) - 2R}{c} \right).$$

Since the transmitted signal is repeated every T_{PRI}, we have $\omega_n = 2\pi n f_0$ where f_0 is the pulse repetition frequency. Equation (8) has been expressed for one T_{PRI}. This equation is valid in all time by assuming an almost periodic h_m with the period of pulse repetition interval which leads to uniform coefficients α_{mn} in every pulse repetition interval. This assumption for a finite number of pulse repetition intervals, will be reasonable if $2T_{CPI}v \ll c\tau$, where T_{CPI} is the time of coherent processing interval. The estimation of velocity for low speed targets (where v is small), needs the duration of coherent processing interval to be long enough. Therefore, in our scenario that the wide range of velocity detection is considered, α_{mn} cannot be assumed constant and varies smoothly in time. Equation (8) can be extended to be valid for every t if $a_n(t)$ is expressed as:

$$a_n(t) = \sum_{m=1}^{M} \sum_{k=1}^{K} Pc_n \alpha_{mn}(t) e^{-j2\pi f_c(2\kappa_m(t))}$$
$$e^{j(\phi_k + \varphi_m + \theta_{nkm})} e^{j2\pi F_d t}. \tag{9}$$

In Appendix A, we find an expression for the correlation function and the spectrum of the time-varying coefficients $a_n(t)$. In there, we demonstrate that the $a_n(t)$ processes are band-limited signals all around the Doppler frequency. So, if a summation of N Fourier components is transmitted as an ultra-short pulse signal as mentioned before, the received signal will be a wideband signal composed of a sum of non-stationary signals each of which acts as a sine wave with a frequency $n f_0$, multiplied by a "smoothly varying" amplitude function. Therefore, the baseband received signal $s_3(t)$ is modeled as a Fourier series with time-varying coefficients $a_n(t)$. Based on what is described in this section, the scattering in UWB radar elongates the pulse width τ and changes the resolution of range gate. Because the number of pulses, time delay δ_m, and the intensity α_{mn} of the signal depends on the target shape and target element, we can use variation of the pulse width as a feature in classification of targets. The scattered signal from a real target and its spectrum are depicted in Figs. 3 and 4, respectively. The experimental test to obtain the signal is discussed in Section 7. In next section, a linear MMSE estimation has been introduced to determine of the mentioned time-varying coefficients.

3   IPCP detector

Let $s(t)$ be the baseband transmitted signal, then, this signal is reflected back from a target with the velocity v at the range of R the received baseband signal is represented as

$$r(t) = e^{j2\pi F_d t}\widehat{s}\left(t - \frac{2R}{c}\right) + n(t), \tag{10}$$

where c is the speed of light, $F_d = \frac{2v}{\lambda}$ is the Doppler frequency, and λ is the wavelength of received signal. The effects of array antennas, anomalies due to the propagation path and effects of target scattering may change $s(t)$ to the pulse train $\widehat{s}(t)$, these effects will be discussed in Section 2. $n(t)$ is a zero mean additive noise which is independent of the transmitted signal.

The IPCP receiver uses the received signal as the reference signal. The receiver model is shown in Fig. 5. The output could be expressed as

$$U(nT_{PRI}) = \int_{nT_{PRI}}^{(n+1)T_{PRI}} r^*(\tau)\, r(\tau - T_{PRI})\, d\tau, \tag{11}$$

where T_{PRI} is the pulse repetition interval and n is the neutral number. Now, using Eq. (10) in Eq. (11), we have

$$U(nT_{PRI}) =$$

$$e^{-j2\pi F_d T_{PRI}} \int_{nT_{PRI}}^{(n+1)T_{PRI}} \widehat{s_3^*}\left(\tau - \frac{2R}{c}\right)\widehat{s}\left(\tau - T_{PRI} - \frac{2R}{c}\right) d\tau \tag{12}$$

$$+ \int_{nT_{PRI}}^{(n+1)T_{PRI}} e^{-j2\pi F_d \tau}\widehat{s^*}\left(\tau - \frac{2R}{c}\right) n(\tau - T_{PRI})\, d\tau \tag{13}$$

$$+ \int_{nT_{PRI}}^{(n+1)T_{PRI}} e^{j2\pi F_d(\tau - T_{PRI})}\widehat{s}\left(\tau - T_{PRI} - \frac{2R}{c}\right) n^*(\tau) d\tau \tag{14}$$

$$+ \int_{nT_{PRI}}^{(n+1)T_{PRI}} n^*(\tau)\, n(\tau - T_{PRI})\, d\tau. \tag{15}$$

Since the noise and signal are uncorrelated, Eqs. (13) and (14) are zero. It can be seen in Eq. (12) that the Doppler frequency exists in a phase factor which is independent of the time sequence. Hence, velocity can not be extracted in the Doppler processing. Since the noise is zero mean and independent of the transmitted signal, there is no information about velocity in Eq. (13) or Eq. (14). Therefore, the velocity information is ignored in the IPCP detector.

4   DPSS-MMSE estimator

Sampling the noisy received waveform $r(t) = s_3(t) + v(t)$ with sampling time T_s, we will obtain L samples where $s_3(t)$ and $v(t)$ are assumed uncorrelated. $v(t)$ accounts for all interferences, clutters, and thermal noise and hence can be considered as colored and non-Gaussian noise. Taking into consideration that f_0 is the only known a priori

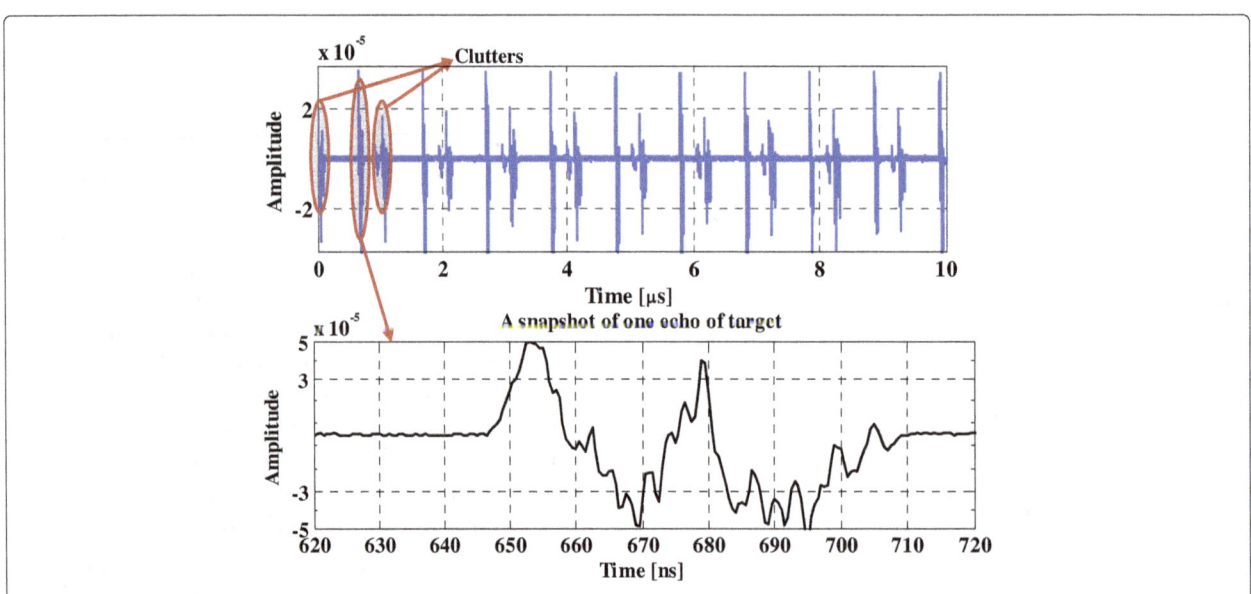

Fig. 3 Backscattered signal from a target. The signal from a target that impinge on the 11-chip barker sequence as the transmitted signal. Chip width is 2 ns and the total pulse width is 22ns. PRI is 1 μs in transmitted signal

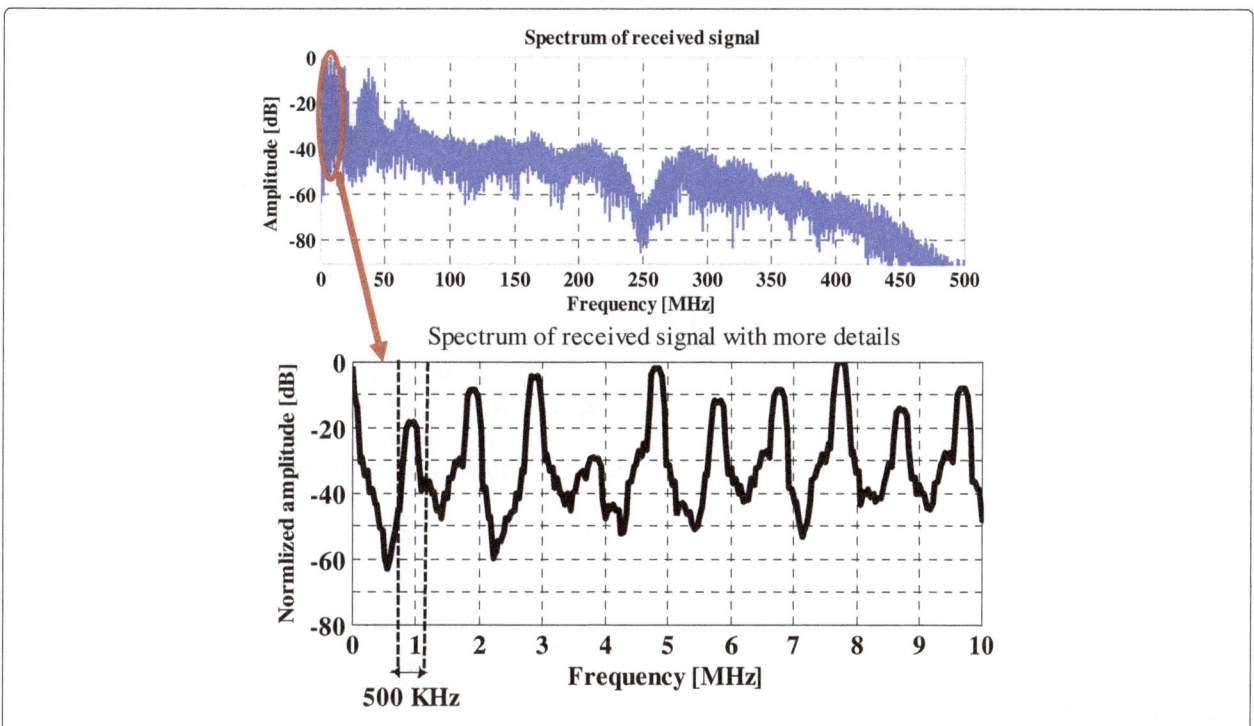

Fig. 4 Spectrum of the backscattered signal from a target. Spectrum of the backscattered signal from a target. The duration of observations in computation of spectrum is 4 PRI s. PRI is 1 μs in the transmitted signal

information, in this section, we will introduce a MMSE estimator for determination of the finite bandwidth signals $a_n(\ell)$, and then, we develop our estimator to detect the presence of the UWB signal and extract the target's range and velocity. In estimating of $a_n(\ell)$, we first expand it by the optimum orthogonal bases. The optimal bases, that is an index-limited sequence with maximum energy concentration in a finite sample interval, is related to the zeroth discrete prolate spheroidal sequence (DPSS) [17]. In accordance with this decomposition, $a_n(\ell)$ can be expressed in terms of DPSS sequences

$$a_n(\ell) = \sum_{m=1}^{\mathcal{M}} c_{mn}\phi_m(\ell) = \Phi(\ell)^T \mathbf{C}_n, \quad \ell = 0, \cdots, L-1$$

$$\Phi(\ell) = [\phi_1(\ell)\, \phi_2(\ell)\, \cdots\, \phi_{\mathcal{M}}(\ell)]^T \quad (16)$$
$$\mathbf{C}_n = [c_{1n}\, c_{2n}\, \cdots\, c_{\mathcal{M}n}]^T,$$

where $\{\phi_m(\ell)\}_{m=1}^{\mathcal{M}}$ are DPSS orthonormal sequences, $\mathbf{C}_n$ is the coefficient vector. $\mathcal{M} < L$ is the order of expansion and depends on the normalized bandwidth of $a_n(t)$, W, and observation length, L,

$$\mathcal{M} = \lceil 2WLT_s \rceil + 1. \quad (17)$$

We use linear estimation model with time-varying weights, which is defined as

$$\hat{a}_n(\ell) = \mathbf{w}_n(\ell)^H \mathbf{r}, \quad (18)$$

where $\mathbf{r}$ is the $L \times 1$ observation vector has been made by sampled received waveform and $\mathbf{w}_n(\ell)$ are time-varying weights. Substituting Eq. (8) in to the sampled received

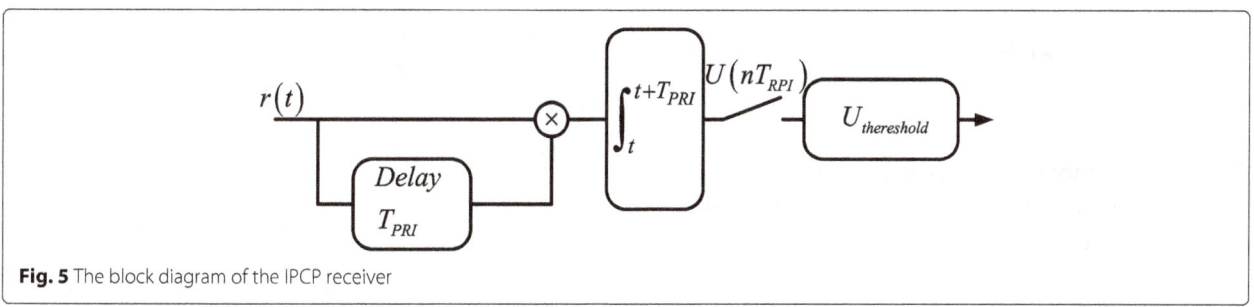

Fig. 5 The block diagram of the IPCP receiver

signal, $r(\ell) = s_3(\ell) + v(\ell)$, the matrix presentation of the observation vector $\mathbf{r}$ will be:

$$\mathbf{r} = \sum_{n=-N/2}^{N/2} \mathbf{F}_n \mathbf{C}_n + v, \qquad (19)$$

where v is the $L \times 1$ noise vector which is uncorrelated to the signal and $\mathbf{F}_n$ is a $L \times \mathcal{M}$ matrix with entries:

$$f_n(i, m) = \left\{ \phi_m(i) e^{j\frac{2\pi n}{T_p} i} \right\}, \begin{cases} 0 \leq i \leq L-1 \\ 1 \leq m \leq \mathcal{M} \end{cases} \qquad (20)$$

Incorporating Eq. (19) into Eq. (18), the expansion of $\hat{a}_n(\ell)$ is achieved:

$$\hat{a}_n(\ell) = \mathbf{w}_n(\ell)^H \sum_{n'=-N/2}^{N/2} \mathbf{F}_{n'} \mathbf{C}_{n'} + \mathbf{w}_n(l)^H v \qquad (21)$$

$$= \mathbf{w}_n(\ell)^H \mathbf{F}_n \mathbf{C}_n + \sum_{n'=-N/2, n' \neq n}^{N/2} \mathbf{w}_n(\ell)^H \mathbf{F}_{n'} \mathbf{C}_{n'} + \mathbf{w}_n(\ell)^H v.$$

The above equation reveals that each estimate is formed of two components: the first depends on the time-varying amplitude at the frequency of interest ω_n, while the second is an error term which depends on all the other components of $r(\ell)$ at frequencies separated from ω_n and noise. The optimal estimator must produce the accurate time-varying amplitude from the first component and minimize the contribution of the error term. In other words, we need to impose the restrictions that:

$$\mathbf{w}_n(\ell)^H \mathbf{F}_n = \Phi(\ell)^H. \qquad (22)$$

Also, we have to minimize the mean-squared error (MSE):

$$\text{MSE}(\ell) = E \left\{ \left| a_n(\ell) - \hat{a}_n(\ell) \right|^2 \right\}. \qquad (23)$$

Minimization of Eq. (23) is performed subject to the constraint (Eq. 22) using the method of Lagrange multipliers, i.e., we have to minimize the cost function:

$$J(\ell) = \mathbf{w}_n(\ell)^H \Theta \mathbf{w}_n(\ell) + \mathbf{w}_n(\ell)^H \mathbf{R}_v \mathbf{w}_n(l) \qquad (24)$$
$$- \left[\mathbf{w}_n(\ell)^H \mathbf{F}_n - \Phi(\ell)^H \right] \lambda,$$

where

$$\Theta = \sum_{n'=-N/2, n' \neq n}^{N/2} \mathbf{F}_{n'} E \left\{ \mathbf{C}_{n'} \mathbf{C}_{n'}^H \right\} \mathbf{F}_{n'}^H, \quad \mathbf{R}_v = E \{ v v^H \} \qquad (25)$$

and λ is a $\mathcal{M} \times 1$ vector of Lagrange multipliers. By setting the first derivative of the cost function to zero, we arrive at optimum least square weights as [23]:

$$\mathbf{w}_n(\ell) = \mathbf{F}_n \Phi(\ell). \qquad (26)$$

By using this model, MMSE is determined as

$$\text{MMSE}(\ell) = \left(1 + \sigma_v^2 \right) \sum_{m=1}^{\mathcal{M}} |\phi_m(\ell)|^2, \qquad (27)$$

where σ_v^2 is the variance of colored or AWG noise. Although, MMSE seems time-dependent in Eq. (27), but it is almost constant in practice and is $\frac{\mathcal{M}}{L} \left(1 + \sigma_v^2 \right)$. By decreasing $\mathcal{M}$ and increasing L, the MMSE will be decreased. Therefore, the suitable selection of L which depends on the bandwidth of $a_n(t)$ alleviate noise as much as possible. The number of L and $\mathcal{M}$ is set by an educated guesstimate. In the two following sections the extraction of range and velocity of the target from estimated time-varying coefficients have been described.

5   Range processing

In this section, we use energy detector to find the range of the target. We can write the received signal as

$$r(t) = \sum_{n=-N/2}^{N/2} r_{\omega_n}(t) + v(t) \qquad (28)$$

$$r_{\omega_n}(t) = a_n(t) e^{j\omega_n t}.$$

And the energy of the received signal in each time range cell is expressed as

$$\mathcal{E}_k = \int_{(k-1)T_g}^{kT_g} |r(t)|^2 dt, \qquad (29)$$

where T_g is the the range cell and $\mathcal{E}_k$ is the energy of the kth range cell. Neglecting the very slight overlap of $r_{\omega_n}(t)$'s, $|r(t)|^2$ can be replaced by:

$$|r(t)|^2 = \sum_{n=-N/2}^{N/2} \left| r_{\omega_n}(t) \right|^2 + |v(t)|^2. \qquad (30)$$

Therefore, the energy of the kth range cell can be expressed as:

$$\mathcal{E}_k \approx \sum_{n=-N/2}^{N/2} \int_{(k-1)T_g}^{kT_g} \left| r_{\omega_n}(t) \right|^2 dt + \varepsilon_k \qquad (31)$$
$$- \sum_{n=-N/2}^{N/2} \int_{(k-1)T_g}^{kT_g} |a_n(t)|^2 dt + \varepsilon_k,$$

where ε_k is the energy of noise in the kth range cell. The detection algorithm can be implemented under two hypotheses test:

$$\begin{cases} r(t) = s_3(t) + v(t) & \mathcal{H}_1 \\ r(t) = v(t) & \mathcal{H}_0, \end{cases}$$

where $\mathcal{H}_1$ shows the presence of the target and $\mathcal{H}_0$ shows the absence of the target. By considering estimated energy in every range cell as decision statistics and based on

Eq. (31), the decision statistics can be presented as the combination of the energy of estimated coefficients:

$$l(\mathbf{a}) = \frac{1}{NL} \sum_{n=-N/2}^{N/2} \sum_{i \in \mathcal{A}^k} |\hat{a}_n(i)|^2, \tag{32}$$

where $l(\mathbf{a})$ is the decision statistics and $\mathcal{A}^k$ is a countable set of the length of $L = \left\lfloor \frac{T_g}{T_s} \right\rfloor$,

$$\mathcal{A}^k \triangleq \{i \in \mathbb{Z} : (k-1)L, \ldots, kL-1\}. \tag{33}$$

Therefore, the decision rule can be given by

$$l(\mathbf{a}) \underset{\mathcal{H}_0}{\overset{\mathcal{H}_1}{\gtrless}} \gamma, \tag{34}$$

where γ is threshold. To define the threshold for a given the probability of false alarm (P_{FA}), we have to determine an analytical expression for the distribution function of $l(\mathbf{a})$ under $\mathcal{H}_0$. This distribution function is approximated as a Gaussian distribution:

$$l(\mathbf{a}) \sim \mathcal{N}(\mu, \Sigma^2), \tag{35}$$

where μ and Σ^2 are mean and variance, respectively. In this approximation, considering large number of elements in observation vector $\mathbf{r}$, central limit theorem is used. When the amount of L is large enough, the approximation of distribution of $l(\mathbf{a})$ under $\mathcal{H}_0$ as a Gaussian is independent of the type of noise (AWGN, colored or non-Gaussian). In Section 7 the Kolmogorov-Smirnov test is used to experimentally show the conformity of this approximation. The statistical average in Eq. (35) for the most general case, where no statistical assumption on the distribution of noise is required, is expressed as:

$$\mu = \frac{1}{NL} \sum_{n=-N/2}^{N/2} \sum_{i=0}^{L-1} \mathbf{w}_n^H(l) \mathbf{R}_v \mathbf{w}_n(l). \tag{36}$$

In this case, variance cannot be evaluated in closed form. But, for *iid* noise, Eq. (36) can be simplified, hence, we are able to evaluate statistical average and variance in closed form as the following:

$$\mu = \frac{\sigma_v^2}{L} \sum_{i=1}^{L} \sum_{m=1}^{M} |\phi_m(i)|^2, \quad \Sigma = \sqrt{\frac{2\sigma_v^4}{L} \sum_{i=1}^{L} \left(\sum_{m=1}^{M} |\phi_m(i)|^2 \right)^2}. \tag{37}$$

This allows the probability of false alarm, P_{FA} to be approximated as

$$P_{FA} = P(l(\mathbf{a})|\mathcal{H}_0) = Q\left(\frac{\gamma - \mu}{\Sigma}\right), \tag{38}$$

where $Q(\cdot)$ is the tail probability function of a zero-mean unit-variance Gaussian random variable. Therefore, for a the given probability of false alarm, we are able to determine the threshold.

The probability of false alarm is decreased, and simultaneously, the probability of detection is increased by increasing the length of L or equivalently the time range cell T_g. We note that $1/T_g$ is the range resolution and it cannot be a large number. The appropriate value for T_g for the purpose of target range detection is the reciprocal of the received signal bandwidth. But, it can be set smaller amount when more range resolution is needed. the amount of this parameter is depend on the scenario of detection.

6 Velocity processing

As previously mentioned in Section 2.3, $a_n(t)$ is a band-limited signal around the Doppler frequency, so the peak of $S_a^{nn}(t; f)$ can determine the Doppler frequency and the velocity of the target can henceforth be determined. Based on Wiener-Kintchin theorem

$$S_a^{nn}(t; f) = \int_{-\infty}^{\infty} R_a^{nn}(t; \tau) e^{j2\pi f \tau} d\tau. \tag{39}$$

Again, based on Eq. (51) we have

$$S_a^{nn}(t; f) = KP^2 |c_n|^2 S_g^{nn}(t; f - F_d). \tag{40}$$

As the Doppler frequency is independent of time, the peak of Eq. (40) when time-averaged, will define the Doppler frequency.

$$\begin{aligned} \overline{S_a^{nn}(f)} &= \lim_{T \to \infty} \frac{1}{T} \int_{-T/2}^{T/2} S_a^{nn}(t, f) dt \\ &= KP^2 |c_n|^2 \overline{S_g^{nn}(f - F_d)}. \end{aligned} \tag{41}$$

Based on [24], the time-averaged signal in Eq. (41) can be considered as a spectral correlation density function at zero cycle frequency whose consistent estimator is the frequency-smoothed cyclic periodogram [24]

$$S_a^{nn}(t,f) = \lim_{T\to\infty} \lim_{\Delta f\to 0} \frac{1}{T} \int_{f-\Delta f/2}^{f+\Delta f/2} \frac{1}{\Delta f} \left|A_T^n(t,\lambda)\right|^2 d\lambda,$$

(42)

where

$$A_T^n(t,\lambda) = \int_{t-T/2}^{t+T/2} a_n(s) e^{-j2\pi\lambda s} ds.$$

(43)

In addition, we define $\mathcal{P}$ as

$$\sqrt{T\Delta f} \left(\widehat{S_a^{nn}(t,f)} - \overline{S_a^{nn}(f)}\right).$$

(44)

In [24], it has been shown that $\mathcal{P}$ is an asymptotically zero-mean complex Gaussian random variable with $\sigma_{\mathcal{P}} = 1$.

To compute $A_T^n(t,\lambda)$, $a_n(iT_{PRI})$, the samples of $a_n(t)$ at sampling time iT_{PRI}, are to be used as the inputs for discrete version of Eq. (43) which it is FFT processing. The resolution of the frequency must be selected such that $\Delta f \geq 1/T$ (in our scenario T is equal coherent processing interval $T_C PI$). So the best resolution for the Doppler frequency can be $1/T$ which, based on Eq. (44), causes maximum variance of the error in the estimation. To reduce the variance of the error we have to increase Δf. By incoherent summation of $S_a^{nn}(t,f)$ over all harmonies we are able to improve the signal-to-noise ratio in the estimation. After this is done, the energy detection could be employed to detect the Doppler frequency. Since, $S_a^{nn}(t,f)$ has been modulated from nf_0 to the baseband, we shall use $S_a^{nn}(t;f - nf_0)$ instead of $S_a^{nn}(t;f)$ in the incoherent summation. Therefore, the result of incoherent summation can be expressed as:

$$S_r(t;f) = \sum_{n=-N/2}^{N/2} S_a^{nn}(t;f - nf_0)$$

(45)

and

$$\overline{S_r(f)} = \sum_{n=-N/2}^{N/2} \overline{S_a^{nn}(f - nf_0)}.$$

(46)

The estimator $\sqrt{T\Delta f}\widehat{S_r(t,f)}$ which is an asymptotically complex normal random variable with the mean of $\sqrt{T\Delta f}\overline{S_r(f)}$ can be computed based on Eq. (42). Based on energy detection, the decision statistics under two hypotheses can be represented as $l(\mathbf{a}) = T\Delta f|\widehat{S_r(t,f)}|^2$.

For white noise assumption under $\mathcal{H}_0$, the mean of the estimator is expressed as:

$$\sqrt{T\Delta f}\,\overline{S_r(f)} = \sqrt{T\Delta f}\sigma_v^2.$$

Therefore, the random variable $l(\mathbf{a})$ has the non-central chi-square distribution, expressed as:

$$f(l(\mathbf{a})|\mathcal{H}_0) = \frac{1}{2} e^{-\left(\frac{l(\mathbf{a})+\lambda}{2}\right)} I_0\left(\sqrt{\lambda l(\mathbf{a})}\right),$$

(47)

where $\lambda = T\Delta f\sigma_v^4$ and $I_0(\cdot)$ is the zeroth order modified Bessel function of the first kind. The probability of false alarm for the threshold γ, hence, is expressed as:

$$P_{FA} = P(l(\mathbf{a}) \geq \gamma|\mathcal{H}_0) = Q_1\left(\sqrt{\lambda}, \sqrt{\gamma}\right),$$

(48)

where $Q_1(.,.)$ is the Marcum Q-function. Again, to decrease the probability of false alarm and to increase the probability of detection simultaneously, Δf should be increased. Increasing Δf would also cause the velocity resolution to increase. Here, it should be taken into account that variations of the Doppler frequency lower than Δf cannot be sensed. As a result, in high-resolution velocity applications, the FFT processing time i.e., T, must be large enough. Since in our novel detection, the spectrum of signal is determined by a non-stationary analysis (Eq. 42), it is still valid for large T.

7 Simulation and results

Computer simulations and an experimental test are carried out to illustrate the performance of the proposed algorithm. In the experimental test, we have produced an ultra-short pulse signal with features mentioned in Table 1 by a signal generator excited with an impulse generator. The generated signal after near $13dB$ amplification is inputted to a horn antenna with an approximately $10°$ beamwidth in both elevation and azimuth. Next, the signal is impinged upon the "remote control Skyartec mini Cessna" plain flying at the distance range of 90 to 130 m. The backscattered signal after being received by the antenna and passing through the circulator will be caught by a fast oscilloscope. A fast switch is used to alleviate ground clutter and to protect the oscilloscope. The output signal of the oscilloscope is logged by a computer at three positions of the target mentioned in the Table 2. Since we have not used array antenna in our experimental test, the received signal model, previously mentioned in Section 2, will be valid for $P = 1$ (P is the number of radiators in the array antenna depicted in Fig. 2). The experimental setup is shown in Fig. 6.

Since the target in the test is near the radar, the received signal level is high. Actually, by measuring the noise and signal level at the input of the oscilloscope we find that the SNR is around 26 to 30dB in various range mentioned in Table 2. To produce multi-target scenarios, we consider this signal as the noiseless signal and combine them to produce a signal refrying to three targets in three position mentioned in Table 2. To produce (non)-Gaussian white (colored) noisy signal, we make the signal under the test by adding noise to the noiseless signal. The colored noise

Table 1 Features of transmitted signal and receiver

Parameter	Info	Parameter	Info
T_{PRI}	1 μs	L	8000
Carrier frequency	3 GHz	CPI	7 ms
Pulse width	2 ns	Bandwidth	500MHz
T_{obs}	4 μs	Velocity resolution	7 m/s
N	40	Range resolution	> 1 m
$\mathcal{M}$	5	Max of range	150 m

is constructed by passing AWGN noise over a low-pass FIR filter. The order of this filter is 9 and its bandwidth is 10MHz. The non-Gaussian noise used in this paper is the Middleton Class-A noise. From [25, 26] the probability of density function for the Class-A noise process is considered as:

$$f_x(x) = e^{-A} \sum_{m=0}^{\infty} \frac{A^m}{m! \sqrt{2\pi \sigma_m^2}} e^{-\frac{x^2}{2\sigma_m^2}},$$

where

$$\sigma_m^2 = \frac{m/A + \Gamma}{1 + \Gamma}.$$

According to this pdf, this noise is thereby a weighted sum of Gaussian distributions. σ_m^2 is the variance of the mth Gaussian distribution. A, which is called impulsive index [25] is set to one in this paper. Γ refers to the portion of the power of each Gaussian part in the total distribution which is set to $\Gamma = 0.01$ in this paper. In this paper, we consider $m = 2$.

In this paper, to drive an equation between false alarm rate and the threshold, i.e., Eq. (35), the hypothesis statistic $l(\mathbf{a})$ is approximated as Gaussian random variable. In this section, the conformity of this approximation is experimentally proved at the first. Based on Eq. (18), $\hat{a}_n(i)$ is a weighted combination of elements of $\mathbf{r}$. Since the length of $\mathbf{r}$, i.e., L, is large, according to central limited theorem $\hat{a}_n(i)$ has the Gaussian distribution. The amount of $L = 8000$ is large enough that the Gaussian distribution is independent of the type of noise (AWGN, colored or non-Gaussian). The Kolmogorov-

Table 2 Targets information

	Velocity	Range
Target #1	+15 m/s^a	70 m
Target #2	-20 m/s	95 m
Target #3	-35 m/s	115 m

[a] plus for close to the radar, and minus for far from the radar

Smirnov test, is a very common and powerful test in comparing the values in the given data vector, such as $\hat{a}_n(i)$, to Gaussian distribution [28]. We use this test for 100 samples of $\hat{a}_n(i)$ in the significance level 15 % (this parameter mentions to the error of unconformity [28]). The test approves the conformity in all types of noise. Therefore, There is no assumption on noise. Since, $\hat{a}_n(i)$ have almost standard normal distribution, $l(\mathbf{a})$ has the chi-squared distribution with N degrees of freedom. The N is large enough to let us approximate the chi-squared distribution with Gaussian distribution. Again, the Kolmogorov-Smirnov test is done for 100 samples of $l(\mathbf{a})$ in the significance level 15 % and the conformity is approved.

In all simulation in this paper, the SNR is considered as the power of the noiseless signal over the power of noise that is added to the noiseless signal. In the first simulation, we have assumed three scenarios with three targets mentioned in Table 2. The noise is assumed to be AWGN for the first, colored Gaussian for the second and non-Gaussian white for the third scenario.

Range and velocity detection are done for each case. Since, the decision statistics and the threshold in dimension of range and velocity are completely different jointly detection is impossible. The detection procedure is done for both range and velocity separately. In the simulations in every hypotheses testing procedure, a target detection is considered as a correct detection if the detection of both range and velocity are correct. Also, in the simulations, a false alarm is considered if a false alarm detection happens in range or velocity and a miss detection is considered if a target is missed in range or in velocity. The hypotheses testing procedure is performed in every cell. In general scenarios in the application considered in this simulation and experimental results, the length of the target is more than one meter in its dimension with maximum length and the distance between two targets can not be less than three meters. Since, the maximum length of target is more than one meter, in the detection algorithm we use the time of range cell, $T_g = 10$ ns, that causes 1.5 m range resolution. Also, we consider 5 ns for time of overlap in the cells. Since, the distance between two targets is not less than three meters, in the implementation of detection algorithm, when the energy of three adjacent cells or two adjacent cells or one cell pass the threshold, we consider it as a one target.

The range-velocity plot for SNR = 2 dB is shown in Fig. 7a. The range and velocity views are shown in Fig. 7b, c, respectively. The energy of the cell for each range cell is computed in range domain, and $\widehat{S_r(t;f)}$ is computed in a CPI time and time is averaged out in velocity domain. In Fig. 7a, peaks which can be seen in

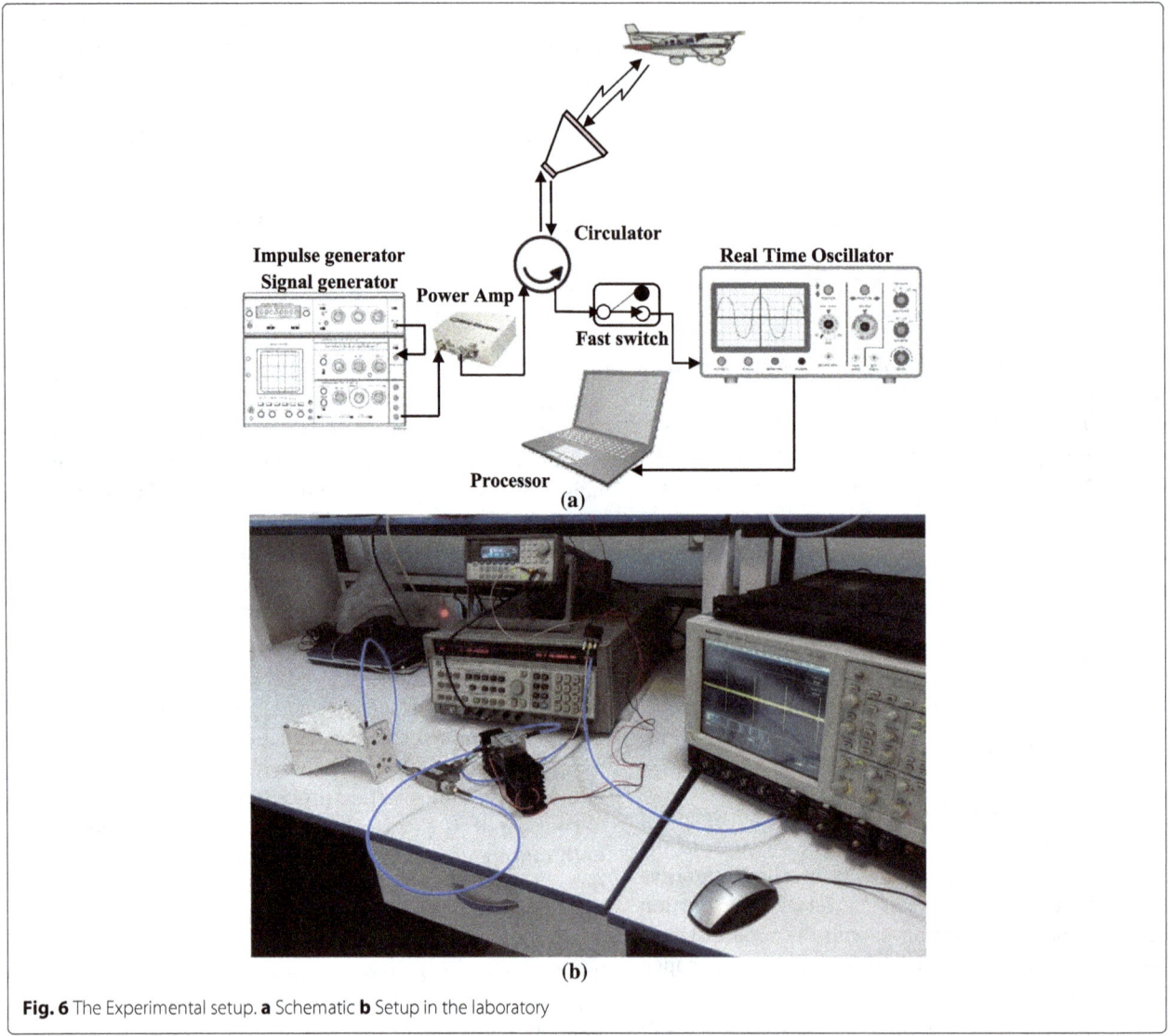

Fig. 6 The Experimental setup. **a** Schematic **b** Setup in the laboratory

the zero velocity are related to non-stationary clutters. In this figure, the robustness of the purposed algorithm in presence of the non-Gaussian and colored noise can be clearly perceived. In Fig. 7b, the peaks of energy happen just in the place of velocity of targets. These peaks, happening in all three noise scenario, are good references for the ability of proposed detection in extraction of velocity.

The time-varying spectrum $\widehat{S_r(t;f)}$ for the target at the velocity of -37 m/s, referring to -744 Hz, is depicted in Fig. 8 as an example of time-varying spectrum. This figure is to provide an evidence that time-varying coefficients have finite band width around Doppler frequency. It is seen that the bandwith of signal is around 350 Hz.

The new detector is compared to the IPCP in two next simulations. In these simulations, the performance of novel detector is surveyed in a single target as well

as three targets in the presence of (non)-Gaussian white (colored) noise. on the other hand, IPCP detector is used only for single target detection in the presence of white Gaussian noise. In the first simulation, the priority of proposed detector is shown by ROC plot for $SNR = 2$ dB in Fig. 9. In this figure, it is seen the superiority of the proposed detector over IPCP in both single and three targets in the presence of AWGN. The probability of detection in the proposed detector in single target is around 0.11 better than IPCP in false alarm rate 10^{-2}. This amount is 0.09 in three targets case. Assuming the probability of detection 0.85 as the applicable performance, it can be seen that the proposed detector is a applicable in all scenarios for noise, when false alarm rate is around 0.06 and $SNR = 2$ dB. In the second simulation (Fig. 10), the probability of detection for various SNR in $P_{fa} = 10^{-2}$ has been regarded. According to this figure, the prob-

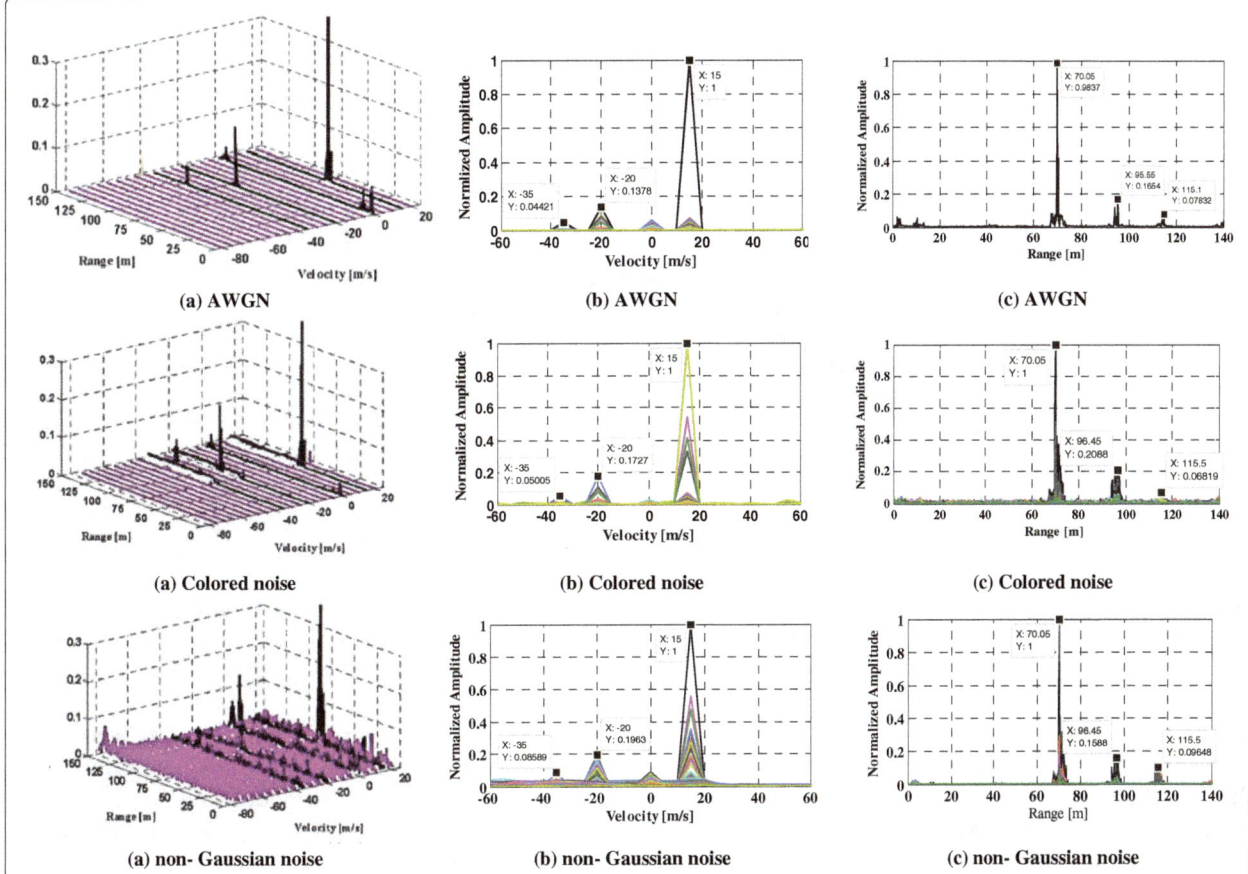

Fig. 7 Range-velocity plot. This plot depicts range-velocity plot for three targets in SNR = 2 dB when noise is AWGN, Gaussian-colored and non-Gaussian white. Energy of the cell for each range cell is computed in range domain, $\widehat{S_r(t;f)}$ is computed in a CPI time and time is averaged out in velocity domain. **a** Range-velocity plot. **b** Velocity view. **c** Range view

ability of detection 0.9 is achieved in the $SNR = -2$ dB in single target detection and $SNR = -0.5$ dB in three targets detection. This performance is achieved in IPCP in $SNR = 3.2$ dB which is around 5.2 dB worse than the performance of novel detector for single target detection and 3.7 dB worse than for three targets detection. Assuming the probability of detection 0.85 as the applicable performance, it is seen that in $SNR = 5$ dB the novel detector is applicable in all scenarios for noise.

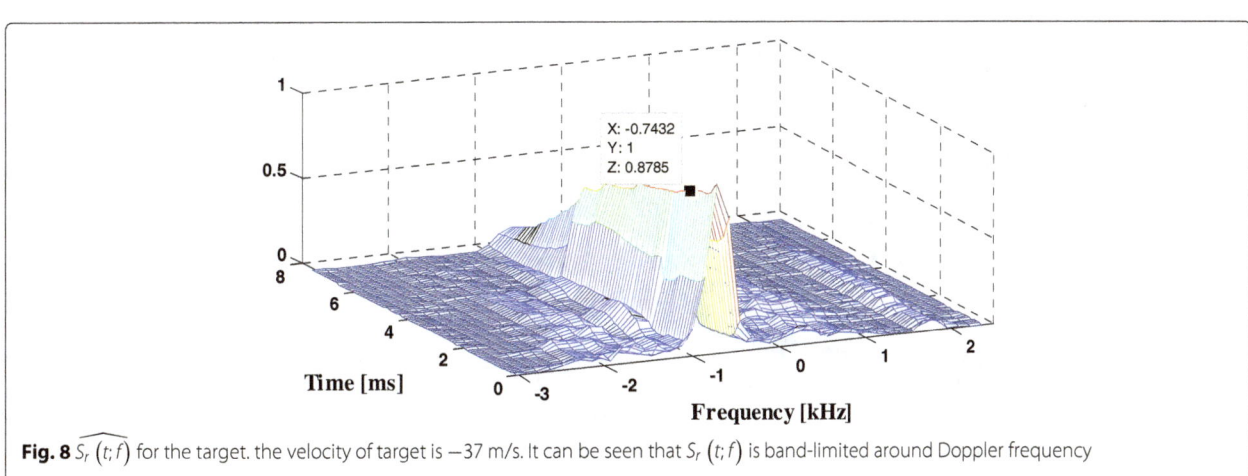

Fig. 8 $\widehat{S_r(t;f)}$ for the target. the velocity of target is −37 m/s. It can be seen that $S_r(t;f)$ is band-limited around Doppler frequency

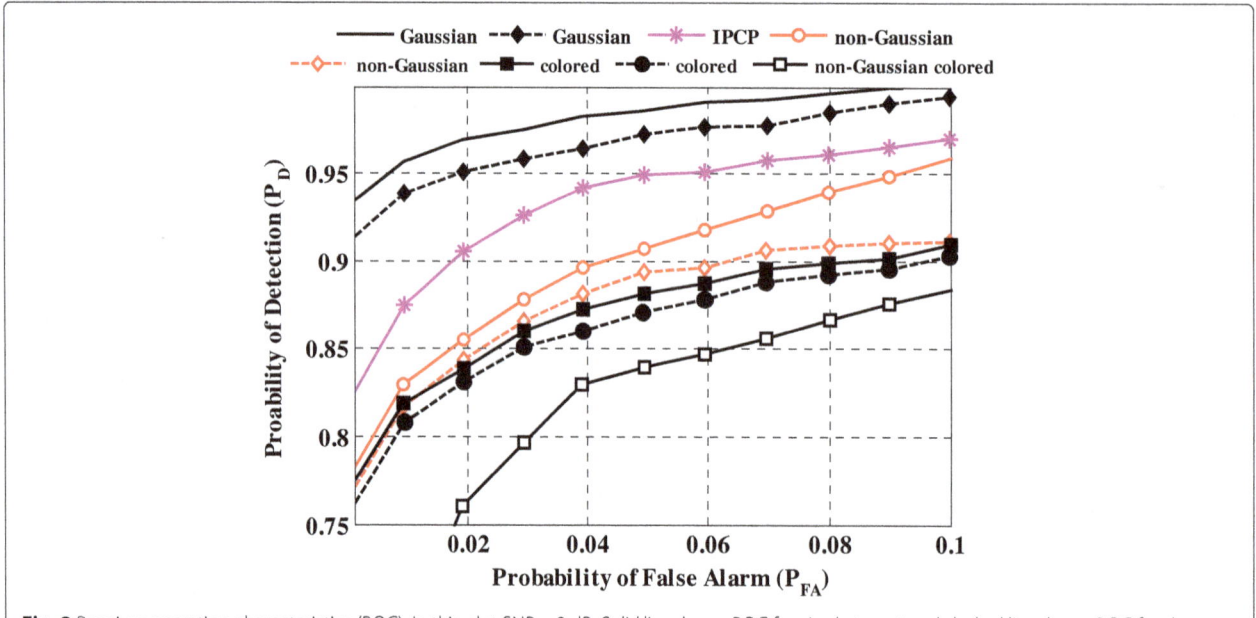

Fig. 9 Receiver operation characteristics (ROC). In this plot SNR = 2 dB. *Solid line* shows ROC for single target and *dashed line* shows ROC for three targets

8   Conclusion

In this paper, we have demonstrated that the receivers in ultra-wideband impulse radar based on IPCP use the received signal as the reference in their correlator which leads to remove Doppler frequency. Also, the Fourier series model with time-varying coefficients is perceived as a convenient model for non-stationary received signal in UWB impulse radar. Accordingly, the innovative MMSE estimator is introduced to evaluate the coefficients. Utilization of the time-varying weights in the estimator makes the blind noise-suppression from non-stationary signals without correlation, conceivable. The robust detection is proposed to extract the range and velocity in accordance with the estimator. Versatile results in comparison to IPCP and robustness to the alteration of noise features do verify the ability of our novel detector to be used in future applications.

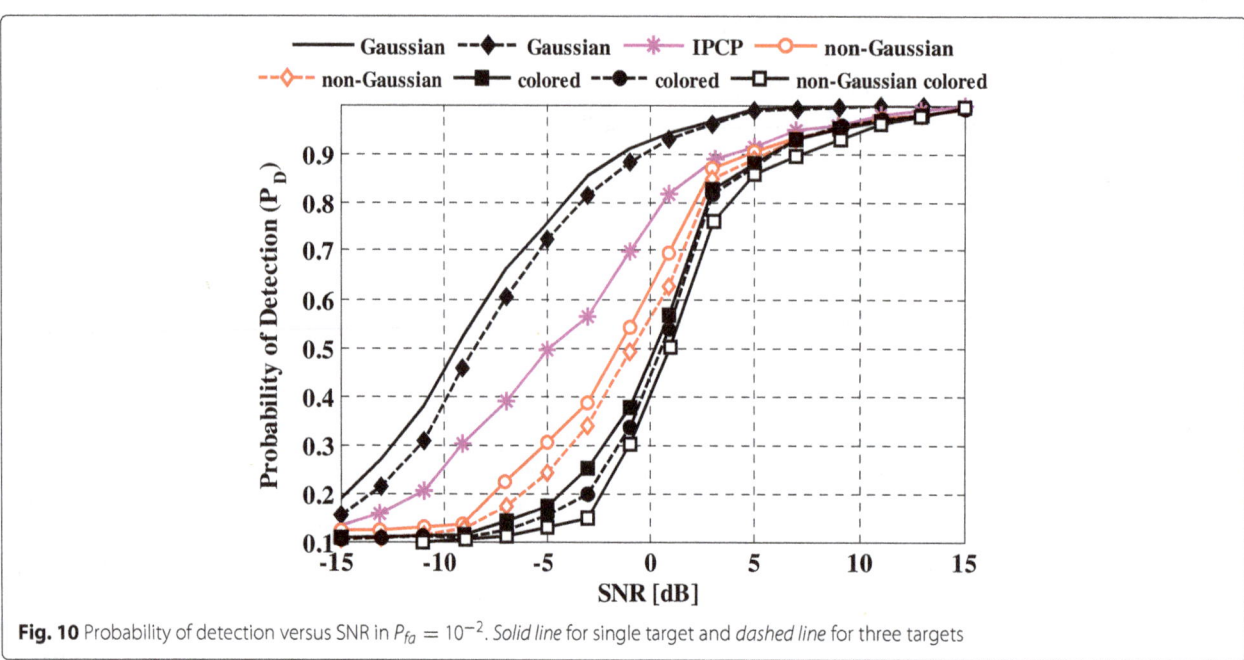

Fig. 10 Probability of detection versus SNR in $P_{fa} = 10^{-2}$. *Solid line* for single target and *dashed line* for three targets

9 Appendix A: Autocorrelation of time-varying coefficients

In this section, we have computed the correlation function and the spectrum of the time-varying coefficients $a_n(t)$

$$R_a^{n_1 n_2}(t;\tau) = E\left\{a_{n_1}(t)\, a_{n_2}^*(t+\tau)\right\}$$

$$= P^2 c_{n1} c_{n2}^* e^{-j2\pi F_d \tau} \sum_{m=1}^{M}\sum_{k=1}^{K}\sum_{\hat{m}=1}^{M}\sum_{\hat{k}=1}^{K} E\left\{g_{mn_1}(t)\, g_{\hat{m}n_2}^*(t+\tau)\right.$$

$$\left. e^{j\left(\varphi_k + \phi_m + \theta_{n_1 km}\right)} e^{-j\left(\varphi_{\hat{k}} + \phi_{\hat{m}} + \theta_{n_2 \hat{k}\hat{m}}\right)}\right\},$$

$$(49)$$

where

$$g_{mn} = \alpha_{mn}(t)\, e^{-j2\pi f_c(2\kappa_m(t))}.$$

Since, we do not have any information about ε_k and δ_m, we can suppose uniform distribution in $[-L_r/2c \ \ L_r/2c]$ and $[-L_t/2c \ \ L_t/2c]$ for ε_k and δ_m, respectively. This assumption would cause uniform distribution in $[-\pi,\pi]$ for $U_k = 2\pi f_c \varepsilon_k$ and $V_m = 2\pi f_c \delta_m$. By this, for $k \neq \hat{k}$ and $m \neq \hat{m}$ the expectation in Eq. (49) can be written as

$$E\left\{g_{mn_1}(t)\, g_{\hat{m}n_2}^*(t+\tau) e^{j\left(\varphi_k + \phi_m + \theta_{n_1 km}\right)} e^{-j\left(\varphi_{\hat{k}} + \phi_{\hat{m}} + \theta_{n_2 \hat{k}\hat{m}}\right)}\right\}$$

$$= \frac{1}{(2\pi)^4}\int_{2\pi}\int_{2\pi}\int_{2\pi}\int_{2\pi} E\left\{g_{mn_1}(t)\, g_{\hat{m}n_2}^*(t+\tau)\, e^{j\left(\theta_{n_1 km} - \theta_{n_2 \hat{k}\hat{m}}\right)}\right\}$$

$$e^{j\left(U_k - U_{\hat{k}} - V_m - V_{\hat{m}}\right)} dU_k\, dU_{\hat{k}}\, dV_m\, dV_{\hat{m}} =$$

$$E\left\{g_{mn_1}(t)\, g_{\hat{m}n_2}^*(t+\tau)\, e^{j\left(\theta_{n_1 km} - \theta_{n_2 \hat{k}\hat{m}}\right)}\right\}\delta\left(k - \hat{k}\right)\delta\left(m - \hat{m}\right).$$

$$(50)$$

For $k = \hat{k}$ and $m = \hat{m}$ the correlation function can be expressed as

$$R_a^{n_1 n_2}(t;\tau) =$$

$$P^2 c_{n1} c_{n2}^* e^{-j2\pi F_d \tau}\sum_{m=1}^{M}\sum_{k=1}^{K} E\left\{g_{mn_1}(t)\, g_{mn_2}^*(t+\tau)\right\}$$

$$\frac{2c}{L_r}\frac{2c}{L_t} e^{j(\omega_{n_1} - \omega_{n_2})\left(\frac{d\cos(\phi) - 2R}{c}\right)}$$

$$\int_{-L_t/2c}^{L_t/2c}\int_{-L_r/2c}^{L_r/2c} e^{j(\omega_{n_1} - \omega_{n_2})(\varepsilon_k - 2\delta_m)}\, d\varepsilon_k\, d\delta_m$$

$$= KP^2 c_{n1} c_{n2}^* e^{-j2\pi F_d \tau}\frac{\sin(\upsilon_r)}{\upsilon_r}\frac{\sin(\upsilon_t)}{\upsilon_t} R_g^{n_1 n_2}(t;\tau),$$

$$(51)$$

where $\upsilon_r = (\omega_{n_1} - \omega_{n_2})L_r/2c$, $\upsilon_t = (\omega_{n_1} - \omega_{n_2})L_t/2c$ and

$$R_g^{n_1 n_2}(t;\tau) = \sum_{m=1}^{M} E\left\{g_{mn_1}(t)\, g_{mn_2}^*(t+\tau)\right\}.$$

$$(52)$$

Let us consider $g_{mn}(t) = b_n g_m(t)$, where b_n are samples of frequency response of the targets and strongly depend on shape and material of the target. Therefore, we have $R_g^{n_1 n_2}(t;\tau) = b_{n_1} b_{n_2}^* R_g(t;\tau)$. In [27], for a typical example, it has been proved that $g_m(t)$ is a band-limited process for long time observations and its bandwidth mostly depends on the rotational speed around the axis which is orthogonal to the plane containing L_t, and also on the bandwidth of transmitted signal.

Competing interests

The authors declare that they have no competing interests.

Author details

[1] Amirkabir University of Technology, Tehran, Iran. [2] University of Washington, Seattle, WA, USA.

References

1. RJ Fontana, Recent system applications of short-pulse ultra-wideband (UWB) technology. IEEE Trans. Microwave Theory Tech. **52**(9), 2087–2104 (2004)
2. IY Immoreev, Practical applications of UWB technology. IEEE Tran. AES. **25**(2), 36–42 (2010)
3. GM Hussain Malek, Ultra wideband impulse radar, An overwiev of the principles. IEEE AES. Mag. **52**(9), 9–14 (1998)
4. J Hu, G Zhu, T Jin, et al, Adaptive through-wall indication of human target with different motions. IEEE Geosci. Remote Sensing Lett. **11**(5), 911–915 (2014)
5. W Guohua, Z Yuxiang, W Siliang, in *Proc. IEEE Int. Conf. Radar.* Detection and localization of high speed moving targets using a short-range UWB impulse radar, (2008)
6. JD Taylor, *Ultra-wideband radar technology. Chapter 10.* (CRC Press, 2001)
7. IY Immoreev, Ultra wideband radars: futures and capabilities. J. Commun. Technol. Electron. **54**(1), 1–26 (2009)
8. IY Immoreev, in *Proc. IEEE Int. Conf. Ultra-Wideband.* Detection of UWB signals reflected from complex targets, (2002)
9. AG Yarovoy, LP Ligthart, J Matuzas, et al, UWB radar for human Benng detection. IEEE AES. Mag. **21**(3), 10–14 (2006)
10. S Singh, Q Liang, et al, Sense through wall human detection using UWB radar. EURASIP J. Wireless Commun. Netw. **2011**(1), 1–11 (2011)
11. Y Xu, S Wu, et al, A novel method for automatic detection of trapped victims by ultrawideband radar. IEEE Trans. Geosci. Remote Sensing. **50**(8), 3132–3142 (2012)
12. JH Lee, MH Jang, S Ko, in *Proc. Eur. Radar Conf.* Measuring the target range and relative velocity in UWB radar for automobile applications, (2011)
13. B Sobhani, E Paolini, A Giorgetti, et al, Target tracking for UWB multistatic radar sensor networks. IEEE J. Sel. Topics Signal Process. **8**(1), 125–136 (2014)
14. B Sobhani, M Mazzotti, et al, in *Proc. IEEE ICUWB.* Multiple target detection and localization in UWB multistatic radars, (2014)
15. GS Gill, in *IEEE Trans. Electromagn. Compat.* Ultra-wideband radar using Fourier synthesized waveforms, vol. 39.2, (1997), pp. 124–131
16. AT Hosseini, H Amindavar, in *Proc. IEEE CIE Int. Conf. on radar.* Cyclostationary detector in ultra Wideband Impulse Radar, (2011)
17. DV De Ville, W Philips, I Lemahieu, On the n-dimensional extension of the discrete prolate spheroidal window. IEEE Signal Process. Lett. **9**(3), 89–91 (2002)

18. RJ Baker, BP Johnson, Applying the Marx bank circuit configuration to power MOSFETs. Electron. Lett. **29**(1), 56 (1993)

19. M Niedzwiecki, *Identification of Time-Varying Processes*. (Wiley, New York, 2000)

20. A Berlinet, C Thomas-Agnan, *Reproducing Kernel Hilbert Spaces in Probability and Statistics*. (Kluwer Academic, Boston, 2004)

21. D Slepian, Prolate spheroidal wave functions, Fourier analysis, and uncertainty–V: the discrete case. Bell Syst. Tech. J. **57**(5), 1371–1430 (1978)

22. HJ Landau, HO Pollak, Prolate spheroidal wave functions, Fourier analysis, and uncertainty–III: the dimension of the space of essentially time-limited and band-limited signals. Bell Syst. Tech. J. **41**(4), 1295-1336 (1962)

23. SM Hosseini, H Amindavar, et al, in *IEEE Int. Workshop on SPAWC*. A New Cyclostationary spectrum sensing approach in cognitive Radio, (2010)

24. WA Gardner, A Napolitano, L Paura, Cyclostationarity: Half a century of research. Eles. J. Signal Proccess. **86**(4), 639–697 (2006)

25. D Middleton, Non-gaussian noise models in signal processing for telecommunications: new methods and results for class a and class b noise models. IEEE Tran. on Infor. Theo. **45**(4), 1129–1149 (1999)

26. RB Ash, C Doleans-Dade, *Probability and measure theory*. (2, ed.) (Academic Press, 2000)

27. BH Borden, ML Mumford, A statistical glint/radar cross section target model. IEEE Tran. AES. **19**(15), 781–785 (1983)

28. FJ Massey, The Kolmogorov-Smirnov test for goodness of fit. J. Am. Stat. Assoc. **46**(253), 68–78 (1951)

7

Motion detection using binocular image flow in dynamic scenes

Qi Min and Yingping Huang[*]

Abstract

Motion detection is a hard task for intelligent vehicles since target motion is mixed with ego-motion caused by moving cameras. This paper proposes a stereo-motion fusion method for detection of moving objects from a moving platform. A 3-dimensional motion model integrating stereo and optical flow has been established to estimate the ego-motion flow. The mixed flow is calculated from an edge-indexed correspondence matching algorithm. The difference between the mixed flow and the ego-motion flow yields residual target motion flow where the intact target is segmented from. To estimate the ego-motion flow, a visual odometer has been implemented. We first extract some feature points in the ground plane that are identified as static points using the height constraint and Harris algorithm. And then, 6 DOF motion parameters of the moving camera are calculated by fitting the feature points into the linear least square algorithm. The approach presented here is tested on substantial traffic videos, and the results prove the efficiency of the method.

Keywords: Motion detection, Stereovision, Optical flow, Ego-motion, Visual odometer

1 Introduction

Detection on moving obstacles like pedestrians and vehicles is of critical importance for autonomous vehicles. Vision-based sensing systems have been used for object detection in many applications including autonomous vehicles, robotics, and surveillance. Compared with the static systems such as the traffic and crowd surveillance, motion detection from a moving platform (vehicle) is more challengeable since target motion is mixed with camera's ego-motion. This paper addresses on this issue and presents a binocular stereovision-based in-vehicle motion detection approach which integrates stereo with optical flow. The approach fully makes use of two pairs of image sequences captured from a stereovision rig, i.e., disparity from left and right pair images and motion fields from consecutive images.

Vision-based motion detection methods can be categorized into three major classes, i.e., temporal difference, background modeling and subtraction, and optical flow. Temporal difference methods [1] readily adapt to sudden changes in the environment, but the resulting shapes of moving objects are often incomplete. Background

modeling and subtraction is mainly used in video surveillance where the background is relatively fixed and static. Its basic idea is to subtract or differentiate the current image from a reference background model [2]. However, the generated background model may not be applicable in some scenes such as gradual or sudden illumination changes and dynamic background (wave trees). To address these issues, a hierarchical background modeling and subtraction [3] and a self-adaptive background matching method [4] have been proposed. Adaptive background models have also been used in autonomous vehicles in an effort to adapt surveillance methods to the dynamic on-road environment. In [5], an adaptive background model was constructed, with vehicles detected based on motion that differentiated them from the background. Dynamic modeling of the scene background in the area of the image where vehicles typically overtake was implemented in [6].

Optical flow, a fundamental machine vision tool, has advantages that directly reflect an accurate estimation of point motion, representing an evident change in position of a moving point. It has been used for motion detection and tracking in defense [7] and abnormal crowd behavior detection in video surveillance [8]. In autonomous vehicles, monocular optical flow has been used to detect

---

[*] Correspondence: huangyingping@usst.edu.cn

School of Optical-Electrical and Computer Engineering, University of Shanghai for Science and Technology, Shanghai 200093, China

head-on vehicle [9], overtaking vehicles in the blind spot [10] and crossing obstacles [11]. In [12], interest points that persisted over long periods of time were detected and tracked using the hidden Markov model as vehicles traveling parallel to the ego vehicle. In [13], optical flow was used to form a spatiotemporal descriptor, which was able to classify the scene as either intersection or non-intersection. The use of optical flow has also heavily been found in stereovision-based motion detection, i.e., stereo-motion fusion method, which benefits from motion cues as well as depth information. There are many different fusion schemes. In [14], Pantilie et al. fuse motion information derived from optical flow into a depth-adaptive occupancy grid (bird-view map) generated from stereovision 3D reconstruction. As an improvement of stereovision-based approach, the method is of benefits to distinguish between static and moving obstacles and to reason about motion speed and direction. Franke and Heinrich [15] propose a depth/flow quotient constraint. Independently moving regions of the image do not fulfill the constraints and are detected. Since the fusion algorithm compare the flow/depth quotient against a threshold function at distinct points only, it is computationally efficient. However, the approach reduces the possibility of carrying out geometrical reasoning and lacks a precise measurement of the detected movements. In addition, the approach is limited with respect to robustness since only two consecutive frames are considered. To get more reliable results, the Kalman filter is equipped to integrate the observations over time. In [16], Rabe et al. employ a Kalman filter to track image points and to fuse the spatial and temporal information so that static and moving pixels can be distinguished before any segmentation is performed. The result is an improved accuracy of the 3D position and an estimation of the 3D motion of the detected moving objects. In [17], Kitt et al. use a sparse set of static image features (e.g., corners) with measured optical flow and disparity and apply the *Longuet-Higgins-Equations* with an implicit extended Kalman filter to recover the ego-motion. The feature points with optical flow and disparity flow not consistent with the estimated ego-motion indicate the existence of independently moving objects. In [18], Bota and Nedevschi focus on fusing stereo and optical flow for multi-class object tracking by designing Kalman filter fitted with static and dynamic cuboidal object models. In [19], interest moving points are first detected and projected on 3D reconstruction ground plane using optical flow and stereo disparity. The scene flow is computed via finite differences for a track up to five 3D positions, and points with a similar scene flow are grouped together as rigid objects in the scene. A graph-like structure connecting all detected interest points is generated, and the resulting edges are removed according to scene flow differences exceeding a certain threshold. The remaining connected components describe moving objects.

A precise recovery of the ego-motion is essential in order to distinguish between static and moving objects in dynamic scenes. One of the methods of ego-motion estimation was to use in-vehicle inertial navigation system (INS) [15]. However, ego-motion from the in-car sensor is not sufficient for a variety of reasons like navigation loss, wheel slip, INS saturation, and calibration errors. Thus, it is ideal to estimate the camera ego-motion directly from the imagery. Ego-motion estimation using monocular optical flow and integrated detection of vehicles was implemented in [20]. Several groups have reported stereo-based ego-motion estimation based on tracking point features. In [18], the concept of 6D vision, i.e., the tracking of interest points in 3D using Kalman filtering, along with ego-motion compensation, was used to identify moving objects in the scene. In [21], vehicle's ego-motion was estimated from computational expensive dense stereo and dense optical flow with the method of iterative learning from all points in the image.

Stereo-motion fusion has been studied in a theoretical manner by Waxman and Duncan [22]. The important result was the relationship between camera's 3D motion and corresponding image velocities with stereo constraints. Our work builds on the basic principles presented in [22] and extends it to dynamic scene analysis. In this work, a mathematical model, integrating optical flow, depth, and camera ego-motion parameters, is firstly derived from Waxman and Duncan's theoretical analysis. Camera's ego-motion is then estimated from the model by using ground feature points, and accordingly ego-motion flow of the image is calculated from the model. A moving target is detected from the difference of the mixed flow and the ego-motion flow.

The main contributions of this work can be summarized as follows: (1) The relationship between optical flow, stereo depth, and camera ego-motion parameters has been established based on Waxman and Duncan's theoretical model. Accordingly, a novel motion detection approach fusing stereo with optical flow sensor has been proposed for in-vehicle environment sensing systems. A visual odometer able to estimate camera's ego-motion has also been proposed. Motion detection using stereo-motion fusion normally identifies image points [16, 19] or features [17] as static or moving and then segment moving objects accordingly. Our method works on the image level, i.e., the difference between the mixed flow image and the ego-motion flow image. (2) Existing motion detection approaches often make some assumptions on object/vehicle motion or scene structure. Our approach can detect moving objects without any constraints on object/vehicle motion or scene structure since the proposed visual odometer can estimate all six motion parameters. (3) When fusing stereo with optical flow, the computational load, accuracy, and comparability (or consistence) between stereo and optical flow calculations are practical issues. Our method uses the edge-indexed method for all calculations and therefore greatly reduces computational load without impact on detection performance, improves calculation accuracy especially on the mixed flow, and provides pixel-wise consistence for all calculations so that the stereo depth, the mixed flow, and the ego-motion flow can be compared pixel by pixel.

2 Approaches

2.1 Overview of the approach

The difficulty of motion detection from a moving camera/vehicle is that the background is also moving and its motion is mixed with target motion. Therefore, the key of motion detection in dynamic scenes is to distinguish the background motion from the target motion. The underlying idea of our approach is to subtract the motion of the camera (ego-motion) from the calculated (mixed) optical flow, that is, a moving target can be detected from the difference between the mixed optical flow and the ego-motion optical flow. Figure 1 gives an overview of the approach.

The mixed flow of the scene is caused by both camera motion and target motion and is obtained from correspondence matching between consecutive images. The ego-motion flow is caused only by camera motion and calculated from a mathematical model derived from Waxman and Duncan's theoretical analysis [22], which indicates the relation between optical flow, depth map, and camera ego-motion parameters. To calculate the ego-motion flow, we need first know the ego-motion parameters of six degree of freedom.

A visual odometer has been implemented for this purpose, in which six motion parameters are estimated by solving a set of equations fitted with a fixed number of feature points using the linear least square method. The feature points are selected as corner points lying on the road surface and determined by using height constraint and Harris corner detection algorithm [23]. Within the two stages, the depth of the image points is provided by the stereovision disparity map. The difference between the mixed flow and the ego-motion flow yields an independent flow which is purely caused by the target motion. The moving target is extracted according to the continuity of the similar independent flow.

To reduce the computational workload and considering that object contour is the most effective cue for object segmentation, all calculations are edge-indexed, i.e., we only conduct calculations on edge points for stereo matching, the mixed flow, and the ego-motion flow calculations. This tactic greatly increases the real-time performance and has no impact on object detection performance.

2.2 The mixed flow

Many methods have been developed to calculate dense optic flow from image sequences [24]. Basically, these approaches can be split into two categories: spatiotemporal gradient-based and correspondence matching techniques. The spatiotemporal gradient-based techniques calculate optic flow based on assumptions including globe smoothness or direction smoothness. Our experiences show that these methods take huge computation cost and are difficult to obtain accurate optical flow in complex traffic scenarios. The correspondence matching-based techniques detect optic flow by searching for correspondence points between consecutive images, therefore are more suitable

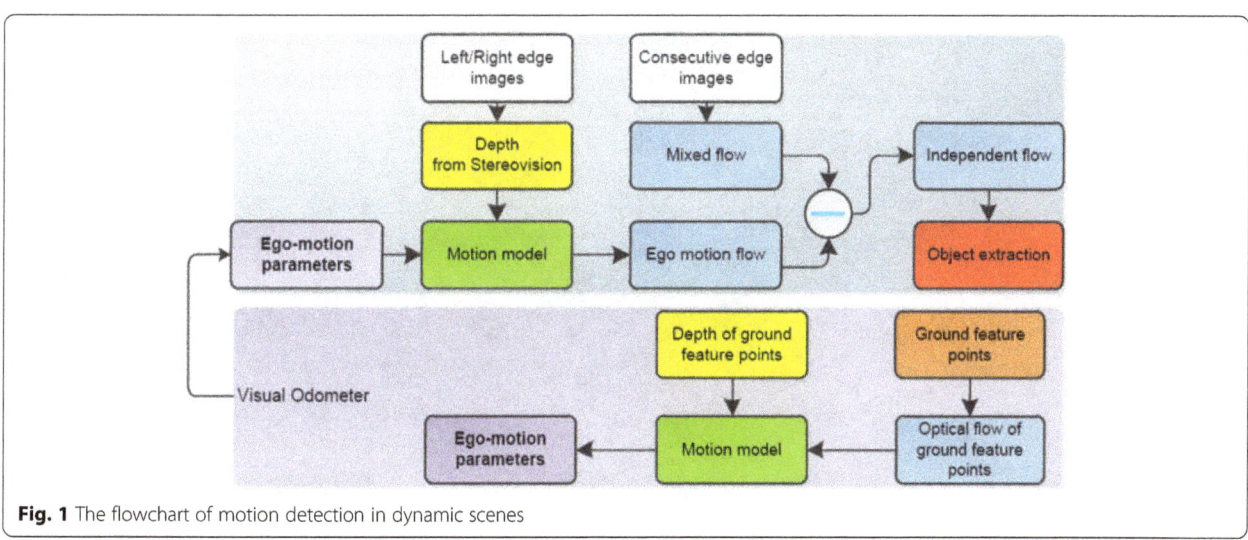

Fig. 1 The flowchart of motion detection in dynamic scenes

for dynamic traffic scene images. In this work, we implement an edge-indexed correspondence matching algorithm based on greyscale similarity to calculate the mixed optical flow. The details of the algorithm can be found in our previous work [25]. A summary is as follows:

Step 1. Generate edge image using Canny operator and use the edge points as seed points to find the correspondence points in next frame.

Step 2. Define the searching range as a square area centered at the seed point and define a rectangular matching window.

Step 3. Use the normalized cross correlation coefficients as a measure of greyscale similarity of two matching windows. The correspondence points are regarded as those with the maximum cross correlation coefficient that must be greater than a predefined threshold.

Step 4. Achieve the sub-pixel estimation of the calculated optical flow along the vertical and horizontal directions by introducing a quadratic interpolation. This is to improve the optical flow resolution so that a higher optical flow accuracy can be achieved.

2.3 3-dimensional motion and ego-motion flow

Ego-motion flow is the optical flow evoked by the moving camera/vehicle, representing the effect of the camera motion. The camera's 3-dimensional motion and planer imaging model is represented in Fig. 2. The origin of the world coordinate system (X, Y, Z) is located at the center of image coordinates (x, y), and the Z-axis is directed along optical axis of the camera. The translational velocity of the camera is $\bar{V} = (V_x, V_y, V_z)$, and the rotational velocity $\bar{W} = (W_x, W_y, W_z)$.

Assuming a point $P(X, Y, Z)$ in space moves to point $P'(X', Y', Z')$, the relation between the point motion and camera motion is as below [22]:

$$\frac{dP}{dt} = -(\bar{V} + \bar{W} \times P) \tag{1}$$

The cost product of the point $P(X, Y, Z)$ and camera's rotational velocity vector can be represented as

$$\bar{W} \times P = \begin{vmatrix} i & j & k \\ W_x & W_y & W_z \\ X & Y & Z \end{vmatrix}$$
$$= (W_y Z - W_z Y)i + (W_z X - W_x Z)j + (W_x Y - W_y X)k \tag{2}$$

where $(i, \ j, \ k)$ denotes the unit vector in the direction of X-, Y-, and Z-axes, $\times$ refers to cross-product. Thus, Eq. (2) can be rewritten as

$$\bar{W} \times P = \begin{bmatrix} W_y Z - W_z Y \\ W_z X - W_x Z \\ W_x Y - W_y X \end{bmatrix} \tag{3}$$

The 3-dimensional velocity $\left(\frac{dX}{dt} \ \frac{dY}{dt} \ \frac{dZ}{dt} \right)$ of the point can be obtained as below:

$$\begin{aligned} dX/dt &= -(V_x + W_y Z - W_z Y) \\ dY/dt &= -(V_y + W_z X - W_x Z) \\ dZ/dt &= -(V_z + W_x Y - W_y X) \end{aligned} \tag{4}$$

For an ideal pinhole camera model, the image point $p(x \ y)$ of the world point $P(X, Y, Z)$ projected in the image plane can be expressed as

$$x = f \frac{X}{Z}, \qquad y = f \frac{Y}{Z} \tag{5}$$

where f denotes the focal length of the stereo camera. The optical flow (u, v) of $P(X, Y, Z)$ can be obtained by

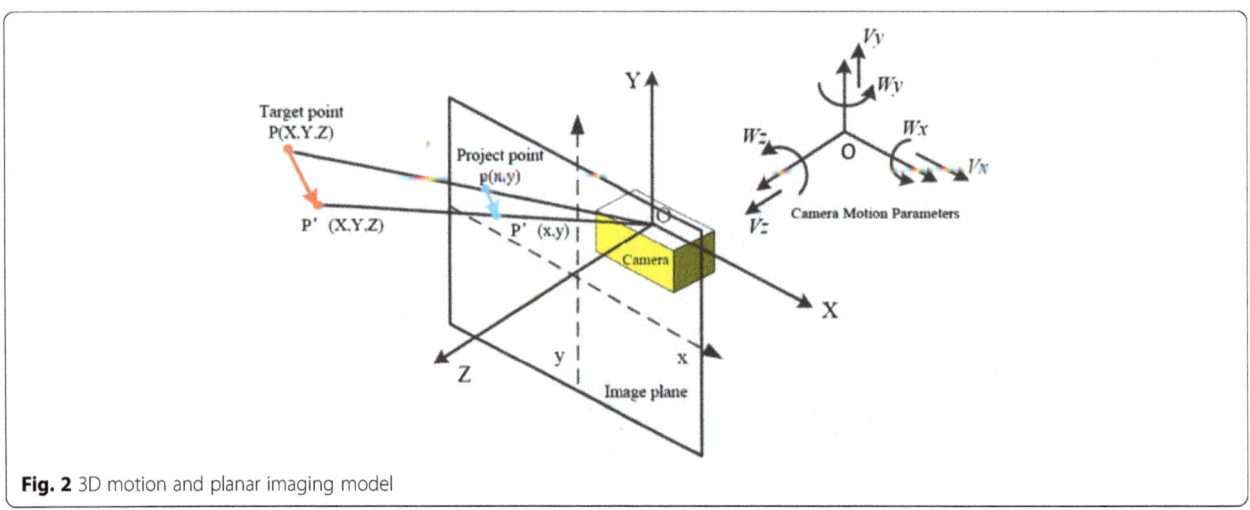

Fig. 2 3D motion and planar imaging model

estimating the derivatives along *X*-axis and *Y*-axis in 2D image coordinates.

$$u = \frac{dx}{dt} = \frac{1}{Z}\left(f\frac{dX}{dt} - x\frac{dZ}{dt}\right)$$
$$v = \frac{dy}{dt} = \frac{1}{Z}\left(f\frac{dY}{dt} - y\frac{dZ}{dt}\right) \tag{6}$$

Integrating Eqs. (4) to (6) yields the following:

$$\begin{bmatrix} u \\ v \end{bmatrix} = -\begin{bmatrix} \frac{f}{Z} & 0 & -\frac{x}{Z} & -\frac{xy}{f} & \frac{f^2+x^2}{f} & -y \\ 0 & \frac{f}{Z} & -\frac{y}{Z} & -\frac{f^2+y^2}{f} & \frac{xy}{f} & x \end{bmatrix} \begin{bmatrix} V_x \\ V_y \\ V_z \\ W_x \\ W_y \\ W_z \end{bmatrix} = A(\bar{V}, \bar{W})^T \tag{7}$$

where $A = \begin{bmatrix} \frac{f}{Z} & 0 & -\frac{x}{Z} & -\frac{xy}{f} & \frac{f^2+x^2}{f} & -y \\ 0 & \frac{f}{Z} & -\frac{y}{Z} & -\frac{f^2+y^2}{f} & \frac{xy}{f} & x \end{bmatrix}$.

Equation (7) indicates the relationship between the ego-motion flow, the depth and the six parameters of the camera motion. It is evident that the ego-motion flow can be calculated from Eq. (7) if the depth and the six motion parameters are known. The depth can be obtained from stereovision as reported in our previous work [26]. Two methods can be used to obtain the motion parameters: one is to use an in-vehicle INS or gyroscope to measure them; the other is to use a visual odometer. However, subject to problems like navigation loss, wheel slip, INS saturation, and calibration errors between the IMU and the cameras, in-vehicle INS may cause inaccurate motion estimation in some cases. Thus, it is ideal to estimate the camera motion directly from the imagery. Ultimately, it could be fused with other state sensors to produce a more accurate and reliable joint estimate of cameral/vehicle motion.

2.4 Visual odometry

It can be known from Eq. (7) that if the ego-motion flow and the depth of six or more points in the scene are known, we can set up a set of equations with six unknown variables, i.e., six camera motion parameters and estimate these variables by solving the equations set using the least square fitting method. The points used for the least square fitting must be assured with accurate optical flow calculation and must not be any moving points.

In this work, the corner points lying on the road surface are selected for this purpose since the ground points are static and the corner points are of good stability and inflexibility to light intensity, therefore possessing relatively accurate optical flow.

2.4.1 Extraction of ground corner points using stereovision and Harris method

Ground points can be determined from the height information that can be obtained from the stereovision as reported in our previous work [26]. The height Y_g of the ground points, namely their *Y*-axis coordinate, depends on the camera installation height H_c, the tilt angle towards the road plane θ, and distance Z_g, as indicated in Eq. (8) and Fig. 3. Those points with *Y*-axis coordinate less than Y_g are regarded as ground points.

$$Y_g = \left(Z_g * \sin\theta - H_c\right)/\cos\theta \tag{8}$$

A corner is defined as a point for which there are two dominant and different edge directions in a local neighborhood of the point. Harris corner points are detected by considering the differential of the corner score with respect to direction [23]. The corner score is referred as autocorrelation. Assuming that a pixel $I(X, Y)$ moves in any directions by small displacements (Vx, Vy), the autocorrelation function is defined as below:

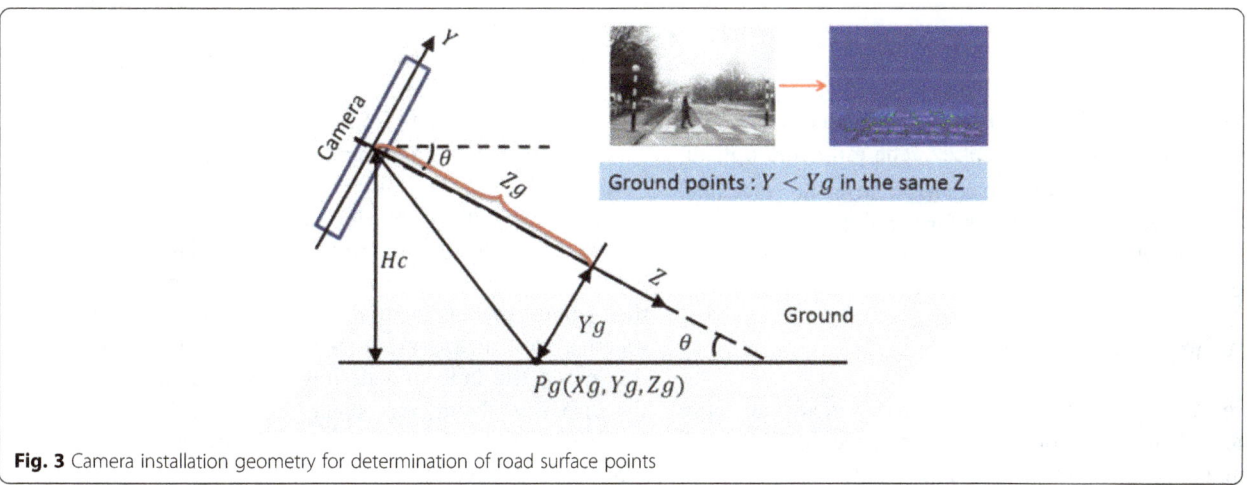

Fig. 3 Camera installation geometry for determination of road surface points

$$
\begin{aligned}
C_{\nabla x, \nabla y} &= \sum_{(x,y) \in W(x,y)} \phi(x,y)[I(x + \nabla x, y + \nabla y) - I(x,y)]^2 \\
&= \sum_{(x,y) \in W(x,y)} \phi(x,y)\left[I_x \nabla x + I_y \nabla y + O(\nabla x^2 + \nabla y^2)\right]^2 \\
&\approx \sum_{(x,y) \in W(x,y)} \phi(x,y)\left([I_x \ I_y]\begin{bmatrix} \nabla x \\ \nabla y \end{bmatrix}\right)^2 \\
&= [\nabla x \ \nabla y] M(x,y)\begin{bmatrix} \nabla x \\ \nabla y \end{bmatrix}
\end{aligned}
\tag{9}
$$

where $\phi(x,y)$ is Gaussian weighting function used here to reduce the impact of noise; $W(x,y)$ denotes window blocks centered at the point; I_x is the gradient in x direction; and I_y is the gradient in y direction. The Sobel convolution kernel ω_x and its transposed form ω_y are used to obtain $I_x = I(X,Y) \otimes \omega_x$, and $I_y = I(X,Y) \otimes \omega_y$. $M(x,y)$ is called the autocorrelation matrix and

$$
M(x,y) = \phi(x,y)\begin{bmatrix} \sum_{W(x,y)} I_x^2 & \sum_{W(x,y)} I_x I_y \\ \sum_{W(x,y)} I_x I_y & \sum_{W(x,y)} I_y^2 \end{bmatrix}
\tag{10}
$$

The corner response function (CRF) can be calculated as.

$$
\text{CRF} = \det(M) - \alpha \cdot (\text{trace}M)^2
\tag{11}
$$

where $\det(M) = \lambda_1 \times \lambda_2$ and $\text{trace}M = \lambda_1 + \lambda_2$, λ_1 and λ_2 denote the eigenvalues of the matrix M, we set $\alpha = 0.04$. The point with CRF bigger than a certain threshold is regarded as a corner point.

2.4.2 Ego-motion parameter estimation using the linear square algorithm

The objective function is defined as the Euclidean distance between the estimated optical flow $(\hat{u}, \hat{v})$ and the true optical flow (u,v).

$$
J = \sum_{n=1}^{N} \|(\hat{u}, \hat{v}) - (u,v)\|^2
\tag{12}
$$

The true optical flow (u,v) is calculated from the method introduced in Section ??. The estimated optical flow $(\hat{u}, \hat{v}) = A(\bar{V}, \bar{W})^T$ is obtained from Eq. (7). The minimum value of the object function is found by setting the gradient to zero and the optimal parameter values are

$$
(\bar{V}, \bar{W}) = (A^T A)^{-1} A^T (u,v)
\tag{13}
$$

where A denotes the coefficient matrix made up with the focal length f of the stereo camera, the depth Z, and the image coordinates as shown in Eq. (7).

2.5 Independent flow and target segmentation

The difference between the mixed flow and the ego-motion flow yields the independent flow which attributes purely to moving targets. This operation ideally cancels out the effects of inter-fame changes caused by vehicle motion and involves a 2D vector difference:

$$
[u_r \ v_r] = [u_m \ v_m] - [u_e \ v_e]
\tag{14}
$$

where $[u_r \ v_r]$ denotes the independent flow in the horizontal and vertical directions, $[u_m \ v_m]$ the mixed flow, and $[u_e \ v_e]$ the ego-motion flow. The synthetic of the two components of the independent flow is calculated as $s = \sqrt{u^2 + v^2}$. Target segmentation is based on the synthetic independent flow.

In theory, the independent flow of the background should be zero. However, the background has some residual independent flow due to calculation errors. The key to distinguish a moving object from the background is to determine a threshold of the independent flow. In this work, we adopt the OTSU algorithm to determine a self-adapting threshold. The algorithm can be described as follows:

1. For a threshold t, $s_{\min} < t < s_{\max}$, define the variance $\varepsilon(t)$ between the moving target's independent flow and the background's independent flow as

$$
\varepsilon(t) = P_o(s_o - t)^2 + P_g(s_g - t)^2
\tag{15}
$$

where s_o denotes the mean of the independent flows of the moving points, $s_o = \dfrac{\sum s_i * p_i}{p_o}$ ($s_i > t$, $i = 1, \ 2, \ 3...$), s_g denotes the mean of the independent flows of the background points, $s_g = \dfrac{\sum s_i * p_i}{p_g}$ ($s_i < t$, $i = 1, \ 2, \ 3...$); p_o denotes the proportion of the points with $s > t$, p_g the proportion of the points with $s < t$, and p_i the proportion of the points with $s < s_i$.

2. Search for the t from $s_{\min}$ to $s_{\max}$ to make $\varepsilon(t)$ maximum and use it as the threshold to segment the moving objects from the background. This process endures a maximum between-class distance.

We cancel out the pixels with the independent flow below the threshold determined above. For the pixels with the independent flow above the threshold, we use the region-growing method to cluster similar potentials together to form the eventual segmentation. Actually, in this work, the independent flow is also combined with the disparity (depth) for object clustering. This tactic is especially useful for separating objects close to each other or with occlusion.

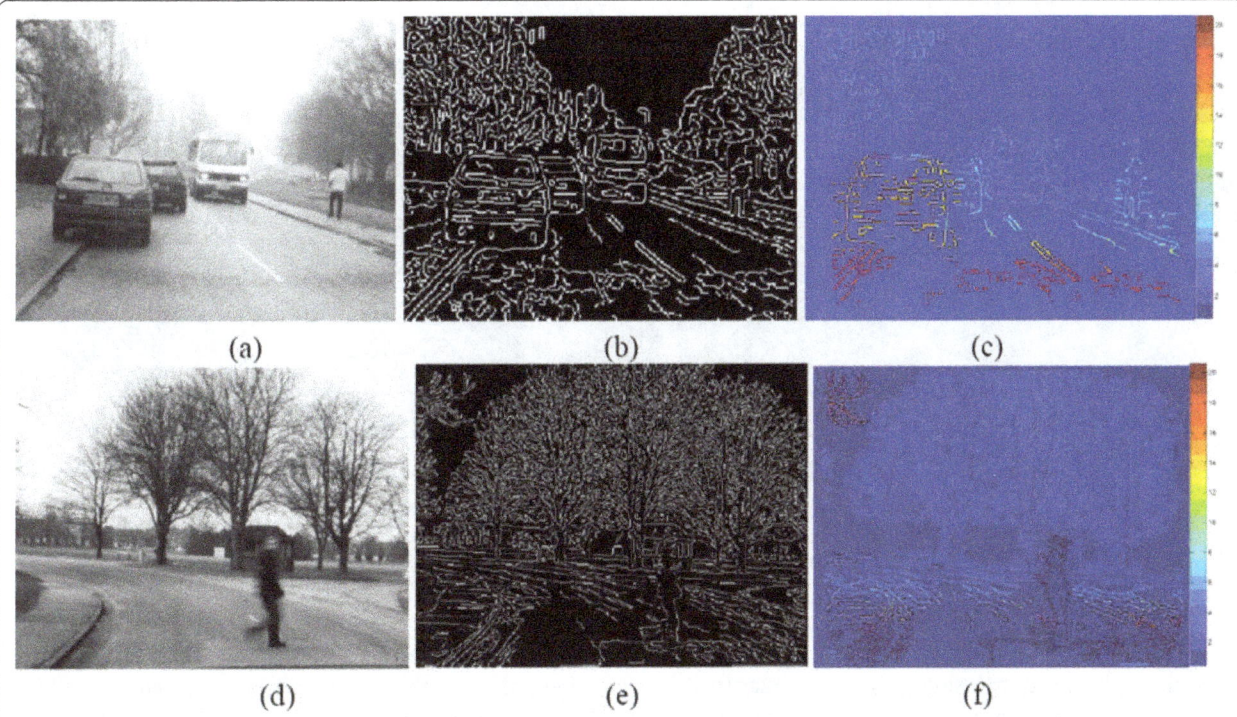

Fig. 4 Traffic scenarios and disparity images. **a** *Left image* at frame 72 of scenario 1. **b** Edge image of scenario 1. **c** Disparity map of scenario 1. **d** *Left image* at frame 72 of scenario 2. **e** Edge image of scenario 2. **f** Disparity map of scenario 2

3 Experiments and results

A VIDERE stereo rig with strict calibration is used to capture images in this work. The two cameras have the image resolution of 640 × 480 pixels, the pixel size of 15 μm, and the baseline of 218.95 mm. The focal length is 16.63 mm. The detection ranges from –8 to 8 m in lateral and from 4 to 50 m in distance. Figure 4a, d shows the left images of two typical traffic scenarios. The first scenario involves a pedestrian, an oncoming coach, and some static obstacles like parked cars and trees, where the equipped vehicle moves in longitudinal direction. In the second scenario, the vehicle is turning in a bend. It helps to evaluate our algorithm when the vehicle undergoes more complex movement.

3.1 Disparity of stereovision

Figure 4b, e shows the edge maps obtained from a Canny detector. The edge points in the left image are

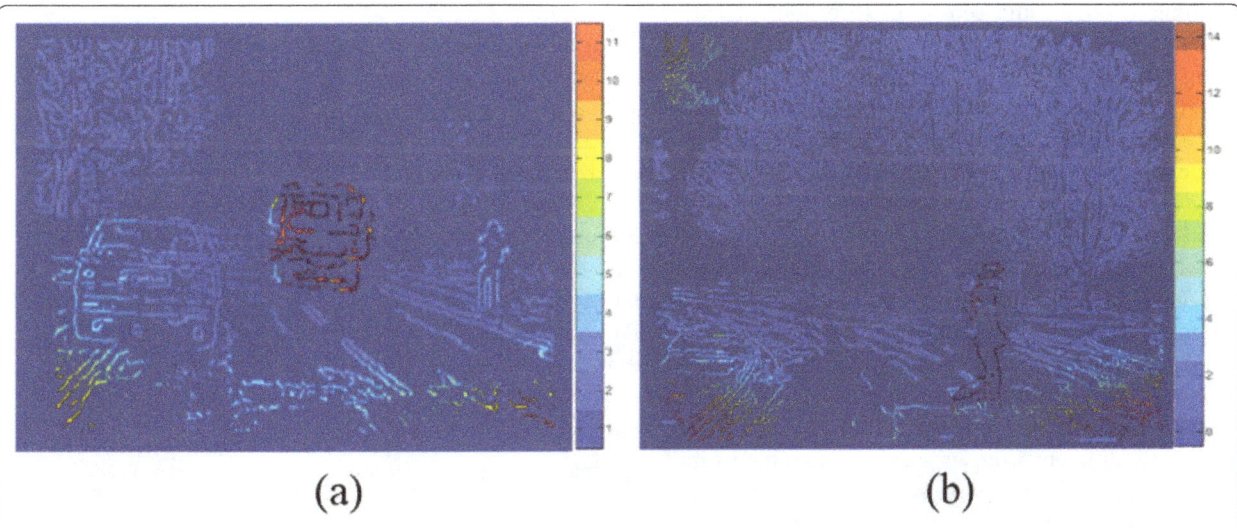

Fig. 5 Mixed flow obtained from the edge-indexed correspondence matching algorithm. **a** The mixed flow of scenario 1. **b** The mixed flow of scenario 2

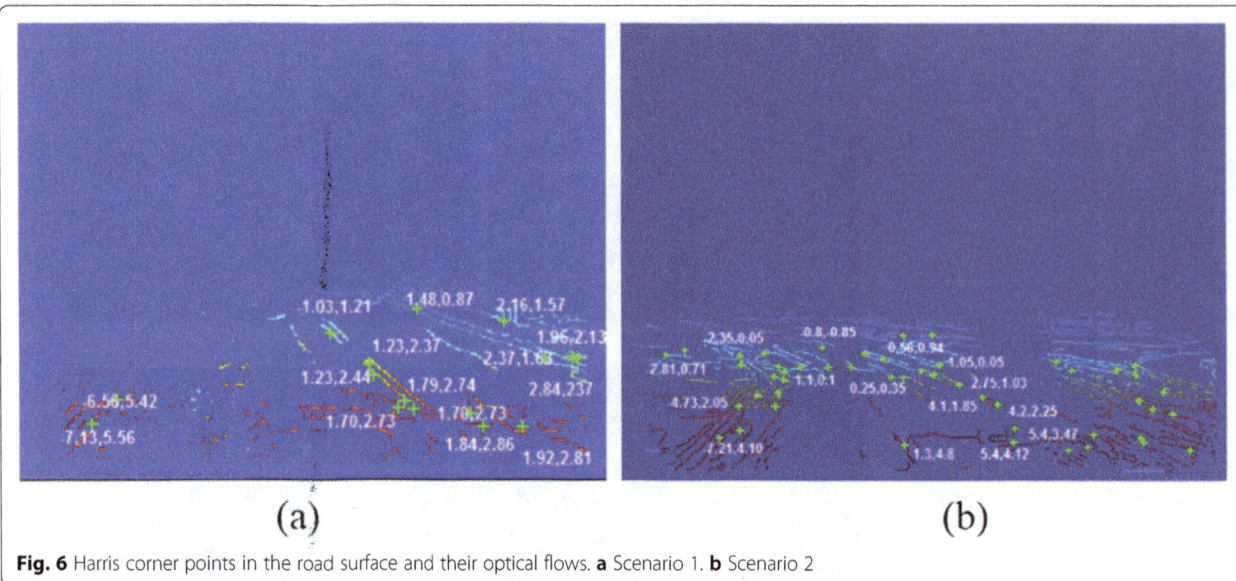

Fig. 6 Harris corner points in the road surface and their optical flows. **a** Scenario 1. **b** Scenario 2

used as seed points to search for the correspondence points in the right image by using greyscale similarity as the measure. The resulting disparity maps are displayed in Fig. 4c, f. A color scheme is used to visualize the disparity. The depth information of the image points can be derived from the disparity map. It should be noted that some points like the trees out of the detection range are not presented in the disparity maps. It is worthy to be noted that contour occluding could be generated due to the different viewpoints of the two cameras and may bring troubles for stereo correspondence matching especially for a short distance with a wider baseline. In our application, we use a relatively short stereo baseline of 218.95 mm, and the detection range is 4 to 50 m. The occluding effect is not significant. In addition, stereo matching depends on the selection of matching windows and setting of threshold of correlation coefficient. The detailed edge-indexed stereo matching procedure can be found in our previous work [26]. All experiments show that the edge-indexed stereo matching can successfully generate an edge-indexed disparity map.

3.2 Mixed flow results

Figure 5 shows the mixed flow obtained using the edge-indexed correspondence matching algorithm described in Section 2.2. It can be noted that even for static objects like parked cars, trees, and ground points, there is obvious motion, which is caused by camera/vehicle's motion. The motion of the pedestrian and the oncoming coach is significantly different from its surroundings due to its own motion. The mixed motion shown in Fig. 5a, b reflects actual movement of the points and will be used for subtraction of motion flows in late stage.

3.3 Visual odometer results

The edge points in the ground surfaces are successfully extracted, as shown in Fig. 6a (scenario 1) and 6b (scenario 2). The Harris corner points are detected and marked with "+" in the figures. For each case, 15 Harris Corner points with higher CRF scores are selected to set up a set of equations for estimating the six ego-motion parameters using the least square fitting method. The results are presented in Table 1. It can be found that for both scenarios, V_z are significant and V_y, W_x, W_y, and W_z are tiny. This is reasonable since the vehicle was moving with a certain speed in a relatively flat road. For scenario 2, V_x is also significant because the vehicle was left turning in a bend. For scenario 1, V_x is equal to 0.17 m/frame, indicating that the vehicle was not strictly moving in longitudinal direction and had a small lateral moving at the moment.

During the video acquisition, a spatial NAV 982 Inertial Navigation System was fitted in the car to measure the ego-motion parameters. Although the INS may lose detection in some cases, the comparison between the effective data of two systems shows that the difference of the results is within 4 %, indicating that our visual odometer is reasonably accurate.

3.4 Ego-motion flow results

The ego-motion flow calculated from Eq. (7) by using above estimated ego-motion parameters is shown in

Table 1 Results of ego-motion estimation

Ego-motion parameters	Scenario 1	Scenario 2
V_x V_y V_z (m/frame)	−0.22 −0.04 227.04	48.61 −0.03 214.25
W_x W_y W_z (rad/frame)	0.12 −0.15 −0.08	0.14 −0.02 0.07

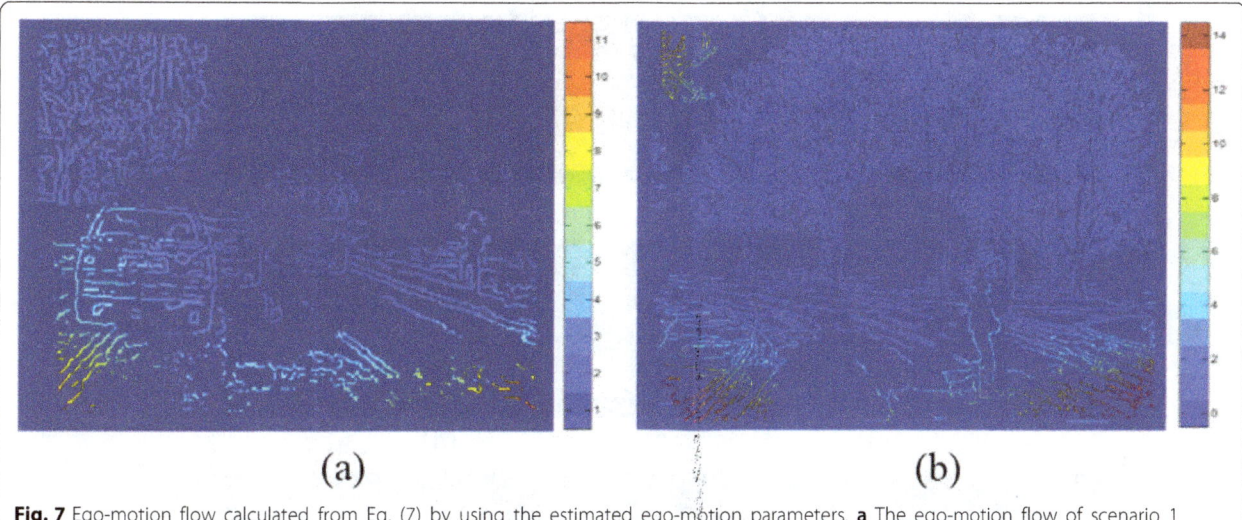

Fig. 7 Ego-motion flow calculated from Eq. (7) by using the estimated ego-motion parameters. **a** The ego-motion flow of scenario 1. **b** The ego-motion flow of scenario 2

Fig. 7. The ego-motion shown in Fig. 7a, b will be used for subtraction of motion flows in late stage.

3.5 Independent flow and motion extraction

The subtraction of Fig. 7a from Fig. 5a is shown in Fig. 8a, while the subtraction of Fig. 7b from Fig. 5b is shown in Fig. 8b. The subtraction yields the independent flow which is purely caused by the target motion. It can be seen that the most of the background have been canceled out and the moving objects are significantly highlighted using the method described in Section 2.5. Furthermore, the pedestrian can be framed according to the continuity of the similar independent flow, as shown in Fig. 8c, d.

3.6 Evaluation of the system

Experiments have also been conducted on the public image database KITTI (Karlsruhe Institute Technology and Toyota Technological Institute) [27]. Figure 9 shows the process of motion detection for one of the

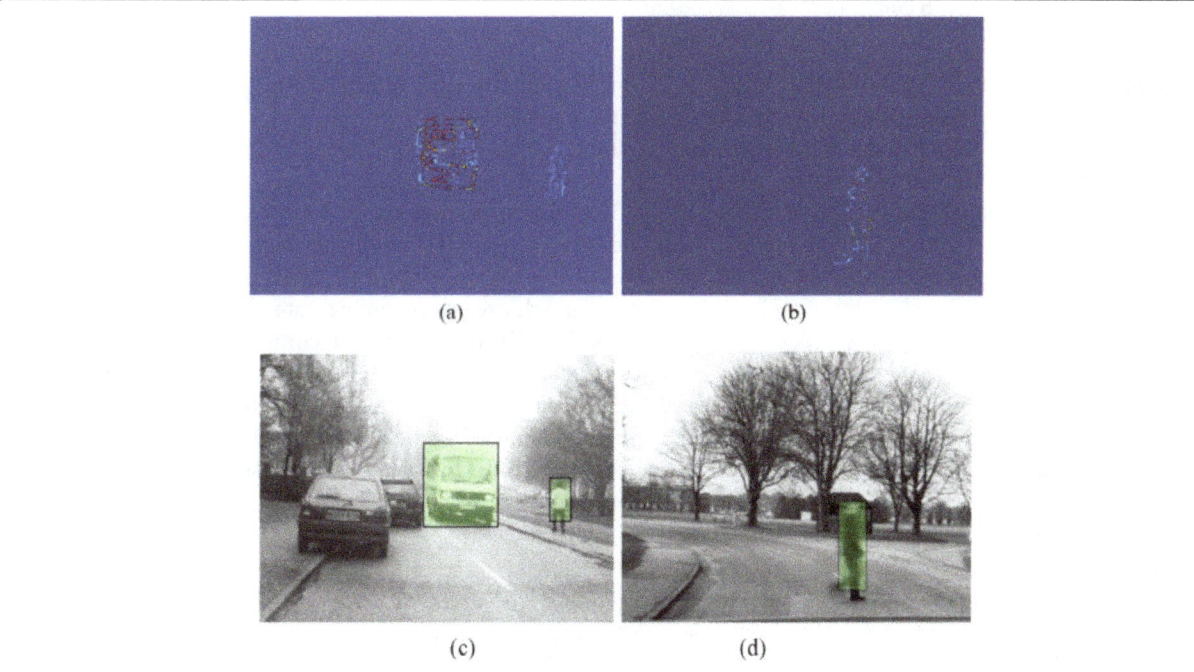

Fig. 8 Independent flow and motion extraction result. **a** Independent flow of scenario 1. **b** Independent flow of scenario 2. **c** Motion extraction in scenario 1. **d** Motion extraction in scenario 2

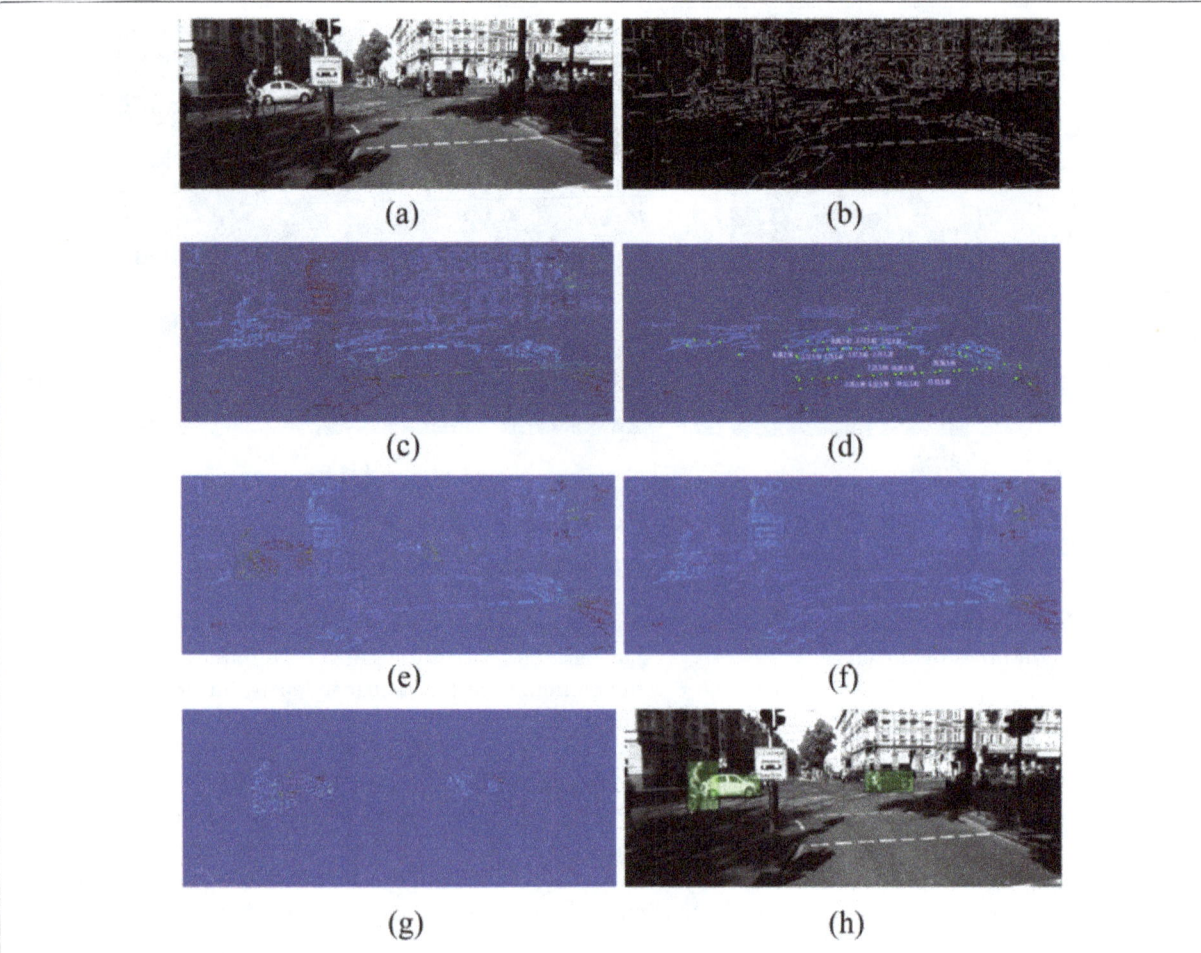

Fig. 9 Detection results for one of KITTI scenarios. **a** Frame 905. **b** Edge image. **c** Disparity image. **d** Feature points in ground plane. **e** Mixed flow. **f** Ego-motion flow. **g** Independent flow. **h** Extraction of the moving objects

scenarios containing multiple moving obstacles. A total of 5000 frames of various scenarios with hand-labeled moving objects including pedestrians and cars have been tested using our approach. In general, *Recall* and *Precision* are usually used to assess the accuracy of object detection.

Recall is defined as follows:

$$Recall = \frac{tp}{tp + fn} \qquad (16)$$

where tp is the total number of true-positively detected objects, fn is the total number of false-negatively detected objects, and $(tp + fn)$ indicates the total number of objects in the ground truth. *Precision* is defined as follows:

$$Precision = \frac{tp}{tp + fp} \qquad (17)$$

where fp is the total number of false-positively detected

objects, and $(tp + fp)$ indicates the total number of the detected objects.

Table 2 lists the performance of our method in terms of detection of pedestrians and cars.

The system is implemented with C++ language in an industrial computer equipped with a 2.40-GHz Intel Dual Core i5 processor and 4 GB of RAM. In general, we can achieve a processing rate of 10–15 frames per second (FPS), depending on complexity of the images. This processing rate includes the stereo pre-processing time. Ideally, it should work at least 25 FPS for a real-time system. But we believe that it will not be a problem to achieve this by using a bespoke image processing hardware in future.

Table 2 Accuracy rate of our method

Object type under detection	Precision	Recall
Pedestrian	94.0 %	92.2 %
Vehicle	94.5 %	93.1 %

Table 3 Comparison with other research work

Category	Approach	Object type under detection	Precision (%)	Recall (%)	Note
Self-adaptive background matching	BBM-based Cauchy distribution [4]	Pedestrian	98.8	88.1	Video surveillance with static camera
		Vehicle	91.3	72.0	
Optical flow	Hidden Markov model (HMM) [12]	Vehicle only	–	86.6	
Stereo-motion fusion	Longuet-Higgins-Equations combined with extended Kalman filter [17]	Pedestrian or car	–	96	Result for feature points detection. The recall definition is slightly different from ours
	Cuboidal object model with extended Kalman filter [18]	Pedestrian or car	–	71.3	Result for object tracking
	Our approach	Pedestrian	94.0	92.2	
		Vehicle	94.5	93.1	

3.7 Comparison with other methods

Table 3 lists the comparison with the some other work reported for moving object detection including applications in video surveillance [4]. It is a hard task to make a uniform comparison with other approaches for two reasons: (1) Evaluation metric used can be different; (2) Many research work do not give statistical accuracy rate. The work most related to our approach can be found in [14–19], which use stereo-motion fusion. However, there are no reports on detection rate or accuracy rate in [14–16, 19]. In [17], authors only give the accuracy for feature point detection rather than accuracy for object detection. Moreover, the accuracy definition is slightly different from ours. In [18], authors only provide result for object tracking.

4 Conclusions

This paper presents a novel motion detection approach using a stereovision sensor for in-vehicle environment sensing system. The relationship between optical flow, stereo depth, and camera ego-motion parameters has been established. Accordingly, a visual odometer has been implemented for estimation of six ego-motion parameters by solving a set of equations fitted with a number of feature points using the linear least square method. The feature points are selected as corner points lying on the road surface and determined by using height constraint and Harris corner detection algorithm. The ego-motion flow evoked by the moving camera/vehicle is calculated from the relational model by using the estimated ego-motion parameters. The mixed flow caused by both camera motion and target motion is obtained from the correspondence matching between consecutive images. The difference between the mixed flow and the ego-motion flow yields the independent flow which attributes purely to the target motion. The moving targets are extracted according to the continuity of the similar independent flow. The approach presented here was tested on substantial complex urban traffic videos. The experimental results demonstrate that the approach can detect moving objects with a correction rate of 93 %. The accuracy of ego-motion estimation is within 4 %, comparing to an in-vehicle INS sensor. The processing rate reaches 10–15 FPS on an industrial computer equipped with a 2.40-GHz Intel Dual Core i5 processor and 4 GB of RAM.

Competing interests
The authors declare that they have no competing interests.

Acknowledgements
This work was sponsored by Specialized Research Fund for the Doctoral Program of Higher Education (Project No. 20133120110006), the National Natural Science Foundation of China (Project No. 61374197), and the Science and Technology Commission of Shanghai Municipality (Project No. 13510502600).

References
1. JE Ha, WH Lee, Foreground objects detection using multiple difference images. Opt. Eng **4**, 047–201 (2010)
2. M.C. arco, F. Michela, B. Domenico, M. Vittorio, Background subtraction for automated multisensor surveillance: a comprehensive review. EURASIP. J. Adv. Signal. Process. 2010, 343057. doi:10.1155/2010/343057
3. L Wei, H Yu, H Yuan, H Zhao, X Xu, Effective background modelling and subtraction approach for moving object detection. IET Computer Vision **9**(1), 13–24 (2015)
4. FC Cheng, SJ Ruan, Accurate motion detection using a self-adaptive background matching framework. IEEE Trans. Intell. Transp. Sys **13**(2), 671–679 (2012)
5. A. Broggi, A. Cappalunga, S. Cattani, P. Zani. In Proceedings of IEEE Intell. Veh. Symp. Lateral vehicles detection using monocular high resolution cameras on TerraMax, (2008), pp. 1143–1148
6. Y Zhu, D Comaniciu, M Pellkofer, T Koehler, Reliable detection of overtaking vehicles using robust information fusion. IEEE Trans. Intell. Transp. Syst **7**(4), 401–414 (2006)
7. L. Kui, D. Qian, Y. He, M. Ben. Optical flow and principal component analysis-based motion detection in outdoor videos. EURASIP. J.Adv. Signal Process. 2010, 680623. doi:10.1155/2010/680623
8. L. Yang, X.F. Li, J. Limin, In Proceedings of the 11th World Congress on intelligent control and automation (WCICA). Abnormal crowd behavior detection based on optical flow and dynamic threshold (2014), pp. 2902-2906

9. E. Martinez, M. Diaz, J. Melenchon, J. Montero, I. Iriondo, J. Socoro, In Proc. IEEE Intell. Veh. Symp.. Driving assistance system based on the detection of head-on collisions (2008); pp. 913–918

10. J Diaz Alonso, E Ros Vidal, A Rotter, M Muhlenberg, Lane-change decision aid system based on motion-driven vehicle tracking. IEEE Trans. Veh. Technol **57**, 2736–2746 (2008)

11. I. Sato, C. Yamano, H. Yanagawa, In Proc. IEEE IV. Crossing obstacle detection with a vehicle-mounted camera (2011), pp. 60–65

12. H Jazayeri, J Cai, Y Zheng, M Tuceryan, Vehicle detection and tracking in car video based on motion model. IEEE Trans. Intell. Transp. Syst. **12**(2), 583–595 (2011)

13. H. Geiger, B. Kitt, In Proc. IEEE Intell. Veh. Symp. Object flow: a descriptor for classifying traffic motion (San Diego, USA, 2010), pp. 287–293

14. D. Pantilie, S. Nedevschi, In Proc. IEEE Conference on Intelligent Transportation Systems. Real-time obstacle detection in complex scenarios using dense stereo vision and optical flow (Funchal, 2010), pp. 439 – 444

15. U Franke, S Heinrich, Fast obstacle detection for urban traffic situations. IEEE Trans. Intell. Transp. Syst **3**(3), 173–181 (2002)

16. C. Rabe, U. Franke, S Gehrig, In Proceedings of IEEE Intelligent Vehicles Symposium. Fast detection of moving objects in complex scenarios (2007), pp. 398–403

17. B. Kitt, B. Ranft, H. Lategahn, In Proc. 13th Int. IEEE Conf. on ITSC. Detection and tracking of independently moving objects in urban environments (2010), pp. 1396–1401

18. S. Bota, S. Nedevschi, In Proc. 14th Int. IEEE Conf. ITSC, Tracking multiple objects in urban traffic environments using dense stereo and optical flow (2011), pp. 791–796

19. P. Lenz, J. Ziegler, A. Geiger, In Proc. IEEE Intell. Veh. Symp.. Roser M. Sparse scene flow segmentation for moving object detection in urban environments (Baden-Baden, Germany, 2011), pp. 926–932

20. T. Yamaguchi, H. Kato, Y. Ninomiya, In Proc. IEEE Intell. Veh. Symp.. Moving obstacle detection using monocular vision (2006), pp. 288–293

21. A. Talukder, L. Matthies, In Proc. IEEE Int. Conf. Intelligent Robots and Systems. Real-time detection of moving vehicles using dense stereo objects from moving and optical flow (2004), pp. 3718-3725

22. AM Waxman, JH Duncan, Binocular image flows: steps towards stereo-motion fusion. IEEE Trans Pattern Anal Mach Intell **8**, 715–729 (1986)

23. C. Harris, M.A. Stephens, In Proceedings of the 4th Alvey Vision Conference. Combined corner and edge detector, (1988), pp. 147–151

24. B McCane, K Novins, D Crannitch, B Galvin, On benchmarking optical flow. Comput Vis Image Underst **84**, 126–143 (2001)

25. Y. Huang, K. Young. Binocular image sequence analysis: integration of stereo disparity and optic flow for improved obstacle detection and tracking. EURASIP. J Adv. Signal. Process. 2008, 843232. doi:10.1155/2008/843232

26. Y Huang, S Fu, C Thompson, Stereovision-based object segmentation for automotive applications. EURASIP J Appl Signal Process **14**, 2322–2329 (2005)

27. KITTI Vision, Available online: http://www.cvlibs.net/datasets/kitti/, Accessed 18 Jul 2015.

8

Online rate adjustment for adaptive random access compressed sensing of time-varying fields

Naveen Kumar[1*], Fatemeh Fazel[2], Milica Stojanovic[2] and Shrikanth S. Naryanan[1]

Abstract

We develop an adaptive sensing framework for tracking time-varying fields using a wireless sensor network. The sensing rate is iteratively adjusted in an online fashion using a scheme that relies on an integrated sensing and communication architecture. As a result, this scheme allows for an implementation that is both energy efficient and robust. The objective is to promote an "active" framework which uses the information extracted from the network data and iteratively adjusts the monitoring process to capture the temporal variations in the monitored field. We propose two metrics based on target detection/tracking for this feedback scheme that seek to trade off between energy efficiency and accuracy of the detection/tracking tasks. Our simulation results suggest that tying target detection with the rate adjustment algorithm ensures that the robustness to changes in the field can be achieved simultaneously with the end goal of accurate target detection. Compared to a baseline method that uses the correlation of the acquired field over time, our method exhibits better performance when the targets of interest have a smaller spatial spread.

Keywords: Sensor networks, Adaptive sensing, Detection, Random access, Compressed sensing, Joint communication and detection

1 Introduction

The emergence of compressed sensing framework presents significant potential for efficient sensing and sampling systems [1–3] by helping to reduce sample complexity under realistic communication constraints. Sensor network technology greatly benefits from the compressed sensing paradigm [4–12]. Consider for example, large-scale networks that are deployed for long-term monitoring of dynamic fields such as the ocean bed that typically need to account for power consumption considerations. An efficient scheme in such cases requires performance optimization that jointly considers both the sensing and communication constraints.

To address this issue, Fazel et al. proposed a random access compressed sensing (RACS) scheme in [13, 14] for energy-efficient reconstruction of sensing fields. The proposed sensing scheme depends on integrating information from the communication and channel access modules into the data acquisition process. The method is simple to implement under realistic communication constraints and requires minimal assumptions on the field. However, the proposed RACS scheme is designed for stationary[1] sensing fields, where the field being monitored is assumed to remain static during sensing. In order to monitor time-varying fields, in [15], the authors employed low-rank matrix recovery to reconstruct the space-time map of the field. However, this approach is offline and entails considerable delays since full recovery can be attained only after the data have been collected over multiple time segments. Moreover, it is assumed that the coherence properties of the underlying field are known a priori and remain unchanged throughout the full sensing duration. This assumption might be justified for the monitoring of natural phenomena, where the field is assumed to be either stationary or changing at a fixed rate. It is also common to find similar assumptions of stationarity in related works in object detection or classification in underwater fields [16, 17]. However, when the field being monitored undergoes a varying rate of change (e.g., when the process

*Correspondence: komathnk@usc.edu
[1] Signal Analysis and Interpretation Laboratory (SAIL), Electrical Engineering, University of Southern California, Los Angeles, CA, USA
Full list of author information is available at the end of the article

is impacted by a target that is moving at an unknown or variable speed), such assumptions may not hold.

In a more recent work, Kerse et al. [18] addressed this problem by proposing to unify target detection with reconstruction using a standard sparse identification technique. Targets are tracked from frame to frame and the authors further suggest that the tracking error could be used as a measure to adjust the sensing rate in turn. While their method does not require any a prioi knowledge of the number of targets or coherence properties of the underlying field, it relies on knowledge of the exact target signatures for both target localization and tracking. In this paper, we propose a rate adjustment method that does not require explicit knowledge of the target signature. Rather, knowledge of the family of models to which the field might belong is adequate as the field model parameters can be jointly estimated.

Similar to the work in [18], we consider that the end goal in sensing is to detect or track targets and incorporate these data processing aspects into the joint sensing and communication scheme. We propose a framework to adjust the sensing rate by estimating different attributes of the field to make an informed decision. Our adaptive rate adjustment procedure for compressed sensing iteratively adjusts the per-node sensing rate to capture the variations in the underlying field. First, we treat the field as piecewise stationary and apply random access compressed sensing within each sensing period. Second, we compute two heuristic metrics that seek to tie in the end goal of target detection/tracking with the rate adjustment scheme. Using the data collected in each segment, the fusion center (FC) relies on a detection algorithm to first determine the current state it is in. Finally, depending on the current state, a control algorithm instructs the FC to change the sensing rate if required.

A high rate of sensing would typically bode well for target detection/tracking, but it is not energy efficient. On the other hand, a low rate of sensing may not necessarily lead to poor performance in detection. Thus, there is a possible trade-off between energy efficiency and the target detection accuracy which the rate adjustment seeks to exploit. We perform simulation experiments using the proposed method and present results that suggest that the proposed sensing rate adjustment method exhibits better performance compared to the baseline method when compared on the following evaluation criteria: (a) mean-squared error of tracking the underlying coherence time and (b) F-score of the target detection accuracy.

The paper is organized as follows. In Section 2, we explain the basic sensing model for the stationary case, and in Section 3, we relax the stationarity assumptions. In Section 4, we describe the sample field model for simulation. In Sections 5 and 6, we describe an adaptive strategy for sensing the time-varying field. Finally, in Sections 7 and 8, we provide results for the simulation.

2 Random sensing network over stationary fields

Consider a grid network consisting of $N = P \times Q$ sensors, with P and Q sensors in the x and y directions, respectively. The underlying assumption is that most signals of interest (natural or man-made) vary smoothly spatially and hence are compressible in the spatial discrete Fourier transform (DFT) basis. We denote the sparsity of the signal by S. The data from the distributed sensors is conveyed to the FC, where a full map of the sensing field is reconstructed. This map can be used for target detection as will be shown in Section 5.

Inspired by the theory of compressed sensing, the architecture proposed in [13, 14] employs *random sensing*, i.e., transmission of sensor data from only a random subset of all the nodes. For a stationary field, each sensor node measures the signal of interest at random time instants—independently of the other nodes—at a rate of λ_1 measurements per second. It then encodes each measurement along with the node's location tag into a packet, which is digitally modulated and transmitted to the FC in a random access fashion. Owing to the random nature of channel access, packets from different nodes may collide, creating interference at the FC, or they may be distorted as a result of the communication noise. A packet is declared erroneous if it does not pass the cyclic redundancy check or a similar verification procedure. Since the recovery is achieved using a randomly selected subset of all the nodes' measurements, we let the FC discard the erroneous packets as long as there are sufficiently many packets remaining to allow for the reconstruction of the field.

The FC thus collects the useful packets over a collection interval of duration T. The interval T is assumed to be much shorter than the coherence time of the process, such that the process can be approximated as fixed during one such interval. Let $R_{xx}(\tau)$ denote the temporal autocorrelation of the process, which quantifies the average correlation between two samples of the process separated by time τ. The coherence time T_{coh} is then defined as the time lag during which the samples of the signal are sufficiently correlated, i.e., $R_{xx}(T_{coh}) = qR_{xx}(0)$, where q is the desired level of the correlation (e.g., $q = 98$ %). We borrow the above random sensing strategy from Fazel et al. [13, 14]. In this strategy, in addition to the compressibility assumption mentioned earlier, we can reconstruct back the field using Eq. (1)

$$\mathbf{y} = \mathbf{R}\boldsymbol{\Psi}\mathbf{v} + \mathbf{z} \tag{1}$$

where $\mathbf{z}$ represents the *sensing noise*, $\boldsymbol{\Psi}$ is the inverse DFT matrix, $\mathbf{v}$ is the sparse vector of Fourier coefficients, and $\mathbf{R}$

is a $K \times N$ matrix—with K corresponding to the number of useful packets collected during T—which models the selection of correct packets. Each row consists of a single one in the position corresponding to the sensor contributing the useful packet. The FC can form $\mathbf{R}$ from the correctly received packets, since they carry the location tag. We emphasize the distinction between the sensing noise $\mathbf{z}$, which arises due to the limitations in the sensing devices, and the communication noise, which is a characteristic of the transmission system. The sensing noise appears as an additive term in Eq. 1, whereas the communication noise results in packet errors and its effect is captured in the matrix $\mathbf{R}$.

The FC then recovers the map of the field using sparse approximation algorithms [19]. It suffices to ensure that the FC collects a minimum number of packets, $N_s = \mathcal{O}$ $(S \log N)$, picked uniformly at random, from different sensors, to guarantee accurate reconstruction of the field with very high probability. The random nature of the system architecture necessitates a probabilistic approach to system design using the notion of sufficient sensing probability [13] denoted by P_s. This is the probability with which full-field reconstruction is guaranteed at the FC. Setting this probability to a desired target value, system optimization under a minimum energy criterion yields the necessary design parameter, i.e., the per-node sensing rate λ_1. The minimum per-node sensing rate can be expressed in terms of the system parameters as shown in [14]

$$\lambda_{1s} = \frac{-1}{2NT_p \frac{b}{b+1}} \cdot W_0 \left(\frac{2NT_p \frac{b}{b+1} e^{\frac{b}{\gamma_0}}}{T} \log \left(1 - \frac{\alpha_s}{N} \right) \right)$$

(2)

where T_p is the packet duration, b is the packet detection threshold, α_s is the average number of packets that need to be collected in one observation interval T to meet the sufficient sensing probability, γ_0 is the nominal received signal-to-noise ratio (SNR), and $W_0(\cdot)$ is the principal branch of the Lambert W function. (More details can be found in [14]). Note that as shown in Fig. 1, λ_{1s} depends on the collection interval T, which in turn must be lower than T_{coh}, for the stationarity assumption to hold.

3 Adaptive sensing

In the above framework, the minimum per-node sensing rate λ_{1s} can be determined based on the properties of the field and is then kept fixed throughout the entire sensing process. However, most fields of interest are seldom stationary, and the coherence properties of a non-stationary dynamic field usually vary significantly

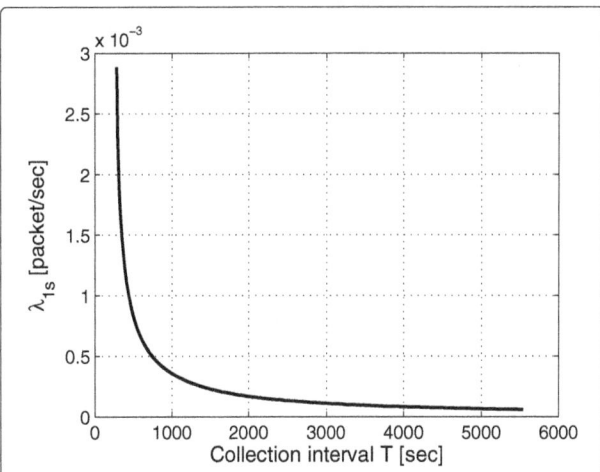

Fig. 1 Per-node sensing rate λ_{1s} vs. the collection interval T, for $N = 1000$ nodes, $S = 10$, $T_p = 0.2$ sec, $b = 4$, $\alpha_s = 244$ packets, and $\gamma_{0=15}$ dB

over time. This calls for the design of an adaptive random sensing network for the monitoring of temporally varying fields. The objective is to transition from passive monitoring where only the map of the sensing field is reconstructed to an active framework where the relevant information is extracted (e.g., target detection/tracking) and exploited to instruct the sensor nodes to adjust their sensing rates. In practice, we assume the coherence properties of a non-stationary field to be piecewise constant over time, such that given prior information about the field at a particular time instant it can be assumed to be stationary within a single coherence time interval. To this end, we employ a detection method, which uses the collected data to determine the attributes of the underlying field. The FC can then use this information to determine any appropriate modifications to the per-node sensing rate, i.e., increase, maintain, or decrease the sensing rate. This cycle of sensing-decision-adjustment is illustrated in Fig. 2.

For a given collection interval T, the corresponding per-node sensing rate can be determined using Eq. (2), as shown in Fig. 1. The proper choice of T however depends on the rate of variations in the field and is adaptively tuned. In particular, we use an approach based on target detection, where we assume that an object/target model of interest is known beforehand. This is a common assumption in most supervised pattern recognition tasks. Given a reconstructed map, the location of targets is first estimated. Using this knowledge, we then estimate the parameters of the object model from the map. These parameters describe the detection system's understanding of the field and can be used to generate the map of the field. Comparing this model-based map with

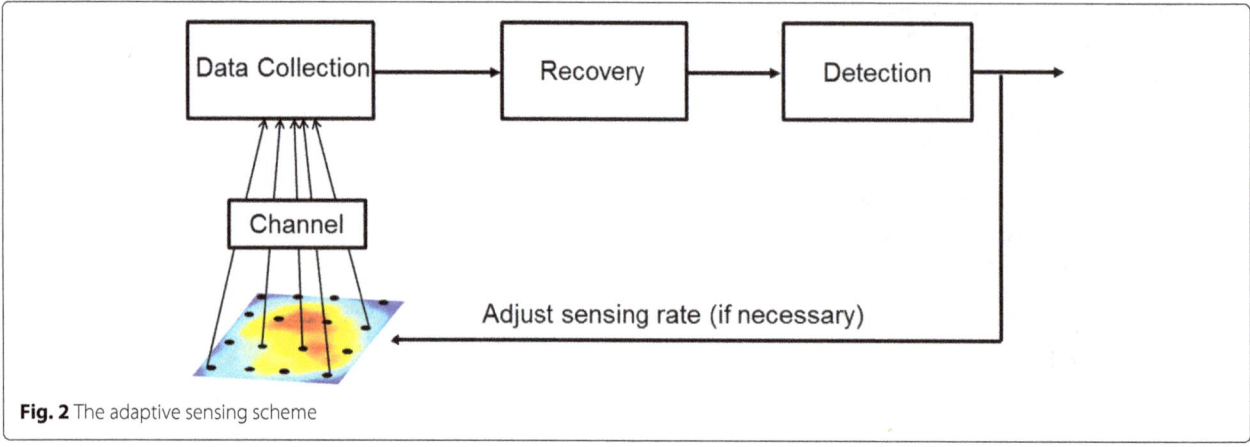

Fig. 2 The adaptive sensing scheme

the observed one reconstructed using sparse approximation algorithms provides us with a reliability metric for reconstruction. In other words, if there is a difference in what the algorithm expects to see and what it sees, it indicates an error in either the acquisition of the field or the algorithm's understanding of the map. In either case, the FC needs to adapt its sensing rate. We specifically discuss our methods in the context of the following two cases:

3.1 Oversensing

This situation corresponds to the case when there occurs redundant sensing because the per-node sensing rate is much larger than the rate of change of the field, i.e., $T << T_{coh}$. Although this case favors reconstruction using the RACS architecture, it leads to a wastage of communication resources. Thus, we seek to lower the sensing rate in this case to an optimal point such that the accuracy of our end goal is not affected. Figure 7 shows the result for of *oversensing* for the example discussed in Section 5. In this case, we devise a scheme to estimate the motion of targets using multiple frames. We use the term frame here and elsewhere in the paper to refer to the field reconstructed from samples acquired within a single sensing time duration.

3.2 Undersensing

In this scenario, the rate of sensing per node is insufficient and the field changes within one collection interval since $T > T_{coh}$. Thus, the per-node sensing rate λ_{1s} needs to be increased. The targets are no longer steady within the duration of a collection interval T leading to blurring, which makes target detection challenging. The estimation task in this case is further complicated by the fact that the packets from different frames may have been collected during the interval T. This scenario leads to a violation of the stationarity assumptions made initially. The reconstructed map is thus blurred because of different packets

originating from different frames (Fig. 3). In this case, we use a specific error metric based on model fitting to estimate the reliability of reconstruction and object detection.

4   Simulation of a sample dynamic field

In this paper, we demonstrate the adaptive monitoring procedure using a sample model for the field. We demonstrate our adaptive scheme to adjust the per node sensing rate in RACS using a simulated example field in this paper. Using a realistic example allows us to control the coherence parameter T_{coh} and monitor the effect of changing the collection interval T in accordance with the adaptive algorithm.

We first describe the example field that serves as our test case. Suppose, a field with M targets is being observed. Each target in the field is assumed to generate a signature (e.g., heat, sound, etc) decaying exponentially with distance from its location. At time t, the process observed by sensor node i at coordinate (x_i, y_i) is given by Eq. (3)

$$u_i(t) = \sum_{m=1}^{M} A_m e^{-p\sqrt{(x_i - a_m(t))^2 + (y_i - b_m(t))^2}} \qquad (3)$$

where $a_m(t)$ and $b_m(t)$ are the coordinates of the mth target at time t, A_m is its strength, and p is the decay rate of the sources. The process then evolves over time as the sources move along random trajectories. Similar models are commonly found in the energy-based localization literature for static sensor networks [20–22] when the targets being detected are not moving.

Initially, the FC has no knowledge of the location of targets or the rate of variation in the field, i.e., the speed at which the targets are moving. It thus instructs the data collection to begin with an initial sensing rate

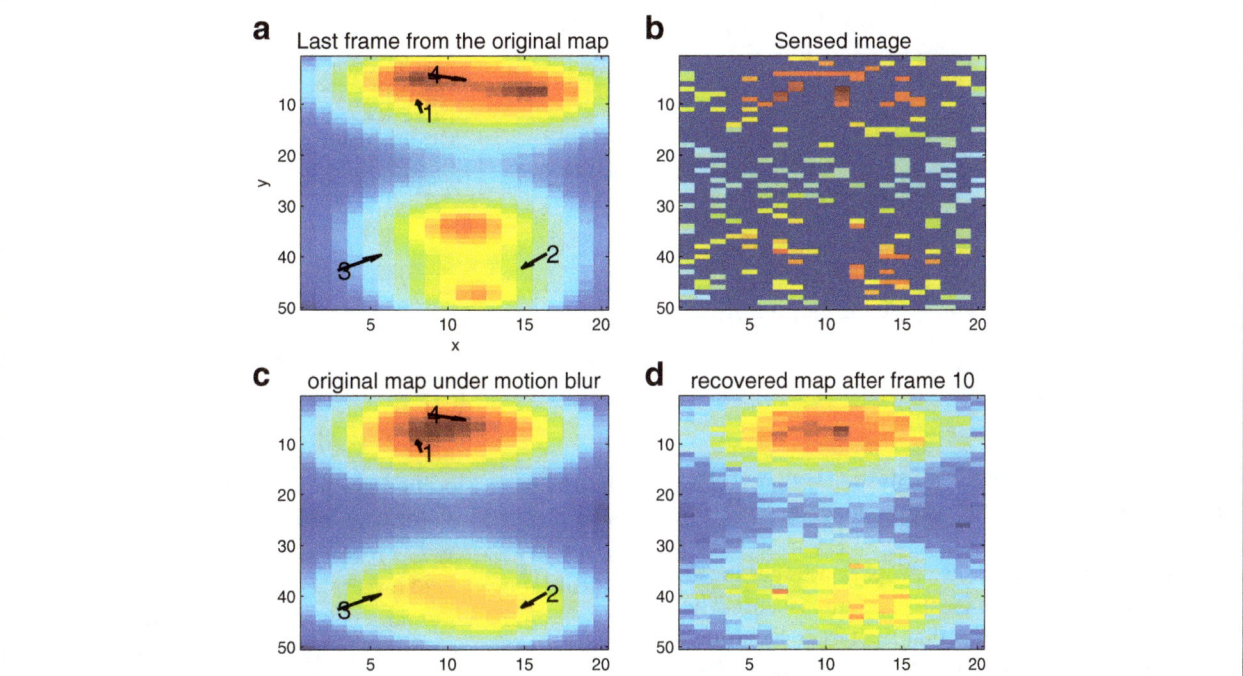

Fig. 3 a–d The effect of the field changing at rate faster than the sensing rate. **a** is the original undistorted map that was sensed at T_{coh} while (**c**) shows the same field obtained through temporal blurring after an observation interval $T = 10 \times T_{coh}$. **d** is the field obtained as a result of sensing at a rate 10 times slower than the field's underlying rate of change. Notice how (**d**) is more similar to the blurred map (**c**) than (**a**). This distortion in the sensed field is an example of undersensing

λ_{1s}^{init}. The initial sensing rate is determined using historical data by setting the desired parameters in Eq. (2). Once the map of the field G^{sa} is recovered using sparse approximation techniques, the FC may now use the rate adjustment algorithm described later to decide if the sensing duration T needs to be adjusted. The sensors employ the adjusted sensing rate in the ensuing sensing duration. In this paper, we discuss metrics for the proposed rate adjustment algorithm for this family of field models, although the framework itself is quite general.

For simulating a map of the field as defined above, we start with a set of randomly chosen parameters. In addition, each target is assigned a random velocity and direction of movement. To simulate the *undersensing* case, we consider that collection occurs over N_b coherence time intervals (referred to as frames) where $T \approx N_b T_{coh}$. The randomly sampled packets are then collected uniformly over the last N_b frames (Fig. 3). This effectively leads to motion blurring of the targets as mentioned before. To simulate the *oversensing* case, the collection interval is reduced to $T' = T_{coh}$ while each target's velocity is scaled by T/T_{coh} to make them appear to be moving slower. To deal with the issue of a finite field size, targets moving out of the field are replaced by new targets starting from the same location assigned a new random velocity and direction.

5 Adaptive field monitoring scheme

In this section, we discuss specific algorithms and metrics extracted using the reconstructed map, which are used for the adaptive rate control of per-node sensing in RACS. Although the algorithms discussed below have been adapted to this particular family of models used in the simulation example, we have attempted to layout broad steps whenever possible and provide a general scheme to be used in such cases. These algorithm stages have been presented in the block diagram shown in Fig. 4.

5.1 Target localization

Before taking any decisions about the current sensing state, we first detect and localize the targets in the image. Similar to [18], by involving target detection in the feedback process, we would like to ensure that the rate of sensing is optimized for the end goal of target detection.

Traditionally, most energy-based localization methods [20–22] assume that the number of targets is known in advance. In addition, they assume a field model for target signature decay allowing for parameter estimation by model fitting. More recently, Kerse et al. proposed a method for direct target localization based on a standard sparse identification technique [18]. The advantage in their method is that target localization can be performed in a single step without first having to reconstruct the field. The number of targets can also be jointly estimated

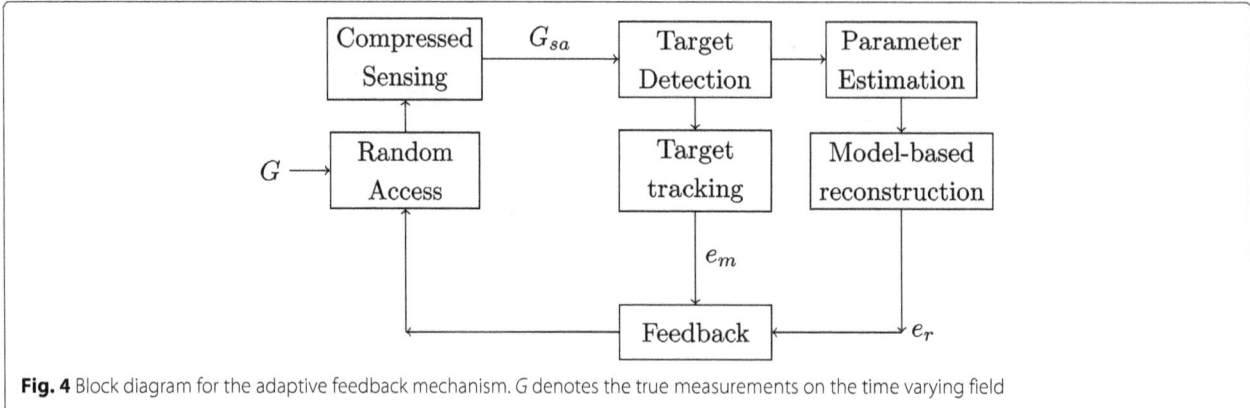

Fig. 4 Block diagram for the adaptive feedback mechanism. G denotes the true measurements on the time varying field

in this process using a sparsity constraint. However, the method still depends on exact knowledge of the field model.

To overcome this drawback, we propose a target localization algorithm based on local gradient ascent. The proposed method only requires that the target signatures be monotonically decaying away from the target. This technique is adapted from the mode seeking mean shift algorithm [23] which is frequently used for unsupervised clustering of data. Consequently, it can be interpreted as searching local peaks in the data histogram. In the task at hand, we are instead interested in finding local peaks in field intensity. We also modify the algorithm slightly to adapt to the discrete search space for this problem. Specifically, we start from a random initial point X on the map. A mean-shift-based gradient ascent technique is then used to update the location of this point in each iteration till a peak is found. Given the current location X, we estimate an intensity weighted mean X_c for locations around X in a window of size W as shown in Eq. (4). Note that the direction from X to X_c gives the direction of gradient ascent, along which X should be updated in the next iteration.

However, in the current problem, we are working on a space of discrete sensor locations. To make the algorithm better suited to such a scenario, we quantize the gradient direction to eight angular bins. Depending on the direction of the gradient, X is then shifted to one of its eight connected points. This process is explained using a schematic in Fig. 5. The iterations are repeated until X converges to a local maximum or reaches a point that has already been visited. The entire procedure is repeated multiple times with new random initial points till the peaks have been discovered. In practice, it is not necessary to traverse all points in the field to discover all the peaks (Fig. 6), and we restrict the algorithm to a fixed number of iterations. The parameter W in this algorithm serves as a mask over which the gradient can be estimated more robustly. The target detection procedure is explained in Algorithm 1.

Algorithm 1: Modified gradient ascent algorithm for target detection on the acquired map. (See Fig. 6)

Input: $P \times Q$ map G^{sa} reconstructed by RACS
Output: Location of targets $\mu_1, \mu_2, \ldots$ where the number of targets is initially unknown

Set maximum number of iterations to M, size of window to W
Cluster index $K \leftarrow 1$
for *iter* $\leftarrow$ 1 **to** M **do**
    Randomly select an initial location X on the map.
    while X hasn't already been visited **do**
        Mark X as visited and belonging to cluster K
        //assuming we're going to find a new target
        Compute the new center of mass X_c within a window of size W around X.
        Compute the direction from X to X_c and quantize it into 8 bins between $\{-\pi/8, \pi/8, 3\pi/8 \ldots\}$ (Fig. 5)
        Update the position of X to one of the 8 adjacent positions based on the quantized direction (Fig. 5)
    end while
    Mark all points in this trail leading upto X as belonging to the same cluster as X
    if X belongs to the new cluster K **then**
        Append X as μ_K to the set of targets
        //new target found
        $K \leftarrow K+1$
    end if
end

$$X_c = \frac{\sum_s I(s)K(s-X)s}{\sum_s I(s)K(s-X)}, s \in P \times Q \text{ grid}$$

$$K(s-x) = \begin{cases} 1 & \text{if } ||s-x||_c \leq W \\ 0 & \text{if otherwise} \end{cases} \tag{4}$$

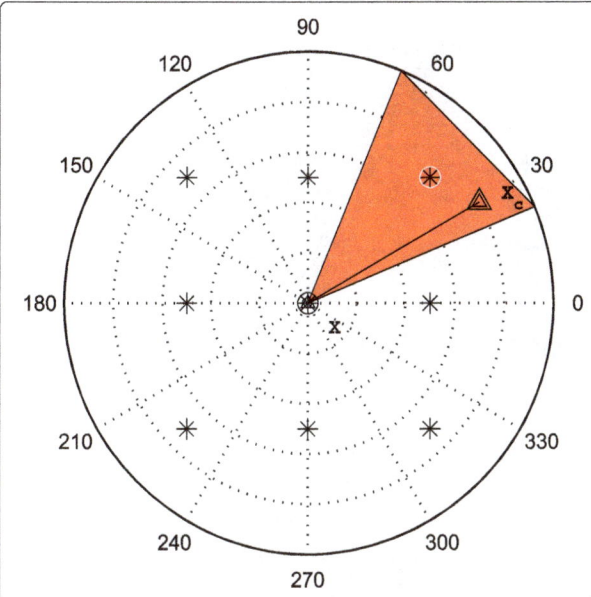

Fig. 5 Modified discrete step update scheme using quantized directions. *Asterisk* (*) indicates points on the grid. *I* denotes the $P \times Q$ grid of intensity values, and $K(.)$ denotes our particular choice of kernel function used for computing mean

5.2 Parameter estimation

Once the positions a_m and b_m for the targets have been identified using the target localization algorithm described above, we try to obtain a minimum mean square error (MMSE) estimate of the field parameters defined in Eq. (3) using a method of comparison by synthesis to match the acquired map G^{sa} against a model. The parameters of model denoted by G^{model} comprise the target locations (a_m, b_m), their respective strengths A_m, and the decay parameters of the target signature (p).

We estimate the parameters using a nonlinear regression [24] technique to optimize an MMSE formulation as

shown in Eq. (5). We use the *nlinfit* function in MATLAB for solving this optimization problem. Since the optimization is not convex, the choice of initial points is important. We use the target localizations estimated earlier and a reasonable value for the decay as our initial point. The estimated parameters are then used to generate a map of the field G^{det} (Fig. 7d). G^{det} represents what the control algorithm expects the field to look like based on the target localizations and its knowledge of the model.

$$\widehat{\mathbf{a}, \mathbf{b}, \mathbf{A}}, p = \underset{a_m, b_m, A_m, p}{\arg\min} \sum_{i=1}^{P} \sum_{j=1}^{Q} \left(\sum_{m=1}^{M} G_{ij}^{model}(a_m, b_m, A_m, p) - G_{ij}^{sa} \right)^2 \tag{5}$$

$$G^{det} = G^{model}(\widehat{\mathbf{a}, \mathbf{b}, \mathbf{A}}, p) \tag{6}$$

5.3 Reliability metric for reconstruction

We obtain a reliability metric for reconstruction based on the error in the model fitting described above. We compare the model-based map G^{det} against the non-model-based map G^{sa} recovered via sparse approximation by projecting both of them on the DFT matrix and decomposing the error into a high-frequency (HF) and low-frequency (LF) term. The intuition is that G^{sa} should be similar to G^{det}, but for any errors resulting from issues in either sensing or localization. A gross mismatch between the coefficients will be captured by the LF term indicating an inaccurate detection while a large value of the HF term (denoted by e_r) is a characteristic of a poor reconstruction due to violation of the assumptions in RACS (Fig. 8a, b). The definition of this error term is shown in Eq. (7) where $\mathcal{F}_\mathcal{H}$ denotes a 2D high-pass filter, and e_r is obtained by convolving it with the difference between the images G^{det} and G^{sa}. The cutoff frequency for the filters can be chosen from a previous estimate of the field. The error metric e_r thus comes in handy in the *undersensing*

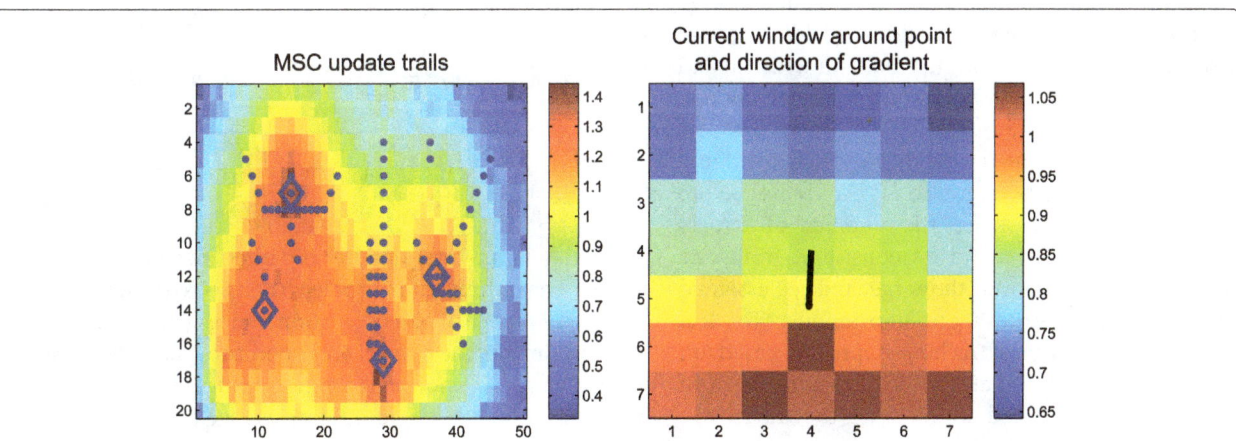

Fig. 6 Figure on the *left* shows a sample map with sequential update steps for each point converging at local peaks (marked by *diamond symbols*). The figure on the *right* shows the window of width $W = 3$ during the current iteration and the general direction of the gradient from the center

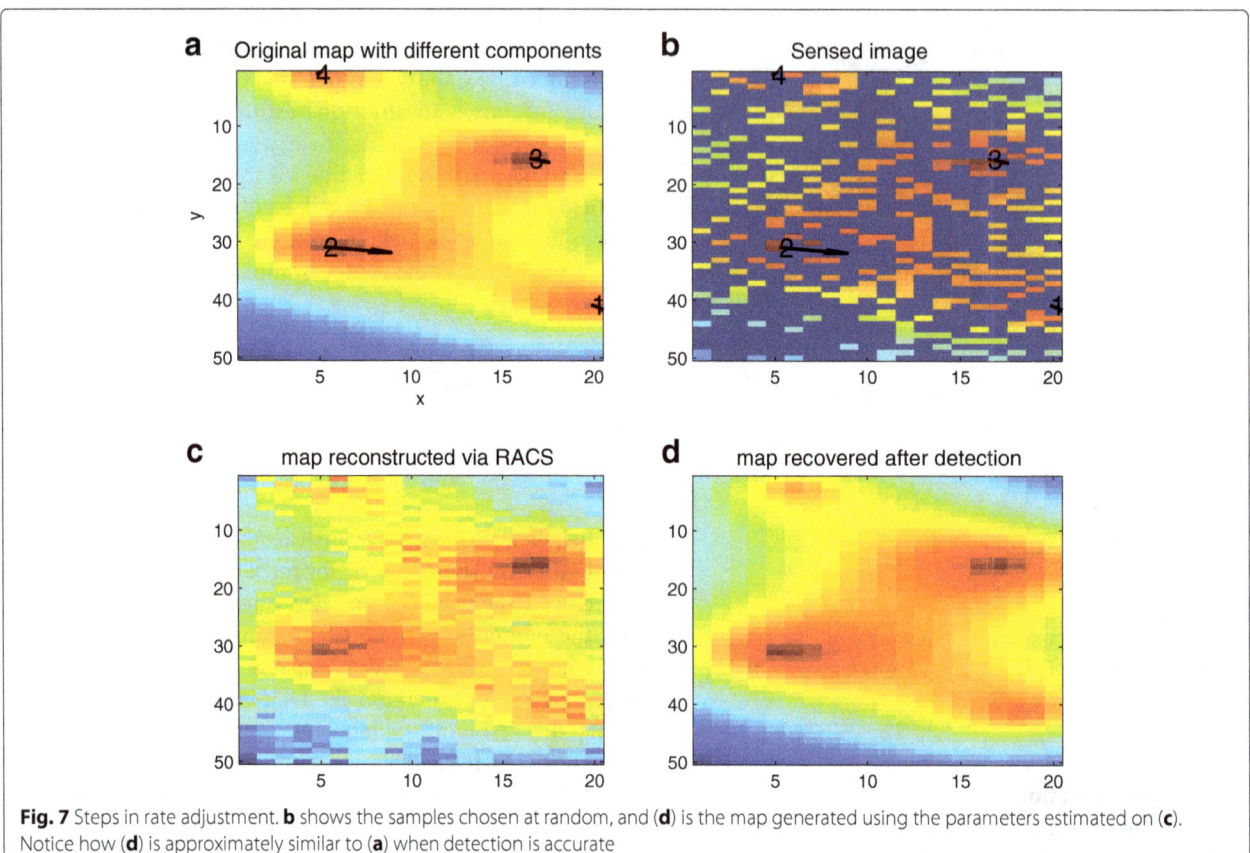

Fig. 7 Steps in rate adjustment. **b** shows the samples chosen at random, and (**d**) is the map generated using the parameters estimated on (**c**). Notice how (**d**) is approximately similar to (**a**) when detection is accurate

case when the acquired map G^{sa} is either blurred or poorly reconstructed.

$$e_r = \mathcal{F}_\mathcal{H} * \left(G^{\text{det}} - G^{sa} \right) \tag{7}$$

5.4 Measuring motion by target tracking

In addition to compensating for the dynamic nature of the field in the *undersensing* case, we would also like to eliminate any redundant sensing. Recall, that in the *oversensing* case, the field changes very slowly, yielding perfect reconstruction and detection. This makes it necessary to monitor the field over multiple frames for detecting any signs of *oversensing*.

To quantify motion in the field, we track the location of detected targets over the last L frames. Note that this is not trivial since the number of targets detected is not guaranteed to be consistent from one frame to another. Moreover, the target indices assigned by the detection algorithm are non-unique. To deal with this issue, we use a tracking approach based on dynamic programming that ensures tracking even if the object is not detected in some of the intermediate frames. More specifically, we recursively minimize the total distance moved by a target over L frames. If the targets are spaced sufficiently apart, this ensures

tracking of the slowest moving target over multiple frames.

Suppose $\left(\hat{a}_i^t, \hat{b}_i^t \right)$ denote the location of the ith target detected in frame t and let d_{ij}^t be defined as the distance between the ith target detected in the frame t and the jth target detected in the frame $t + 1$. Then the net distance $D(n_1, n_2, \ldots, n_L)$ moved by a target over L frames and detected at the indices $\{n_1, n_2, \ldots, n_L\}$ is given by Eq. (8).

$$D(n_1, n_2, \ldots, n_L) = \sum_{t=1}^{L-1} d_{n_t n_{t+1}}^t; \quad 1 \le n_t \le M_t \tag{8}$$

$$d_{ij}^t = \sqrt{\left(a_i^t - a_j^{t+1} \right)^2 + \left(b_i^t - b_j^{t+1} \right)^2} \tag{9}$$

where M_t is the total number of targets detected at each frame. The optimization problem then boils down to finding the sequence of object location indices $\{n_1, n_2, \ldots, n_L\}$ that minimizes the sum of total point-to-point distances over L frames. To normalize over the length of the temporal window, we use the minimum average distance moved as a tracking-based metric as shown in Eq. (10). This metric (denoted by e_m) is expected to be low when the targets move slowly in the *oversensing* case while it is expected

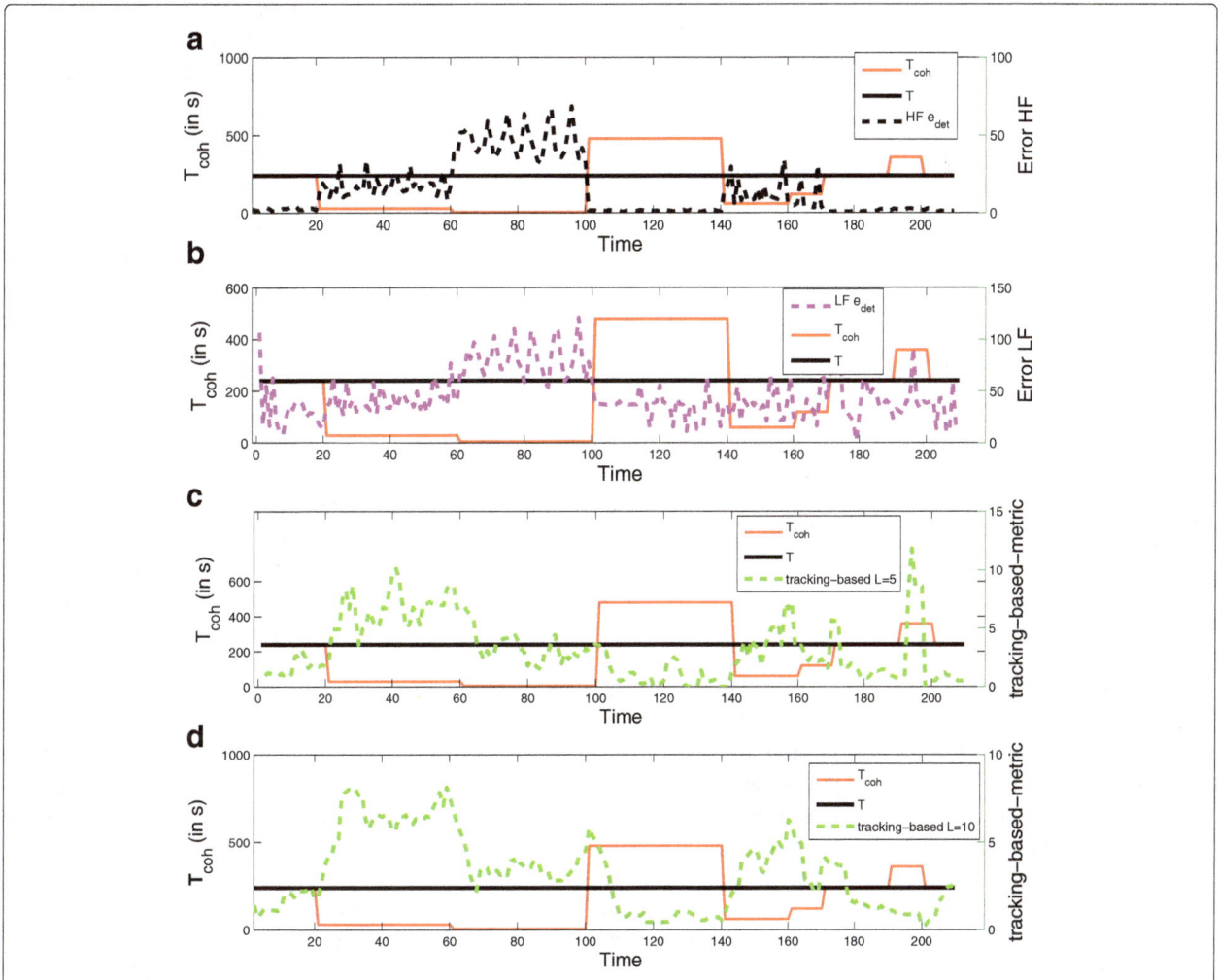

Fig. 8 a HF and **b** LF components of the model-based error and the minimum average motion per frame for (**c**) L=5 and (**d**) L=10 cases. The system is in open-loop control, i.e., the error metrics are not being compensated for which can be noted by the flat T line

to be much higher when the objects get blurred in the *undersensing* case.

$$e_m = \frac{1}{L} \min_{n_1,\ldots,n_L} D(n_1,\ldots,n_L) \qquad (10)$$

6 Feedback algorithm for rate adjustment

After computing the error metrics e_r and e_m, a control algorithm is used to determine if the current sensing rate needs to be changed. This information is fed back to the FC which makes any necessary changes to the sensing rate, thereby establishing a closed-loop control. The objective of control is to minimize the reconstruction error-based metric e_r at the same time ensuring that the targets move significantly from frame to frame as indicated by the tracking error metric e_r. This ensures that the system is in a sweet spot between *undersensing* and *oversensing*.

In this paper, we propose a dual threshold feedback scheme to keep the system in this "optimal" state. We define two parameters: a lower threshold th_m for the metric e_m and an upper threshold th_r for e_r, respectively. In the proposed scheme, the value of these two parameters is tuned adaptively. The control algorithm itself consists of two modes: an "adjust" mode and a "calibrate" mode. The system's current mode is decided based on which of the four system states (A–D) it currently is in. These system states are shown in Fig. 9.

In the adjust mode (states A and D in Table 1), the collection interval T is incremented or decremented by a scalar update ($\kappa > 1$) depending on the value of the metrics. In state A, the condition $e_m \le th_m$ indicates *oversensing* and hence the algorithm would decide to increase T, thereby decreasing sensing rate. In state D, the condition $e_r \ge th_r$ indicates that the system is *undersensing* which would result in the control decision to decrease T

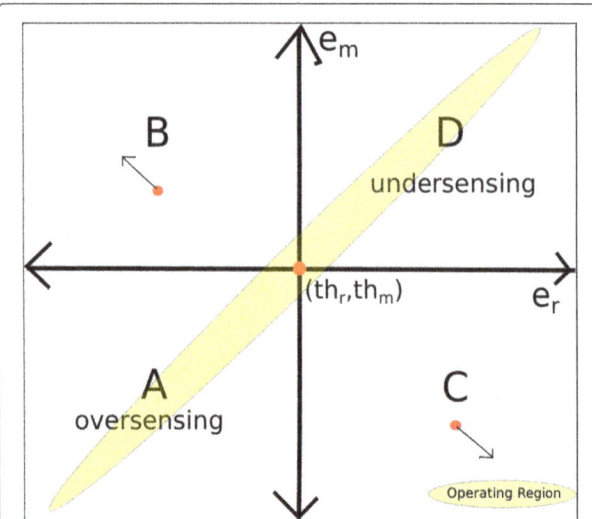

Fig. 9 Four states of the control algorithm as determined by the error metrics e_r and e_m based on the thresholds th_r and th_m. The ideal operating zone is shown by the *shaded region*

thereby increasing the sensing rate. The mutually opposing nature of these two conditions gives rise to a "stable" or buffer region for T (see Table 1) where it is not updated. In this mode, we only adapt the width of the buffer region by adjusting the threshold values such that for the current value of T, the control is just within the stable zone. Note, that it might not be feasible to guess a value for th_m or th_r making this adaptive approach necessary.

In practice, the thresholds can be initialized to extreme values such that the buffer region is wide enough and the system is guaranteed to start in state B. In subsequent intervals, the thresholds are updated in the "calibration" mode to shrink this buffer region. The ideal operating zone, centered at (th_r, th_m), is shown by a shaded region in Fig. 9. If the system steps out of this operating zone, the thresholds are adjusted in a direction such that the current system state is contained within the operating zone. In the calibration mode, the thresholds are adjusted by

predefined steps α and β such that the system stays within the operating zone. In other words, the aim of this adaptive control is to obtain the tightest stable region of control for T.

To provide an intuition of how the scheme works, consider the graphical illustration of the proposed feedback algorithm in Fig. 9. At any time instant, the system state can be represented as a point on the $e_r e_m$ plane. The relative position of this point with respect to the thresholds th_r and th_m is then used to decide the next course of action. In general, the algorithm tries to maintain a narrow buffer region of control by keeping the operating point (e_r, e_m) close to (th_r, th_m). This can be achieved by either updating the sensing time period T (states A and D) or by adapting the thresholds (states B and C).

By observing how the states of the algorithm transition, we present an argument to show that the proposed algorithm is bounded-input bounded-output (BIBO) stable. This means that for a finite coherence time T_{coh}, the controlled variable viz. the sensing time period T is always bounded. This can be shown by considering each of the four states of operation A, B, C, and D of the proposed algorithm above.

Suppose, we are currently in state A (oversensing) and the feedback algorithm responds by increasing T. This leads to an increase in both e_m and e_r since fewer samples are sensed per unit time. This, in turn, prevents the system from staying in state A and is most likely to cause a transition to states B or C. Similarly, if the system is in state D (undersensing), the algorithm responds by decreasing T. This leads to the net effect that both e_r and e_m decrease, taking the system out of state D. When the system is in states B or C, the thresholds are varied so that operating point is maintained close to the adapted thresholds. This ensures that the sensing time period T and in turn the error metrics e_r and e_m cannot grow unbounded for a given finite T_{coh}. As an example, note the state transitions in Fig. 10 which shows the internal states of the algorithm for a simulated control scenario.

Table 1 Control feedback rules ("adjust" mode: A, D ; "calibrate" mode: B, C)

	$e_m < th_m$		$e_r > th_r$	
	Oversensing	**Stable region**	**Undersensing**	
State	**A**	**B**	**C**	**D**
Condition	$e_m < th_m$ $e_r < th_r$	$e_m > th_m$ $e_r < th_r$	$e_m < th_m$ $e_r > th_r$	$e_r > th_r$ $e_m > th_m$
Decision	$T \uparrow$	*Narrow*	*Widen*	$T \downarrow$
Update	$T = \kappa T$	$th_r = th_r - \alpha$ $th_m = th_m + \beta$	$th_r = th_r + \alpha$ $th_m = th_m - \beta$	$T = T/\kappa$

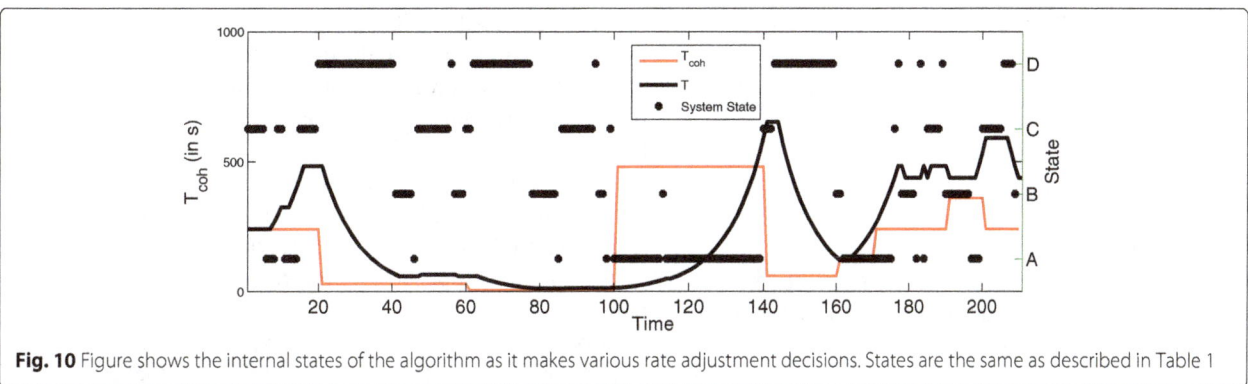

Fig. 10 Figure shows the internal states of the algorithm as it makes various rate adjustment decisions. States are the same as described in Table 1

7 System operation

We presented an example of the control system in an open-loop operation earlier in Fig. 8 which showed the value of metrics e_r and e_m for a particular T_{coh} profile. Figure 8c, d suggests that a larger L might be related to a slower response time of e_m as can be seen around $t = 80$, 100 when the field's T_{coh} changes. The error metrics based on model fitting (Fig. 8a, b) are computed from the current frame and thus respond instantaneously to the changes in field coherence. It is also worth noting here that the HF errors (spatial high-frequency component) typically correspond to the reconstruction error in the *undersensing* case, while the LF error serves as a sanity check for detection.

In this section, we present an example of closed-loop control by using the feedback mechanism (Table 1) discussed earlier to compensate for the error metrics as shown in Fig. 11. Note that the collection time interval T now attempts to trace the underlying hidden parameter T_{coh} as the system tries to keep both e_r and e_m within bounds. Figure 10 provides a better insight into the system by showing the feedback algorithm's underlying internal state. Recall that states A and D correspond to oversensing

and undersensing respectively while in states B and C, the thresholds are adapted to achieve a tight buffer region of stability.

8 Simulation results

We run simulations with $M = 4$ targets and different velocity and decay parameters, using the proposed feedback algorithm for sensing rate adjustment. We evaluate the performance of the algorithm using two different criteria. First, the average mean squared error (MSE) between the estimated sampling duration T (blue line in Fig. 11) and the coherence time ground truth T_{coh} (red line in Fig. 11) is computed as a measure of the algorithm's ability to track and adapt to changes in coherence time. Secondly, we calculate the F-score measure for target detection in each frame. This is a reasonable criterion for our task since the end goal of such an application would be target detection and tracking. To calculate the F-score of target detection in a frame, we check how many of the localized targets correspond to actual target locations. F-score weights both the precision and recall of target detection as follows

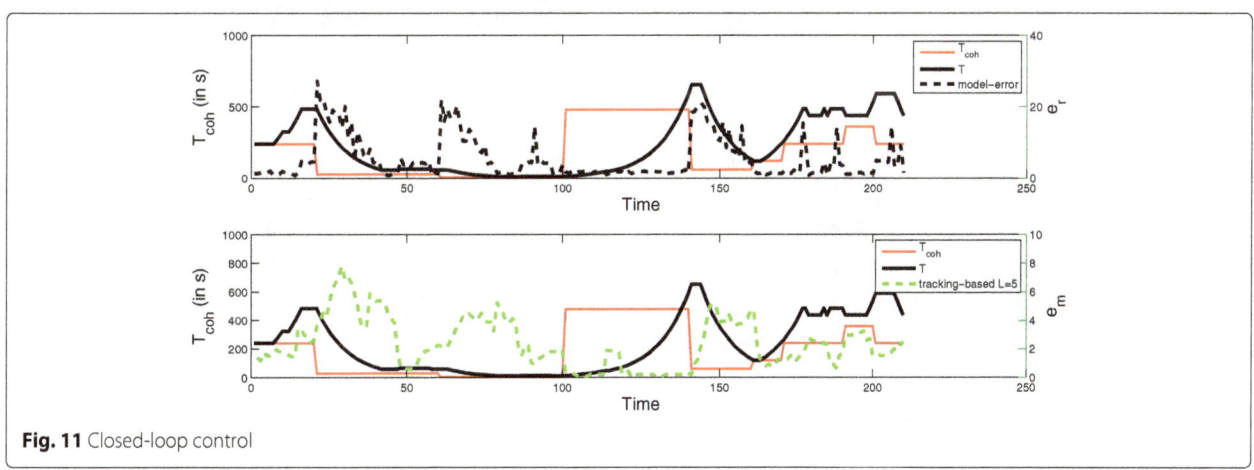

Fig. 11 Closed-loop control

$$\text{Precision} = \frac{\text{\# targets that were correctly localized}}{\text{\# targets that were localized}}$$

$$\text{Recall} = \frac{\text{\# targets that were correctly localized}}{\text{\# targets actually present}}$$

$$F = \frac{2.\text{Precision}.\text{Recall}}{\text{Precision} + \text{Recall}} \qquad (11)$$

8.1 Baseline algorithm

We compare the results against a baseline method based on correlation of the currently acquired field G^{sa} with the field acquired in the previous cycle as suggested in [15]. Control decisions are taken depending on two *fixed* high (δ_h) and low (δ_l) thresholds on the correlation value. T is incremented if correlation exceeds δ_h and decremented if correlation falls below δ_l. T is not updated if correlation lies between these two thresholds. The sampling duration T is changed using a scaling parameter κ as earlier.

Figure 12 shows the results of our simulation for the baseline and proposed methods. The results, averaged for each run, are shown for different parameter settings of velocity and decay p as defined in Eq. (3). Remember that a high F-score indicates accurate target detection while a low MSE denotes accurate tracking of the underlying coherence time (T_{coh}).

Figure 12a shows the results for the baseline method using correlation as a feedback metric. From the MSE and F-score plots, it is evident that the baseline algorithm performance drops with increase in decay parameter p. This is expected since the targets with a lesser spread do not affect the correlation metric significantly. In contrast,

for our proposed feedback algorithm, the performance improves with increase in the decay parameter p since targets with lesser spread are easier to localize and hence field reconstruction is more accurate (Fig. 12b). A similar trend can be seen for average MSE which is related inversely to the target detection F-score. In other words, this might indicate that poor tracking of the underlying T_{coh} is related to poor target detection, thereby making a case for the proposed sensing rate adjustment in this paper.

Since velocity for each target is assigned uniformly at random in the simulation, we control the velocity range that can be assigned to each target as a parameter instead. We note that increase in velocity has a positive influence on the performance of both the algorithms. This can be easily explained by considering that an increase in velocity of the targets makes it easier to discriminate between the oversensing, adequate sensing, and undersensing states. This holds for both the algorithms, and hence, we notice a general increase in F-score and decrease in MSE with increase in velocity. Finally, we also note that our proposed detection-based feedback mechanism outperforms the baseline correlation-based method in terms of the accuracy of target detection as indicated by the higher F-score. The average target detection F-score obtained for the baseline algorithm was 0.32 while our proposed method achieved an average F-score of 0.45.

8.2 Simulations with measurement noise

Next, we test the noise robustness of our proposed method by adding synthetic sensor or measurement noise

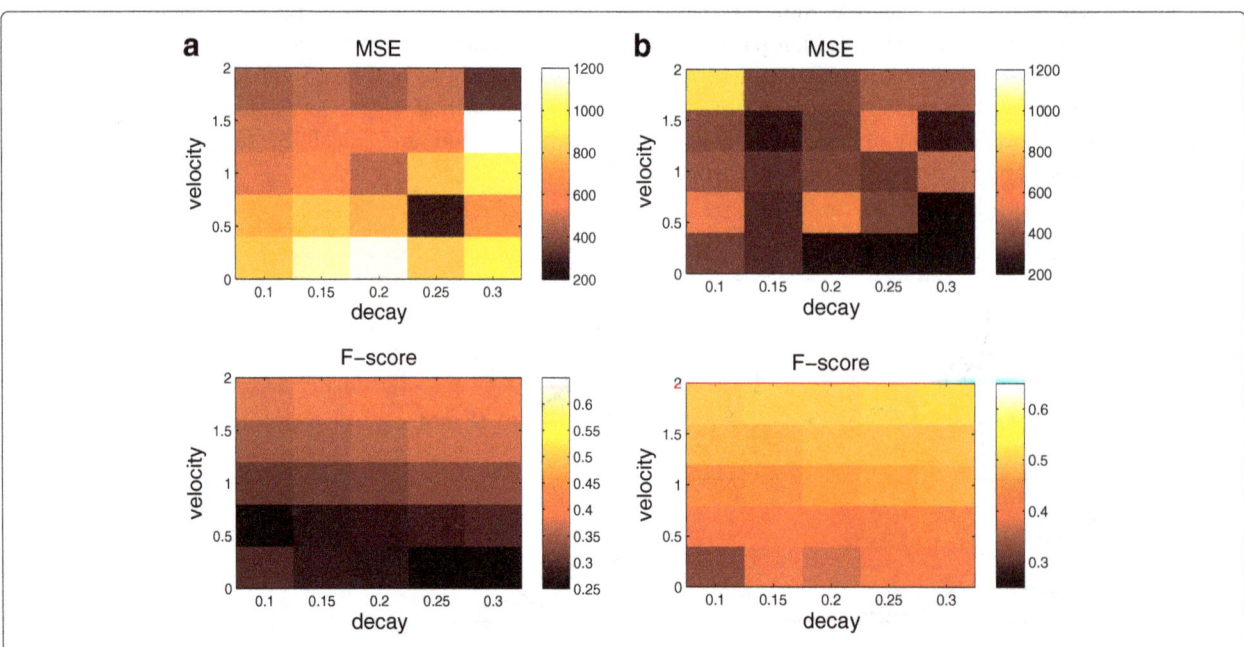

Fig. 12 The values for averaged MSE and F-score evaluation criteria for (**a**) correlation baseline mechanism and the (**b**) proposed feedback algorithm at different velocity and decay parameter settings

to the field. We simulate measurement noise by adding zero mean white Gaussian noise at different amplitudes (σ_{noise}) while keeping the original field's intensity the same. Results are averaged on full runs for different noise level, and the F-score for target detection accuracy is shown in Fig. 13. Other simulation factors such as velocity-scale and decay were held constant at 1 and 0.3 respectively for the purposes of this simulation.

The results show a clear degradation in F-score for both the proposed detection-based and baseline correlation-based methods, as the SNR decreases. However, we note that for the baseline algorithm, the F-score initially increases with decrease in SNR hinting at suboptimality in processing. On further investigation, we found this peculiar trait to be an artifact of the control algorithm and the choice of metric used in the baseline scheme. As noise is added to the field value of the correlation metric drops, which leads the FC to confuse the current state as undersensing. This inference leads to a monotonic decrease in sampling duration T, temporarily increasing performance at the cost of energy efficiency. The target detection F-score peaks around $\sigma_{\text{noise}} = 0.05$ beyond which the accuracy of target-detection algorithm is significantly affected. On the other hand, this artifact does not affect the proposed method significantly, since the metric used for our control algorithm is based directly on target detection. As a result, we note that our proposed method exhibits a more graceful degradation in performance with increase in noise without compromising on energy efficiency.

9 Conclusions

In this work, we present a rate adjustment scheme for random access compress sensing (RACS) to monitor and compensate for the rate of change in time-varying fields. Adjustment of sensing rate for RACS is motivated by the trade-off between energy efficiency and the accuracy of tracking targets of interest. Although direct estimation of the coherence time might seem to be the best approach for sensing a time varying field, we note that it is not directly related to our end goal of target detection in the field. We also observe that the reconstruction error in RACS has a complex relation to the current coherence time of the field and factors like the position of targets, their velocity, etc. Thus, by making the algorithm depend on the detection and localization of targets, we ensure that the rate adjustment is tied in to our actual objective.

In this paper, we propose two unsupervised metrics to inform the rate adjustment scheme. A model-based field reconstruction is done after target localization for each acquired frame, and the model fit error e_r is used as a metric to detect cases when the FC is undersensing. To account for oversensing in operation, we define a measure for the motion of detected targets in the past L frames. This motion based metric e_m indicates any redundancy in sensing and can be used to decrease the collection time interval T if needed.

We proposed a dual-threshold-based feedback mechanism using these error metrics for rate sensing adjustments. The technique assumes minimum prior knowledge and adapts the threshold in an online fashion using reasonable assumptions about the field. We also show that the proposed control mechanism is BIBO stable. In addition, we compare our results against a baseline algorithm that uses temporal correlation of the acquired fields and show that the proposed rate adjustment mechanism performs better on an average in terms of target detection accuracy even in the presence of noise.

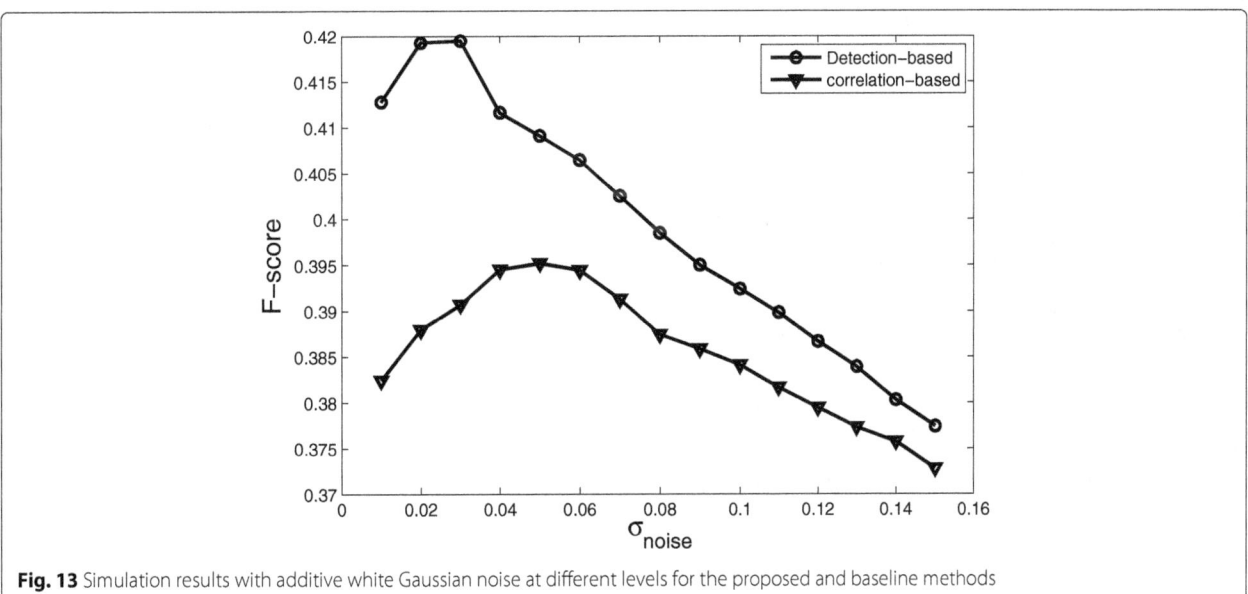

Fig. 13 Simulation results with additive white Gaussian noise at different levels for the proposed and baseline methods

Endnote

[1] We use the term *stationary* in this paper to refer to fields that are temporally static during the sensing duration.

Competing interests

The authors declare that they have no competing interests.

Acknowledgements

This work was supported by ONR and NSF.

Author details

[1] Signal Analysis and Interpretation Laboratory (SAIL), Electrical Engineering, University of Southern California, Los Angeles, CA, USA. [2] Department of Electrical and Computer Engineering, Northeastern University, Boston, MA, USA.

References

1. EJ Candes, J Romberg, T Tao, Stable signal recovery from incomplete and inaccurate measurements. Comm. Pure Appl. Math. **59**, 1207–1223 (2005)
2. EJ Candes, J Romberg, T Tao, Robust uncertainty principles: exact signal reconstruction from highly incomplete frequency information. IEEE Trans. Inform. Theory. **52**, 489–509 (2006)
3. E Candes, J Romberg, Sparsity and incoherence in compressive sampling. Inverse Problems. **23**, 969–985 (2006)
4. WU Bajwa, J Haupt, AM Sayeed, R Nowak, in *5th Int. Conf. Information Processing in Sensor Networks (IPSN'06)*. Compressive wireless sensing, (2006), pp. 134–142
5. WU Bajwa, J Haupt, AM Sayeed, R Nowak, Joint source-channel communication for distributed estimation in sensor networks. IEEE Trans. Inform. Theory. **53**(10), 3629–3653 (2007)
6. CT Chou, R Rana, W Hu, in *IEEE 34th Conference on Local Computer Networks (LCN)*. Energy efficient information collection in wireless sensor networks using adaptive compressive sensing, (2009), pp. 443–450
7. C Luo, F Wu, J Sun, CW Chen, in *Proceedings of the 15th Annual International Conference on Mobile Computing and Networking*. MobiCom. Compressive data gathering for large-scale wireless sensor networks, (2009), pp. 145–156
8. J Meng, H Li, Z Han, in *43rd Annual Conference on Information Sciences and Systems (CISS)*. Sparse event detection in wireless sensor networks using compressive sensing, (2009), pp. 181–185
9. R Masiero, G Quer, D Munaretto, M Rossi, J Widmer, M Zorzi, in *Proceedings of the 28th IEEE Conference on Global Telecommunications*. GLOBECOM'09. Data acquisition through joint compressive sensing and principal component analysis, (2009), pp. 1271–1276
10. R Masiero, G Quer, M Rossi, M Zorzi, in *ICUMT'09*. A Bayesian analysis of compressive sensing data recovery in wireless sensor networks, (2009), pp. 1–6
11. S Lee, S Pattem, M Sathiamoorthy, B Krishnamachari, A Ortega, in *GeoSensor Networks*. Lecture Notes in Computer Science. Spatially-localized compressed sensing and routing in multi-hop sensor networks, vol. 5659 (Springer, 2009), pp. 11–20
12. D Motamedvaziri, V Saligrama, D Castanon, in *Communication, Control, and Computing (Allerton), 2010 48th Annual Allerton Conference On*. Decentralized compressive sensing, (2010), pp. 607–614
13. F Fazel, M Fazel, M Stojanovic, Random access compressed sensing for energy-efficient underwater sensor networks. IEEE J. Selected Areas Commun. (JSAC). **29**(8), 1660–1670 (2011)
14. F Fazel, M Fazel, M Stojanovic, Random access compressed sensing over fading and noisy communication channels. Wireless Commun. IEEE Trans. **12**(5), 2114–2125 (2013)
15. F Fazel, M Fazel, M Stojanovic, in *Information Theory and Applications Workshop (ITA), 2012*. Random access sensor networks: field reconstruction from incomplete data (IEEE, 2012), pp. 300–305
16. S Reed, Y Petillot, J Bell, An automatic approach to the detection and extraction of mine features in sidescan sonar. Oceanic Eng. IEEE J. **28**(1), 90–105 (2003)
17. N Kumar, QF Tan, SS Narayanan, in *Acoustics, Speech and Signal Processing (ICASSP), 2012 IEEE International Conference On*. Object classification in sidescan sonar images with sparse representation techniques (IEEE, 2012), pp. 1333–1336
18. K Kerse, F Fazel, M Stojanovic, in *Signals, Systems and Computers, 2013 Asilomar Conference On*. Target localization and tracking in a random access sensor network (IEEE, 2013), pp. 103–107
19. JA Tropp, SJ Wright, Computational methods for sparse solution of linear inverse problems. Proc. IEEE, Special Issue, Appl. Sparse Representation and Compressive Sensing. **98**(6), 948–958 (2010)
20. D Li, YH Hu, Energy-based collaborative source localization using acoustic microsensor array. EURASIP J. Appl. Signal Process. **2003**, 321–337 (2003)
21. X Sheng, Y-H Hu, Maximum likelihood multiple-source localization using acoustic energy measurements with wireless sensor networks. Signal Process. IEEE Trans. **53**(1), 44–53 (2005)
22. W Meng, W Xiao, L Xie, An efficient em algorithm for energy-based multisource localization in wireless sensor networks. Instrum. Meas. IEEE Trans. **60**(3), 1017–1027 (2011)
23. Y Cheng, Mean shift, mode seeking, and clustering. Pattern Anal. Mach. Intell. IEEE Trans. **17**(8), 790–799 (1995)
24. PW Holland, RE Welsch, Robust regression using iteratively reweighted least-squares. Commun. Statistics-Theory Methods. **6**(9), 813–827 (1977)

9

An incremental learning algorithm for the hybrid RBF-BP network classifier

Hui Wen, Weixin Xie, Jihong Pei[*] and Lixin Guan

Abstract

This paper presents an incremental learning algorithm for the hybrid RBF-BP (ILRBF-BP) network classifier. A potential function is introduced to the training sample space in space mapping stage, and an incremental learning method for the construction of RBF hidden neurons is proposed. The proposed method can incrementally generate RBF hidden neurons and effectively estimate the center and number of RBF hidden neurons by determining the density of different regions in the training sample space. A hybrid RBF-BP network architecture is designed to train the output weights. The output of the original RBF hidden layer is processed and connected with a multilayer perceptron (MLP) network; then, a back propagation (BP) algorithm is used to update the MLP weights. The RBF hidden neurons are used for nonlinear kernel mapping and the BP network is then used for nonlinear classification, which improves classification performance further. The ILRBF-BP algorithm is compared with other algorithms in artificial data sets and UCI data sets, and the experiments demonstrate the superiority of the proposed algorithm.

Keywords: Radial basis function (RBF), Back propagation (BP), Incremental learning, Hybrid, Neural network

1 Introduction

In the field of pattern recognition and data mining, various methods and models are proposed to solve different problems. Existing methods can be divided into two levels including data level and algorithmic level. The data levels are mainly concerned with various sampling techniques [1]. The algorithmic level tried to apply or improve varieties of existing traditional learning algorithms such as fuzzy clustering [2], Markovian jumping system [3–5], k-nearest neighbors [6], and neural network, where single-layer feed-forward networks (SLFNs) have been intensively studied in the past several decades and applied to solve various problems in different fields, such as image recognition [7], signal processing [8], disease prediction [9], and industrial fault diagnosis [10]; in particular, radial basis function (RBF) neural networks offer an effective mechanism for nonlinear mapping and classification. In a typical RBF network, the number of hidden neurons is assigned a priori [11, 12], which leads to poor adaptability for different sample sets. Several sequential learning algorithms have been

proposed to determine proper sizes of RBF network architectures. A resource allocation network (RAN) for constructing the RBF network is proposed in [13], which uses the novelty of incoming data as the learning strategy. A RAN algorithm based on an extended Kalman filter (RANEKF) is proposed in [14], which uses the extended Kalman filter algorithm instead of the least mean squares (LMS) algorithm. In [15], a minimal resource allocation network (MRAN) is proposed, which is allowed for the deletion of the previous center. The deletion strategy is based on the overall contribution of each hidden unit to the network output. A sequential learning algorithm for growing and pruning the RBF (GAP-RBF) is proposed in [16, 17]; this algorithm uses the significance of neurons as the learning strategy. In [18], a Gaussian mixture model (GMM) to approximate the generalized growing and pruning evaluation formula is proposed; the GMM can be used for problems with a high-dimensional probability density distribution. In [19], an error correction (ErrCor) algorithm is used for function approximation; this algorithm can achieve a desired error rate with fewer RBF units. Other methods have also been established to identify

* Correspondence: jhpei@szu.edu.cn
ATR Key Lab of National Defense, Shenzhen University, 518060 Shenzhen, China

a proper architecture while maintaining a desired accuracy [20–22].

Support vector machines (SVMs), which are maximal margin classifiers, can also be used to train SLFNs. RBFs and SVMs differ in that at the output layer, a SVM employs convex optimization to find an optimal linear classifier, whereas the output weights of RBF network are typically estimated by a linear least squares algorithm, such as the LMS or recursive least squares (RLS) algorithm. Regarding other training SLFNs, extreme learning machines (ELMs) are proposed in [23]; ELMs choose random hidden neuron parameters and calculate the output weights with the least squares algorithm. This method can achieve a fast training speed. Subsequently, an online sequence extreme learning machine (OS-ELM) algorithm that can learn one by one and data blocks of the input samples is proposed in [24]. In ELMs, the number of hidden nodes is assigned a priori, and many nonoptimal nodes may exist; thus, in [25–28], several types of growing and pruning techniques based on ELMs are proposed to effectively estimate the number of hidden neurons.

All of the algorithms for training SLFNs consist of two stages: (1) suitable feature mapping and (2) output weight adjustment. To train SLFNs efficiently, in this paper, a potential function is introduced in the feature mapping stage to train the sample space, and an incremental learning method of constructing RBF hidden neurons is proposed. Note that although the sequence learning RBF algorithms can also generate RBF hidden neurons automatically, because of the lack of global information in the sample space, the adaptability of complex sample space may be poor. In contrast to GAP-RBF, the proposed method does not require an assumption that the input samples obey a unified distribution. Furthermore, it does not need to fit the input sample distribution, such as the algorithm proposed in [18]. The proposed method utilizes global information about each class of training sample space and can generate RBF hidden neurons incrementally to adapt the sample space. By using a potential function to measure the density in each class of training sample space, the corresponding RBF hidden neurons that cover different sample areas can be established. The center of the Gaussian kernel function can be determined by learning the density of different regions in the training sample space. Once the width is given, a hidden neuron is generated and introduced into the RBF network, and a mechanism for eliminating the potentials of original samples is presented. This mechanism is ready for the next learning step, and thus, the RBF centers and number of hidden neurons can be effectively estimated. In this way,

a suitable network size for RBF hidden layer that matches the complexity of the sample space can be built up. Thus, the proposed method solves the problem of dimension change from sample space mapping to feature space, and it reduces the restrictions on the sample sets, which is adaptable to more complex sample sets.

In this paper, a hybrid RBF-BP network architecture is designed for the output weight adjustment stage to further improve the generalization and classification performance. The output of the original RBF hidden layer is processed and connected with a new hidden layer, which means that the output of the original RBF hidden layer, the new hidden layer, and the output layer consists of a multilayer perceptrons (MLPs), and the output of the original RBF hidden layer is the input of the MLPs. Once the network architecture is established, a back propagation (BP) algorithm is used to update the weights of the MLPs. In the hybrid RBF-BP network, the RBF hidden neurons are used for nonlinear kernel mapping, the complexity of sample space is mapped onto the dimension of the BP network input layer, and the BP network is then used for nonlinear classification. The nonlinear kernel mapping can improve the separability of sample spaces, and a nonlinear BP classifier can then supply a better classification surface. In this manner, the improved network architecture combines the local response characteristics of the RBF network with the global response characteristics of the BP network, which simplifies the neuron number selection in the BP network hidden layer while further reducing the dependence on space mapping in the RBF hidden layer.

The incremental learning algorithm for the hybrid RBF-BP (ILRBF-BP), which is a batch learning algorithm, is proposed by combining the proposed incremental learning algorithm with the hybrid RBF-BP network architecture. In this paper, the performance of the ILRBF-BP algorithm is compared with other well-known learning algorithms, such as back propagation based on stochastic gradient descent (SGBP) [29], the RBF algorithm based on k-means clustering (KM-RBF) [12], GAP-RBF, SVM, and an ELM, on artificial data sets. To measure the unique features of the proposed method, the k-means clustering learning algorithm based on the hybrid RBF-BP network (KMRBF-BP) is also compared with ILRBF-BP on artificial data sets. Because SGBP and KM-RBF are not suitable for considering more complex problems, for multi-class data sets, in addition to batch learning algorithms, such as SVM and ELM, other well-known sequential algorithms, such as MRAN, GAP-RBFN, and OS-ELM, are also compared with the ILRBF-BP algorithm. The results indicate that the ILRBF-BP algorithm can provide a higher classification accuracy with comparable complexity.

The remainder of this paper is organized as follows. Section 2 describes the principal ideas of the ILRBF-BP, followed by a summary of the algorithm. Section 3 presents the experimental results and performance comparisons with other existing batch and sequential algorithms. Section 4 provides the conclusions of this study.

2 Main concepts of the ILRBF-BP algorithm

In this section, the main concepts of the ILRBF-BP algorithm are described. First, we provide the problem definition of the basic RBF network and then present the incremental learning algorithm for constructing RBF hidden neurons. Then, a hybrid RBF-BP network architecture is designed, and the ILRBF-BP algorithm is summarized. Finally, a method of adjusting the output saturation for multi-class classification problem is proposed.

2.1 Problem definition

For a RBF network, the output can be given by

$$F(\mathbf{x}) = \sum_{k=1}^{K} \omega_k \phi_k(\mathbf{x}) \tag{1}$$

where

$$\phi_k(\mathbf{x}) = \exp\left(-\frac{1}{2\sigma_k^2}\|\mathbf{x}-\mu_k\|^2\right) \tag{2}$$

where K is the number of RBF hidden neurons; $\phi_k(\mathbf{x})$ is the response of the kth hidden node for an input vector $\mathbf{x}$, where $\mathbf{x} \in R^t$; ω_k is its connecting weight to the output node, which determines the classification surface; and μ_k and σ_k are the center and width of the kth hidden node, respectively, where $k = 1, 2, \ldots K$.

A RBF network can localize the input sample space, which maps input samples to the interior of the hypercube, and the localized area is near a vertex. The dimension of the hypercube is the number of RBF hidden neurons. Thus, when going through the RBF network, an input vector $\mathbf{x} \in R^t$ can be denoted as $f: R^t \rightarrow (0, 1]^K$. Figure 1 shows the results of mapping input samples going through the RBF hidden neurons, where the number of RBF hidden neurons is set as $K = 3$. In Fig. 1, we assume that every input sample vector is near the center of a RBF hidden neuron and that there is no overlap area covered by different RBF hidden neurons.

Figure 1 illustrates that in a RBF network, to achieve good training algorithms, an effective method

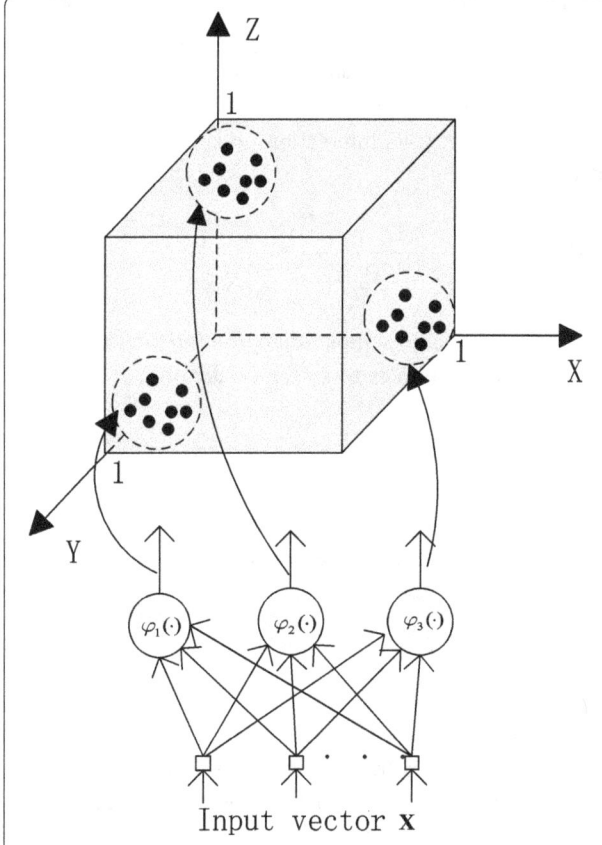

Fig. 1 Result of mapping input samples going through RBF hidden neurons

of mapping the input sample space should be established, which means completing the estimation of the parameter set $\{K, \mu_k, \sigma_k\}_{k=1}^{K}$. Then, an effective classification surface is needed, which depends on output weight adjustment.

2.2 Incremental learning algorithm for constructing RBF hidden neurons

In the fields of data mining and pattern recognition, potential functions can be used for density clustering and image segmentation (IS) [30]. Several methods of constructing potential function are proposed in [31]; here, we choose the potential function

$$\gamma(\mathbf{x_1}, \mathbf{x_2}) = \frac{1}{1 + T \cdot d^2(\mathbf{x_1}, \mathbf{x_2})} \tag{3}$$

where $\gamma(\mathbf{x_1}, \mathbf{x_2})$ represents the interaction potential of two points $\mathbf{x_1}, \mathbf{x_2}$ in the input sample space, $d(\mathbf{x_1}, \mathbf{x_2})$ represents the distance measure, and T is a constant, which can be regarded as the distance weighting factor.

Given a training sample set S, where a specific label $\mathbf{y_i}, \mathbf{y_i} \in \{\mathbf{y_i}; i = 1, 2, \ldots h\}$ is attached to each sample vector $\mathbf{x}$

in S, h is the number of pattern class. Let S_i denote the set of feature vectors that are labeled $\mathbf{y_i}$, $S_i = \{\mathbf{x_1^i}, \mathbf{x_2^i}, ..., \mathbf{x_{N_i}^i}\}$, where N_i is the number of training samples in the ith pattern class. Thus, $S = \cup_{i=1}^{h} S_i$, $S_i \cap S_j = \varnothing, \forall\, i \neq j$. For a pair of samples $(\mathbf{x_u^i}, \mathbf{x_v^i})$ in S_i, its interaction potential can be denoted as

$$\gamma(\mathbf{x_u^i}, \mathbf{x_v^i}) = \frac{1}{1 + T \cdot d^2(\mathbf{x_u^i}, \mathbf{x_v^i})} \tag{4}$$

Let $\mathbf{x_v^i}$ be the baseline sample; then, the interaction potential of all other samples to $\mathbf{x_v^i}$ can be denoted as

$$\rho(\mathbf{x_v^i}) = \sum_{u=1, u \neq v}^{N_i} \gamma(\mathbf{x_u^i}, \mathbf{x_v^i}) \tag{5}$$

Therefore, the potentials of each sample in S_i is given by

$$\rho^i = \left\{\rho(\mathbf{x_1^i}), \rho(\mathbf{x_2^i}), ..., \rho(\mathbf{x_{N_i}^i})\right\} \tag{6}$$

The potentials can be used to measure the density of different regions in the pattern class. Potentials are relatively large in the dense region, whereas they are relatively small in the sparse region. Once the potentials of each sample in S_i are given, the sample with the maximum potential can be selected, where it is assumed the sample is $\mathbf{x_p^i}$, that is,

$$\rho\left(\mathbf{x_p^i}\right) = \max\left\{\rho(\mathbf{x_1^i}), \rho(\mathbf{x_2^i}), ..., \rho\left(\mathbf{x_{N_i}^i}\right)\right\} \tag{7}$$

In a RBF network, the activation response of hidden neurons has local characteristics. The sample space is divided into different subspaces by establishing different Gaussian kernel functions. To generate valid Gaussian kernel functions, we find the most densely region in the sample space and then establish a Gaussian kernel to cover the region. For that purpose, the sample with the maximum potential is chosen as the center of Gauss kernel function, which is given below.

$$\mu_k = \mathbf{x_p^i} \tag{8}$$

where k refers to the number of RBF hidden neurons generated. To simplify the calculation, the width is fixed and selected by cross validation.

When a hidden neuron is established, it is necessary to eliminate the potentials of the region to find the next center in the remaining samples. This process can be updated by

$$\rho_{new}\left(\mathbf{x_v^i}\right) = \rho\left(\mathbf{x_v^i}\right) - \rho\left(\mathbf{x_p^i}\right) \cdot \exp\left(-\frac{1}{2\sigma_k^2} ||\mathbf{x_v^i} - \mathbf{x_p^i}||^2\right),$$

$$v = 1, 2, ...N_i$$

$$\tag{9}$$

where $\mathbf{x_p^i}$ is the center of the current hidden neuron. For the potential value update process, Eq.(9) shows when a sample $\mathbf{x_v^i}$ is close to the center $\mathbf{x_p^i}$, the potential value of $\mathbf{x_v^i}$ is attenuated fast, whereas when a sample $\mathbf{x_v^i}$ is far away from the center, the potential value of $\mathbf{x_v^i}$ is attenuated slowly. When meeting the inequality

$$\max\left\{\rho_{new}\left(\mathbf{x_1^i}\right), \rho_{new}\left(\mathbf{x_2^i}\right), ..., \rho_{new}\left(\mathbf{x_{N_i}^i}\right)\right\} > \delta \tag{10}$$

a new hidden neuron is introduced into the RBF network and is ready to search for the next center; otherwise, the algorithm of constructing RBF hidden neurons in the current pattern class is over, where δ is a threshold.

The above process is called the incremental learning algorithm of constructing RBF hidden neurons. Figure 2 shows a schematic diagram of generating RBF hidden neurons incrementally, where the serial numbers in the training sample space represent the regions covered by different RBF hidden neurons. These covered regions transition from dense to sparse. The incremental learning algorithm of constructing RBF hidden neurons is summarized in Algorithm 1.

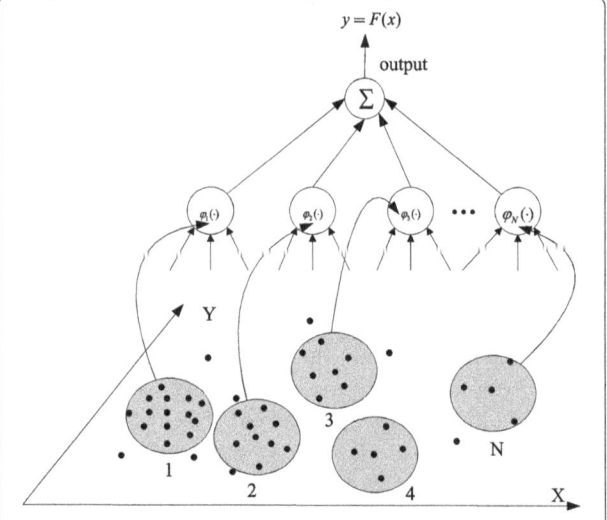

Fig. 2 Illustrative diagram of incrementally generating RBF hidden neurons

---

Algorithm 1. Incremental learning algorithm of constructing RBF hidden neurons

Initialize the number of RBF hidden neurons $k = 0$. Given the width σ and the distance weighting factor T. Given training samples $S = \bigcup_{i=1}^{h} S_i$, $S_i = \{\mathbf{x}_1^i, \mathbf{x}_2^i, ..., \mathbf{x}_{N_i}^i\}$. For each pattern class S_i, do

1. Compute the potential value of each sample according to Eq.(5);

2. Determine the maximum potential value of each sample according to Eqs.(6) and (7);

3. The number of RBF hidden neurons k counts plus 1. Use Eq.(8) to allocate a new hidden neuron center;

4. Eliminate the sample potential value of the region according to Eq.(9);

5. Set iteration termination condition

   If $\max\{\rho_{new}(\mathbf{x}_1^i), \rho_{new}(\mathbf{x}_2^i), ..., \rho_{new}(\mathbf{x}_{N_i}^i)\} > \delta$

       Go to Step 2.

   Else

       The process of learning current pattern class is over. Go to learn other pattern classes.

   EndIf

2.3 Hybrid RBF-BP network architecture

As noted above, in a typical RBF network, the output weights are typically estimated by a linear least squares algorithm, such as the LMS or RLS algorithm. In this section, we transform the linear least squares algorithm into a nonlinear algorithm. When classifying a problem, a nonlinear algorithm can supply a better classification surface to adapt the sample space. For that purpose, a hybrid RBF-BP network architecture is designed. The output of the RBF hidden neurons is processed and connected with a MLPs network, and then, the nonlinear BP algorithm is used to update the weights of the MLPs. The architecture of the hybrid RBF-BP network is shown in Fig. 3, which consists of four components:

1. The input layer, which consists of t source nodes, where t is the dimensionality of the input vector

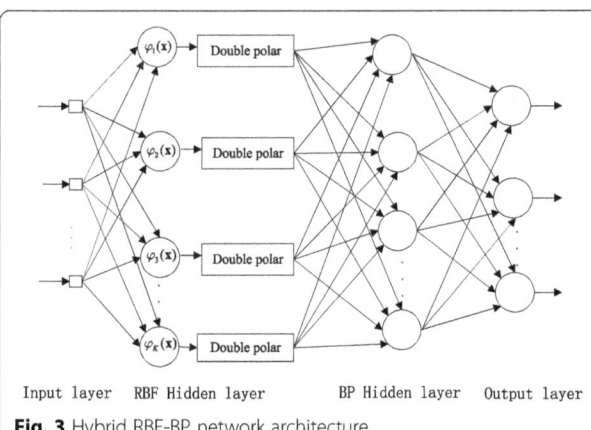

Input layer   RBF Hidden layer   BP Hidden layer   Output layer

Fig. 3 Hybrid RBF-BP network architecture

2. The RBF hidden layer, which consists of a group of Gaussian kernel functions:

$$\phi_k(\mathbf{x}) = \exp\left(-\frac{1}{2\sigma_k^2}\|\mathbf{x}-\mu_k\|^2\right), \quad k = 1, 2, ...K \quad (11)$$

where μ_k and σ_k are the center and width of the hidden neuron, respectively, and K is the number of hidden neurons.

3. The BP hidden layer, which consists of the neurons between the RBF hidden layer and output layer. The induced local field $v_j^{(l)}$ for neuron j in layer l of the BP network is

$$v_j^{(l)} = \sum_i \omega_{ji}^{(l)} y_i^{(l-1)} \quad (12)$$

where $y_i^{(l-1)}$ is the output signal of the neuron i in the previous layer $l-1$ of the BP network and $\omega_{ji}^{(l)}$ is the synaptic weight of neuron j in layer l that is fed from neuron i in layer $l-1$. Assuming the use of a sigmoid function, the output signal of neuron j in layer l is

$$y_j^{(l)} = \phi_j(v_j) = a\tanh(bv_j) \quad (13)$$

where a and b are constants.

If neuron j is in the first BP network hidden layer, i.e., $l = 1$, set

$$y_j^{(0)} = g_j(\mathbf{x}) \quad (14)$$

where $g_j(\mathbf{x})$ is the double polar output of $\phi_j(\mathbf{x})$ and can be denoted as

$$g_j(\mathbf{x}) = 2\cdot\phi_j(\mathbf{x})-1 \qquad (15)$$

4. The output layer. Set L is the depth of the BP network, note the depth of the BP network is equal to the sum of the BP network input layer, the hidden layer, and the output layer, i.e., if $l = 1$, then $L = 3$, and the output can be given as

$$o_j = y_j^{(L)} \qquad (16)$$

In Fig. 3, the double polar processing can ensure the validity of the BP network input. The hybrid RBF-BP network architecture is designed such that the RBF network has good stability, where the activation response in the RBF hidden neurons has local characteristics and maps the output value between 0 and 1. Thus, the original samples including outliers will be limited to a finite space. When the results of mapping the RBF hidden neurons are processed and used for the input of the BP network, the convergence rate of the BP algorithm can be increased and local minima can be avoided. For a BP network, the activation response in hidden neurons has global characteristics, especially those regions not fully displayed in the training set. Therefore, the hybrid RBF-BP network architecture is a reasonable model; it provides a new strategy that combines the local characteristics of the RBF network with the global characteristics of the BP network. In addition, the hybrid network simplifies the number of neurons in the BP hidden layer while further reducing the dependence on space mapping in the RBF hidden layer.

A single hidden layer MLP neural network with an input-output mapping can provide an approximate realization of any continuous mapping [32]. Combined with the above discussion, in the hybrid network, we set the number of BP network hidden layers as $l = 1$.

Combining the proposed incremental learning algorithm with the hybrid RBF-BP network architecture, the incremental learning RBF-BP (ILRBF-BP) algorithm is summarized in Algorithm 2.

Algorithm 2: The ILRBF-BP algorithm

1. Assign random initialized weights to each layer of the MLP network, initialize the BP iteration step m, set $a = 1.716$, $b = 2/3$.

2. Use the incremental learning algorithm of constructing RBF hidden neurons proposed in Algorithm 1.

3. Use Eqs.(11) and (15) to compute $g_j(\mathbf{x})$, let $g(\mathbf{x})$ be the input of the BP network, where

$$g(\mathbf{x}) = (g_1(\mathbf{x}), g_2(\mathbf{x}), ..., g_K(\mathbf{x})).$$

4. Forward compute the BP network. Use Eqs.(12)-(14) and Eq.(16). Compute the error signal

$$e_j = d_j - o_j$$

where d_j is the jth element of the desired response vector $\mathbf{d}$.

5. Backward compute the BP network. Compute the local gradients of the network, which is denoted by

$$\delta_j^{(l)} = \begin{cases} e_j^{(L)}\phi_j'\left(v_j^{(L)}\right) & \text{for neuron } j \text{ in output layer } L \\ \varphi_j'(v_j^{(l)})\sum_k \delta_k^{(l+1)}\omega_{kj}^{(l+1)} & \text{for neuron } j \text{ in MLP hidden layer } l \end{cases}$$

where $\phi_j'(\square)$ is the differentiation with respect to the argument. Adjust the synaptic weights of the network in layer l of MLP as shown below.

$$\omega_{ji}^{(l)}(m+1) = \omega_{ji}^{(l)}(m) + \alpha\left[\omega_{ji}^{(l)}(m-1)\right] + \eta\delta_j^{(l)}(m)y_i^{(l-1)}(m)$$

where α is the momentum constant and η is the learning rate.

6. Iteration. Iterate the forward and backward computations in Steps 4 and 5 by presenting new epochs of training examples to the network until the chosen stopping criterion is met.

2.4 Adjustment of the output label values

The ILRBF-BP algorithm can handle binary problems and multi-class problems. For multi-class classification problems, suppose that the observation data set is given as $\{\mathbf{x_n}, \mathbf{y_n}\}_{n=1}^{N}$, where $\mathbf{x_n} \in R^t$ is an t-dimensional observation features and $\mathbf{y_n} \in R^h$ is its coded class label. Here, h is the total number of classes, which is equal to the number of output hidden neurons. If the observation data $\mathbf{x_n}$ is assigned to the class label c, then the cth element of $\mathbf{y_n} = [y_1, ..., y_c, ... y_h]^T$ is 1 and other elements are –1, which can be denoted as follows:

$$y_j = \left\{ \begin{array}{ll} 1 & \text{if } j = c \\ -1 & \text{otherwise} \end{array} j = 1, 2, ..., h \right\} \tag{17}$$

The output tags of the ILRBF-BP classier are $\hat{\mathbf{y}}_\mathbf{n} = [\hat{y}_1, ..., \hat{y}_c, ... \hat{y}_h]^T$, where

$$\hat{y}_j = \text{sgn}(o_j), \quad j = 1, 2, ...h \tag{18}$$

According to the coding rules, only one output tag value is 1 and the other value is –1. If this condition is not met, the output tag is saturated and must be adjusted. Therefore, we set an effective way to correct the saturation problem in the learning process, which can be denoted as the pseudo code in Algorithm 3.

3 Performance evaluation of the ILRBF-BP algorithm

In this section, we evaluate the performance of the ILRBF-BP algorithm using two artificial classification problems from [33] and three classification problems from the UCI machine learning repository [34]. The artificial binary data sets, including the Double-moon and Twist problems are used to measure the unique features of ILRBF-BP and the main advantages of the results over others. Table 1 provides a description of the classifying data sets, where Double-moon, Twist, and IS are well-balanced data sets and Heart and vehicle classification (VC) are imbalanced data sets. For balanced data sets, the numbers of training samples in each class are identical. For the heart problem, the numbers of training samples in classes 1 and 2 are 33 and 40, respectively. For the VC problem, the numbers of training samples in classes 1–4 are 119, 118, 98, and 89, respectively.

The performance of ILRBF-BP is compared with other well-known batch and sequential learning algorithms, such as SGBP, KM-RBF, KMRBF-BP, SVM and ELM, MRAN, GAP-RBF, and OS-ELM on different data sets. Note that the number of SGBP, KM-RBF, KMRBF-BP, ELM, and OS-ELM hidden neurons is selected manually. When changing the number of hidden neurons several times, the one with the lowest overall testing error is selected as the suitable number of hidden neurons. For multi-class problems, the method of adjusting output saturation problems is used. All simulations in each algorithm are performed ten times and are conducted in the MATLAB 2013 environment on an Intel(R) Core(TM) i5, 3.2 GHZ CPU with 4G of RAM. The simulations for the SVM are carried out using the popular LIBSVM package in C [35].

Algorithm 3 Method of adjusting the output saturation problem

Given an observation data set $\{\mathbf{x_n}, \mathbf{y_n}\}_{n=1}^{N}$, for every input vector $\mathbf{x_n}$,

While j £ h

    If the number of $\hat{y}_j == -1$ is equal to h

        Set $\max(o_j) = 1$ and hold other output values fixed.

    EndIf

    If the number of $\hat{y}_j == 1$ is more than 1

        Set $\max(o_j) = 1$ and the other output values are -1.

    EndIf

Endwhile

Table 1 Descriptions of the classifying data sets

Data sets	No. of features	No. of classes	No. of training	No. of testing	Attribute	Sources
Double-moon	2	2	200~2000	4000	Balance	Artificial
Twist	2	2	200~2000	4000	Balance	Artificial
Heart	13	2	73	230	Imbalance	UCI
IS	19	7	210	2100	Balance	UCI
VC	18	4	424	422	Imbalance	UCI

3.1 Performance measures

In this paper, the overall and average per-class classification accuracies are used to measure performance. The confusion matrix Q is used to obtain the class-level performance and global performance of the various classifiers. Class-level performance is measured by the percentage classification (η_i), which is defined as

$$\eta_i = \frac{q_{ii}}{N_i^T} \tag{19}$$

where q_{ii} is the number of correctly classified samples and N_i^T is the number of samples for the class $\mathbf{y_i}$ in the training/testing data set. The overall (η_o) and average per-class (η_a) classification accuracies are defined as

$$\eta_o = 100 \times \frac{1}{N^T} \sum_{i=1}^{h} q_{ii} \tag{20}$$

$$\eta_a = 100 \times \frac{1}{h} \sum_{i=1}^{h} \eta_i \tag{21}$$

where h is the number of classes and N^T is the number of training/testing samples.

3.2 Performance comparison

3.2.1 Artificial binary data sets: Double-moon problem

The prototype and data set of the Double-moon classification problem are shown in Fig. 4a, b, respectively,

where $r = 10$, $\omega = 6$ and $d = -6$. The main parameters of distance weighting factor, width, incremental learning threshold, number of BP hidden neurons, and momentum constant in ILRBF-BP are set as $T = 1$, $\sigma = 3$, $\delta = 0.01$, $M = 5$, and $\alpha = 0$, respectively. Figure 4c shows the classification results for the testing samples under these parameters. The classification results illustrate that the proposed algorithm can provide a superior classification surface. Figure 5 shows using different width parameters to cover the training sample space, where each cover generates a RBF hidden neuron and the number of RBF hidden neurons is increased incrementally, the bold lines represent the first coverage region in each pattern class. In Fig. 5, with the increase of the width parameter, the corresponding region covered each RBF hidden neuron is increased accordingly, which will affect the location of the next center, thus generates different RBF hidden neurons. Though the number of RBF hidden neurons has changed, ILRBF-BP still can effectively cover each class of training samples. Thus, the incremental learning algorithm based on potential function clustering is feasible. ILRBF-BP can be well adapted to the sample space, which is an effective algorithm to incrementally generate RBF hidden neurons for the Double-moon problem.

Figure 6a, b demonstrates that when the number of training samples has changed, KMRBF-BP needs less

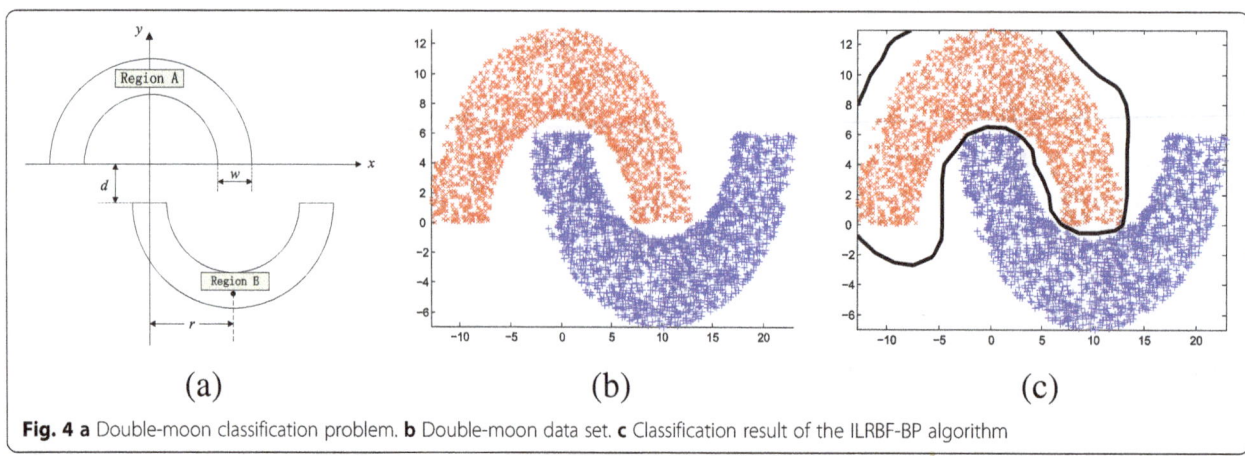

Fig. 4 **a** Double-moon classification problem. **b** Double-moon data set. **c** Classification result of the ILRBF-BP algorithm

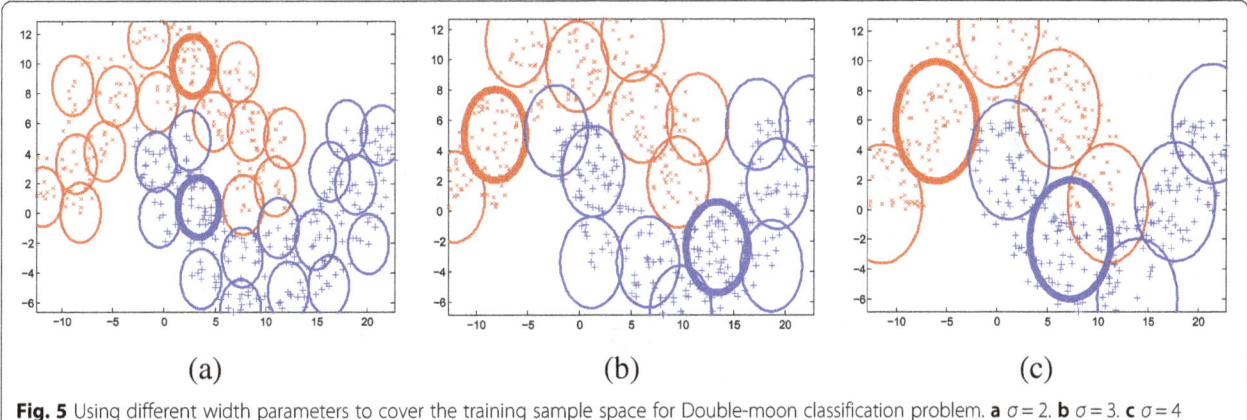

Fig. 5 Using different width parameters to cover the training sample space for Double-moon classification problem. **a** $\sigma = 2$. **b** $\sigma = 3$. **c** $\sigma = 4$

number of RBF hidden neurons than KM-RBF. When the number of training samples is more than 500, KMRBF-BP can get a higher classifying accuracy than KM-RBF. These results show that the hybrid RBF-BP network architecture is effective, which can improve the classifying accuracy and reduce the dependence on the original sample space mapping. In GAP-RBF and ILRBF-BP, the number of RBF hidden neurons is generated automatically. ILRBF-BP needs less number of RBF hidden neurons than GAP-RBF, and the overall testing accuracy outperforms GAP-RBF. The classifying accuracy of ILRBF-BP is comparable with SVM and KMRBF-BP and outperforms ELM and KM-RBF. Note that the number of KM-RBF and KMRBF-BP is selected manually. When changing the number of hidden neurons several times, the one with the highest overall testing accuracy is selected as the suitable number of hidden neurons. As ILRBF-BP utilizes global information about each class of training sample space, it can generate RBF hidden neurons incrementally to adapt the sample space, and the hybrid RBF-BP network architecture improves the network performance further.

3.2.2 Artificial binary data sets: Twist problem

The prototype and data set for the twist classification problem are shown in Fig. 7a, b, respectively, where $d_1 = 0.2$, $d_2 = 0.5$ and $d_3 = 0.8$. Compared to the Double-moon problem, the twist classification problem is more complex and can thus be used to evaluate the classification performance of the different algorithms. The main parameters of distance weighting factor, width, incremental learning threshold, number of BP hidden neurons, and momentum constant in ILRBF-BP are set as $T = 200$, $\sigma = 0.15$, $\delta = 0.01$, $M = 5$, and $\alpha = 0$, respectively. Figure 7c shows the classification results for the testing samples under these parameters. The classification results illustrate that the proposed algorithm still provides a superior classification surface for the Twist classification problem. Figure 8 shows using different width parameters to cover the training sample space, where each cover generates a RBF hidden neuron. In Fig. 8, the bold lines represent the first coverage region, which denote the most dense region in each pattern class. Although there are some overlap in different coverage regions, ILRBF-BP still can effectively cover each class of training

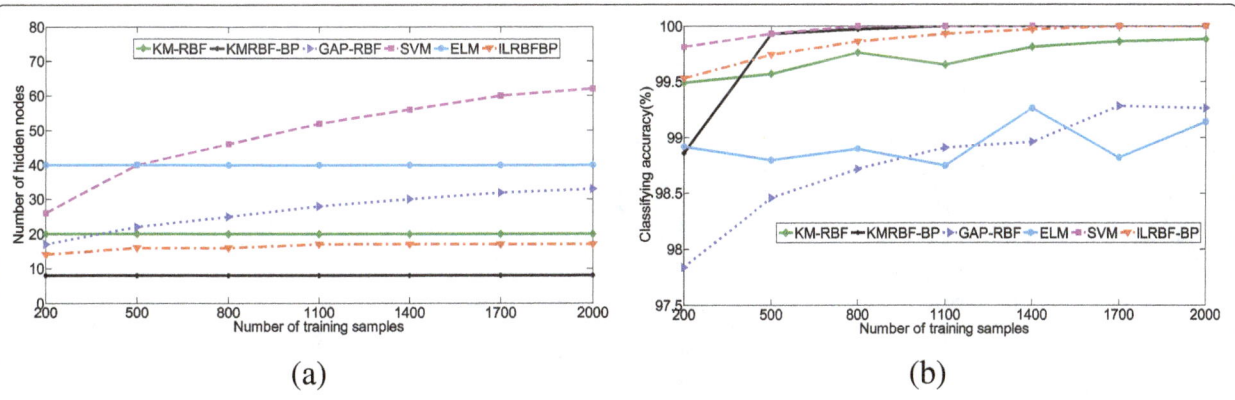

Fig. 6 Performance comparisons between ILRBF-BP and other algorithms on the Double-moon problem. **a** Number of training samples—number of hidden neurons. **b** Number of training samples—classifying accuracy

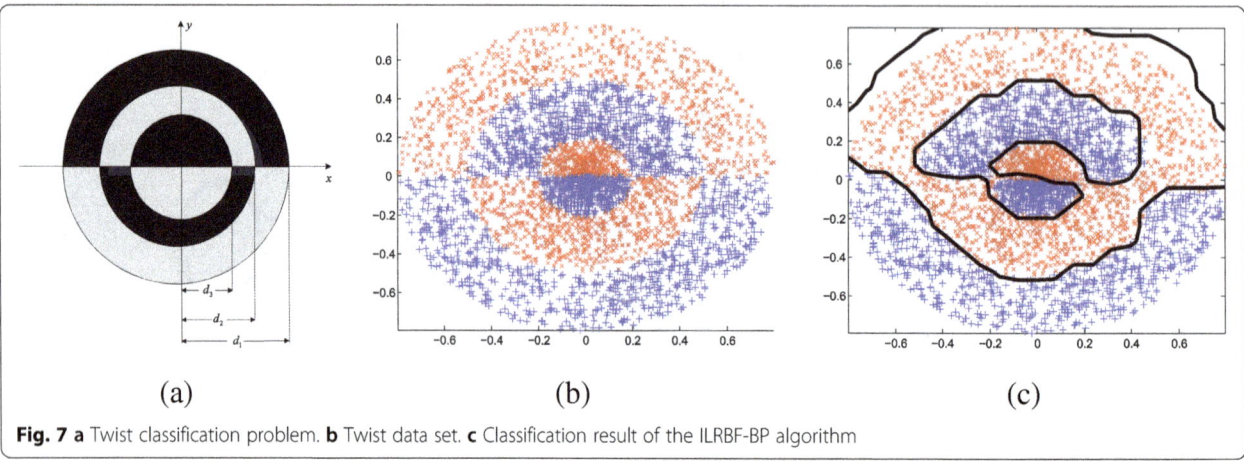

Fig. 7 a Twist classification problem. **b** Twist data set. **c** Classification result of the ILRBF-BP algorithm

samples and generate corresponding RBF hidden neurons incrementally.

Figure 9a, b demonstrates that when the number of training samples has changed, KMRBF-BP needs less number of RBF hidden neurons than KM-RBF and can get a higher classifying accuracy. Thus, the hybrid RBF-BP network architecture improves the classifying accuracy and reduces the dependence on the original sample space mapping. Note that in KM-RBF and KMRBF-BP, when the number of training samples is changed, the number of RBF hidden neurons has to be adjusted manually; otherwise, it will lead to a poor classification accuracy. Compared to KM-RBF and KMRBF-BP, ILRBF-BP can adapt the training sample space well; when the number of training samples is changed, the number of RBF hidden neurons in ILRBF-BP is changed accordingly and can get a higher classifying accuracy. Compared to GAP-RBF, ILRBF-BP can better adapt to the change of sample space. The classifying accuracy of ILRBF-BP outperforms GAP-RBF as well as SVM and ELM. Thus, the incremental learning algorithm based on potential function is effective, which utilizes global information about each class of

training sample space to construct RBF hidden neurons incrementally, and the hybrid RBF-BP network architecture improves the network performance further.

3.2.3 UCI binary data set: Heart problem

In this section, the Heart problem in the UCI binary data set is used to evaluate the performance of the ILRBF-BP algorithm. In the Heart problem, the sample distribution values of each dimension are between 0 and 1, and the main parameters of distance weighting factor, width, incremental learning threshold, number of BP hidden neurons, and momentum constant in ILRBF-BP are set as $T = 1$, $\sigma = 1.2$, $\delta = 0.001$, $M = 5$, and $\alpha = 0.1$, respectively. As noted above, the Heart problem is an imbalanced classification problem. This, in addition to the overall testing η_o, the average testing η_a is also used to measure the performance of each algorithm.

The performance comparisons between ILRBF-BP and the other batch learning algorithms are shown in Table 2. For the Heart problem, the overall and average testing accuracy of ILRBF-BP are clearly higher than those of SGBP, and the proposed algorithm outperforms ELM

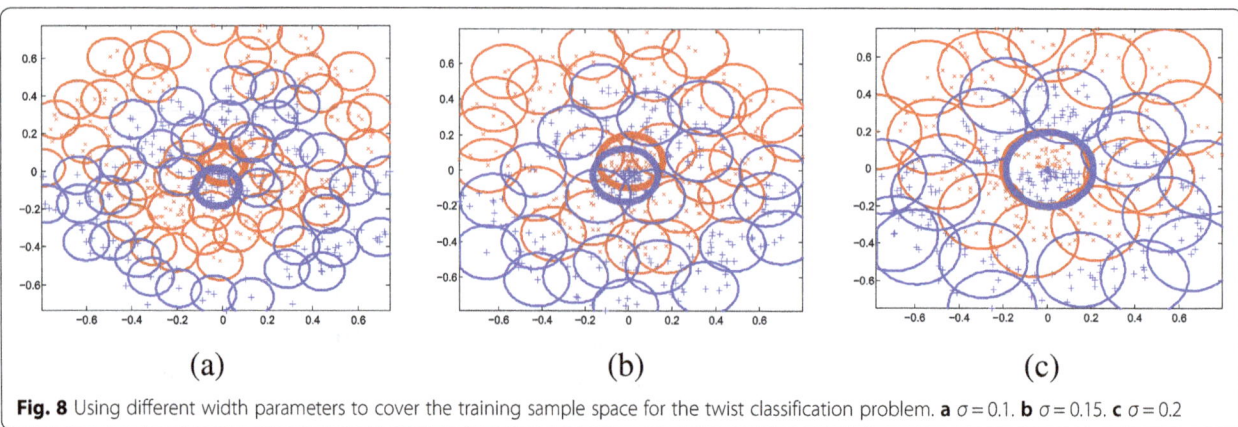

Fig. 8 Using different width parameters to cover the training sample space for the twist classification problem. **a** $\sigma = 0.1$. **b** $\sigma = 0.15$. **c** $\sigma = 0.2$

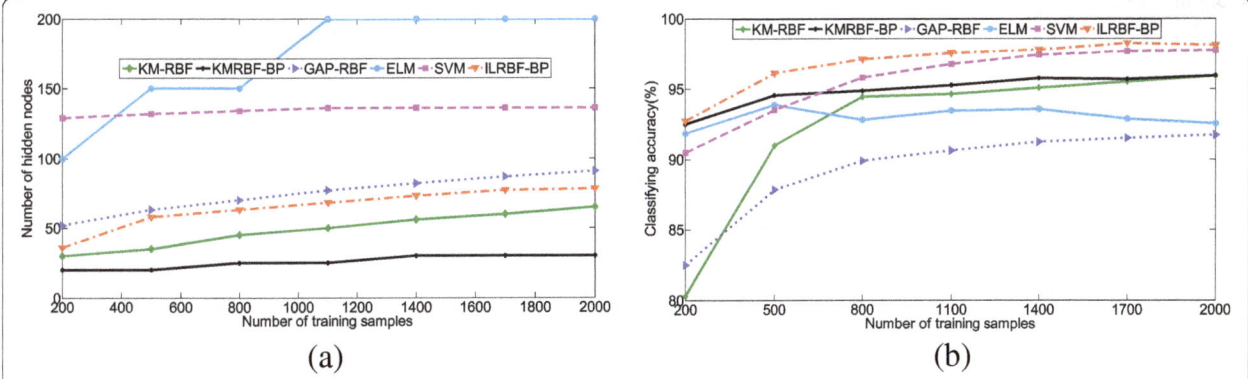

Fig. 9 Performance comparisons between ILRBF-BP and other algorithms on the Twist problem. **a** Number of training samples—number of hidden neurons. **b** Number of training samples—classifying accuracy

and KM-RBF by approximately 2.5–5 %. The average testing accuracy of ILRBF-BP is 1.74 % lower than that of the SVM; however, the overall testing accuracy is approximately 3 % higher than that of the SVM, and fewer hidden neurons are needed.

3.2.4 UCI multi-class data sets: IS and VC problems

In this section, the IS and VC problems are used to evaluate the performance of the ILRBF-BP algorithm. The output saturation is adjusted for the multi-class classifying problem in the ILRBF-BP algorithm. For the IS problem, the sample distribution range in each dimension is different, so the inputs of each algorithm are scaled appropriately between 0 and +1. The main parameters of distance weighting factor, width, incremental learning threshold, number of BP hidden neurons, and momentum constant in ILRBF-BP are set as $T = 1$, $\sigma = 0.3$, $\delta = 0.001$, $M = 8$, and $\alpha = 0.2$, respectively. The IS problem is a well-balanced data set; the number of training samples in each class is 30, and the overall testing η_o is used to measure performance of each algorithm. For the VC problem, the sample distribution values of each dimension are between −1 and +1, and the main parameters of distance weighting factor, width, incremental learning threshold, number of BP hidden neurons, and

momentum constant in ILRBF-BP are set as $T = 1$, $\sigma = 0.4$, $\delta = 0.001$, $M = 9$, and $\alpha = 0.1$, respectively. The number of training samples in each class is 119, 118, 98, and 89. The VC problem is a highly imbalanced data set, where the strong overlap between the classes influences the performance of each algorithm. The overall testing η_o and average testing η_a are used to measure the performance of each algorithm.

Table 3 shows the performance comparisons for the IS and VC problems. For the IS problem, the overall testing accuracy of ILRBF-BP is approximately 5–6 % higher than those of MRAN and GAP-RBF and approximately 0.9–1.3 % higher than those of OS-ELM, SVM, and ELM. For the VC problem, the overall and average testing accuracies of ILRBF-BP are approximately 9–11 % higher than those of MRAN and GAP-RBF and approximately 1.2–2.5 % higher than those of the SVM, ELM, and OS-ELM. The number of RBF hidden neurons and training time of ILRBF-BP are the greatest because the strong overlap of sample space increases the number of RBF hidden neurons and learning time, which yields a higher classification accuracy.

3.3 Analysis of the parameters in the ILRBF-BP algorithm

In this section, the parameter selection for the ILRBF-BP algorithm is discussed, which mainly refers to the

Table 2 Performance comparison for the Heart problem

Method	N_H neurons	Training time(s)	Training η_o	Testing η_o	Testing η_a
SGBP	7	0.95	95.01	46.09	48.42
KM-RBF	7	0.78	82.19	75.22	75.30
SVM	39[a]	0.08	100	77.39	81.81
ELM	10	0	87.67	77.83	77.66
ILRBF-BP	11 and 5[b]	0.66	91.78	80.43	80.07

[a]Support vectors
[b]RBF and BP hidden neurons

Table 3 Performance comparisons for the IS and VC problems

Data sets	Method	N_H neurons	Training time(s)	Testing η_o	Testing η_a
IS	SVM	96[a]	11.61	90.62	–
	MRAN	78	11.68	85.82	–
	GAP-RBF	87	5.77	86.34.	–
	ELM	49	0	90.23	–
	OS-ELM	100	0.01	90.67	–
	ILRBF-BP	77 and 8[b]	2.09	91.57	–
VC	SVM	234[a]	10.74	68.72	67.99
	MRAN	105	10.38	60.24	60.02
	GAP-RBF	81	9.87	58.94	58.17
	ELM	300	0.09	68.01	67.39
	OS-ELM	300	0.12	68.95	67.56
	ILRBF-BP	258 and 9[b]	11.53	70.17	69.43

[a]Support vectors
[b]RBF and BP hidden neurons

distance weighting factor T, width σ and number of BP hidden neurons.

3.3.1 Selection of distance weighting factor T

In this paper, parameter T is used for distance weighting, which can be used to control the interaction potential between two samples. By changing T, the nonlinear mapping of the potential γ can be achieved.

To determine a proper choice of T, in this paper, the standard deviation is considered to measure the impact on T. Here, the Twist classification problem is used in the experiment. Given the number of training samples is 500 and testing samples is 4000; other parameters are given as follows:

1) Twist 1: Set $d_1 = 0.2$, $d_2 = 0.5$, and $d_3 = 0.8$,the standard deviation in each dimension is 0.3281 and 0.3196, respectively. The width parameter is set as $\sigma = 0.1$.

2) Twist 2: Set $d_1 = 2$, $d_2 = 5$, and $d_3 = 8$,the standard deviation in each dimension is 3.2744 and 3.2689, respectively. The width parameter is set as $\sigma = 1$

Figure 10a shows that when the samples are not normalized, for the Twist 2 sample set, the standard deviation of each dimension is relatively large; with the increase of T, the classification performance is reduced. For the Twist 1 sample set, the standard deviation of each dimension is relatively small and the sensitivity of classifying accuracy on T is reduced; however, when the T is selected as 200, the maximum classification accuracy is achieved. Thus, the choice of T should be inversely proportional to the standard deviation of each dimension, that is, $T \propto 1 / \max i = 1, 2...t\{\alpha_i\}$, where α_i is the standard deviation of ith dimension and t is the sample dimension. Figure 10b further indicates that when the samples are normalized to $[-1, 1]$, the dependence on T is reduced and a relatively stable classification accuracy can be achieved.

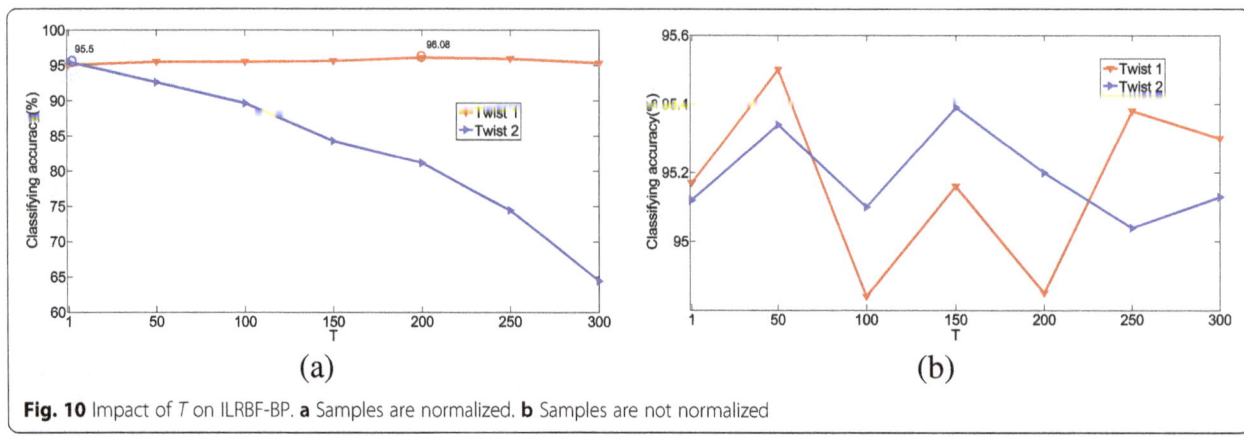

Fig. 10 Impact of T on ILRBF-BP. **a** Samples are normalized. **b** Samples are not normalized

In this paper, for the Double-moon data set, the maximum standard deviation of two dimensions is 8.6448, so a small T should be provided and T is set as $T = 1$. For the Twist data set, the maximum standard deviation of two dimensions is 0.3281, and T is set as $T = 200$.

In high-dimensional space, the sample distribution is often relatively sparse. The sample dimension is considered to be inversely proportional to T, thus $T \propto 1/t$. In this paper, for the IS classification problem, the input values in each dimension are scaled appropriately between 0 and +1. For the Heart and VC classification problems, the values in each dimension are between −1 and 1. Thus, the impact of standard deviation on T is eliminated. Taken into account the dimension information, for the IS, Heart, and VC classification problems, a small T should be provided and T is set as $T = 1$.

3.3.2 Impact of the width σ on ILRBF-BP

The width parameter σ can be used to control the classification accuracy and generalization performance in a RBF network. In the ILRBF-BP algorithm, the width is fixed and selected by cross validation. To reduce the range of the width parameter value selection, we conduct preprocessing for the sample space. If the sample distribution values of each dimension vary considerably, such as in the IS data set, the inputs to each algorithm are scaled appropriately between 0 and +1, whereas the inputs to each algorithm remain unchanged in the Heart and VC data sets.

In the proposed incremental learning algorithm, using a potential function approach to construct RBF hidden neurons incrementally has to complete the effective coverage of the training sample space. As the samples in high-dimensional space are relatively sparse, if the width is too small, it may lead to establish the corresponding Gaussian kernel at each sample, and the proposed incremental learning algorithm is invalid. The reason is that although the potential value of each sample in the training sample space is measured, in the process of eliminating the

potential value of the sample, the generated RBF hidden neurons do not cover other samples, which will lead to a failure of Eq. (9), and excessive RBF hidden neurons will lead to the redundancy of the network architecture, which affects the classification performance of the BP network. Thus, in the proposed ILRBF-BP algorithm, an effective kernel width parameter should be provided, which can generate proper RBF hidden neurons to cover the sample space. Note that the number of generated RBF hidden neurons should not be close to the number of the training samples; otherwise, the proposed algorithm is invalid.

Figure 11a, b shows the impact of width on the overall classification accuracy and the number of RBF hidden neurons, respectively. Figure 11 illustrates that for the Heart and VC data sets, when the width parameter is small, such as $\sigma = 0.1$ and $\sigma = 0.2$, the overall classification accuracy is poor, and effective coverage of the input sample space is not achieved.

When the value of the width parameter is in a suitable range, the number of generated RBF hidden neurons will change, but a relatively stable classification accuracy can be achieved. For the proposed ILRBF-BP algorithm, once the width is given, it can learn the sample space automatically, and the changes in the width parameter will affect the coverage of RBF hidden neurons and generate different RBF hidden neurons. Thus, the incremental learning strategy can counteract the effect of the width to some extent.

4 Impact of the number of BP hidden neurons on ILRBF-BP

In the hybrid RBF-BP network architecture, the nonlinear BP algorithm is used to adjust the weights of the MLPs, which further improves the classification result. However, this method results in an increase in the number of parameters to be selected, especially the selection of the number of BP hidden neurons. For this problem, we conduct experiments on the UCI data sets and discuss the results.

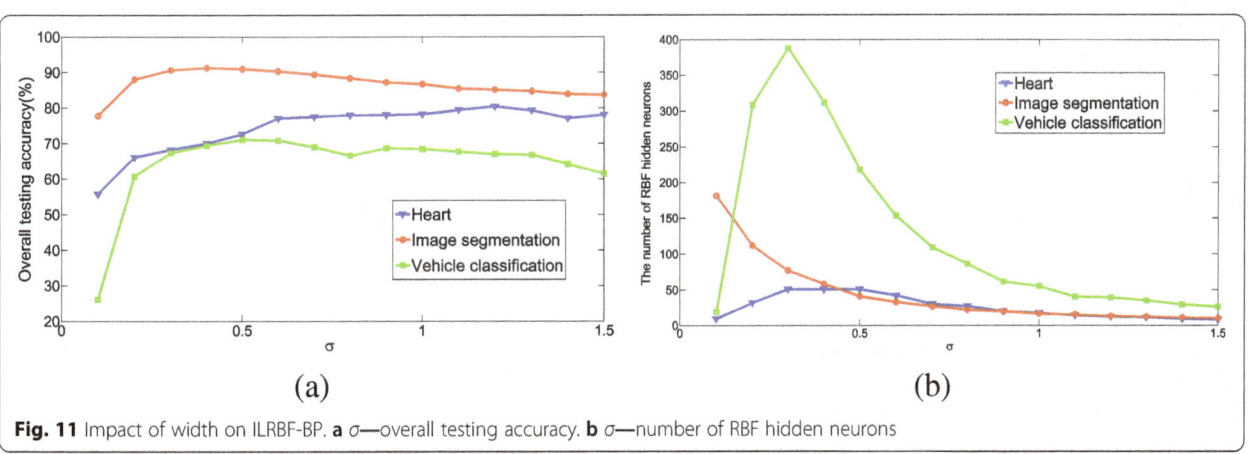

Fig. 11 Impact of width on ILRBF-BP. **a** σ—overall testing accuracy. **b** σ—number of RBF hidden neurons

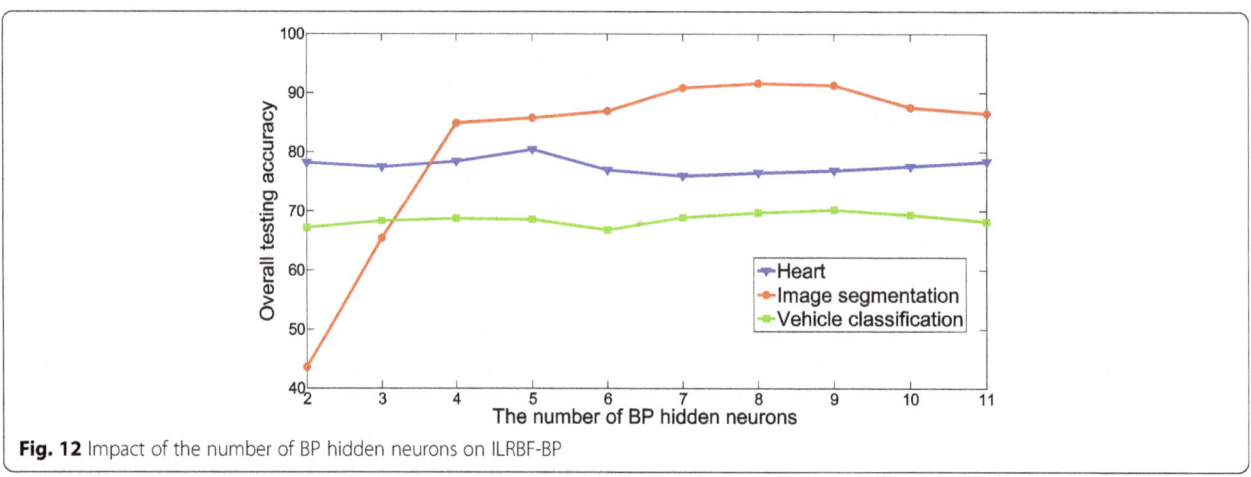

Fig. 12 Impact of the number of BP hidden neurons on ILRBF-BP

Figure 12 shows the impact of the number of BP hidden neurons on ILRBF-BP. For the Heart, IS, and VC problems, when the number of BP hidden neurons is greater or equal to 4, the overall classification accuracy does not change considerably. For the hybrid RBF-BP network, the mapping results of RBF hidden neurons are processed and used for the input of BP network, which improves the stability of the BP network and effectively avoids falling into local minima for the BP algorithm. Thus, the dependence on the number of BP hidden neurons is reduced. When the sample set is more complex, the momentum term can be used to improve the BP algorithm further.

5 Conclusions

In this paper, an incremental learning algorithm for the hybrid RBP-BP (ILRBF-BP) network classifier is proposed. The ILRBF-BP algorithm uses a potential function to measure the density of the training sample space and incrementally generates RBF hidden neurons, enabling the effective estimation of the center and number of RBF hidden neurons. In this way, a suitable network size for RBF hidden layer that matches the complexity of the sample space can be built up. A hybrid RBF-BP network architecture is designed to improve classification performance further, which shows good stability and generalization performance. The hybrid network simplifies the selection of the number of neurons in the BP hidden layer while further reducing the dependence on space mapping in the RBF hidden layer.

The performance of the ILRBF-BP algorithm has been compared with other batch learning algorithms, such as SGBP, KM-RBF, SVM, and ELM, and sequential learning algorithms, such as MRAN, GAP-RBF, and OS-ELM, in artificial data sets and UCI data sets. The method of adjusting output label values is used to prevent the

output saturation problem for multi-class classification. Experiments demonstrate the superiority of the ILRBF-BP algorithm.

In the future, we will focus on the optimization of kernel width and imbalanced data classification problems. In the ILRBF-BP algorithm, the width is fixed and selected by cross validation and the adjustment of width parameter will affect the location of next center, as well as the network size. Therefore, it is necessary to design an adaptive width adjustment to adapt to the different regions of the sample space. In addition, for the imbalanced data classification problem, the samples in the boundary regions contain more classification information, thus how to measure and select these samples is particularly important. Further studies are needed to address these concerns.

Competing interests
The authors declare that they have no competing interests.

Acknowledgements
The authors thank the support provided by the National Science Foundation of China (No. 61331021, U1301251) and the Shenzhen Science and Technology Plan Project (JCYJ20130408173025036). The authors would like to thank the Editor-in-Chief, the Associate Editor, and the Anonymous Reviewers for their helpful comments and suggestions which have greatly improved the quality of presentation.

References
1. M Lin, K Tang, X Yao, Dynamic sampling approach to training neural networks for multiclass imbalance classification. IEEE Trans Neural Netw and Learning Systems **24**(4), 647–660 (2013)
2. L-Q Li, W-X Xie, Intuitionistic fuzzy joint probabilistic data association filter and its application to multitarget tracking. Signal Process **96**, 433–444 (2014)
3. Y-L Wei, J-B Qiu, HR Karimi, M Wang, H-infinity model reduction for continuous-time Markovian jump systems with incomplete statistics of mode information. Int J Syst Sci **45**(7), 1496–1507 (2014)
4. Y-L Wei, J-B Qiu, HR Karimi, M Wang, Filtering design for two-dimensional Markovian jump systems with state-delays and deficient mode information. Inform Sci **269**, 316–331 (2014)

5. Y-L Wei, J-B Qiu, HR Karimi, M Wang, A new design of H∞ filtering for continuous-time Markovian jump systems with time-varying delay and partially accessible mode information. Signal Process **93**(9), 2392–2407 (2013)

6. F-Y Meng, X Li, J-H Pei, A feature point matching based on spatial order constraints bilateral-neighbor vote. IEEE Trans Image Process **24**(11), 4160–4171 (2015)

7. L-X Guan, W-X Xie, J-H Pei, Segmented minimum noise fraction transformation for efficient feature extraction of hyperspectral images. Pattern Recogn **48**(10), 3216–3226 (2015)

8. HC Nejad, O Khayat, B Azadbakht, M Mohammadi, Using feed forward neural network for electrocardiogram signal analysis in chaotic domain. J Intelligent and Fuzzy Systems **27**(5), 2289–2296 (2014)

9. CH Weng, CK Huang, RP Han, Disease prediction with different types of neural network classifiers. Telematics Inform **33**(2), 277–292 (2016)

10. C Lu, N Ma, ZP Wang, Fault detection for hydraulic pump based on chaotic parallel RBF network. EURASIP J on Advances in Signal Processing **49**, (2011). doi: 10.1186/1687-6180-2011-49

11. J Moody, CJ Darken, Fast learning in networks of locally-tuned processing. Neurocomputing **1**(2), 281–294 (1989)

12. D Lowe, Characterising complexity by the degrees of freedom in a radial basis function network. Neurocomputing **19**(1-3), 199–209 (1998)

13. J Platt, A resource-allocating network for function interpolation. Neural Comput **3**(2), 213–225 (1991)

14. V Kadirkamanathan, M Niranjan, A function estimation approach to sequential learning with neural networks. Neural Comput **5**(6), 954–975 (1993)

15. L Yingwei, N Sundararajan, P Saratchandran, A sequential learning scheme for function approximation using minimal radial basis function. Neural Comput **9**(2), 461–478 (1997)

16. G-B Huang, P Saratchandran, N Sundararajan, An efficient sequential learning algorithm for growing and pruning RBF (GAP-RBF) networks. IEEE Trans Syst Man Cybern B Cybern **34**(6), 2284–2292 (2004)

17. G-B Huang, P Saratchandran, N Sundararajan, A generalized growing and pruning RBF (GAP-RBF) neural network for function approximation. IEEE Trans Neural Netw **16**(1), 57–67 (2005)

18. M Bortman, M Aladjem, A growing and pruning method for radial basis function networks. IEEE Trans Neural Netw **20**(6), 1030–1045 (2009)

19. H Yu, PD Reiner, T Xie, T Bartczak, BM Wilamowski, An incremental design of radial basis function networks. IEEE Trans Neural Netw and Learning Systems **2**(10), 1793–1803 (2014)

20. S Suresh, D Keming, HJ Kim, A sequential learning algorithm for self-adaptive resource allocation network classifier. Neurocomputing **73**(16-18), 3012–3019 (2010)

21. T Xie, H Yu, J Hewlett, P Rózycki, B Wilamowski, Fast and efficient second-order method for training radial basis function networks. IEEE Trans Neural Netw and Learning Systems **23**(4), 609–619 (2012)

22. C Constantinopoulos, A Likas, An incremental training method for the probabilistic RBF network. IEEE Trans Neural Netw **17**(4), 966–974 (2006)

23. G-B Huang, Q-Y Zhu, C-K Siew, A new learning scheme of feedforward neural, in *Proceedings of International Joint Conference on Neural Networks (IJCNN 2004)*, pp. 985–99

24. N-Y Liang, G-B Huang, P Saratchandran, N Sundararajan, A fast and accurate online sequential learning algorithm for feedforward networks. IEEE Trans Neural Netw **17**(6), 1411–1423 (2006)

25. G-B Huang, L CHEN, C-K Siew, Universal approximation using incremental constructive feedforward networks with random hidden nodes. IEEE Trans Neural Netw **17**(4), 879–892 (2006)

26. G-B Huang, L CHEN, Convex incremental extreme learning machine. Neurocomputing **70**(16-18), 3056–3062 (2007)

27. G-B Huang, L Chen, Enhanced random search based incremental extreme learning machine. Neurocomputing **71**(16-18), 3460–3468 (2008)

28. G Feng, G-B Huang, Q Lin, Error minimized extreme learning machine with growth of hidden nodes and incremental learning. IEEE Trans Neural Netw **20**(8), 1352–1357 (2009)

29. Y LeCun, L Bottou, GB Orr, K-R Müller, Efficient backprop. Lecture Notes Comput Sci **1524**, 9–50 (1998)

30. J-H Pei, W-X Xie, Adaptive multi thresholds image segmentation based on potential function clustering. Chinese J Computers **22**(7), 758–762 (1999)

31. OA Bashkerov, EM Braverman, IB Muchnik, Potential function algorithms for pattern recognition learning machines. Autom Remote Control **25**(5), 692–695 (1964)

32. G Cybenko, Approximation by superpositions of a sigmoidal function. Mathematics of Control, Signal, and Systems **2**, 303–314 (1989)

33. S Hayin, *Neural networks and learning machines. Third Edition* (China Machine Press, China, 2009), pp. 61–63

34. C Blake, C Merz, *UCI repository of machine learning databases* (Department of Information and Computer Sciences, University of California, Irvine, 1998). available at http://archive.ics.uci.edu/ml/

35. C-C Chang, C-J, *LIBSVM: a library for support vector machines* (Department of Computer Science and Information Engineering, National Taiwan University, Taiwan, 2003). available at http://www.csie.ntu.edu.tw/~cjlin/libsvm/index.html

10

Quantization in zero leakage helper data schemes

Joep de Groot[1,4], Boris Škorić[2], Niels de Vreede[3] and Jean-Paul Linnartz[1*]

Abstract

A helper data scheme (HDS) is a cryptographic primitive that extracts a high-entropy noise-free string from noisy data. Helper data schemes are used for preserving privacy in biometric databases and for physical unclonable functions. HDSs are known for the guided quantization of continuous-valued sources as well as for repairing errors in discrete-valued (digitized) sources. We refine the theory of helper data schemes with the zero leakage (ZL) property, i.e., the mutual information between the helper data and the extracted secret is zero. We focus on quantization and prove that ZL necessitates particular properties of the helper data generating function: (1) the existence of "sibling points", enrollment values that lead to the same helper data but different secrets and (2) quantile helper data. We present an optimal reconstruction algorithm for our ZL scheme, that not only minimizes the reconstruction error rate but also yields a very efficient implementation of the verification. We compare the error rate to schemes that do not have the ZL property.

Keywords: Biometrics, Fuzzy extractor, Helper data, Privacy, Secrecy leakage, Secure sketch

1 Introduction

1.1 Biometric authentication: the noise problem

Biometrics have become a popular solution for authentication or identification, mainly because of their convenience. A biometric feature cannot be forgotten (like a password) or lost (like a token). Nowadays identity documents such as passports nearly always include biometric features extracted from fingerprints, faces, or irises. Governments store biometric data for forensic investigations. Some laptops and smart phones authenticate users by means of biometrics.

Strictly speaking, biometrics are not secret. In fact, fingerprints can be found on many objects. It is hard to prevent one's face or iris from being photographed. However, storing biometric features in an unprotected, open database. Introduces both security and privacy risks. Security risks include the production of fake biometrics from the stored data, e.g., rubber fingers [1, 2]. These fake biometrics can be used to obtain unauthorized access to services, to gain confidential information or to leave fake evidence at crime scenes. We also mention two privacy risks. (1) Some biometrics are known to reveal diseases and disorders of the user. (2) Unprotected storage allows for cross-matching between databases.

These security and privacy problems cannot be solved by simply encrypting the database. It would not prevent *insider attacks*, i.e., attacks or misuse by people who are authorized to access the database. As they legally possess the decryption keys, database encryption does not stop them.

The problem of storing biometrics is very similar to the problem of storing passwords. The standard solution is to store *hashed* passwords. Cryptographic hash functions are one-way functions, i.e., inverting them to calculate a secret password from a public hash value is computationally infeasible. Even inside attackers who have access to all the hashed passwords cannot deduce the user passwords from them.

Straightforward application of this hashing method to biometrics does not work for biometrics, however. Biometric measurements are noisy, which causes (small) differences between the digital representation of the enrollment measurement and the digitized measurement during verification. Particularly if the biometric value lies near a quantization boundary, a small amount of noise can

*Correspondence: j.p.linnartz@tue.nl
[1] Signal Processing Systems group, Department of Electrical Engineering, Eindhoven University of Technology, 5600 MB, Eindhoven, The Netherlands
Full list of author information is available at the end of the article

flip the discretized value and trigger an avalanche of bit flips at the output of the hash.

1.2 Helper data schemes

The solution to the noise problem is to use a helper data scheme (HDS) [3, 4]. A HDS consists of two algorithms, Gen and Rep. In the enrollment phase, the Gen algorithm takes a noisy (biometric) value as input and generates not only a secret but also public data called *helper data*. The Rep algorithm is used in the verification phase. It has two inputs: the helper data and a fresh noisy (biometric) value obtained from the same source. The Rep algorithm outputs an estimator for the secret that was generated by Gen.

The helper data makes it possible to derive the (discrete) secret reproducibly from noisy measurements, i.e., to perform error correction, while not revealing too much information about the enrollment measurement. The noise-resistant secret can be hashed as in the password protection scheme.

1.3 A two-stage approach

We describe a commonly adopted two-stage approach for real-valued sources, as for instance presented in ([5], Chap. 16). The main idea is as follows. A first-stage HDS performs quantization (discretization) of the real-valued input. Helper data is applied in the "analog" domain, i.e., before quantization. Typically, the helper data consists of a 'pointer' to the center of a quantization interval. The quantization intervals can be chosen at will, which allows for optimizations of various sorts [6–8].

After the first stage, there is typically still some noise in the quantized output. A second-stage HDS employs digital error correction techniques, for instance the code offset method (also known as Fuzzy Commitment) [3, 9] or a variant thereof [10, 11].

Such a two-stage approach is also common practice in communication systems that suffer from unreliable (wireless) channels: the signal conditioning prior to the quantization involves optimization of signal constellations and multidimensional transforms. The discrete mathematical operations, such as error correction decoding,

are known to be effective only for sufficiently error-free signals. According to the asymptotic Elias bound ([12], Chap. 17), at bit error probabilities above 10 % one cannot achieve code rates better than 0.5. Similarly, in biometric authentication, optimization of the first stage appears essential to achieve adequate system performance. The design of the first stage is the prime motivation, and key contribution, of this paper.

Figure 1 shows the data flow and processing steps in the two-stage helper data scheme. In a preparation phase preceding all enrollments, the population's biometrics are studied and a transform is derived (using well known techniques such as principal component analysis or linear discriminant analysis [13]). The transform splits the biometric vector $\underline{x}$ into scalar components $(x_i)_{i=1}^M$. We will refer to these components x_i as features. The transform ensures that they are mutually independent, or nearly so.

At enrollment, a person's biometric $\underline{x}$ is obtained. The transform is applied, yielding features $(x_i)_{i=1}^M$. The Gen algorithm of the first-stage HDS is applied to each feature independently. This gives continuous helper data $(w_i)_{i=1}^M$ and short secret strings $s_1, \ldots, s_M$ which may or may not have equal length, depending on the signal-to-noise ratio of the features. All these secrets are combined into one high-entropy secret k, e.g., by concatenating them after Gray-coding. Biometric features are subject to noise, which will lead to some errors in the reproduced secret $\hat{k}$; hence, a second stage of error correction is done with another HDS. The output of the second-stage Gen algorithm is discrete helper data r and a practically noiseless string c. The hash $h(c\|z)$ is stored in the enrollment database, along with the helper data $(w_i)_{i=1}^M$ and r. Here z is salt, a random string to prevent easy cross-matching.

In the authentication phase, a fresh biometric measurement $\underline{y}$ is obtained and split into components $(y_i)_{i=1}^M$. For each i independently, the estimator $\hat{s}_i$ is computed from y_i and w_i. The $\hat{s}_i$ are combined into an estimator $\hat{k}$, which is then input into the 2nd-stage HDS reconstruction together with r. The result is an estimator $\hat{c}$. Finally, $h(\hat{c}\|z)$ is compared with the stored hash $h(c\|z)$.

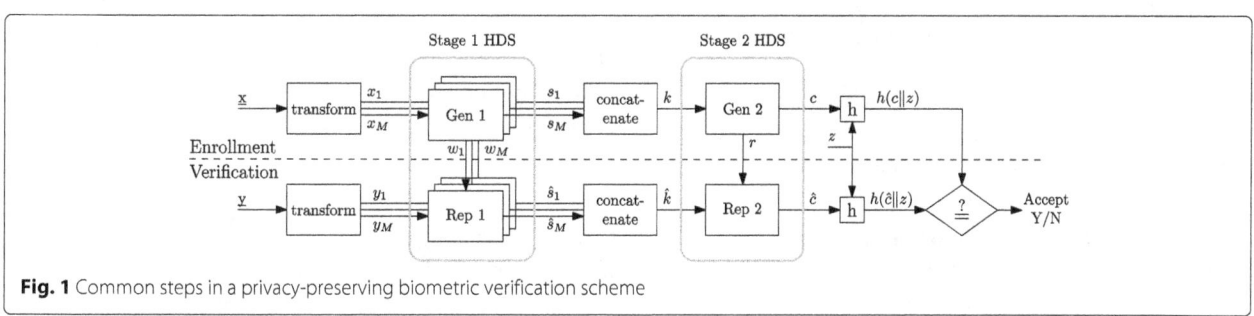

Fig. 1 Common steps in a privacy-preserving biometric verification scheme

1.4 Fuzzy extractors and secure sketches

Special algorithms have been developed for HDSs [4, 6, 8, 9]: Fuzzy extractors (FE) and secure sketches (SS). The FE and SS are special cases of the general HDS concept. They have different requirements,

- *Fuzzy extractor*
  The probability distribution of s given w has to be (nearly) uniform.
- *Secure sketch*
  s given w must have high entropy, but does not have to be uniform. Typically, s is equal to (a discretized version of) x.

The FE is typically used for the extraction of cryptographic keys from noisy sources such as physical unclonable functions (PUFs) [14–16]. Some fixed quantization schemes support the use of a fuzzy extractor, provided that the quantization intervals can be chosen such that each secret s is equiprobable, as in [17].

The SS is very well suited to the biometrics scenario described above.

1.5 Security and privacy

In the HDS context, the main privacy question is how much information, and *which* information, about the biometric $\underline{x}$ is leaked by the helper data. Ideally, the helper data would contain just enough information to enable the error correction. Roughly speaking, this means that the vector $\underline{w} = (w_i)_{i=1}^{M}$ consists of the noisy "least significant bits" of $\underline{x}$, which typically do not reveal sensitive information since they are noisy anyway. In order to make this kind of intuitive statement more precise, one studies the information-theoretic properties of HDSs. In the system as sketched in Fig. 1, the mutual information[1] $I(C; \underline{W}, R)$ is of particular interest: it measures the leakage about the string c caused by the fact that the attacker observes $\underline{w}$ and r. By properly separating the "most significant digits" of $\underline{x}$ from the "least significant digits", it is possible to achieve $I(C; \underline{W}, R) = 0$. We call this zero secrecy leakage or, more compactly, zero leakage (ZL).[2] HDSs with the ZL property are very interesting for quantifying privacy guarantees: if a privacy-sensitive piece of a biometric is fully contained in c, and not in $(\underline{w}, r)$, then a ZL HDS based database reveals *absolutely nothing* about that piece.[3]

We will focus in particular on schemes whose first stage has the ZL property for each feature separately: $I(S_i; W_i) = 0$. If the transform in Fig. 1 yields independent features, then automatically $I(S_j; W_i) = 0$ for all i, j, and the whole first stage has the ZL property.

1.6 Contributions and organization of this paper

In this paper, we zoom in on the first-stage HDS and focus on the ZL property in particular. Our aim is to minimize reconstruction errors in ZL HDSs that have scalar input $x \in \mathbb{R}$. We treat the helper data as being real-valued, $w \in \mathbb{R}$, though of course w is in practice stored as a finite-precision value.

- We show that the ZL constraint for continuous helper data necessitates the existence of "Sibling Points", points x that correspond to different s but give rise to the same helper data w.
- We prove that the ZL constraint for $x \in \mathbb{R}$ implies "quantile" helper data. This holds for uniformly distributed s as well as for non-uniform s. Thus, we identify a simple quantile construction as being the generic ZL scheme for all HDS types, including the FE and SS as special cases. It turns out that the continuum limit of a FE scheme of Verbitskiy et al. [7] precisely corresponds to our quantile HDS.
- We derive a reconstruction algorithm for the quantile ZL FE that minimizes the reconstruction errors. It amounts to using a set of optimized threshold values, and is very suitable for low-footprint implementation.
- We analyze, in an all-Gaussian example, the performance (in terms of reconstruction error rate) of our ZL FE combined with the optimal reconstruction algorithm. We compare this scheme to fixed quantization and a likelihood-based classifier. It turns out that our error rate is better than that of fixed quantization, and not much worse than that of the likelihood-based classifier.

The organization of this paper is as follows. Section 2 discusses quantization techniques. After some preliminaries (Section 3), the sibling points and the quantile helper data are treated in Section 4. Section 5 discusses the optimal reconstruction thresholds. The performance analysis in the Gaussian model is presented in Section 6.

2 Related work on biometric quantization

Many biometric parameters can be converted by a principal component analysis (PCA) into a vector of (near)independent components [18]. For this reason, most papers on helper data in the analog domain can restrict themselves to a one-dimensional quantization, e.g., [4, 6, 18]. Yet, the quantization strategies differ, as we will review below. Figure 2 shows the probability density function (PDF) of the measurement y in the verification phase and how the choice of quantization regions in the verification phase affects the probability of erroneous reconstruction (shaded area) in the various schemes.

2.1 Fixed quantization (FQ)

The simplest form of quantization applies a uniform, fixed quantization grid during both enrollment and verification. An example for $N = 4$ quantization regions is depicted

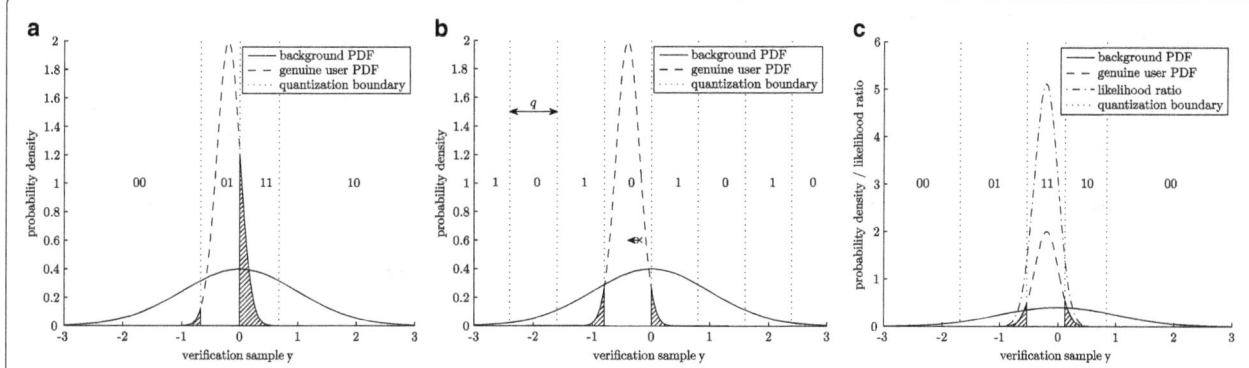

Fig. 2 Examples of adaptating the genuine user PDF in the verification phase. FQ does not translate the PDF; QIM centers the PDF on a quantization interval; LQ uses a likelihood ratio to adjust the quantization regions. **a** Fixed equiprobable quantization. **b** Quantization Index Modulation. **c** Multi-bits based on likelihood ratio [6]

in Fig. 2a. An unfavorably located genuine user pdf, near a quantization boundary, can cause a high reconstruction error.

The inherently large error probability can be mitigated by "reliable component" selection [17]. Only components x_i far away from a boundary are selected; the rest are discarded. The indices of the reliable components constitute the helper data. Such a scheme is very inefficient, as it wastes resources: features that are unfavorably located w.r.t. the quantization grid, but nonetheless carry information, are eliminated. Furthermore, the helper data leaks information about the biometric, since the intervals have unequal width and therefore unequal probabilities of producing reliable components [19].

2.2 Quantization index modulation (QIM)

QIM borrows principles from digital watermarking [20] and writing on dirty paper [21]. QIM has quantization intervals alternatingly labeled with 's'0" and "1" as the values for the secret s. The helper data w is constructed as the distance from x to the middle of a quantization interval; adding w to y then offsets the pdf so that the pdf is centered on the interval (Fig. 2b), yielding a significantly lower reconstruction error probability than FQ.

The freedom to choose quantization step sizes allows for a trade-off between reconstruction performance and leakage [4]. The alternating labeling was adopted to reduce leakage but sacrifices a large part of the source's entropy.

2.3 Likelihood-based quantization (LQ)

At enrollment, the LQ scheme [6] allocates N quantization regions as follows. The first two boundaries are chosen such that they yield the same probability of y given x, and at the same time enclose a probability mass $1/N$ on the background distribution (the whole population's distribution). Subsequent quantization intervals are

chosen contiguous to the first and again enclose a $1/N$ probability mass. Finally, the probability mass in the tails of the background distribution is added up as a wrap-around interval, which also holds a probability mass of $1/N$. Since the quantization boundaries are at fixed probability mass intervals, it suffices to communicate a single boundary t as helper data to the verification phase.

In LQ, the secret s is not equiprobable. The error rates are low, but the revealed t leaks information about s.

2.4 Dynamic detection-rate-based bit allocation

In [22], Lim et al. proposed dynamic genuine interval search (DGIS) as an improvement of the bit allocation scheme of Chen et al. [23]. The resulting scheme has some similarity to our approach in that they both determine discretization intervals per user and store these intervals as helper data. However, their scheme is motivated solely by optimization of the detection rate, whereas in our scheme the optimization is subject to the zero leakage restriction. Applying the DGIS method introduces some additional leakage to the underlying bit allocation scheme. Furthermore, DGIS performs its search for the optimal discretization intervals using a sliding window algorithm, which in general will not succeed in finding the exact optimum. In contrast, in our scheme, we analytically derive the optimal solution from the background distribution.

3 Preliminaries
3.1 Notation
Random variables are denoted with capital letters and their realizations in lowercase. The notation $\mathbb{E}$ stands for expectation. Sets are written in calligraphic font. We zoom in on the one-dimensional first-stage HDS in Fig. 1. For brevity of notation the index $i \in \{1, \ldots, M\}$ on x_i, w_i, s_i, y_i and $\hat{s}_i$ will be omitted.

The probability density function (PDF) or probability mass function (PMF) of a random variable A is denoted as f_A, and the cumulative distribution function (CDF) as F_A. We consider $X \in \mathbb{R}$. The helper data is considered continuous, $W \in \mathcal{W} \subset \mathbb{R}$. Without loss of generality we fix $\mathcal{W} = [0, 1)$. The secret S is an integer in the range $S = \{0, \ldots, N - 1\}$, where N is a system design choice, typically chosen according to the signal to noise ratio of the biometric feature. The helper data is computed from X using a function g, i.e., $W = g(X)$. Similarly, we define a quantization function Q such that $S = Q(X)$. The enrollment part of the HDS is given by the pair Q, g. We define quantization regions as follows,

$$A_s = \{x \in \mathbb{R} : Q(x) = s\}. \tag{1}$$

The quantization regions are non-overlapping and cover the complete feature space, hence form a partitioning:

$$A_s \cap A_t = \emptyset \quad \text{for } s \neq t; \qquad \bigcup_{s \in \mathcal{S}} A_s = \mathbb{R}. \tag{2}$$

We consider only quantization regions that are contiguous, i.e., for all s it holds that A_s is a simple interval. In Section 5.3, we will see that many other choices may work equally well, *but not better*; our preference for contiguous A_s regions is tantamount to choosing the simplest element Q out of a whole equivalence class of quantization functions that lead to the same HDS performance. We define quantization boundaries $q_s = \inf A_s$. Without loss of generality, we choose Q to be a monotonically increasing function. This gives $\sup A_s = q_{s+1}$. An overview of the quantization regions and boundaries is depicted in Fig. 3.

In a generic HDS, the probabilities $\mathbb{P}[S = s]$ can be different for each s. We will use shorthand notation

$$\mathbb{P}[S = s] = p_s > 0. \tag{3}$$

The quantization boundaries are given by

$$q_s = F_X^{-1}\left(\sum_{t=0}^{s-1} p_t\right), \tag{4}$$

where F_X^{-1} is the inverse CDF. For a Fuzzy extractor, one requires $p_s = 1/N$ for all s, in which case (4) simplifies to

$$q_s^{\text{FE}} = F_X^{-1}\left(\frac{s}{N}\right). \tag{5}$$

3.2 Zero leakage

We will work with a definition of the zeroleakage property that is a bit stricter than the usual formulation [7], which pertains to mutual information. This is necessary in order to avoid problems caused by the fact that W is a continuum variable (e.g., pathological cases where some property does not hold on measure-zero subsets of $\mathcal{W}$),

*Definition 3.1. We call a helper data scheme **Zero Leakage** if and only if*

$$\forall_{\mathcal{V} \subseteq \mathcal{W}} \quad \mathbb{P}[S = s | W \in \mathcal{V}] = \mathbb{P}[S = s]. \tag{6}$$

In words, we define the ZL property as independence between S and W. Knowledge about W has no effect on the adversary's uncertainty about S. ZL implies $I(S; W) = 0$ or, equivalently, $H(S|W) = H(S)$. Here H stands for Shannon entropy, and I for mutual information (see, e.g., ([24], Eq. (2.35)–(2.39)).

3.3 Noise model

It is common to assume a noise model in which the enrollment measurement x and verification measurement y are both derived from a hidden 'true' biometric value z, i.e., $X = Z + N_e$ and $Y = Z + N_v$, where N_e stands for the noise in the enrollment measurement and N_v for the noise in the verification measurement. It is assumed that N_e and N_v are mutually independent and independent of X and Y. The N_e, N_v have zero mean and variance σ_e^2, σ_v^2 respectively. The variance of z is denoted as σ_Z^2. This is a very generic model. It allows for various special cases such as noiseless enrollment, equal noise at enrollment, and verification, etc.

It is readily seen that the variance of X and Y is given by $\sigma_X^2 = \sigma_Z^2 + \sigma_e^2$ and $\sigma_Y^2 = \sigma_Z^2 + \sigma_v^2$. The correlation coefficient ρ between X and Y is defined as $\rho = (\mathbb{E}[XY] - \mathbb{E}[X]\mathbb{E}[Y])/(\sigma_X \sigma_Y)$ and it can be expressed as $\rho = \sigma_Z^2/(\sigma_X \sigma_Y)$.

For zero-mean X, Y, it is possible to write

$$Y = \lambda X + R, \quad \text{with } \lambda = \frac{\sigma_Y}{\sigma_X}\rho \quad \text{and } \text{Var}(R) = \sigma^2 \overset{\text{def}}{=} \sigma_Y^2(1 - \rho^2), \tag{7}$$

where R is zero mean noise. This way of expressing Y is motivated by the first and second order statistics, i.e., the variance of Y is $\lambda^2 \sigma_X^2 + \sigma^2 = \sigma_Y^2$, and the correlation between X and Y is $\mathbb{E}[XY]/(\sigma_X \sigma_Y) = \lambda \sigma_X^2/(\sigma_X \sigma_Y) = \rho$.

In the case of Gaussian Z, N_e, N_v, the X and Y are Gaussian, and the noise R is Gaussian as well.

From (7), it follows that the PDF of y given x (i.e., the noise between enrollment and verification) is centered on λx and has variance σ^2. The parameter λ is called the attenuation parameter. In the "identical conditions" case $\sigma_e = \sigma_v$ it holds that $\lambda = \rho = \sigma_Z^2/(\sigma_Z^2 + \sigma_v^2)$. In the

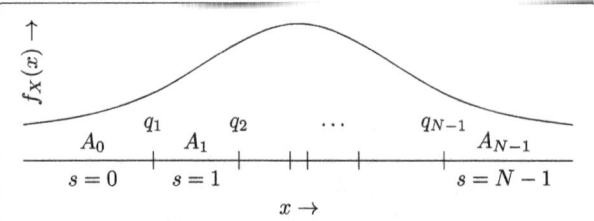

Fig. 3 Quantization regions A_s and boundaries q_s. The locations of the quantization boundaries are based on the distribution of x, such that secret s occurs with probability p_s

"noiseless enrollment" case $\sigma_e = 0$ we have $\lambda = 1$ and $\rho = \sigma_Z/\sqrt{\sigma_Z^2 + \sigma_v^2}$.

We will adopt expression (7) as our noise model.

In Section 5.1, we will be considering a class of noise distributions that we call *symmetric fading noise*.

Definition 3.2. Let X be the enrollment measurement and let Y be the verification measurement, where we adopt be model of Eq. (7). Let $f_{Y|X}$ denote the probability density function of Y given X. The noise is called symmetric fading noise *if for all x, y_1, y_2 it holds that*

$$|y_1 - \lambda x| < |y_2 - \lambda x| \implies f_{Y|X}(y_1|x) > f_{Y|X}(y_2|x). \quad (8)$$

Equation (8) reflects the property that small noise excursions are more likely than large ones, and that the sign of the noise is equally likely to be positive or negative. Gaussian noise is an example of symmetric fading noise.

4 Zero leakage: quantile helper data

In Section 4.1, we present a chain of arguments from which we conclude that, for ZL helper data, it is sufficient to consider only functions g with the following properties: (1) covering $\mathcal{W}$ on each quantization interval (surjective); (2) monotonically increasing on each quantization interval. This is then used in Section 4.2 to derive the main result, Theorem 4.8: Zero Leakage is equivalent to having helper data obeying a specific quantile rule. This rule makes it possible to construct a very simple ZL HDS which is entirely generic.

4.1 Why it is sufficient to consider monotonically increasing surjective functions g

The reasoning in this section is as follows. We define *sibling points* as points x in different quantization intervals but with equal w. We first show that for every w, there must be at least one sibling point in each interval (surjectivity); then, we demonstrate that having more than one is bad for the reconstruction error rate. This establishes that each interval must contain exactly one sibling point for each w. Then, we show that the ordering of sibling points must be the same in each interval, because otherwise the error rate increases. Finally, assuming g to be differentiable yields the monotonicity property.

The verifier has to reconstruct x based on y and w. In general this is done by first identifying which points $x \in \mathbb{R}$ are compatible with w, and then selecting which one is most likely, given y and w. For the first step, we introduce the concept of *sibling points*.

Definition 4.1. (Sibling points): Two points $x, x' \in \mathbb{R}$, with $x \neq x'$, are called Sibling Points if $g(x) = g(x')$.

The verifier determines a set $\mathcal{X}_w = \{x \in \mathbb{R}|g(x) = w\}$ of sibling points that correspond to helper data value w. We write $\mathcal{X}_w = \cup_{s \in \mathcal{S}}\mathcal{X}_{sw}$, with $\mathcal{X}_{sw} = \{x \in \mathbb{R}|Q(x) = s \wedge g(x) = w\}$. We derive a number of requirements on the sets $\mathcal{X}_{sw}$.

Lemma 4.2. ZL implies that

$$\forall_{w \in \mathcal{W}, s \in \mathcal{S}} \quad \mathcal{X}_{sw} \neq \emptyset. \quad (9)$$

Proof: see Appendix A1. Lemma 4.2 tells us that there is significant leakage if there is not at least one sibling point compatible with w in each interval A_s, for all $w \in \mathcal{W}$. Since we are interested in zero leakage, we will from this point onward consider only functions g such that $\mathcal{X}_{sw} \neq \emptyset$ for all s, w.

Next, we look at the requirement of low reconstruction error probability. We focus on the *minimum distance* between sibling points that belong to different quantization intervals.

Definition 4.3. The minimum distance between sibling points in different quantization intervals is defined as

$$D_{\min}(w) = \min_{s,t \in \mathcal{S}: \, s<t} |\min \mathcal{X}_{tw} - \max \mathcal{X}_{sw}|, \quad (10)$$

$$D_{\min} = \min_{w \in \mathcal{W}} D_{\min}(w). \quad (11)$$

We take the approach of maximizing $D_{\min}$. It is intuitively clear that such an approach yields low error rates given the noise model introduced in Section 3.3. The following lemma gives a constraint that improves the $D_{\min}$.

Lemma 4.4. Let $w \in \mathcal{W}$ and $\mathcal{X}_{sw} \neq \emptyset$ for all $s \in \mathcal{S}$. The $D_{min}(w)$ is maximized by setting $|\mathcal{X}_{sw}| = 1$ for all $s \in \mathcal{S}$.

Proof: see Appendix A.2. Lemma 4.4 states that each quantization interval A_s should contain exactly one point x compatible with w. From here onward we will only consider functions g with this property.

The set $\mathcal{X}_{sw}$ consists of a single point which we will denote as x_{sw}. Note that g is then an *invertible* function on each interval A_s. For given $w \in \mathcal{W}$, we now have a set $\mathcal{X}_w = \cup_{s \in \mathcal{S}}x_{sw}$ that consists of one sibling point per quantization interval. This vastly simplifies the analysis. Our next step is to put further constraints on g.

Lemma 4.5. Let $x_1, x_2 \in A_s$ and $x_3, x_4 \in A_t$, $s \neq t$, with $x_1 < x_2 < x_3 < x_4$ and $g(x_1) = w_1, g(x_2) = w_2$. Consider two cases,

1. $g(x_3) = w_1$; $g(x_4) = w_2$
2. $g(x_4) = w_1$; $g(x_3) = w_2$.

Then it holds that

$$\min_{w \in \{w_1, w_2\}} D_{\min}^{\text{case } 2}(w) \leq \min_{w \in \{w_1, w_2\}} D_{\min}^{\text{case } 1}(w). \tag{12}$$

Proof: see Appendix A.3. Lemma 4.5 tells us that *the ordering of sibling points should be the same in each quantization interval*, for otherwise the overall minimum distance $D_{\min}$ suffers. If, for some s, a point x with helper data w_2 is higher than a point with helper data w_1, then this order has to be the same for all intervals.

The combination of having a preserved order (Lemma 4.5) together with g being invertible on each interval (Lemma 4.4) points us in the direction of "smooth" functions. If g is piecewise differentiable, then we can formulate a simple constraint as follows.

Theorem 4.6. (sign of g' equal on each A_s) : *Let $x_s \in A_s, x_t \in A_t$ be sibling points as defined in Def. 4.1. Let g be differentiable in x_s and x_t. Then having sign $g'(x_s) =$ sign $g'(x_t)$ leads to a higher D_{min} than sign $g'(x_s) \neq$ sign $g'(x_t)$.*

Proof: see Appendix A.4.

If we consider a function g that is differentiable on each quantization interval, then (1) its piecewise invertibility implies that it has to be either monotonously increasing or monotonously decreasing on each interval, and (2) Theorem 4.6 then implies that it has to be either increasing on *all* intervals or decreasing on all intervals.

Of course there is no reason to assume that g is piecewise differentiable. For instance, take a piecewise differentiable g and apply a permutation to the w-axis. This procedure yields a function g_2 which, in terms of error probabilities, has exactly the same performance as g, but is not differentiable (nor even continuous). Thus, there exist huge equivalence classes of helper data generating functions that satisfy invertibility (Lemma 4.4) and proper ordering (Lemma 4.5). This brings us to the following conjecture, which allows us to concentrate on functions that are easy to analyze.

Conjecture 4.7. Without loss of generality we can choose the function g to be differentiable on each quantization interval $A_s, s \in \mathcal{S}$.

Based on Conjecture 4.7, we will consider only functions g that are monotonically *increasing* on each interval. This assumption is in line with all (first stage) HDSs [4, 6, 8] known to us.

4.2   Quantile helper data

We state our main result in the theorem below.

*Theorem 4.8. (**ZL is equivalent to quantile relationship between sibling points**): Let g be monotonously increasing on each interval A_s, with $g(A_0) = \cdots = g(A_{N-1}) = W$. Let $s, t \in \mathcal{S}$. Let $x_s \in A_s, x_t \in A_t$ be sibling points as defined in Def. 4.1. In order to satisfy Zero Leakage we have the following necessary and sufficient condition on the sibling points,*

$$\frac{F_X(x_s) - F_X(q_s)}{p_s} = \frac{F_X(x_t) - F_X(q_t)}{p_t}. \tag{13}$$

Proof: see Appendix A.5.

Corollary 4.9. (ZL FE sibling point relation): Let g be monotonously increasing on each interval A_s, with $g(A_0) = \cdots = g(A_{N-1}) = W$. Let $s, t \in \mathcal{S}$. Let $x_s \in A_s, x_t \in A_t$ be sibling points as defined in Def. 4.1. Then for a Fuzzy Extractor we have the following necessary and sufficient condition on the sibling points in order to satisfy Zero Leakage,

$$F_X(x_s) - \frac{s}{N} = F_X(x_t) - \frac{t}{N}. \tag{14}$$

Proof. Immediately follows by combining Eq. (13) with the fact that $p_s = 1/N$ $\forall s \in \mathcal{S}$ in a FE scheme, and with the FE quantization boundaries given in Eq. (5).   □

Theorem 4.8 allows us to define the enrollment steps in a ZL HDS in a very simple way,

$$s = Q(x)$$
$$w = g(x) = \frac{F_X(x) - F_X(q_s)}{p_s}. \tag{15}$$

Note that $w \in [0, 1)$, and $F_X(q_s) = \sum_{t=0}^{s-1} p_t$. The helper data can be interpreted as a quantile distance between x and the quantization boundary q_s, normalized with respect to the probability mass p_s in the interval A_s. An example of such a function is depicted in Fig. 4. For a specific distribution, e.g., a standard Gaussian distribution, the helper data generation function is depicted in Fig. 5. In the FE case, Eq. (15) simplifies to

$$F_X(x) = \frac{s + w}{N}; \qquad w \in [0, 1) \tag{16}$$

and the helper data generation function becomes

$$w = g^{\text{FE}}(x) = N \cdot F_X(x) - s. \tag{17}$$

Equation (16) coincides with the continuum limit of the Fuzzy extractor construction by Verbitskiy et al. [7]. A similar equation was later independently proposed for uniform key generation from a noisy channel by Ye et al. [25]. Equation (15) is the *simplest* way to implement

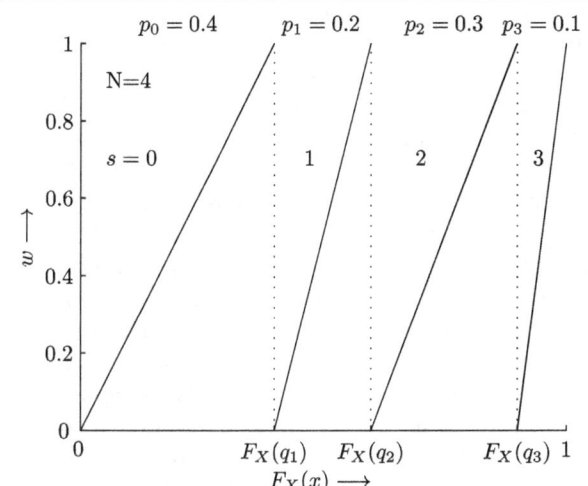

Fig. 4 Example of helper data generating function g for $N = 4$ on quantile x, i.e., $F_X(x)$. The probabilities of the secrets do not have to be equal; in this case, we have used $(p_0, \ldots, p_3) = (0.4, 0.2, 0.3, 0.1)$

an enrollment that satisfies the sibling point relation of Theorem 4.8. However, it is not the *only* way. For instance, by applying any invertible function to w, a new helper data scheme is obtained that also satisfies the sibling point relation (13) and hence is ZL. Another example is to store the whole set of sibling points $\{x_{tw}\}_{t \in S}$; this contains exactly the same information as w. The transformed scheme can be seen as merely a different representation of the "basic" ZL HDS (15). Such a representation may have various advantages over (15), e.g., allowing for a faster reconstruction procedure, while being completely equivalent in terms of the ZL property. We will see such a case in Section 5.3.

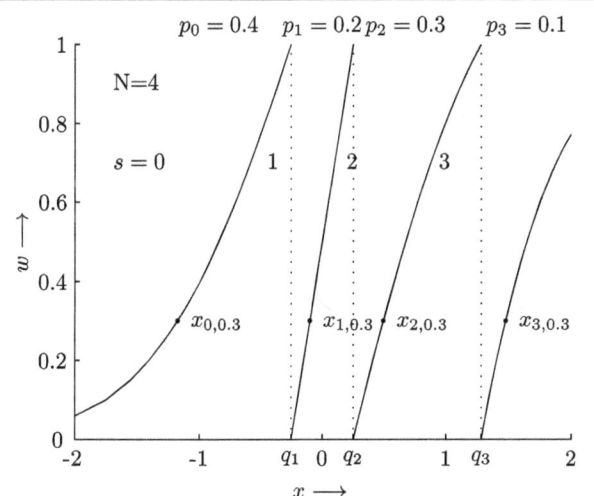

Fig. 5 Example of a helper data generating function g for a standard Gaussian distribution, i.e., $x \sim \mathcal{N}(0, 1)$, and $N = 4$. Sibling points x_{sw} are given for $s \in \{0, \ldots, 3\}$ and $w = 0.3$

5 Optimal reconstruction

5.1 Maximum likelihood and thresholds

The goal of the HDS reconstruction algorithm $\mathrm{Rep}(y, w)$ is to reliably reproduce the secret s. The best way to achieve this is to choose the most probable $\hat{s}$ given y and w, i.e., a maximum likelihood algorithm. We derive optimal decision intervals for the reconstruction phase in a Zero Leakage Fuzzy Extractor.

Lemma 5.1. Let $\mathrm{Rep}(y, w)$ be the reproduction algorithm of a ZL FE system. Let g_s^{-1} be the inverse of the helper data generation function for a given secret s. Then optimal reconstruction is achieved by

$$\mathrm{Rep}(y, w) = \arg\max_{s \in S} f_{Y|X}\left(y | g_s^{-1}(w)\right). \tag{18}$$

Proof: see Appendix A.6. To simplify the verification phase we can identify thresholds τ_s that denote the lower boundary of a decision region. If $\tau_s \leq y < \tau_{s+1}$, we reconstruct $\hat{s} = s$. The $\tau_0 = -\infty$ and $\tau_N = \infty$ are fixed, which implies we have to find optimal values only for the $N - 1$ variables $\tau_1, \ldots, \tau_{N-1}$ as a function of w.

Theorem 5.2. Let $f_{Y|X}$ represent symmetric fading noise. Then optimal reconstruction in a FE scheme is obtained by the following choice of thresholds

$$\tau_s = \lambda \frac{g_s^{-1}(w) + g_{s-1}^{-1}(w)}{2}. \tag{19}$$

Proof. In case of symmetric fading noise we know that

$$f_{Y|X}(y|x) = \varphi(|y - \lambda x|), \tag{20}$$

with φ some monotonic decreasing function. Combining this notion with that of Eq. (18) to find a point $y = \tau_s$ that gives equal probability for s and $s - 1$ yields

$$\varphi\left(|\tau_s - \lambda g_{s-1}^{-1}(w)|\right) = \varphi\left(|\tau_s - \lambda g_s^{-1}(w)|\right). \tag{21}$$

The left and right hand side of this equation can only be equal for equal arguments, and hence

$$\tau_s - \lambda g_{s-1}^{-1}(w) = \pm\left(\tau_s - \lambda g_s^{-1}(w)\right). \tag{22}$$

Since $g_s^{-1}(w) \neq g_{s-1}^{-1}(w)$ the only viable solution is Eq. (19). $\square$

Instead of storing the ZL helper data w according to (15), one can also store the set of thresholds $\tau_1, \ldots, \tau_{N-1}$. This contains precisely the same information, and allows for quicker reconstruction of s: just a thresholding operation on y and the τ_s values, which can be implemented on computationally limited devices.

5.2 Special case: 1-bit secret

In the case of a one-bit secret s, i.e., $N = 2$, the above ZL FE scheme is reduced to storing a single threshold τ_1.

It is interesting and somewhat counterintuitive that this yields a threshold for verification that does not leak information about the secret. In case the average of X is zero, one might assume that a positive threshold value implies $s = 0$. However, both $s = 0$ and $s = 1$ allow positive as well as negative τ_1, dependent on the relative location of x in the quantization interval.

5.3 FE: equivalent choices for the quantization

Let us reconsider the quantization function $Q(x)$ in the case of a Fuzzy extractor. Let us fix N and take the $g(x)$ as specified in Eq. (16). Then, it is possible to find an infinite number of different functions Q that will conserve the ZL property and lead to exactly the same error rate as the original scheme. This is seen as follows. For any $w \in [\,0, 1)$, there is an N-tuplet of sibling points. Without any impact on the reconstruction performance, we can permute the s-values of these points; the error rate of the reconstruction procedure depends only on the x-values of the sibling points, not on the s-label they carry. It is allowed to do this permutation for every w independently, resulting in an infinite equivalence class of Q-functions. The choice we made in Section 3 yields the simplest function in an equivalence class.

6 Example: Gaussian features and BCH codes

To benchmark the reproduction performance of our scheme, we give an example based on Gaussian-distributed variables. In this example, we will assume all variables to be Gaussian distributed, though we remind

the reader that our scheme specifies optimal reconstruction thresholds even for non-Gaussian distributions.

We compare the reproduction performance of our ZL quantization scheme with Likelihood-based reproduction (ZLQ-LR) to a scheme with (1) fixed quantization (FQ), see Section 2.1, and (2) likelihood classification (LC). The former is, to our knowledge, the only other scheme sharing the zero secrecy leakage property, since it does not use any helper data. An example with $N = 4$ intervals is depicted in Fig. 6a. LC is not an actual quantization scheme since it requires the enrollment sample to be stored in-the-clear. However, a likelihood based classifier provides an optimal trade-off between false acceptance and false rejection according to communication theory [24] and should therefore yield the lowest possible error rate. Instead of quantization boundaries, the classifier is characterized by decision boundaries as depicted in Fig. 6b.

A comparison with QIM cannot be made since there the probability for an impostor to guess the enrolled secret cannot be made equal to $1/N$. This would result in an unfair comparison since the other schemes are designed to possess this property. Moreover, the QIM scheme allows the reproduction error probability to be made arbitrary small by increasing the quantization width at the cost of leakage.

Also, the likelihood based classification can be tuned by setting the decision threshold. However, for this scheme, it is possible to choose a threshold such that an impostor will have a probability of $1/N$ to be accepted, which corresponds to the $1/N$ probability of guessing the enrolled secret in a FE scheme. Note that for a likelihood classifier, there is no enrolled secret since this is not a quantization scheme.

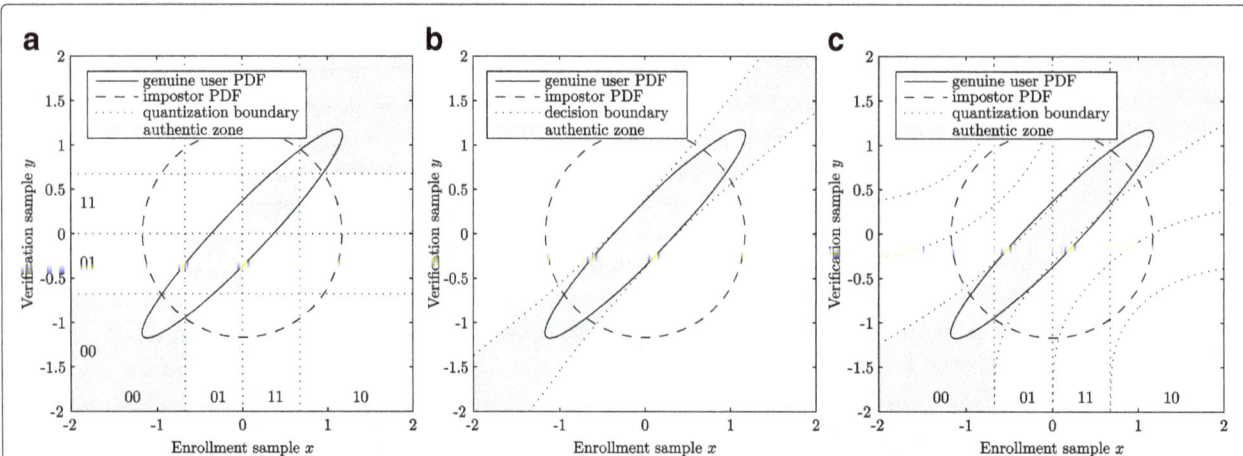

Fig. 6 Quantization and decision patterns based on the genuine user and impostor PDFs. Ideally the genuine user PDF should be contained in the authentic zone and the impostor PDF should have a large mass outside the authentic zone. Fifty percent probability mass is contained in the genuine user and impostor PDF ellipse and circle. The genuine user PDF is based on a 10 dB SNR. **a** Fixed equiprobable quantization (FQ). **b** Likelihood classification (LC). **c** Zero leakage quantization scheme with likelihood based reproduction (ZLQ-LR)

As can be seen from Fig. 7, the reproduction performance for a ZL scheme with likelihood based reproduction is always better than that of a fixed quantization scheme. However, it is outperformed by the likelihood classifier. Differences are especially apparent for features with a higher signal-to-noise ratio. In these regions, the fixed quantization struggles with a inherent high error probability, while the ZL scheme follows the LC.

In a good quantization scheme, the gap between $I(X; Y)$ and $I(S; \hat{S})$ must be small. For a Gaussian channel, standard expressions are known from ([24], Eq. (9.16)). Figure 8 shows that a fixed quantization requires a higher SNR on order to converge to the maximum number of bits, whereas the ZLQ-LR scheme directly reaches this value.

Finally, we consider the vector case of the two quantization schemes discussed above. We concluded that FQ has a larger error probability, but we now show how this relates to either false rejection or secret length when combined with a code offset method [3].

We assume i.i.d. features and therefore we can calculate false acceptance rate (FAR) and false rejection rate (FRR) based on a binomial distribution. In practice, features can be made (nearly) independent, but they will in general not be identically distributed. However, results will be similar. Furthermore we assume the error correcting code can be applied such that its error correcting properties can be fully exploited. This implies that we have to use a Gray code to label the extracted secrets before concatenation.

We used 64 i.i.d. features, each having a SNR of 17 dB, which is a typical average value for biometric features

[8, 17]. From these features, we extract 2 bits per feature on which we apply BCH codes with a code length of 127. (We omit one bit). For analysis, we have also included the code (127, 127, 0), which is not an actual code, but represents the case in which no error correction is applied.

Suppose we want to achieve a target FRR of $1 \cdot 10^{-3}$, the topmost dotted line in Fig. 9, then we require a BCH (127, 92, 5) code for the ZLQ-LR scheme, while a BCH (127, 15, 27) code is required for the FQ scheme. This implies that we would have a secret key size of 92 bits versus 15 bits. Clearly, the latter is not sufficient for any security application. At the same time, due to the small key size, FQ has an increased FAR.

7 Conclusions

In this paper, we have studied a generic helper data scheme (HDS) which comprises the Fuzzy extractor (FE) and the secure sketch (SS) as special cases. In particular, we have looked at the zero leakage (ZL) property of HDSs in the case of a one-dimensional continuous source X and continuous helper data W.

We make minimal assumptions, justified by Conjecture 4.7: we consider only monotonic $g(x)$. We have shown that the ZL property implies the existence of sibling points $\{x_{sw}\}_{s \in S}$ for every w. These are values of x that have the same helper data w. Furthermore, the ZL requirement is equivalent to a quantile relationship (Theorem 4.8) between the sibling points. This directly leads to Eq. (15) for computing w from x. (Applying any reversible function to this w yields a completely equivalent helper data

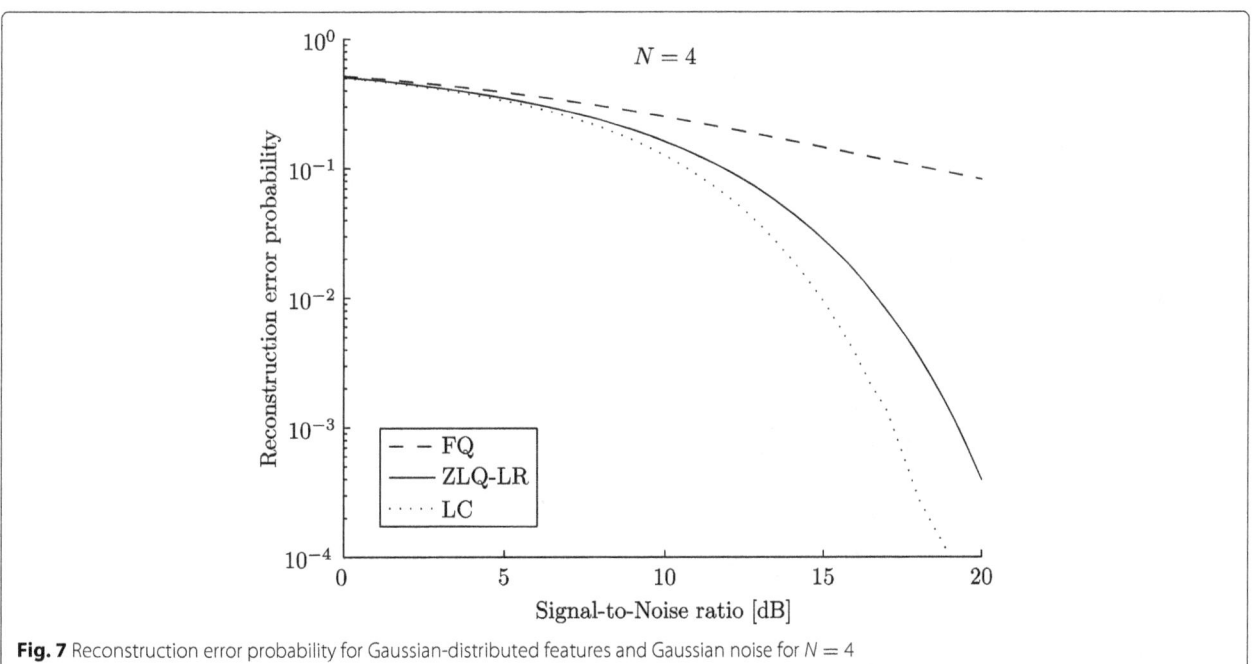

Fig. 7 Reconstruction error probability for Gaussian-distributed features and Gaussian noise for $N = 4$

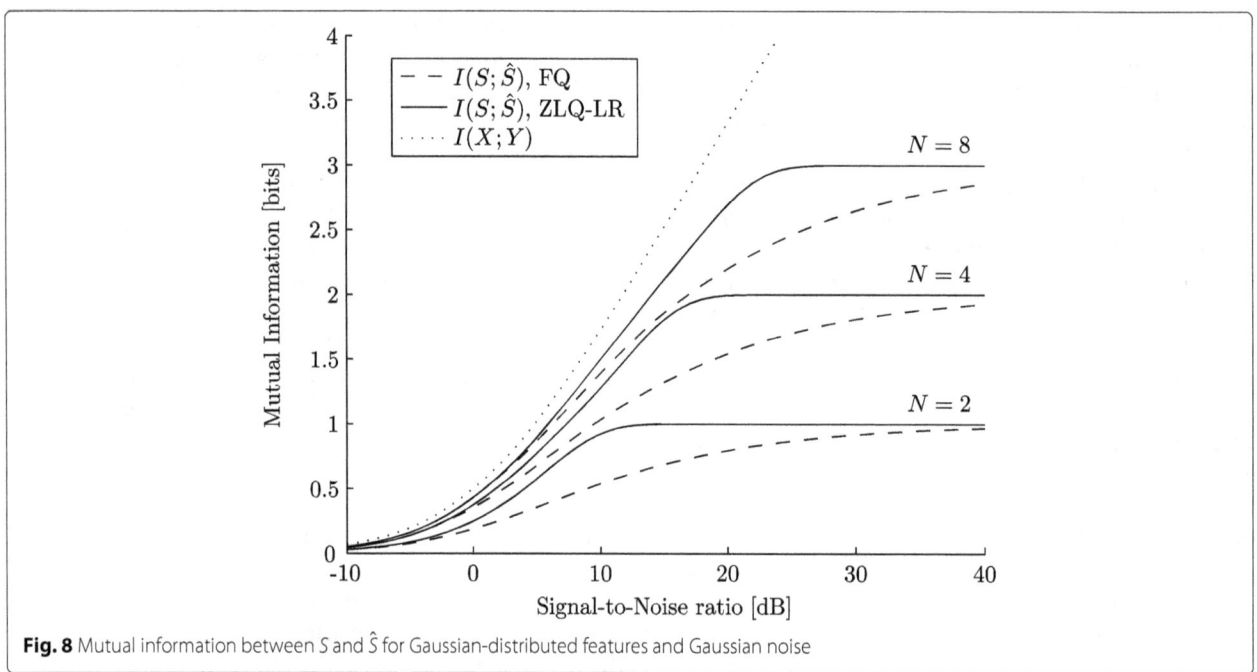

Fig. 8 Mutual information between S and $\hat{S}$ for Gaussian-distributed features and Gaussian noise

system.) The special case of a FE ($p_s = 1/N$) yields the $m \to \infty$ limit of the Verbitskiy et al. [7] construction.

We have derived reconstruction thresholds τ_s for a ZL FE that minimize the error rate in the reconstruction of s (Theorem 5.2). This result holds under very mild assumptions on the noise: symmetric and fading. Equation (19) contains the attenuation parameter λ, which follows from the noise model as specified in Section 3.3.

Finally, we have analyzed reproduction performance in an all-Gaussian example. Fixed quantization struggles with inherent high error probability, while the ZL FE with optimal reproduction follows the performance of the optimal classification algorithm. This results in a larger key size in the protected template compared to the fixed quantization scheme, since an ECC with a larger message length can be applied in the second stage HDS to achieve the same FRR.

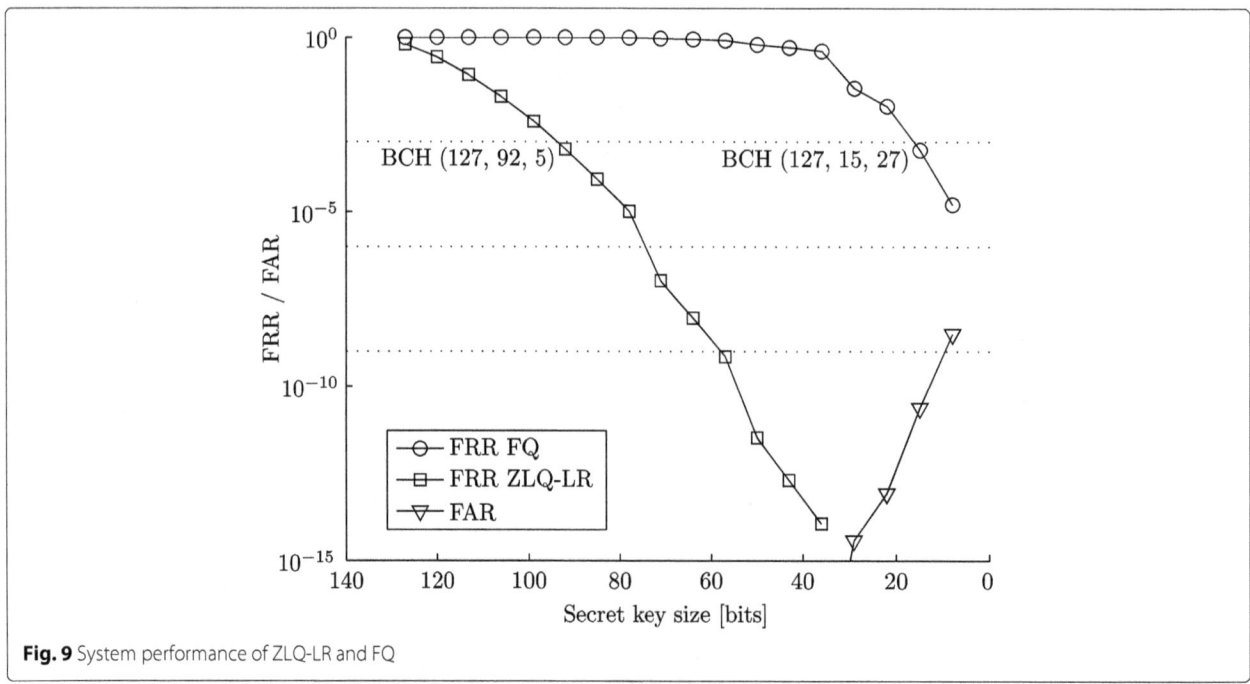

Fig. 9 System performance of ZLQ-LR and FQ

In this paper, we have focused on arbitrary but known probability densities. Experiments with real data are beyond the scope of this paper, but have been reported in [26, 27]. A key finding there was that modeling the distributions can be problematic, especially due to statistical outliers. Even so, improvements were obtained with respect to earlier ZL schemes. We see modeling refinements as a topic for future research.

Endnotes

[1] For information-theoretic concepts such as Shannon entropy and mutual information we refer to e.g. [24].

[2] This concept is not new. Achieving zero mutual information has always been a (sometimes achievable) desideratum in the literature on fuzzy vaults/fuzzy extractors/secure sketches for discrete and continuous sources.

[3] An overall Zero-Leakage scheme can be obtained in the final HDS stage even from a leaky HDS by applying privacy amplification as a post-processing step. However, this procedure discards substantial amounts of source entropy, while in many practical applications it is already a challenge to achieve reasonable security levels from biometrics without privacy protection.

Appendix
Appendix A: Proofs
A.1 Proof of Lemma 4.2

Pick any $w \in \mathcal{W}$. The statement $w \in \mathcal{W}$ means that there exists at least one $s' \in \mathcal{S}$ such that $\mathcal{X}_{s'w} \neq \emptyset$. Now suppose there exists some $s'' \in \mathcal{S}$ with $\mathcal{X}_{s''w} = \emptyset$. Then knowledge of w reveals information about S, i.e. $\mathbb{P}[S = s|W = w] \neq \mathbb{P}[S = s]$, which contradicts ZL.

A.2 Proof of Lemma 4.4

Let g be such that $|\mathcal{X}_{sw}| > 1$ for some s, w. Then choose a point $\bar{x} \in \mathcal{X}_{sw}$. Construct a function g_2 such that

$$g_2(x) \begin{cases} = g(x) & \text{if } x = \bar{x} \text{ or } x \notin \mathcal{X}_{sw} \\ \neq g(x) & \text{otherwise} \end{cases}. \tag{23}$$

The $D_{\min}(w)$ for g_2 cannot be smaller than $D_{\min}(w)$ for g.

A.3 Proof of Lemma 4.5

We tabulate the $D_{\min}(w)$ values for case 1 and 2,

	case 1	case 2
w_1	$x_3 - x_1$	$x_4 - x_1$
w_2	$x_4 - x_2$	$x_3 - x_2$

The smallest of these distances is $x_3 - x_2$.

A.4 Proof of Theorem 4.6

Let $0 < \varepsilon \ll 1$ and $0 < \delta \ll 1$. Without loss of generality we consider $s < t$. We invoke Lemma 4.5 with $x_1 = x_s$,

$x_2 = x_s + \varepsilon, x_3 = x_t, x_4 = x_t + \delta$. According to Lemma 4.5 we have to take $g(x_2) = g(x_4)$ in order to obtain a large $D_{\min}$. Applying a first order Taylor expansion, this gives

$$g(x_s) + \varepsilon g'(x_s) + \mathcal{O}(\varepsilon^2) = g(x_t) + \delta g'(x_t) + \mathcal{O}(\delta^2). \tag{24}$$

We use the fact that $g(x_s) = g(x_t)$, and that ε and δ are positive. Taking the sign of both sides of (24) and neglecting second order contributions, we get sign $g'(x_s) = $ sign $g'(x_t)$.

A.5 Proof of Theorem 4.8

The ZL property is equivalent to $f_W = f_{W|S}$, which gives for all $s \in \mathcal{S}$

$$f_W(w) = f_{W|S}(w|s) = \frac{f_{W,S}(w,s)}{p_s}, \tag{25}$$

where $f_{W,S}$ is the joint distribution for W and S. We work under the assumption that $w = g(x)$ is a monotonous function on each interval A_s, fully spanning $\mathcal{W}$. Then for given s and w there exists exactly one point x_{sw} that satisfies $Q(x) = s$ and $g(x) = w$. Furthermore, conservation of probability then gives $f_{W,S}(w,s) \, dw = f_X(x_{sw}) \, dx_{sw}$. Since the right hand side of (25) is independent of s, we can write $f_W(w)dw = p_s^{-1} f_X(x_{sw})dx_{sw}$ for *any* $s \in \mathcal{S}$. Hence for any $s, t \in \mathcal{S}, w \in \mathcal{W}$ it holds that

$$\frac{f_X(x_{sw})dx_{sw}}{p_s} = \frac{f_X(x_{tw})dx_{tw}}{p_t}, \tag{26}$$

which can be rewritten as

$$\frac{dF_X(x_{sw})}{p_s} = \frac{dF_X(x_{tw})}{p_t}. \tag{27}$$

The result (13) follows by integration, using the fact that A_s has lower boundary q_s.

A.6 Proof of Lemma 5.1

Optimal reconstruction can be done by selecting the most likely secret given y, w,

$$\text{Rep}(y,w) = \arg\max_{s \in \mathcal{S}} f_{S|Y,W}(s|y,w) = \arg\max_{s \in \mathcal{S}} \frac{f_{Y,S,W}(y,s,w)}{f_{Y,W}(y,w)}. \tag{28}$$

The denominator does not depend on s, and can hence be omitted. This gives

$$\text{Rep}(y,w) = \arg\max_{s \in \mathcal{S}} f_{S,Y,W}(s,y,w) \tag{29}$$

$$= \arg\max_{s \in \mathcal{S}} f_{Y|S,W}(y|s,w) f_{W|S}(w|s) p_s. \tag{30}$$

We constructed the scheme to be a FE with ZL, and therefore $p_s = 1/N$ and $f_{W|S}(w|s) = f_W(w)$. We see that both p_s and $f_{W|S}(w|s)$ do not depend on s, which implies they can be omitted from Eq. (30), yielding $\text{Rep}(y,w) = \arg\max_{s \in \mathcal{S}} f_{Y|S,W}(y|s,w)$. Finally, knowing S and W is equivalent to knowing X. Hence $f_{Y|S,W}(y|s,w)$ can be replaced by $f_{Y|X}(y|x)$ with x satisfying $Q(x) = s$ and $g(x) = w$. The unique x value that satisfies these constraints is $g_s^{-1}(w)$.

Competing interests

The authors declare that they have no competing interests.

Author details

[1]Signal Processing Systems group, Department of Electrical Engineering, Eindhoven University of Technology, 5600 MB, Eindhoven, The Netherlands. [2]Security and Embedded Networked Systems group, Department of Mathematics and Computer Science, Eindhoven University of Technology, 5600 MB Eindhoven, The Netherlands. [3]Discrete Mathematics group, Department of Mathematics and Computer Science, Eindhoven University of Technology, 5600 MB Eindhoven, The Netherlands. [4]Genkey Solutions B.V., High Tech Campus 69, 5656 AG Eindhoven, The Netherlands.

References

1. T van der Putte, J Keuning, in *Proceedings of the Fourth Working Conference on Smart Card Research and Advanced Applications on Smart Card Research and Advanced Applications*. Biometrical fingerprint recognition: don't get your fingers burned (Kluwer Academic Publishers, Norwell, MA, USA, 2001), pp. 289–303
2. T Matsumoto, H Matsumoto, K Yamada, S Hoshino, Impact of artificial "gummy" fingers on fingerprint systems. Opt. Secur. Counterfeit Deterrence Tech. **4677**, 275–289 (2002)
3. A Juels, M Wattenberg, in *CCS '99: Proceedings of the 6th ACM Conf on Comp and Comm Security*. A fuzzy commitment scheme (ACM, New York, NY, USA, 1999), pp. 28–36. doi:10.1145/319709.319714. http://doi.acm.org/10.1145/319709.319714
4. J-P Linnartz, P Tuyls, in *New Shielding Functions to Enhance Privacy and Prevent Misuse of Biometric Templates*, ed. by J Kittler, Mark Nixon. Audio- and Video-Based Biometric Person Authentication: 4th International Conference, AVBPA 2003 Guildford, UK, June 9–11, 2003 Proceedings (Springer Berlin Heidelberg, Berlin, Heidelberg, 2003), pp. 393–402. doi:10.1007/3-540-44887-X_47. http://dx.doi.org/10.1007/3-540-44887-X_47
5. P Tuyls, B Škorić, T Kevenaar, *Security with Noisy Data: Private Biometrics, Secure Key Storage and Anti-Counterfeiting*. (Springer, Secaucus, NJ, USA, 2007)
6. C Chen, RNJ Veldhuis, TAM Kevenaar, AHM Akkermans, in *Proc. IEEE Int. Conf. on Biometrics: Theory, Applications, and Systems*. Multi-bits biometric string generation based on the likelihood ratio (IEEE, Piscataway, 2007)
7. EA Verbitskiy, P Tuyls, C Obi, B Schoenmakers, B Škorić, Key extraction from general nondiscrete signals. Inform. Forensics Secur. IEEE Trans. **5**(2), 269–279 (2010)
8. JA de Groot, J-PMG Linnartz, in *Proc. IEEE Int. Conf. Acoust., Speech, Signal Process*. Zero leakage quantization scheme for biometric verification, (Piscataway, 2011)
9. Y Dodis, L Reyzin, A Smith, in *Fuzzy Extractors: How to Generate Strong Keys from Biometrics and Other Noisy Data*, ed. by C Cachin, JL Camenisch. Advances in Cryptology - EUROCRYPT 2004. International Conference on the Theory and Applications of Cryptographic Techniques, Interlaken, Switzerland, May 2-6, 2004. Proceedings (Springer Berlin Heidelberg, Berlin, Heidelberg, 2004), pp. 523–540. doi:10.1007/978-3-540-24676-3_31 http://dx.doi.org/10.1007/978-3-540-24676-3_31
10. AV Herrewege, S Katzenbeisser, R Maes, R Peeters, A-R Sadeghi, I Verbauwhede, C Wachsmann, in *Reverse Fuzzy Extractors: Enabling Lightweight Mutual Authentication for PUF-Enabled RFIDs*, ed. by AD Keromytis. Financial Cryptography and Data Security: 16th International Conference, FC 2012, Kralendijk, Bonaire, Februray 27-March 2, 2012, Revised Selected Papers (Springer Berlin Heidelberg, Berlin, Heidelberg, 2012), pp. 374–389. doi:10.1007/978-3-642-32946-3_27. http://dx.doi.org/10.1007/978-3-642-32946-3_27

11. B Škorić, N de Vreede, The spammed code offset method. IEEE Trans. Inform. Forensics Secur. **9**(5), 875–884 (2014)
12. F MacWilliams, N Sloane, *The Theory of Error Correcting Codes*. (Elsevier, Amsterdam, 1978)
13. JL Wayman, AK Jain, D Maltoni, D Maio (eds.), *Biometric Systems: Technology, Design and Performance Evaluation*, 1st edn. (Spring Verlag, London, 2005)
14. B Škorić, P Tuyls, W Ophey, in *Robust Key Extraction from Physical Uncloneable Functions*, ed. by J Ioannidis, A Keromytis, and M Yung. Applied Cryptography and Network Security: Third International Conference, ACNS 2005, New York, NY, USA, June 7-10, 2005. Proceedings (Springer Berlin Heidelberg, Berlin, Heidelberg, 2005), pp. 407–422. doi:10.1007/11496137_28. http://dx.doi.org/10.1007/11496137_28
15. GE Suh, S Devadas, in *Proceedings of the 44th Annual Design Automation Conference*. DAC '07. Physical unclonable functions for device authentication and secret key generation (ACM, New York, NY, USA, 2007), pp. 9–14
16. DE Holcomb, WP Burleson, K Fu, Power-Up SRAM state as an identifying fingerprint and source of true random numbers. Comput. IEEE Trans. **58**(9), 1198–1210 (2009)
17. P Tuyls, A Akkermans, T Kevenaar, G-J Schrijen, A Bazen, R Veldhuis, in *Practical Biometric Authentication with Template Protection*, ed. by T Kanade, A Jain, and N Ratha. Audio- and Video-Based Biometric Person Authentication: 5th International Conference, AVBPA 2005, Hilton Rye Town, NY, USA, July 20-22, 2005. Proceedings (Springer Berlin Heidelberg, Berlin, Heidelberg, 2005), pp. 436–446. doi:10.1007/11527923_45. http://dx.doi.org/10.1007/11527923_45
18. EJC Kelkboom, GG Molina, J Breebaart, RNJ Veldhuis, TAM Kevenaar, W Jonker, Binary biometrics: an analytic framework to estimate the performance curves under gaussian assumption. Syst. Man Cybernetics, Part A: Syst. Hum. IEEE Trans. **40**(3), 555–571 (2010). doi:10.1109/TSMCA.2010.2041657
19. EJC Kelkboom, KTJ de Groot, C Chen, J Breebaart, RNJ Veldhuis, in *Biometrics: Theory, Applications, and Systems, 2009. BTAS '09. IEEE 3rd International Conference On*. Pitfall of the detection rate optimized bit allocation within template protection and a remedy, (2009), pp. 1–8. doi:10.1109/BTAS.2009.5339046
20. B Chen, GW Wornell, Quantization index modulation: a class of provably good methods for digital watermarking and information embedding. Inform. Theory IEEE Trans. **47**(4), 1423–1443 (2001). doi:10.1109/18.923725
21. MHM Costa, Writing on dirty paper (corresp.) IEEE Trans. Inform. Theory. **29**(3), 439–441 (1983)
22. M-H Lim, ABJ Teoh, K-A Toh, Dynamic detection-rate-based bit allocation with genuine interval concealment for binary biometric representation. IEEE Trans. Cybernet., 843–857 (2013)
23. C Chen, RNJ Veldhuis, TAM Kevenaar, AHM Akkermans, Biometric quantization through detection rate optimized bit allocation. EURASIP J. Adv. Signal Process. **2009**, 29–12916 (2009). doi:10.1155/2009/784834
24. TM Cover, JA Thomas, *Elements of Information Theory*, 2nd edn. (John Wiley & Sons, Inc., 2005)
25. C Ye, S Mathur, A Reznik, Y Shah, W Trappe, NB Mandayam, Information-theoretically secret key generation for fading wireless channels. IEEE Trans. Inform. Forensics Secur. **5**(2), 240–254 (2010)
26. JA de Groot, J-PMG Linnartz, in *Proc. WIC Symposium on Information Theory in the Benelux*. Improved privacy protection in authentication by fingerprints (WIC, The Netherlands, 2011)
27. JA de Groot, B Škorić, N de Vreede, J-PMG Linnartz, in *Security and Cryptography (SECRYPT), 2013 International Conference On*. Diagnostic category leakage in helper data schemes for biometric authentication (IEEE, Piscataway, 2013), pp. 1–6

11

Advanced flooding-based routing protocols for underwater sensor networks

Elvin Isufi[1]*, Henry Dol[2] and Geert Leus[1]

Abstract

Flooding-based protocols are a reliable solution to deliver packets in underwater sensor networks. However, these protocols potentially involve all the nodes in the forwarding process. Thus, the performance and energy efficiency are not optimal. In this work, we propose some advances of a flooding-based protocol with the goal to improve the performance and the energy efficiency. The first idea considers the node position information in order to reduce the number of relays that may apply flooding. Second, a network coding-based protocol is proposed in order to make a better use of the duplicates. With network coding, each node in the network recombines a certain number of packets into one or more output packets. This may give good results in flooding-based protocols considering the high amount of packets that are flooded in the network. Finally, a fusion of both ideas is considered in order to exploit the benefits of both of them.

Keywords: Underwater communications, Flooding-based routing, Network coding, Geographical routing, Implicit acknowledgement

1 Introduction

Starting from the first underwater telephone, developed by the Naval Underwater Sound Laboratory (USA) [1], many research efforts have been put in underwater communications for both civil and military applications [2]. Acoustic technology is mostly preferred for communication distances that exceed about a hundred meters. However, in contrast to terrestrial radio-frequency communications, underwater acoustic links are characterized by long propagation delays, low data rates, limited and variable bandwidth, and high bit error rates [3, 4]. According to [5], the attenuation of an underwater acoustic link increases exponentially with the distance, and in [6], it has been shown that we also pay in terms of bandwidth for greater transmission ranges. In several underwater communication scenarios, we may be interested in reaching distances longer than the range of a direct transmission link, or large areas need to be covered. In these cases, an extension to an underwater acoustic communication *network* is required (i.e., multihop transmission instead of a single direct transmission), which brings benefits in terms of energy and capacity.

On the other hand, underwater acoustic networks bring new issues, and thus efficient routing protocols are required to determine the path that the packets must follow to reach the destination. We redirect the reader to [3, 7] and references therein for a deeper analysis and overview about underwater routing protocols and to [8] for design guidelines about opportunistic routing. Irrespective of the routing protocol, or the application scenario, one of the main goals is to obtain a high packet delivery ratio (PDR) while keeping the end-to-end delay and energy consumption limited. One class of routing protocols, which can be used in underwater scenarios, consists of flooding-based protocols. These protocols may be preferred in networks where the nodes are generally not static, and when the communication links face outages, meaning that continuously updating the routing table may reduce the overall throughput. The performance of flooding-based protocols starts degrading when the network becomes overloaded by traffic. Thus, in order to avoid this situation, and to reduce the number of collisions, the number of duplicates that are flooded in the network must be kept limited.

*Correspondence: e.isufi-1@tudelft.nl
[1] Faculty of Electical Engineering, Mathematics and Computer Science, Delft University of Technology, 2628 CD, Delft, The Netherlands
Full list of author information is available at the end of the article

In [9], a duplicate reduction flooding-based protocol (called *Dflood*) has been proposed[1]. Here, the nodes use some backoff time for each information packet, which is continuously adapted when a duplicate is overheard. Even considering these duplicate reduction policies applied by Dflood, it may happen that in certain scenarios, the number of packets forwarded by the network, i.e., *the energy consumption* is still high. For more details about the Dflood protocol, we redirect the reader to [9] and [10], where more information is provided for all protocol layers. In this work, we will mostly focus on advancing the performance (improve the PDR), and the energy efficiency of Dflood.

The first idea is to reduce *the number of duplicates* that are forwarded by the network, using node position information. Our motivation comes from the fact that Dflood does not distinguish between a source node that is closer to the destination and another one that might be further. As a result, whatever the physical position of the source, the entire network is prospected to take part in the relaying process. In large networks, and when the sources are close to the destination, the resulting amount of packets forwarded by the network may be unacceptable. To reduce the energy consumption, we propose an enhancement of Dflood exploiting the location information of the node that is transmitting and the final destination position. This information has been used in other approaches such as [11] and [12]. The main difference with the last approach [12] is that our nodes are equipped with omnidirectional hydrophones, so all the neighbours within the transmission range can receive the transmitted packet. The way in which we use location information is the main difference with the first approach [11]. Instead of creating a straight pipe from the source node to the destination, and to allow only the nodes inside the pipe to take part in the forwarding, we consider the possibility to involve also other nodes when necessary. More details will be provided in the next section. Furthermore, we have extended our approach with the use of an implicit acknowledgement (ACK). This is an efficient retransmission strategy, where the ACK is not explicitly sent by the receiver node. More specifically, if the sender node, within a certain interval of time, does not overhear the transmission of that specific packet from one of its neighbours (remember that the nodes are equipped with omnidirectional hydrophones), it will retransmit another copy. This strategy helps improving the PDR, as well as reducing the end-to-end delay and energy consumption with respect to the other automatic-repeat-request (ARQ) schemes.

Our second contribution is motivated by the fact that the number of replicas that are flooded in the network and received by the destination is still high. Instead of trying to reduce them further, we have considered the possibility that these packets may share the information of more than one packet. Basically speaking, we propose network coding (NC) [13], in order to have more information flooding in the network than replicas. With NC, each node in the network is allowed to recombine a certain number of packets into one or more output packets, instead of simply forwarding each of them. It is clear that this increases the robustness of the transmission, especially in a flooding fashion. *Linear* NC has been proposed recently, where the output packets are linear combinations of the packets presented in the node's buffer. It has also been used in underwater sensor networks to potentially increase the PDR, the energy efficiency, and the throughput, as well as to reduce the end-to-end delay [14–16]. It has shown promising results in both simulations [17], and real experimental trials [18, 19]. This motivated us to consider NC also in flooding-based protocols, where many packets flood in the network and the sharing of information between them will potentially improve the information at the destination. However, we should be aware of the fact that being inspired by a flooding-based approach, the number of replicas must be kept limited in order to be energy-efficient. For this reason, we propose to fuse linear NC with Dflood in order to gain the benefits from both approaches, i.e., more robustness from the NC and a reduced energy consumption using the Dflood idea. Our idea does not simply consist of using NC on top of Dflood. Instead, we have reviewed the latter's rules and changed them in order to exploit the potential of NC. This is the main difference with the approach in [14]. The key differences with [15] and [16] are, respectively, that our approach is developed for any scenario with a sink node (not only for a chain topology), and that the packets arrive at the next relay node with a different and variable delay. Next, we have considered such an approach in conjunction with geographical information, as discussed earlier. For the considered scenario, simulation results show an increment of up to 10 % of the PDR, with respect to the original Dflood, for low packet error rates (PERs) and a low traffic load. At the same time, the energy efficiency of the network is increased.

The rest of this paper is organized as follows. In the next section, we will present our geographical protocol; the NC-based solution will be explained in Section 3; in Section 4, we will show the simulation results; and the work will be concluded in Section 5.

2 Geographical Dflood
2.1 Proposed protocol
Geographical Dflood (GDflood) uses the node position information in order to reduce the number of relays that take part in the forwarding process. With GDflood, the transmitting node makes use only of its own position and the final destination position. Using this information during the relaying, it is more likely that only those nodes

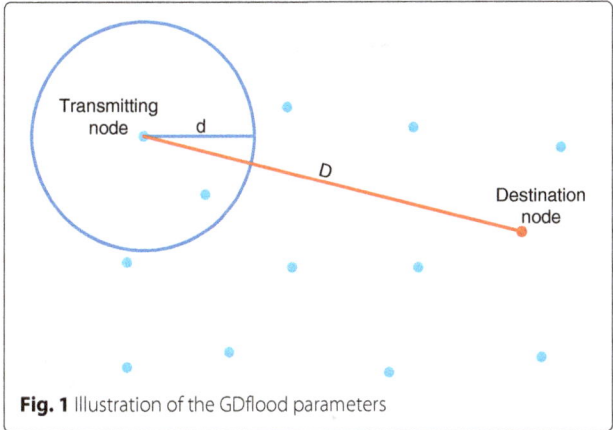

Fig. 1 Illustration of the GDflood parameters

that are closer to the destination take part in the forwarding process. GDflood uses the same network header, as Dflood, composed by *a source address, a destination address, a sequence number,* and *a hop count.* To distinguish between different packets, the first three fields form a *unique ID* in the network.

We assume that in a single transmission (i.e., 1 hop), a node can send a packet to its neighbors within a distance d. When a node has a packet to send, it first calculates the distance D, between itself and the destination, as illustrated in Fig. 1. Then, this distance is quantized into hop counts

$$D_{\text{HC}} = \lceil D/d \rceil, \tag{1}$$

where $\lceil \cdot \rceil$ indicates the ceiling operator. The transmitting node puts the value $D_{\text{HC}} + f_{\text{n}}$ in the hop count field, where $f_{\text{n}} \leq 0$ is a common term for all nodes that influences the number of relays that may consider the packet for the next transmission. The role of this parameter is to keep the number of relays limited (see later on). Only if the node is the source node, it will consider a redundancy factor $f_{\text{s}} \geq 0$, for its own packets. The hop count of these packets will be $D_{\text{HC}} + f_{\text{n}} + f_{\text{s}}$ (see also later on). After setting the hop count, the node sends the packet immediately to the MAC layer. Depending on the MAC protocol used, if the packet is not directly transmitted but sent back to the upper layers, the above mechanism has to be repeated. This is important because the moving nodes change continuously their position, and as a consequence D_{HC} changes. GDflood is explained in detail below in relation to the role that a specific node has in the transmission, i.e., if it is *the source* of the packet, *the destination* node, or *a relay* node, see also Fig. 2.

Source node: In case the node is the source node, the packets to be transmitted are received from the upper layers. The source will schedule the packet directly, so the backoff time, T_{back}, in this case is the next available time

instant. When this time is reached, the source node calculates the D_{HC}, as explained before enters as hop count $D_{\text{HC}} + f_{\text{n}} + f_{\text{s}}$, and sends the packet down to the MAC layer.

Destination node: In case a node receives a packet intended for itself, i.e., if it is the destination node, it will immediately broadcast a receive notification (RN). The nodes that receive the RN will stop relaying the packet with the same unique ID. The RN packet will not be forwarded.

Relay node: When the node is a relay, it will first check if the packet has been received before or not. In case the packet is a duplicate, and already transmitted, it will not be forwarded anymore.

In case the packet is a duplicate but scheduled to be transmitted, the node will apply some duplicate policies. First, it will check if it has to stop taking part in the relaying process. This is done by uniformly drawing a random number $\rho \in [0,1]$ and by checking if

$$n_{\text{d}} > N_{\text{dupl}} - \rho, \tag{2}$$

where n_{d} is the number of duplicates collected till that moment, and N_{dupl} is defined as the maximum number of duplicates (in practice, it can be defined also as a rational number). If the test fails, the relay node will add a T_{dupl} to the scheduled time for that packet.

In case the packet is not a duplicate, the relay node will check if it can take part in the forwarding. This is done by checking if $D_{\text{HC}} \leq$ hop count, where D_{HC} is the distance in hop counts of the relay node to the destination. In case the node is allowed to take part in the relaying process, the packet will be scheduled to be transmitted after a random time T_{back}, uniformly drawn in $[T_{\text{min}}, T_{\text{max}}]$. The new packet will have $D_{\text{HC}} + f_{\text{n}}$ as new hop count value.

Considering the behaviour of the relay nodes, it should be clear that a positive value of f_{n} will increase the hop count of the newly forwarded packet with respect to the received one. In this way, at each forwarding stage, more nodes will get involved and this will cause flooding in the whole network. As a result, we always consider a negative f_{n}. In contrast, the parameter f_{s} is useful to be positive, especially in those scenarios where the source node is characterized by low connectivity. In other words, in the first hop transmission, the source node might also need those neighbours that are not closer to the destination than itself. In this way, more relays will participate in the first transmissions, and their number will reduce while we get closer to the destination in the next stages.

2.2 Implicit ACK

Considering the fact that the underwater sensors are mostly equipped with omnidirectional hydrophones, and that the information is broadcast to the nodes that are within the transmission range, an implicit ACK strategy

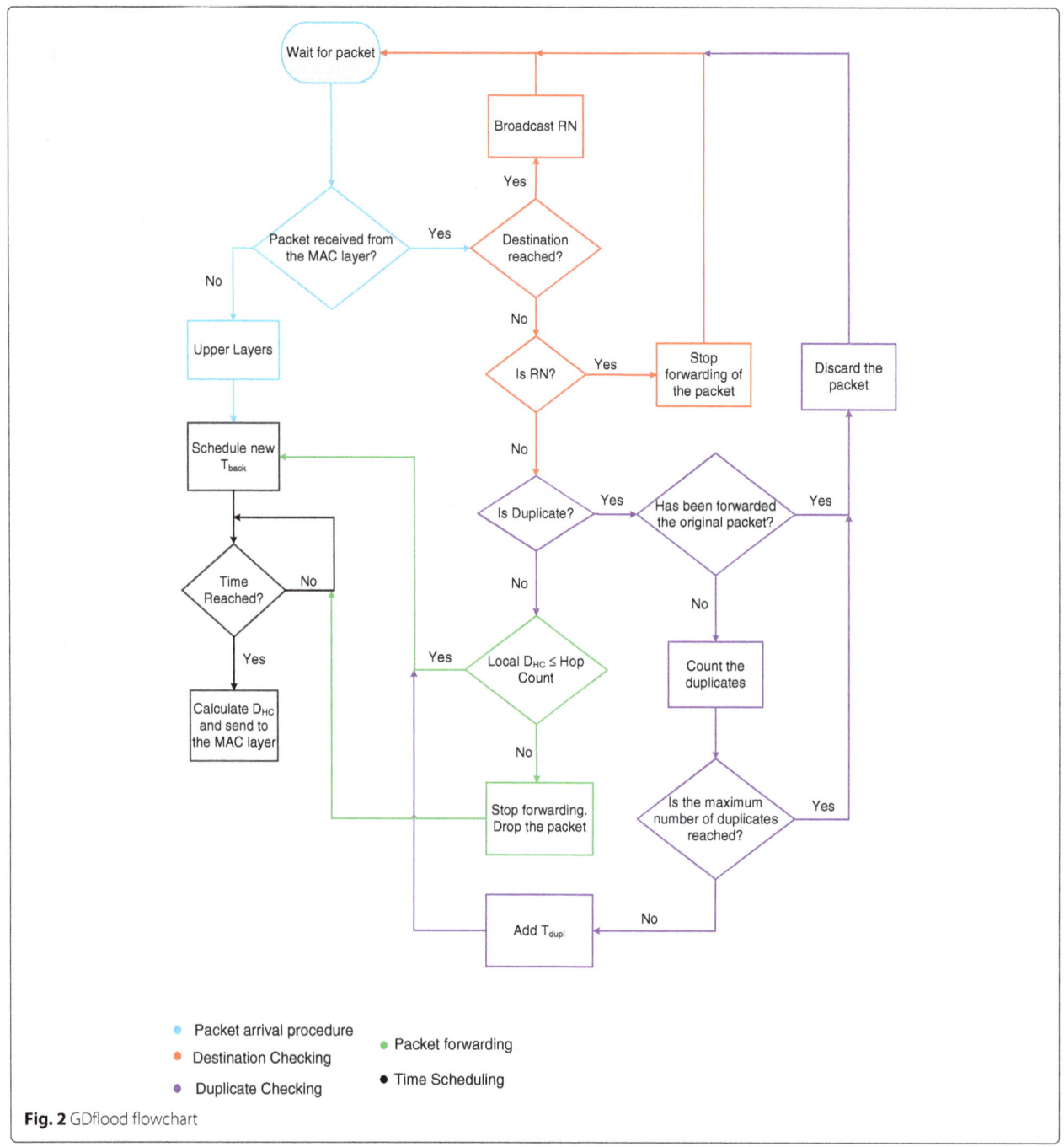

Fig. 2 GDflood flowchart

can be easily adopted. This way of re-transmission can be used by the nodes to detect if a packet has been successfully received by one of its neighbours. In this manner, we use a link-by-link ACK without the need of transmitting special packets. If, after some time from the moment of transmitting the packet, the node does not overhear any forwarding by one of its neighbors, it will retransmit another copy. This procedure can be repeated by all the nodes till the packet reaches the destination. To be energy-efficient, the number of retransmissions must be kept limited. For each packet forwarded, each node in the network starts a local timer, called the ACK time. If during this time, the node under consideration overhears a transmission of that packet, by any of its neighbors, it will not retransmit other copies. Otherwise, at the end of the ACK time, the node will retransmit another copy and will start another ACK time. This procedure will be done for a fixed number of re-transmissions. In [10], a random selection of the implicit ACK time has been proposed for the Dflood protocol. We will use the same approach in order to have a direct comparison between the two protocols. The ACK time is selected as

$$\tau = \mathrm{rand}\left(\left[\frac{t_{\mathrm{delay}}}{2}, t_{\mathrm{delay}}\right]\right), \tag{3}$$

where τ is the ACK time, t_{delay} is the delay time relative to the retransmissions (retx.), and rand $[a, b]$ selects a random number, uniformly drawn in $[a, b]$.

3 Network coding Dflood

In this section, we will explain our network coding (NC) approach. With NC, the nodes encode the incoming packets into one or more output packets instead of using the classic store and forward approach. In this way, the original information is shared among the encoded packets, and thus the destination is more likely to receive a piece of information instead of replicas of the same packet. We will use *linear* NC due to its effectiveness and applicability in underwater sensor networks. With linear NC, the output packets are a linear combination of the packets present in the node's buffer. Together with the Dflood idea, we aim to keep the number of packets that are flooded in the network limited, yet improving the PDR.

3.1 Linear network coding background

3.1.1 Encoding

Let us consider a network where a source node has a set of g packets to send $\mathbf{X} = [\mathbf{x}_1, \ldots, \mathbf{x}_g]^T$, where $\mathbf{x}_i \in \{0, 1\}^{L_b \times 1}$, $(\cdot)^T$ indicates the transpose operator and L_b is the number of bits that each packet is composed of. We interpret s consecutive bits as a symbol over the field $\mathbb{F}_{2^s}$, so each packet consists of L_b/s symbols. In linear NC, the source node linearly combines the g original packets into h encoded packets $\mathbf{Y}_1 = [\mathbf{y}_{11}, \ldots, \mathbf{y}_{1h}]^T$, with $h \geq g$. The operations are carried out over the field $\mathbb{F}_{2^s}$ [20]. For each encoded packet $\mathbf{y}_{1i}$, the node selects the respective encoding vector $\mathbf{e}_{1i}$ over the field $\mathbb{F}_{2^s}$, [21], and the output encoded packets are obtained as

$$\mathbf{Y}_1 = \mathbf{E}_1 \mathbf{X}, \tag{4}$$

where $\mathbf{E}_1$ $(h \times g)$ is the matrix that contains as the ith row the encoding vector $\mathbf{e}_{1i}^T$. Note that both $\mathbf{Y}_1$ and $\mathbf{E}_1$ will be transmitted. With NC, not only the source, but also the relay nodes can re-encode the received packets. Let us consider a relay node that receives a set of encoded packets $(\mathbf{e}_{11}, \mathbf{y}_{11}), (\mathbf{e}_{12}, \mathbf{y}_{12}), \ldots, (\mathbf{e}_{1k}, \mathbf{y}_{1k})$, with $k \leq h$. The relay node will consider the received packets $\tilde{\mathbf{Y}}_1 = [\mathbf{y}_{11}, \ldots, \mathbf{y}_{1k}]^T$ and the corresponding encoding matrix $\tilde{\mathbf{E}}_1 = [\mathbf{e}_{11}, \ldots, \mathbf{e}_{1k}]^T$ to re-encode these packets into l output packets, $\mathbf{Y}_2 = [\mathbf{y}_{21}, \ldots, \mathbf{y}_{2l}]^T$, with $l \leq k$. It is clear that $\tilde{\mathbf{Y}}_1$ and $\tilde{\mathbf{E}}_1$ are, respectively, submatrices composed of k rows, of $\mathbf{Y}_1$ and $\mathbf{E}_1$. For each output packet $\mathbf{y}_{2i}$, the relay

node selects locally a set of coefficients $\mathbf{e}'_{2i}$, in $\mathbb{F}_{2^s}$, and the output packets are computed as

$$\mathbf{Y}_2 = \mathbf{E}'_2 \tilde{\mathbf{Y}}_1, \tag{5}$$

where $\mathbf{E}'_2$ is the matrix that contains as the ith row the local re-encoding vector $\mathbf{e}'^T_{2i}$. Substituting the relationship of $\tilde{\mathbf{Y}}_1$ with the original information $\mathbf{X}$ in (5), we obtain a direct relationship of the new information vectors with the original packets

$$\mathbf{Y}_2 = \mathbf{E}'_2 \tilde{\mathbf{E}}_1 \mathbf{X} = \mathbf{E}_2 \mathbf{X}. \tag{6}$$

The generic ith output packet $\mathbf{y}_{2i}$ and the corresponding encoding vector $\mathbf{e}^T_{2i}$, which will be transmitted, are obtained as

$$\mathbf{y}^T_{2i} = \mathbf{e}'^T_{2i} \tilde{\mathbf{E}}_1 \mathbf{X} = \mathbf{e}^T_{2i} \mathbf{X} \tag{7}$$

This procedure can be repeated by all the nodes in the network until the packets reach the destination.

3.1.2 Decoding

In order to decode the data, the destination needs a sufficient number of packets. Let us assume for simplicity that the sink receives a set of m packets after passing n stages $(\mathbf{e}_{n1}, \mathbf{y}_{n1}), (\mathbf{e}_{n2}, \mathbf{y}_{n2}), \ldots, (\mathbf{e}_{nm}, \mathbf{y}_{nm})$. In order to decode these packets and retrieve the original information, i.e., $\mathbf{X} = [\mathbf{x}_1, \ldots, \mathbf{x}_g]^T$, the node must solve the linear system

$$\tilde{\mathbf{Y}}_n = \tilde{\mathbf{E}}_n \mathbf{X}, \tag{8}$$

where $\tilde{\mathbf{Y}}_n = [\mathbf{y}_{n1}, \ldots, \mathbf{y}_{nm}]^T$ contains the received packets and $\tilde{\mathbf{E}}_n = [\mathbf{e}_{n1}, \ldots, \mathbf{e}_{nm}]^T$ contains the encoding vectors. This linear system can be solved only if the destination collects enough packet, i.e., $n \geq g$, and at least g combinations must be independent. This means that the rank of the matrix $\tilde{\mathbf{E}}_n$ must be g. Considering the computational cost, the inversion of a matrix is related to the cubic power of its rank. Thus, the source groups the packets in so-called generations, and the encoding process is limited only to the packets of the same generation [21].

Another key element for decoding the information are the encoding coefficients. In [22], a randomized selection of the encoding coefficients has been proposed. Each node in the network selects randomly the coefficients, uniformly distributed in $\mathbb{F}_{2^s}$. This is a liked strategy since it helps to have an independent and decentralized network. The benefits are enhanced in underwater sensor networks where transmitting the information from a central unit may take too long. However, selecting randomly the coefficients may lead to linearly dependent combinations, which happens with a probability related to the field size s, [22]. However, [21] has shown that in practice, $s = 8$ is sufficient to have a full rank decoding matrix with very high probability. Considering that the coefficients are chosen locally at each node, the encoding vectors must be included in the packet headers. Obviously, this brings an increment to the overhead, which grows linearly with the generation size. This is because the higher the generation size, the longer the dimension of each encoding vector in order to contain all the entries selected for encoding the information packets into one specific output packet.

3.2 Proposed protocol

Also here, we have considered the same network header as Dflood, with the note that the sequence number is now called generation number. Thus, the *unique* ID refers to the packets of the same generation. So, more packets will hold the same ID. However, this does not give any problems, since we are interested in receiving g independent packets, instead of receiving each of them independently. After decoding, the packets can be sorted in the same order as produced by the application layer of the transmitter. This can be done using the information of the upper layers, or by splitting the b bits of the generation number in two sub-fields. The first one with

$$b_1 = \lceil \log_2 g \rceil \tag{9}$$

bits will be used to sort the packets inside the generation, and the second with

$$b_2 = b - b_1 \tag{10}$$

bits are used for the effective generation number.

We have used the same terminology as in previous scientific works about NC. The *innovative packets* are those which are not a linear combination of the packets present in the node buffer[2]. On the other hand, *non-innovative packets* are those packets that can be expressed as a linear combination of the packets that the node keeps. In order to exploit the benefits of NC, we have changed the Dflood rules, and adapted them into a generation fashion. In this way, we aim to reduce the number of non-innovative packets. On top of that, we want to transmit the necessary amount of innovative packets considering that we are working in a flooding fashion. In the following, we will explain how each node acts in the network, mainly focusing on the novelty of this approach.

Source node: The source node will start transmitting when it receives g packets from the upper layers. It will encode them into h output packets and send them down to the MAC layer with hop count equal to 1. In this way, the source performs NC to the desired group of packets. There might also be cases when there are less than g packets to transmit. In this case, the upper layers may inform the encoder to process a smaller group of packets. This will save time and keep the end-to-end delay limited. To contrast harsh environments, it is preferable to have $h > g$. This is because the first hop is the bottleneck of the overall transmission. If all the source's 1-hop neighbours do not receive g independent packets, the overall

transmission will result in a waste of energy since the destination cannot retrieve the original information.

Destination node: The destination, on the other side, will not broadcast an RN for each packet received. Instead, it will inform its neighbours when g-independent packets are received and the original information is recovered.

Relay node: The relay node's behaviour is more complicated. Each time it receives a packet of a generation for the first time, i.e., an innovative packet, it will schedule it for forwarding after the relative backoff time, still uniformly drawn in $[T_{\min}, T_{\max}]$. If during the waiting time, the relay receives an innovative packet with the same ID, it will encode this packet together with the packet in the buffer into two new ones. One of them will replace the packet scheduled to be transmitted, and meanwhile the other will be scheduled after the relative backoff time. This process continues whenever an innovative packet with the same ID arrives. So if a new innovative packet arrives, it will be re-encoded together with all the packets with the same ID in the buffer. The number of new re-encoded packets is equal to the number of packets in the buffer not yet transmitted (these will replace the packets in the buffer not yet transmitted) plus one (this packet will receive a new relative backoff time). Note that, in this approach, we assume that even when a packet is transmitted, it will not be removed from the buffer and it will be used for re-encoding packets of the same ID that arrive later.

It may happen that the backoff time expires without receiving an innovative packet. In order to keep the end-to-end delay limited, and to allow also more nodes to take part in the forwarding, the packet will be transmitted by the relay node without extra encoding.

In this approach, we have considered also a maximum number of times that a packet will be encoded in the network, $H_{\max}$. So the relay will create a new coded packet, if the received one is innovative and with hop count lower than or equal to $H_{\max}$. As before, all the packets in the buffer will be encoded together with the new one received. In case the hop count is higher than $H_{\max}$, the received innovative packet will be considered only to update the scheduled packets and a new one will not be created. In case a new packet is created, the hop count of this packet will be increased by one with respect to the packet received.

The *relay node* will apply the duplicate policy when non-innovative packets are received. First of all, when it has forwarded g packets with the same ID, it will drop any non-innovative packet received. In case not all the packets of a generation are forwarded, and a non-innovative packet is received, the forwarding will be delayed by T_{dupl}. In the NC-Dflood case, there are many ways to do this. We propose two: (1) delay by T_{dupl} only the first packet to be sent, with the same ID and (2) delay by T_{dupl} all the packets scheduled to be sent with the same ID. Depending on the specific scenario, one should simulate both cases and choose the solution that gives the best performance. We refer the reader to [23] for more details on how the selection is done for this particular case.

As before, the number of non-innovative packets will be counted in n_d. The relay will try for each received non-innovative packet to quit the forwarding for that generation, if

$$n_d > N_{dupl} - \rho. \tag{11}$$

In this case, ρ is drawn uniformly in $[0, R]$, and for N_{dupl}, we have proposed two approaches. In the first one, N_{dupl} is related by a constant c to the generation size,

$$N_{dupl} = c \cdot g, \tag{12}$$

and $R = g$; in the second approach, it is related to the number of independent packets that the relay has collected till that moment,

$$N_{dupl} = c \cdot \text{rank}(\tilde{\mathbf{E}}) \tag{13}$$

and $R = \text{rank}(\tilde{\mathbf{E}})$.

Finally, we will extend this approach with position information, as in GDflood, in order to reduce even more the energy consumption in the network. In this case, the hop count of the packets is substituted by the quantized distance to the destination, together with the factors f_n and f_s. In the latter case, each node in the network that receives a packet will apply the duplicate reduction policy if it is a non-innovative packet, or will use it for re-encoding the packets present in the buffer if it is an innovative packet. But, a new encoded packet will be created if the D_{HC} local of this node is lower than or equal to the hop count value. Now it is more clear why the parameter f_n can generally not have a positive value. If all the nodes put $f_n > 0$, then an avalanche effect will be created and all the nodes will put a hop count value larger than the actual hop count contained in the packet. So, the whole network will be involved. The parameter $f_s > 0$ can be useful in those cases when the source node has low connectivity. In the NC case, this is even more useful to contrast also the bottleneck effect that is created in the first hop transmission.

4 Simulations

4.1 Simulation setup

4.1.1 Considered scenario and traffic model

To compare the protocol performance, we have considered the scenario illustrated in Fig. 3 composed of 22 sea bottom nodes and 1 autonomous underwater vehicle (AUV). For simplicity, we have considered a regular network grid with an intra-node distance of 3 km. The red line is the trajectory of the AUV, which makes a round trip from checkpoint A to B and back with a speed of 4 knots. We have considered three source nodes, node 1,

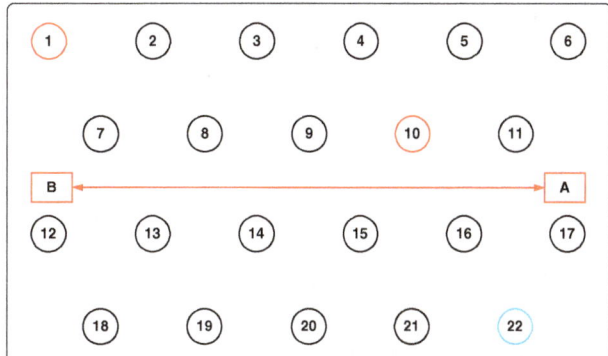

Fig. 3 Network topology. The nodes are deployed in a regular grid, with an intra-node distance of 3 km in a area of 20.25 × 15 km. The network is composed of 22 sea bottom nodes and one moving node

10, and the AUV, and one destination, node 22. We have compared our solutions with the Dflood protocol. For the Dflood and GDflood, we have assumed the same traffic rate at the application layer. The arrival times of the packets are generated from a Poisson process with the λ parameter defined as

$$\lambda = L_b/r, \tag{14}$$

where L_b is assumed to be 160 bits and r (in bits per second) is a simulation parameter that influences the traffic introduced in the network by the application layer of the source nodes. If NC is applied, λ is assumed to be

$$\lambda = (L_b \cdot g)/r \tag{15}$$

since, in this case, it is considered that the application layer produces g packets and sends them down to the MAC layer. The assumption to make the arrival times between sets g times bigger, is done to ensure the same traffic for both cases. However, we have to note that the NC simulations will be affected by more interferences considering that the packets are sent in a shorter time interval.

4.1.2 Physical and MAC layer

For our simulations, we have assumed a PER $= p$, common to all the links present in the network. In this way, the evaluated performance can be for different physical transmission schemes. The transmission data rate of the physical layer is considered in such a way that the packet duration is one second. This assumption is considered for simplicity, but nevertheless, it is justified in underwater communications, since the packet lengths are very short and the bit rates are low [24], e.g., in [9], a packet length of 160 bits and a bit rate of 200 bits/s is considered for Dflood. The transmission power is assumed to offer a PER $= p$ for links up to 3 km. Farther nodes receive a packet erroneously with probability one.

In the MAC layer, a simple unslotted ALOHA protocol is considered. We did not consider any carrier sensing

Table 1 Protocol parameters used in the simulations. N/A means that the parameter is not applicable for that protocol

Protocol	T_{min}	T_{max}	T_{dupl}	N_{dupl}	H_{max}	f_n	f_s	No. of retx	t_{delay}
Dflood	0 s	50 s	35 s	2.5	N/A	N/A	N/A	N/A	N/A
GDflood	0 s	50 s	35 s	2.5	N/A	0	2	N/A	N/A
Dflood-ACK	0 s	50 s	35 s	2.5	N/A	N/A	N/A	2	50 s
GDflood-ACK	0 s	50 s	35 s	2.5	N/A	−1	1	2	80 s
NC-Dflood	0 s	70 s	30 s	2.5·g	15	N/A	N/A	N/A	N/A
NC-GDflood	0 s	70 s	30 s	2.5·g	15	0	1	N/A	N/A

or end-to-end ACKs. We have assumed this protocol due to its simplicity and suitability in today's underwater sensor networks, which are generally characterized by low traffic rates. Also a carrier sensing may be proposed for a short time before transmitting; however, it will not prevent all collisions. We leave this aspect for future work. Here, we have assumed a simple interference model. A total destructive interference is considered if two or more neighboring nodes are transmitting in the same time interval. This time interval consists of the time it takes for the packet to travel through the medium (a sound speed of 1500 m/s is assumed) plus the time required to receive the full packet. None of these nodes will corectly receive the packet intended for them. Also, the nodes that are neighbors, with the two overlapping transmitting nodes, will not receive any of these packets. To illustrate this, with reference to Fig. 3, let us consider that only the nodes 1 and 2 are transmitting in the same time interval. According to the considered interference model, the nodes 1, 2, and 7 will not receive correctly the packet, the transmission to the nodes 3 and 8 will not be affected by any interference. We have made these assumptions about the interference model in order to simulate the *worst case* scenario. In a real scenario, the performance will be better than the simulated one, because the interference is not always destructive. In some cases, the forward error correction bits can recover erroneous ones. In some other cases, it may happen that the channel buffers the packets from different nodes, and thus avoids the interference from different ongoing transmissions. With the latter assumption for the interference model, in the NC case, we have assumed that the network layer sends the packets down to the MAC layer with a backoff time drawn uniformly from [1, 10]. This is done with the goal to avoid interference that may affect all the packets of the same generation, if they are transmitted sequentially.

4.2 Simulation results

To evaluate the performance, we have considered three evaluation criteria: the PDR, the end-to-end delay, and the average number of packets forwarded by the network (Av. PKT), for each information packet produced by the source node. The protocol parameters are selected heuristically

by a parameter scanning, as a trade-off between the three evaluation criteria, paying most attention to the PDR. The selected parameters are shown in Table 1. We first determined the Dflood parameters by simulations, considering $p = 0.1$ and $r = 1$ bit per second. The backoff time interval is the one that mostly influences the PDR and end-to-end delay. These parameters (i.e., T_{min} and T_{max}) are selected in order to ensure a high value of the PDR (close to the saturation). The duplicate parameters, T_{dupl} and N_{dupl}, are selected in such a way to keep the PDR as high as possible and to reduce the average number of packets. These parameters have less influence on the PDR and end-to-end delay, compared to T_{min} and T_{max}. For more details on the sensitivity to the three evaluation criteria, see [23]. Using the same approach, we find also the parameters for the other protocols.

For the GDflood protocol, we have used the same parameters as for Dflood[3], and an inaccuracy is considered when the distance D is measured. We have assumed

$$D = D^{real} + u, \qquad (16)$$

where u is an error uniformly distributed in [−100, 100] m, and D^{real} is the real distance to the destination. When the implicit ACK is not used, the "best" parameters for GDflood are $f_n = 0$ and $f_s = 2$. From simulation results, we have found that, when we use the implicit ACK strategy, the "best" values are $f_n = -1$ and $f_s = 1$. In Dflood and GDflood, a good trade-off occurs when a maximum of two retransmissions is considered with t_{delay} equal to 50 and 80 s, respectively.

In the case of NC-Dflood, different values of the parameters with respect to Dflood give a better performance. More specifically, T_{min} and T_{max} are considered 0 s and 70 s respectively; a value of 30 s added only to the first scheduled packet is considered for T_{dupl}; for N_{dupl} the trade-off value is 2.5 · g common to all the nodes in the network; and H_{max} is selected to be equal to 15. The value of H_{max} is found using the same approach as used for T_{dupl} and N_{dupl}. These values are selected for a generation size $g = 2$ and $h = 3$. It is obvious that for a larger g, the performance improves, up to a certain point, but the end-to-end delay and the Av. PKT increases as well. This is because, for larger generation sizes, the destination needs more packets to retrieve the information, which effectively increases the end-to-end delay. Also, larger generations cause higher energy consumption since less non-innovative packets are received at the relays, and if we try to reduce also the innovative information, the PDR will be affected. Meanwhile, an extra packet, during the encoding by the source nodes, is enough to improve the PDR without having big consequences for the Av. PKT. For higher values, the PDR starts saturating, and the energy consumption increases. In case the geographical position is used, we have found $f_n = 0$ and $f_s = 1$, respectively, as the best values keeping the other parameters same as in NC-Dflood.

Simulation results for different values of PER p and the traffic rate r are shown in Figs. 4 and 5. First, we have anal-ysed the performance of our approaches and Dflood for different PERs, p. In this case, the traffic rate r of the appli-cation layer is considered 1 bit per second. In order to save space, we have shown the average PDR of all the transmis-sions, the end-to-end delay of the transmission of node 1 (considering that it is the farthest from the destination), and the mean of the Av. PKTs for the three singular trans-missions in Fig 4a–c, respectively. From the first figure, we can see that Dflood without implicit ACK has the worst performance for a p smaller than 0.5. For reasonable small values of p, we can see that the retransmission strategy and NC offers values of the PDR that are close to the maxi-mum. The GDflood protocol, even without implicit ACK, has a better PDR than Dflood since it reduces the number of collisions. The drawback is an increment of the end-to-end delay per packet, compared to the latter. As we can see from Fig. 4b, all the approaches result in a higher end-to-end delay compared to Dflood. In the case when the implicit ACK is used, this is obvious since the increment of the PDR is due to the retransmission strategy, which retransmits copies after a certain time. In the NC case, the selected parameters, which ensure the desired PDR value, also result in a higher end-to-end delay per packet. In Fig. 4c, we can see that the use of the implicit ACK in the Dflood protocol leads to a waste of energy, considering the high number of packets that are forwarded for each origi-nal one. Meanwhile, in the case of GDflood, the use of the retransmission strategy is more energy efficient. The use of the geographical position reduces the number of nodes that take part in the forwarding, and thus protocols that make use of it (including also NC-GDflood) flood fewer packets. For low values of p, we can reach values that are even lower than the pure Dflood protocol.

Without getting too much into detail, we can see that the same trend is kept also for different values of r, see Fig. 5a–c. In these simulations, we have assumed $p = 0.1$. We can see that the implicit ACK strategy improves the

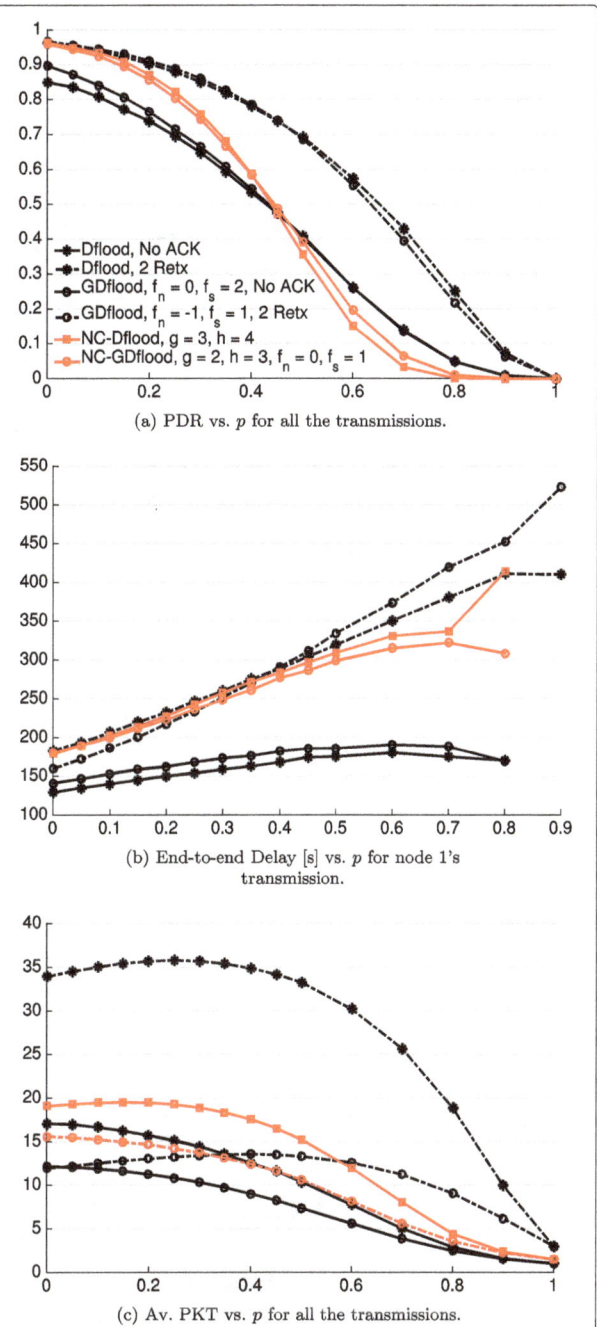

(a) PDR vs. p for all the transmissions.

(b) End-to-end Delay [s] vs. p for node 1's transmission.

(c) Av. PKT vs. p for all the transmissions.

Fig. 4 Performance comparison of proposed protocols with Dflood for different values of packet error rate, p. **a** PDR vs. p for all the transmissions. **b** End-to-end Delay [s] vs. p for node 1's transmission. **c** Av. PKT vs. p for all the transmissions

PDR, but at the same time a higher end-to-end delay is obtained. For the Dflood protocol, the implicit ACK is also worse in terms of energy, since the network floods too many packets. The use of the implicit ACK becomes more useful when the traffic increases, and thus the interference becomes more predominant. In those cases, by allowing a retransmission, we increase the possibility that a packet

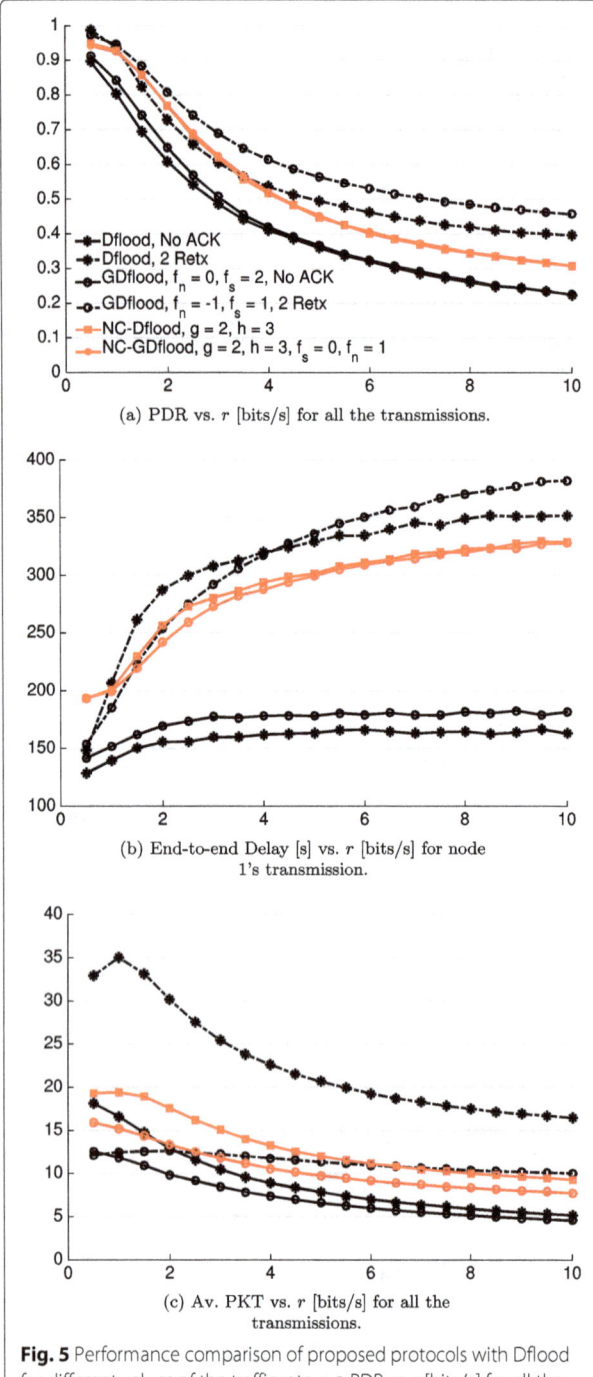

(a) PDR vs. r [bits/s] for all the transmissions.

(b) End-to-end Delay [s] vs. r [bits/s] for node 1's transmission.

(c) Av. PKT vs. r [bits/s] for all the transmissions.

Fig. 5 Performance comparison of proposed protocols with Dflood for different values of the traffic rate, r. **a** PDR vs. r [bits/s] for all the transmissions. **b** End-to-end Delay [s] vs. r [bits/s] for node 1's transmission. **c** Av. PKT vs. r [bits/s] for all the transmissions

reaches the destination. The NC approach still ensures high PDRs, especially for low traffic rates, matching those of Dflood and GDflood with the implicit ACK. It also ensures the same end-to-end delay, but more packets are transmitted compared to GDflood. It is worth to mention that the use of geographical information does not improve

the PDR when the traffic increases. It seems that the limited number of relay nodes selected by GDflood is affected by the high interference present in the network. As we can see from Fig. 5a, when the traffic increases, and thus more collisions happen, all the protocols' performances degrade. This has direct implications on the end-to-end delay and average number of packets forwarded, as shown in Fig. 5b, c, respectively. However, most of today's underwater networks are characterized by sparse traffic, where the proposed protocols perform better.

4.3 Discussion

In evaluating the performance of the proposed protocols, we have been relying on simulation results. This is because the main goal of the paper is to propose to the reader an alternative way of using flooding in underwater sensor networks, and to illustrate its benefits in a basic scenario. However, in this subsection we discuss some challenges of the proposed approaches from a practical point of view.

For the GDflood and NC-GDflood, we consider that each node must know its own position and the final destination position. In underwater sensor networks, precise positioning is a challenge, especially for moving nodes [25]. However, our approach does not need precise information about the nodes' positions. Indeed, this information is only required to calculate the distance in hop counts between the transmitting node and the final destination. Considering that the position accuracy of such an estimate is about some tens of meters, or even hundreds of meters, and the transmitting distance is several kilometres, in the worst case, the D_{HC} will deviate by 1 from the true value. For this reason, in (16), we also consider an error on the true distance between the transmitting node and final destination.

Our next point considers the encoding coefficients required to perform the linear random NC, which must be transmitted alongside with the encoded packets. We want to point out that the benefits that NC brings with respect to Dflood (specially for low PER p and traffic rate r) come with a low extra information in the packet header. Indeed, the improved performance is achieved by only considering pairs of packets (i.e., $g = 2$) in the encoding process[4]. Thus, we may conclude that the use of NC is beneficial with a relative low price to pay in a flooding fashion.

From an implementation point of view, our protocols require a lot of data to be stored, e.g., the protocol's parameters, the already forwarded packets, the packets needed for applying NC and so on. In all these considerations, we assume that there are no memory constraints at the relay nodes. This is because most underwater sensors are bigger than terrestrial ones, and there is enough space to include a large memory.

Another point is the dependence of the protocols' parameters on the network topology. In this work, we consider a simple network topology, which satisfies the RACUN project requirements, mainly to illustrate that the proposed protocols might improve the performance of flooding-based protocols. However, we find it difficult to find a closed-form relationship between the protocols' parameters and the underlying topology, and thus we found their value with a scanning approach by simulations. It is obvious that for a different topology, their optimal value is different in order to ensure the highest performance. Indeed, they have to be tuned to the particular topology in order to achieve the desired performance. A challenging topology results when the AUV is the destination node. Geography-based protocols may suffer since the destination position changes continuously over time. In this case, the nodes may be informed about the operating sector of the AUV, and consider a reference point to calculate D_{HC}, e.g., this might be the center of the sector, or it can be improved if the AUV trajectory is know to the transmitting nodes. For the other protocols, this aspect is irrelevant as long as the AUV stays in the network coverage area.

5 Conclusions

In this work, we have proposed some advanced upgrades of a flooding-based protocol (Dflood), for underwater sensor networks. Our first idea was to incorporate the node position information in the relaying process. In this way, the participation of the nodes farther from the destination can be avoided. Simulation results show that a considerable amount of energy can be saved and an improvement of the PDR can be achieved as well. The price to pay is the end-to-end delay, which increases with respect to the original protocol. The use of an implicit ACK ensures that the PDR remains closer to the maximum for low values of the packet error rate, but our geographic approach outperforms the standard one in terms of energy consumption.

The second idea was to use network coding in order to flood more informative packets in the network than replicas. Considering the high amount of packets that are flooded in the network, the sharing of information between them will bring more information to the destination. Our proposed protocol, NC-Dflood, increases the PDR of the transmission with respect to Dflood (no ACK), with an increment in the end-to-end delay and energy consumption in a single transmission. An increment in the PDR, in this case, means that less original packets will be retransmitted in a second try in order to deliver all the information to the destination. Thus, the gap in the end-to-end delay and energy consumption is potentially reduced with respect to Dflood. Finally, we have used this approach also with geographical information and better results are obtained, even for a single transmission.

As future work, we consider the implementation of these ideas in real scenarios, with real communication channels, and different MAC protocols. We will also consider the implicit ACK strategy in the network coding case. Lastly, inspired by terrestrial wireless network coding schemes, it will be very interesting to use instantly-decodable random network coding [26, 27] in order to retrieve the original information packet before completing the full rank decodable problem.

Endnotes

[1] One more rule must be added to the rules in [9], which was inadvertently left out: "Forwarding is delayed by a time T_{dupl}, when a duplicate is received (with hop count greater than that of the original reception)" [10].

[2] Sometimes, the innovative packets are defined as those which increase the degrees of freedom, since they bring new information to that node.

[3] This is done in order to compare the GDflood performance with the "best" performance of Dflood. The protocol's parameters can be tuned to offer a better performance. Here, we have played only with f_n and f_s to achieve the desired performance of GDflood.

[4] Note that the protocol requires 1 extra byte for each encoded packet to be included in the packet header.

Competing interests

The authors declare that they have no competing interests.

Acknowledgements

The authors kindly acknowledge Paolo Casari (IMDEA Networks, Madrid, Spain) and Roald Otnes (FFI, Horten, Norway) for their support. This work has been supported by the European Defence Agency (EDA) project RACUN [28]. Part of this work has been presented in [29].

Author details

[1] Faculty of Electical Engineering, Mathematics and Computer Science, Delft University of Technology, 2628 CD, Delft, The Netherlands. [2] Acoustics and Sonar Dept., TNO, 2597 AK, The Hague, The Netherlands.

References

1. A Quasi, W Konrad, Underwater Acoustic Communications. IEEE Commun. Mag. **20**, 24–30 (1982)
2. IF Akyildiz, D Pompili, T Melodia, Underwater Acoustic Sensor Networks: Research Challenges. Ad Hoc Netw. Elsevier. **3**, 257–279 (2005)
3. R Otnes, et al., Underwater Acoustic Networking Techniques, Springer Briefs in Electrical and Computer Engineering (2012). doi:10.1007/978-3-642-25224-2-1
4. M Stojanovic, JC Preisig, Underwater Acoustic Communication Channels: Propagation Models and Statistical Characterization. IEEE Commun. Mag. **47**, 84–89 (2009)
5. R Urick, *Principles of Underwater Sound*. (McGraw-Hill, 1983)
6. EM Sozer, M Stojanovic, JG Proakis, Underwater Acoustic Networks. IEEE J. Oceanic Eng. **25**, 72–83 (2000)
7. G Han, J Jiang, N Bao, L Wan, M Guizani, Routing Protocols for Underwater Wireless Sensor Networks. IEEE Commun. Mag. **53**, 72–78 (2015)
8. RWL Coutinho, A Boukerche, LFM Vieira, AAF Loureiro, Design Guidelines for Opportunistic Routing in Underwater Networks. IEEE Commun. Mag. **54**, 40–48 (2016)
9. R Otnes, S Haavik, *Duplicate Reduction with Adaptive Backoff for a Flooding-Based Underwater Network Protocol*. (IEEE Oceans, Norway, 2013)

10. R Otnes, PA van Walree, H Buen, H Song, Underwater Acoustic Network Simulation with Lookup Tables from Physical-Layer Replay. IEEE J. Oceanic Eng. **40**, 822–840 (2015)

11. P Xie, JH Cui, L Lao, VBF, in *Proc. of IFPF Networking*. Vector-Based Forwarding Protocol for Underwater Sensor Networks, (Canada, 2005)

12. JM Jornet, M Stojanovic, M Zorzi, On Joint Frequency and Power Allocation in a Cross-Layer Protocol for Underwater Acoustic Networks. IEEE J. Oceanic Eng. **35**, 936–947 (2010)

13. R Ahlswede, N Cai, SR Li, RW Yeng, Network Information Flow. IEEE Trans. Information Theory. **46**, 1204–1216 (2000)

14. Z Guo, P Xie, JH Cui, B Wang, *On applying Network Coding to Underwater Sensor Networks*. (WUWNET'06, LA California USA, 2006)

15. DE Lucani, M Medrad, M Stojanovic, *Network Coding Schemes for Underwater Networks*. (WUWNET'07, Canada, 2007)

16. N Chirdchoo, M Chitre, WS Soh, *A Study on Network Coding in Underwater Networks*. (IEEE Oceans, Washington USA, 2010)

17. Z Guo, B Wang, P Xie, W Zeng, JH Cui, Efficient Error Recovery with Network Coding in Underwater Sensor Networks, Ad Hoc Networks, Elsevier. **7**, 791–802 (2009)

18. V Kebkal, K Kebkal, O Kebkal, *Experiments with Network Coding in Dynamic Underwater Acoustic Channel*. (IEEE UComms, Italy, 2014)

19. C Meanville, A Miyajan, A Alharbi, M Haining, *Network Coding in Underwater Sensor Networks*. (IEEE Oceans, Norway, 2013)

20. C Fragouli, JY Le Boudec, J Widmer, Network Coding: An Instant Primer, LCA-Report-2005-010 (2005). Available online from, www.infospace.epfl.ch

21. PA Chou, Y Wu, K Jain, in *Proc. of Allerton Conference on Communication*. Practical Network Coding (Control and Computing, Illinois USA, p. 2003

22. T Ho, R Koetter, M Medrad, DR Krager, M Effors, in *International Symposium on Information Theory (ISIT)*. The Benefits of Coding Over Routing in a Randomized Setting, Japan, 2003)

23. E Isufi, *Network Coding for Flooding-Based Routing in Underwater Sensor Networks, Master Thesis*. (University of Perugia, Italy, 2014)

24. S Basagni, C Petrioli, R Petroccia, M Stojanovic, Optimized Packet Size Selection in Underwater Wireless Sensor Network Communications. IEEE J. Oceanic Eng. **37**, 321–337 (2012)

25. H Yan, Z Shi, JH Cui, *DBR: Depth-Based Routing for Underwater Sensor Networks*. (IFIP Networking'08, Singapore, 2008)

26. Y Mingchao, P Sadeghi, N Aboutorab, Performance characterization and transmission schemes for instantly decodable network coding in wireless broadcast. EURASIP J. Adv. Signal Process. **2015**, 1–17 (2015)

27. A Douik, S Sorour, TY Al-Naffouri, MS Alouini, Instantly Decodable Network Coding for Real-Time Device-to-Device Communications. EURASIP J. Adv. Signal Process. **2016**, 1–14 (2016)

28. J Kalwa, *The RACUN-Project: Robust Acoustic Communications in Underwater Networks - an Overview*. (IEEE Oceans, Spain, 2011)

29. E Isufi, G Leus, H Dol, in *Proceedings of the International Conference on Underwater Networks and Systems (WUWNET'14)*. Network Coding for Flooding-Based Routing in Underwater Sensor Networks (ACM, Italy, 2014). http://dl.acm.org/citation.cfm?id=2674570&CFID=600486126&CFTOKEN=90052992. doi:10.1145/2671490.2674570

12

Singular spectrum-based matrix completion for time series recovery and prediction

Grigorios Tsagkatakis[1*], Baltasar Beferull-Lozano[2] and Panagiotis Tsakalides[1,3]

Abstract

Big data, characterized by huge volumes of continuously varying streams of information, present formidable challenges in terms of acquisition, processing, and transmission, especially when one considers novel technology platforms such as the Internet-of-Things and Wireless Sensor Networks. Either by design or by physical limitations, a large number of measurements never reach the central processing stations, making the task of data analytics even more problematic. In this work, we propose Singular Spectrum Matrix Completion (SS-MC), a novel approach for the simultaneous recovery of missing data and the prediction of future behavior in the absence of complete measurement sets. The goal is achieved via the solution of an efficient minimization problem which exploits the low rank representation of the associated trajectory matrices when expressed in terms of appropriately designed dictionaries obtained by leveraging the theory of Singular Spectrum Analysis. Experimental results in real datasets demonstrate that the proposed scheme is well suited for the recovery and prediction of multiple time series, achieving lower estimation error compared to state-of-the-art schemes.

1 Introduction

The dynamic nature of Big Data, a feature termed velocity, is a critical aspect of massive data streams from a signal processing viewpoint [1]. Due to the high velocity of the input streams, measurements may be missing with a high probability. This phenomenon can be attributed to three factors, namely: (a) intentionally collecting a subset of the measurements for efficiency purposes; (b) unintentional subsampling due to desynchronization; and (c) missing measurements due to communications errors including packet drops, outages, and congestion. To elaborate on these factors, we consider data streams associated with the Internet-of-Things (IoT) paradigm and we focus on Wireless Sensor Networks (WSNs) since WSNs can serve as an enabling platform for IoT applications [2, 3]. In the context of IoT/WSNs, one source of missing measurements is attributed to *intentional subsampling*, a scenario where the designer/operator reduces the sampling rate of the sensing infrastructure in order to increase the lifetime of the network. The relationship between sampling rate and lifetime is governed by the limited energy availability that typically characterizes WSNs. While efficient compression and aggregation schemes can be employed to reduce power consumption, reducing the number of measurements is the most efficient approach to achieve this goal [4].

Even when a specific sampling rate is selected, *desynchronization* between nodes inevitably leads to a reduction of the network-wide sampling rate, since nodes that were supposed to sample at the same time instance end up acquiring measurements at different instances [5]. This issue is also closely related to the quantization of the sampling time, as measurements that were collected in succession can be mapped to different sampling instances, introducing missing measurements for particular time slots. In addition to energy consumption and desynchronization, missing measurements can also be attributed to *network outages* and *packet losses*, which are frequent in WSNs deployed in harsh and cluttered environments, causing a large number of packets to fail in reaching their destination.

*Correspondence: greg@ics.forth.gr
[1] Institute of Computer Science, Foundation for Research & Technology - Hellas (FORTH), Crete, Greece
Full list of author information is available at the end of the article

In this work, we investigate a novel paradigm in distributed data acquisition and centralized reconstruction and forecasting. The proposed sampling, reconstruction, and prediction scheme assumes that only *a small number of randomly selected nodes* acquire measurements during each sampling instance, while nodes that are not in the sampling group enter a low-power state. Because of the sampling scheme, in addition to missing data due to packet losses, the base station only observes a subset of the entire collection of measurements. To address this issue, we propose the so-called *Singular Spectrum Matrix Completion (SS-MC)* scheme, a formal approach for the recovery of missing values and the forecasting of future ones from a single or multiple time series measurements. The proposed SS-MC scheme builds upon the recently proposed framework of Matrix Completion (MC) [6, 7] for the recovery of low-rank matrices from a minimal set of measurements by extending the low-rank matrix recovery framework to the estimation of missing measurements from appropriately generated trajectory matrices and combines it with the Singular Spectrum Analysis framework for exploiting the information encoded in training data. Figure 1 presents a visual overview of the proposed reconstruction scheme, where incomplete trajectory matrices are recovered, providing accurate estimations of past and future measurements. In short, the key novelties of this work include the following:

- A novel efficient paradigm for estimating missing measurements which extents the recently developed framework of low-rank matrix recovery by exploiting inherent correlations without the need for explicit models.

- The proposed SS-MC scheme is an integrated approach for accurately predicting future values even when only a limited number of past measurements is available. This is radical departure from traditional time series forecasting schemes which assume the full availability of historical data.

- The proposed scheme can naturally handle a single or multiple time series sources extending traditional estimation approaches that operate strictly on either single or multi-source data.

- The performance of the proposed method against state-of-the-art techniques is evaluated on real data acquired by a distributed sensor network, which serves as an illustrative example of a Big Data application.

The rest of the paper is organized as follows: Section 2 presents an overview of state-of-the-art methods for energy-efficient data collection. Sections 3 and 4 provide the description of the two theoretical models we consider in this work, namely time series modeling via Singular Spectrum Analysis and missing measurement estimation via the Matrix Completion framework. Section 5 introduces SS-MC, our proposed recovery and prediction method, including the mathematical formulation as well as an efficient optimization approach based on Augmented Lagrange Multipliers. The performance of the proposed scheme is experimentally validated against state-of-the-art methods in Section 6 and the paper concludes in Section 7.

2 Related work

Designing efficient techniques for minimizing the cost of continuous data collection by exploiting data correlations

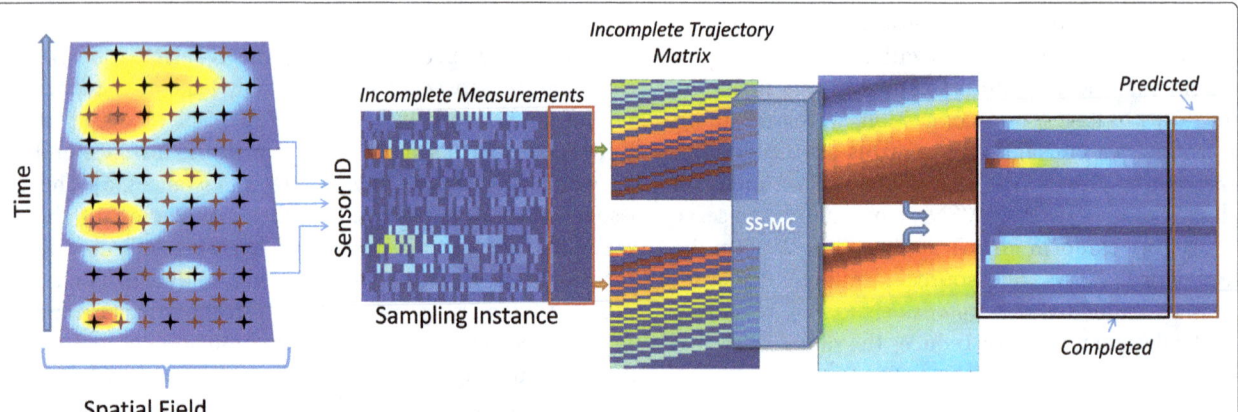

Fig. 1 Overview of the proposed sampling, recovery, and prediction scheme. On the *left*, the three images correspond to the spatial field at three different time instances, where the *star symbols* indicate the sensing nodes. During each sampling instance, *red stars* indicate sampling sensors while *black stars* indicate non-sampling sensors. The figure in the *center* depicts the incomplete measurement matrix where rows correspond to measurements from a specific sensor and columns to different sampling instances. The *red square* over the right part of the matrix highlights that in addition to missing value estimation, our system can generate a number of instances (columns) corresponding to future predictions. Individual sensor measurements are transformed to trajectory matrices that are introduced to the proposed SS-MC framework. The SS-MC algorithm produces completed trajectory matrices that can be joined to generate a fully completed (past and future) measurement matrix

has been extensively studied from multiple aspects and different perspectives in the context of WSNs [8]. Jindal and Psounis [8] presented a method for inferring the spatial correlation of WSN data and for generating synthetic data using a statistical tool called variagriam. Estimating the sampling field at a given location, based on the available sensor data at other additional locations is a common approach for energy efficient sampling. Data imputation and interpolation techniques, such as *Nearest Neighbors Imputation* and *Kriging,* are two very efficient schemes for estimating unavailable data [9]. While in interpolation, one seeks the value of the field in a location where no sensors are present, imputation approaches try to estimate the value at the sensor location at a time instance where sampling did not take place. Kriging relies on the semi-variogram, a statistical tool developed by geo-statisticians [10] in order to estimate the value of a field at a specific location, given prior knowledge about the inherent correlations of data from neighboring nodes. In k-Nearest Neighbors, this objective is reached by using a weighted nearest neighbor interpolation, where the weight corresponding to each sample is based on statistical information indicating the degree of spatial dependence in the field [11].

Another line of work for data imputation exploits probabilistic models for estimating the missing entries. In [12], an Expectation Maximization (EM) algorithm is presented which estimates the parameters of the probability distribution of the data by iteratively maximizing the likelihood of the available data as a function of these parameters. In order to increase the robustness of the process, the authors proposed the regularized EM (RegEM) where a regularization term is added during the inversion of the correlation matrix in order to increase the robustness of the algorithm when more variables are present than data records. RegEM is currently one of the state-of-the-art data imputation techniques, and its performance is compared against the proposed and other schemes in the experimental section.

Data compression has also been extensively explored in the context of energy-efficient data collection in WSNs, based on the premise that data processing is less demanding in terms of energy consumption compared to transmission; hence, energy reduction can be achieved. For example, the recently proposed framework of Compressed Sensing (CS), a state-of-the-art signal sampling and compression scheme, was investigated for WSN data acquisition and aggregation [13, 14] exploiting the sparsity of the sampled data when expressed in an appropriate basis [15]. Distributed compression schemes such as Distributed Source Coding [16] have also been proposed for compressing WSN measurements in densely deployed networks, since utilizing side information from neighboring nodes can dramatically reduce communication cost.

The sparse characteristics of correlated datasets have also been recently considered for transmission of EEG signals [17, 18]. Although sparsity and CS-based methods can have a dramatic reduction in transmission power, typically in these scenarios, the signals are first fully sampled and then compressed.

While the CS framework requires a particular form of sampling (incoherent sampling), the related paradigm of low-rank matrix recovery (MC) assumes a random sampling of the matrix entries. Due to the intuitive sampling, the MC framework has been considered for a variety of signal recovery problems including collaborative spectrum sensing [19], sensor localization [20, 21], and image reconstruction problems [22, 23] among others. MC has been recently explored as a sampling scheme for WSNs [24–27]. In [24], the authors investigated the scenario where sensors lie on a uniform rectangular grid and random sub-sampling is taking place by each sensor. Our work bares some similarities with this line of work; however, we do not pose specific deployment constraints and we allow the sensors to occupy any location in the sensed region. Furthermore, our work differs significantly in the exploitation of prior knowledge in the form of a *dictionary,* which is utilized during the reconstruction stage. The utilization of the singular spectrum dictionary allows for the incorporation of prior knowledge regarding the data generation process which can significantly improve the reconstruction performance [25]. Furthermore, the proposed scheme is able to predict future measurements in addition to estimating missing past ones.

Low-rank recovery was also recently considered in [28] where the authors employ MC for the recovery of under-sampled correlated EEG signals. Our work in this paper investigates different extensions of MC-based recovery by considering trajectory matrices and singular spectrum dictionaries. We develop a generative model where the sampled data can be jointly represented as a low-rank linear combination of dictionary elements, spanning the subspace where data is lying. A similar situation was recently explored, leading to the low rank representations (LRR) framework [29] where the objective is to identify a low rank matrix which can accurately represent the source data. LRR has been considered for subspace clustering problems [30]; however, only fully populated matrices were considered.

In the context of Big Data, matrix and tensor data recovery via an online rank minimization process [31] was recently proposed for scalable imputation of missing data. This was achieved by low-dimensional subspace tracking through the minimization of a weighted least squares regression, regularized with a nuclear norm. While this work bares resemblance to our work, our generative model does not require a fixed bilinear factorization due to a pre-specified rank, while it exploits the subspace identified by the SSA for simultaneous missing past measurement imputation and future predictions.

3 Analysis of time series data

Singular Spectrum Analysis (SSA) is a model-free method for time series analysis and forecasting which has been widely exploited in the analysis of environmental, economical, and computer network data [32, 33]. The basic assumption underlying SSA is that one can approximate a time series $\mathbb{M}_i$ of length K from L lagged samples, by considering the spectral analysis of specialized matrices, called trajectory matrices. Embedding at sampling instance T, the first step of SSA, involves the process of generating a trajectory matrix $\mathbf{M}_i = \{\mathbf{m}_{i,t} | t = T - L : T\} \in \mathbb{R}^{K \times L}$ of lag L measurement vectors, where each vector $\mathbf{m}_{i,t'} = \{m_{i,t'} | t' = t - K : t\}$ encodes the measurements corresponding to a sampling window of length K for sensor i. The length K of the time window and the lag L are two critical parameters encoding important aspects of the underlying data.

In SSA, once the trajectory matrix of the time series has been generated, the subsequent step involves the spectral analysis of the lag-covariance matrix. Formally, given the matrix $\mathbf{M}_i$, the lag-covariance matrix defined as $\mathbf{C}_i = \mathbf{M}_i\mathbf{M}_i^T$ can be used for extracting the eigenvectors of $\mathbf{C}$ which define an L-dimensional subspace where the time series $\mathbb{M}_i$ resides, while the associated eigenvalues encode the variance along the direction of the associated eigenvector. Alternatively, one can apply the SVD decomposition to the original trajectory matrix $\mathbf{M}_i$ in which case the outputs are two matrices containing the right and left singular vectors $\mathbf{U}$ and $\mathbf{V}$ and a diagonal matrix $\mathbf{\Sigma}$ containing the singular values. Given the SVD decomposition, the trajectory matrix $\mathbf{M}_i$ can be expressed as the sum of rank-1 matrices given by $\mathbf{M}_j = \sum_j \sqrt{\lambda_j}\mathbf{u}_j\mathbf{v}_j^T$, where each collection $(\lambda_j, \mathbf{u}_j, \mathbf{v}_j)$ is called eigentriple.

Given the eigenvectors extracted via the SSA, one can project and reconstruct the time series or perform prediction by employing two steps, eigentriple grouping and diagonal averaging. Eigentriple grouping aims at arranging the eigentripes in sets in order to separate additive components that are exactly or approximate separable, facilitating the analysis of the eigenvectors. Diagonal averaging aims at translating the recovered trajectory matrix into a time series according to

$$\hat{m}[k] = \begin{cases} \frac{1}{k}\sum_{m=1}^{K} m^*[m, k-m+1], \\ \qquad \text{for } T-L-K \le k < L \\ \frac{1}{L}\sum_{m=1}^{L} m^*[m, k-m+1], \quad \text{for } L \le k < K \quad (1) \\ \frac{1}{T-K+1}\sum_{m=k-K+1}^{T-K+1} m^*[m, k-m+1], \\ \qquad \text{for } K \le k < T \end{cases}$$

where $m^*[i,j] = m[i,j]$ for $L < K$ and $m^*[i,j] = m[j,i]$ otherwise.

It is worth noting that SSA has also been considered in situations when a number of measurements are missing. A straightforward approach, also employed here, is to estimate the eigenvectors and eigenvalues using only the available measurements during the lag-covariance matrix generation [34]. SSA has also been considered when missing measurements are present [35, 36]; however, the proposed methods differ from our work in that we exploit prior knowledge in the form of a dictionary. Furthermore, the proposed scheme is able to perform missing value estimation, either past or future, while there is no constraint associated with the structure of the missing measurements.

In addition to the analysis of time series, SSA can also be used as a forecasting mechanism. In recurrent forecasting SSA, the time series of known measurements and unknown components is transformed to its Hankel form and the linear recurrent relation coefficients are utilized for forecasting the future values. While typical SSA considers the trajectory matrices associated with a single time series, the Multivariate Singular Spectrum Analysis (MSSA) method has been proposed for handling multiple time series [37–39]. In this work, we consider a simple extension of SSA where instead of analyzing a single trajectory matrix, we consider a compound trajectory matrix generated by the concatenation of S individual matrices, i.e., $\mathbf{M} = [\mathbf{M}_1, \mathbf{M}_2, \ldots, \mathbf{M}_S] \in \mathbb{R}^{S(K \times L)}$. Introducing multiple sources of data can have a dramatic impact in performance as will be shown in the experimental results, with at most linear increase in computational complexity.

4 Low-rank matrix completion

The low-rank approximation of a given matrix is a frequent problem in data analysis [40]. The rank of the matrix indicates the number of linearly independent columns (or rows), and thus it is a indicator of the degree of linear correlation that exists within the data. There are multiple reasons that justify the need for such an analysis. For example, prior knowledge regarding the linear correlation of the data may suggest that the requested matrix is low rank. In other situations, noise in the data artificially increases the rank of the matrix, so reducing the rank effectively amounts to a denoising process. Assuming without loss of generality that $S = 1$, given a noisy $(K \times L)$ matrix $\mathbf{M}$, the objective of low-rank approximation is to identify a matrix $\mathbf{X}$ such that:

$$\underset{\mathbf{X}}{\text{minimize}} \ \text{rank}(\mathbf{X})$$
$$\text{subject to } \|\mathbf{X} - \mathbf{M}\|_F < \epsilon \quad (2)$$

where ϵ is the approximation error, related to the noise power. By utilizing the SVD decomposition $\mathbf{M} = \mathbf{USV}^T$, a low-rank approximation matrix $\mathbf{X}$ can be found by $\mathbf{X} = \mathbf{U}\mathcal{T}(\mathbf{S})\mathbf{V}^T$, where $\mathcal{D}_\tau(\mathbf{S}) = diag([\,\sigma_i(\mathbf{S}) - \tau]_+)$ is a thresholding operator that selects only the elements with values greater than τ from the diagonal matrix $\mathbf{S}$ and sets the rest to zero. The effect of this process is that only a small number of singular values are kept for the low-rank approximation $\mathbf{X}$ of $\mathbf{M}$.

The rank of the matrix is a key property in the recently proposed framework of Matrix Completion (MC) where one tries to estimate the $(K \times L)$ entries of the matrix $\mathbf{M}$ from a smaller number of q entries, where $q \ll (K \times L)$. According to MC, such a recovery is possible provided the matrix is characterized by a small rank (compared to its dimensions) and enough randomly selected entries of the matrix are acquired [6, 41]. More specifically, one can recover an accurate approximation $\mathbf{X}$ of the matrix $\mathbf{M}$ from a small number of entries by solving the minimization problem:

$$\begin{array}{c} \underset{\mathbf{X}}{\text{minimize}} \ \text{rank}(\mathbf{X}) \\ \text{subject to} \ \mathcal{P}_\Omega(\mathbf{X}) = \mathcal{P}_\Omega(\mathbf{M}) \end{array} \qquad (3)$$

where $\mathcal{P}_\Omega$ is a random sampling operator which records only a small number of entries from the matrix $\mathbf{M}$, i.e.,

$$\mathcal{P}_\Omega(\mathbf{M}) = \begin{cases} m_{ij}, & \text{if } ij \in \Omega \\ 0, & \text{otherwise} \end{cases} \qquad (4)$$

where Ω is the sampling set. In the context of WSN for example, the set Ω specifies the collection of sensors that are active at each specific sampling instance. In general, to solve the MC problem, the sampling operator $\mathcal{P}$ must satisfy the modified restricted isometry property, which is the case when uniform random sparse sampling is employed in both rows and columns of matrix $\mathbf{M}$ [42]. The incoherence of sampling introduced by $\mathcal{P}$ with respect to $\mathbf{M}$ guarantees that recovery is possible from a limited number of measurements.

Although solving the above problem will generate a low-rank matrix consistent with the observations, rank minimization is an NP-hard problem. Fortunately, a relaxation of the above problem was shown to produce very accurate approximations, by replacing the rank constraint by the tractable nuclear norm, which represents the convex envelope of the rank [6]. The minimization in Eq. (4) can then be reformulated as:

$$\begin{array}{c} \underset{\mathbf{X}}{\text{minimize}} \ \|\mathbf{X}\|_* \\ \text{subject to} \ \mathcal{P}_\Omega(\mathbf{X}) = \mathcal{P}_\Omega(\mathbf{M}) \end{array} \qquad (5)$$

where the nuclear norm is defined as $\|\mathbf{X}\|_* = \sum \|\sigma_i\|_1$, i.e., the sum of absolute values of the singular values. Candès and Tao showed that under certain conditions the nuclear norm minimization in Eq. (5) can estimate the same matrix as the rank minimization in Eq. (3) with high probability provided $q \geq CK^{6/5} r log(K)$ randomly selected

entries of the rank r matrix are acquired [7] (assuming $K \geq L$).

To solve the nuclear norm minimization problem, various approaches have been proposed including Singular Value Thresholding [43] and the Augmented Lagrange Multipliers [44], among others. We review the technique based on the ALM due to its exceptional performance in terms of both processing complexity and reconstruction accuracy and since it is used as a basis for the extended scheme we discuss next.

To express the MC problem in Eq. (5) in the ALM form, we reformulate it as:

$$\begin{array}{c} \underset{\mathbf{X},\mathbf{E}}{\text{minimize}} \ \|\mathbf{X}\|_* \\ \text{subject to} \ \mathbf{X} + \mathbf{E} = \mathbf{M} \\ \mathcal{P}_\Omega(\mathbf{E}) = 0 \end{array} \qquad (6)$$

The additional variable $\mathbf{E}$ is introduced in order to encode the unknown values in the trajectory matrix $\mathbf{M}$, by restricting the estimation error on the recorded values only. The optimization encoded in Eq. (6) can be expressed in an augmented Lagrangian form by defining the Lagrangian function:

$$\begin{aligned} \mathcal{L}(\mathbf{X}, \mathbf{E}, \mathbf{Y}, \mu) = \|\mathbf{X}\|_* &+ tr(\mathbf{Y}^T(\mathbf{X} - \mathbf{M} + \mathbf{E})) \\ &+ \frac{\mu}{2}\|\mathbf{X} - \mathbf{M} + \mathbf{E}\|_F^2 \end{aligned} \qquad (7)$$

where $\mathbf{Y}$ is the Lagrange multiplier matrix associated to the first equality constraint and μ is the penalty parameter. Minimization of the problem in Eq. (7) involves an iterative process, where a sequential minimization over all variables, i.e., $\mathbf{X}, \mathbf{E}$, and $\mathbf{Y}$, takes place at each iteration. This method of iteratively minimizing over each variable is refereed to as the Alternating Directions Method of Multipliers (ADMM) [45, 46].

One of the key characteristics of MC is the minimal conditions that are imposed for successful recovery, namely the incoherence of sampling and the low rank of the recovered matrix. While a minimal set of requirements is beneficial in situations where limited prior information is available, when such information exists introducing additional constraints can lead to a significantly better recovery. In this section, we exploit the temporal dynamic that time series exhibit in order to enhance typical MC with an additional dictionary which encodes past behavior in a proposed SS-MC framework.

5 The SS-MC algorithm

We consider the truncated trajectory matrices $\mathbf{M}$ formed by concatenating the individual trajectory matrices according to the MSSA approach. The objective of this work is to consider a generative model that produces the time series Hankel matrices $\mathbf{M}$ according to the factorization $\mathbf{M} = \mathbf{DL}$ where $\mathbf{M}$ may correspond to a single or multiple sources. In both cases, our key assumption is that given a full rank dictionary matrix $\mathbf{D}$ obtained through training data, the coefficient matrix $\mathbf{L}$ is approximately low rank, i.e., the number of significant singular vectors is much smaller than the ambient dimensions of the matrix.

To apply the low-rank representation scheme on matrices with missing data, the introduction of the random sub-sampling operator is necessary. Our proposed sampling scheme is a combination of MC and reduced rank multivariate linear regression and it seeks a low-rank presentation coefficient matrix $\mathbf{L}$ from a small number of measurements $\mathcal{P}_\Omega(\mathbf{M})$. Based on this generative model, our proposed Singular Spectrum Matrix Completion (SS-MC) formulation is given by:

$$\underset{\mathbf{L}}{\text{minimize}} \;\; rank(\mathbf{L})$$
$$\text{subject to} \;\; \mathcal{P}_\Omega(\mathbf{M}) = \mathcal{P}_\Omega(\mathbf{DL}) \qquad (8)$$

where $\mathbf{D}$ is a dictionary of elementary atoms that span a low-rank data-induced subspace. Figure 2 presents an example of a real trajectory matrix (left), the representations coefficients $\mathbf{L}$ (center), and the singular value distribution of the coefficients (right).

5.1 Efficient optimization

Similarly to MC optimization, the problem in Eq. (8) is NP-hard due to the rank in the objective function and thus it cannot be solved efficiently for reasonably sized data. A remedy to this problem is to replace the rank constraint with the nuclear norm constraint, thus solving:

$$\underset{\mathbf{L}}{\text{minimize}} \;\; \|\mathbf{L}\|_*$$
$$\text{subject to} \;\; \mathcal{P}_\Omega(\mathbf{M}) = \mathcal{P}_\Omega(\mathbf{DL}) \qquad (9)$$

A key novelty of our work is that in addition to the low rank of the matrix, during the recovery, we employ a dictionary for modeling the generative process that produces the sensed data, as it can be seen in Eq. (9).

The problem in Eq. (9) can be transformed to a semidefinite programming problem and solved using interior point methods [47, 48]. However, utilizing such off-the-shelf solvers introduces a very high algorithmic complexity which renders them impractical, even for moderately sized scenarios. Motivated by the requirements for a data collection mechanism that is both accurate and efficient, we reformulate the SS-MC problem in an Augmented Lagrangian form. By utilizing the ALM formulation for SS-MC, we can achieve efficient recovery, tailored to the specific properties of the problem. Introducing the intermediate dummy variables $\mathbf{Z}$ and $\mathbf{E}$, Eq. (9) can be written as:

$$\underset{\mathbf{L,Z,E}}{\text{minimize}} \;\; \|\mathbf{L}\|_*$$
$$\text{subject to} \;\; \mathbf{M} = \mathbf{DZ} + \mathbf{E}$$
$$\mathbf{Z} = \mathbf{L}$$
$$\mathcal{P}_\Omega(\mathbf{E}) = 0 \qquad (10)$$

where $\mathbf{L}, \mathbf{Z},$ and $\mathbf{E}$ are the minimization variables. The extra variable $\mathbf{Z}$ is introduced in order to decouple the minimization variables by separating the $\mathbf{L}$ variable in the objective function with the $\mathbf{Z}$ variable in the first constraint. Similar to the ALM formulation for MC in Eq. (7), $\mathbf{E}$ is introduced in order to account for the missing entries in $\mathbf{M}$. More specifically, the constraint on the error matrix $\mathbf{E}$ is applied only on the available data via the sampling operator $\mathcal{P}$. The ALM form of Eq. (10) is an unconstrained minimization given by:

$$\mathcal{L}(\mathbf{L}, \mathbf{Z}, \mathbf{E}, \mathbf{Y_1}, \mathbf{Y_2}, \mu) = \|\mathbf{L}\|_* + tr\left(\mathbf{Y}_1^T (\mathcal{P}_\Omega(\mathbf{M} - \mathbf{DZ}))\right)$$
$$+ tr\left(\mathbf{Y}_2^T (\mathbf{Z} - \mathbf{L})\right)$$
$$+ \frac{\mu}{2}(\|\mathcal{P}_\Omega(\mathbf{M} - \mathbf{DZ})\|_F^2 + \|\mathbf{Z} - \mathbf{L}\|_{F^2}) \qquad (11)$$

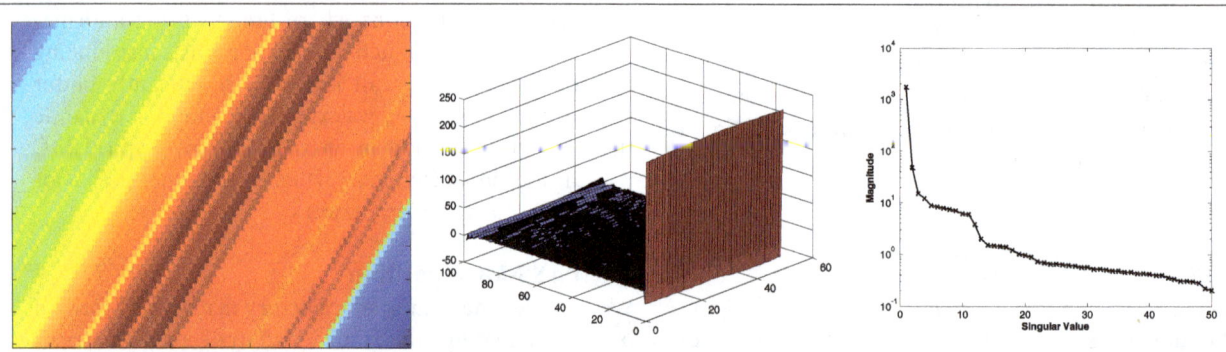

Fig. 2 Example of the generative process of a real trajectory matrix from the Intel-Berkeley dataset. The matrix on the *left* is the Hankel matrix generated by a sensor for a given set of window and lag values. By utilizing the SSA-based dictionary, the mapping of the Hankel matrix results in an extremely low rank representation matrix shown in the *middle*, where a small number of singular values capture most of the signal energy, as shown in the *right* figure

where $\mathbf{Y}_1$ and $\mathbf{Y}_2$ are Lagrange multiplier matrices. The solution can be found by iteratively minimizing Eq. (11) with respect to each of the variables via an ADMM approach. Formally, the minimization problem with respect to $\mathbf{L}$ is given by:

$$\mathbf{L}^{(k+1)} = \min_{\mathbf{L}} \mathcal{L}\left(\mathbf{L}^{(k)}, \mathbf{Z}^{(k)}, \mathbf{E}^{(k)}, \mathbf{Y}_1^{(k)}, \mathbf{Y}_2^{(k)}, \mu^{(k)}\right)$$

$$= \min \|\mathbf{L}\|_* + tr\left(\mathbf{Y}_2^{\mathrm{T}}(\mathbf{Z}-\mathbf{L})\right) + \frac{\mu}{2}\left(\|\mathbf{Z}-\mathbf{L}\|_F^2\right)$$

$$= \min \frac{1}{\mu}\|\mathbf{L}\|_* + \frac{1}{2}\|\mathbf{L}-(\mathbf{Z}+\mathbf{Y}_2/\mu)\|_F^2 . \quad (12)$$

The sub-problem in Eq. (12) is a nuclear norm minimization problem and can be solved very efficiently by the Singular Value Thresholding operator [43]. The minimization with respect to $\mathbf{Z}$ is given by:

$$\mathbf{Z}^{(k+1)} = \min_{\mathbf{Z}} \mathcal{L}\left(\mathbf{L}^{(k+1)}, \mathbf{Z}^{(k)}, \mathbf{E}^{(k)}, \mathbf{Y}_1^{(k)}, \mathbf{Y}_2^{(k)}, \mu^{(k)}\right)$$

$$= \min tr\left(\mathbf{Y}_1^T(\mathcal{P}_\Omega(\mathbf{M}) - \mathcal{P}_\Omega(\mathbf{DZ}))\right) + tr\left(\mathbf{Y}_2^T(\mathbf{Z}-\mathbf{L})\right)$$

$$+ \frac{\mu}{2}\left(\|\mathcal{P}_\Omega(\mathbf{M}-\mathbf{DZ})\|_F^2 + \|\mathbf{Z}-\mathbf{L}\|_F^2\right). \quad (13)$$

Calculating the gradient of the expression in Eq. (13), we obtain:

$$\frac{\partial \mathcal{L}}{\partial \mathbf{Z}} = \mathbf{D}^T\mathbf{Y}_1 - \mathbf{Y}_2 + \mu\left(\mathbf{D}^T(\mathbf{M}-\mathbf{E}-\mathbf{DZ}) - \mathbf{Z} + \mathbf{L}\right) \quad (14)$$

which after setting it equal to zero provides the update equation for Eq. (14) given by:

$$\mathbf{Z}^{(k+1)} = \left(\mathbf{I}+\mathbf{D}^T\mathbf{D}\right)^{-1}\left(\mathbf{D}^T\left(\mathbf{M}^{(k)}-\mathbf{E}^{(k)}\right) + \mathbf{L}^{(k)}\right.$$

$$\left. + \left(\mathbf{D}^T\mathbf{Y}_1^{(k)} - \mathbf{Y}_2^{(k)}\right)/\mu^{(k)}\right). \quad (15)$$

Furthermore, the augmented Lagrangian in Eq. (11) has to be minimized with respect to $\mathbf{E}$, i.e.,

$$\mathbf{E}^{(k+1)} = \min_{\mathcal{P}_\Omega(\mathbf{M})=0} \mathcal{L}\left(\mathbf{L}^{(k+1)}, \mathbf{Z}^{(k+1)}, \mathbf{E}^{(k)}, \mathbf{Y}_1^{(k)}, \mathbf{Y}_2^{(k)}, \mu^{(k)}\right)$$

$$= \min_{\mathcal{P}_\Omega(\mathbf{M})=0} \mathbf{Y}_1 + \mu(\mathbf{E}-\mathbf{M}+\mathbf{DZ}) \quad (16)$$

which provides the update equation for Eq. (16) that is given by:

$$\mathbf{E}^{(k+1)} = \mathcal{P}_{\bar{\Omega}}\left(\mathbf{M}-\mathbf{DZ}^{(k+1)} + \frac{1}{\mu^{(k)}}\mathbf{Y}_1^k\right) \quad (17)$$

where the notation $\mathcal{P}_{\bar{\Omega}}$ is used to restrict the error estimation only on the measurements that do not belong to the sampling set. Last, we perform updates on the two Lagrange multipliers $\mathbf{Y}_1$ and $\mathbf{Y}_2$. The steps at each iteration of the optimization are shown in Algorithm 1.

Algorithm 1: Singular Spectrum Matrix Completion (SS-MC)

Input: The subsampled trajectory matrix $\mathbf{M}_{ij}, (i,j) \in \Omega$,

The dictionary of examples $\mathbf{D}$,

The error tolerance *threshold*,

The maximum number of iterations *limit*.

Output: The representation coefficients matrix $\mathbf{L}$ and the estimated matrix $\mathbf{X}=\mathbf{DL}$.

1: **initialization**
  $\mathbf{L}^0 = \mathbf{0}$, $\mathbf{E}^{(0)} = \mathbf{0}$, $\mathbf{Z}^{(0)} = \mathbf{0}$, $k=0$, $\alpha=1.1$

2: **while** *error* $\geq$ *threshold* or *iterations* $\leq$ *limit* **do**

3: Minimize with respect to $\mathbf{L}$ to obtain $\mathbf{L}^{(k+1)}$

$$(\mathbf{U}, \mathbf{S}, \mathbf{V}) = SVD(\mathbf{Z}+\mathbf{Y}_2/\mu)$$
$$\mathbf{L}^{(k+1)} = \mathbf{U}\mathcal{D}_\tau(\mathbf{S})\mathbf{V}^{\mathrm{T}}$$

4: Minimize with respect to $\mathbf{Z}$ to obtain $\mathbf{Z}^{(k+1)}$

$$\mathbf{Z}^{(k+1)} = (\mathbf{I}+\mathbf{D}^{\mathrm{T}}\mathbf{D})^{-1}(\mathbf{D}^{\mathrm{T}}(\mathbf{M}-\mathbf{E})+\mathbf{L}$$
$$+(\mathbf{D}^{\mathrm{T}}\mathbf{Y}_1 - \mathbf{Y}_2)/\mu)$$

5: Minimize with respect to $\mathbf{E}$ to obtain $\mathbf{E}^{(k+1)}$

$$\mathbf{E}^{(k+1)} = \mathcal{P}_{\bar{\Omega}}\left(\mathbf{M}-\mathbf{DZ}+\frac{1}{\mu^{(k)}}\mathbf{Y}_1\right)$$

6: Update the Lagrangian multipliers

$$\mathbf{Y}_1^{(k+1)} = \mathbf{Y}_1^{(k)} + \mu^{(k)}(\mathbf{M}-\mathbf{DZ}^{(k+1)} - \mathbf{E}^{(k+1)})$$
$$\mathbf{Y}_2^{(k+1)} = \mathbf{Y}_2^{(k)} + \mu^{(k)}(\mathbf{Z}^{(k+1)} - \mathbf{L}^{(k+1)})$$

set $k \longleftarrow k+1$

7: **end while**

Due to its numerous applications, the ADMM method has been extensively studied in the literature for the case of two variables [45, 46] where it has been shown that under mild conditions regarding the convexity of the cost functions, the two-variables ADMM converges at a rate $\mathcal{O}(1/r)$ [49]. Although extending the convergence properties to a larger number of variables has not been shown in general, recently the convergence properties of ADMM for a sum of two or more non-smooth convex separable functions subject to linear constraints were examined [50].

The proposed minimization scheme in Eq. (11) satisfies a large number of the constraints suggested in [50] such as the convexity of each sub-problem, the strict convexity and continuous differentiability of the nuclear norm, the full rank of the dictionary, and the size of the step for the dual update α, while empirical evidence suggests that the closed form solution of each sub-problem allows the SS-MC algorithm to converge to an accurate solution in a small number of iterations.

5.2 Singular spectrum dictionary

In this work, we investigate the utilization of prior knowledge for the efficient reconstruction of severely undersampled time series data. To model the data, we follow a generative scheme where the full collection of acquired measurements is encoded in the trajectory matrix $\mathbf{M} \in \mathbb{R}^{K \times L}$. $\mathbf{M}$ is assumed to be generated from a combination of a dictionary $\mathbf{D} \in \mathbb{R}^{K \times K}$ and a coefficient matrix $\mathbf{L} \in \mathbb{R}^{K \times L}$ according to $\mathbf{M} = \mathbf{DL}$, where we assume that $K \leq L$. This particular factorization is related to SVD by $\mathbf{M} = \mathbf{DL} = \mathbf{U}(\mathbf{SV}^T)$ where the orthonormal matrix $\mathbf{D} = \mathbf{U}$ is a basis for the subspace associated with the column space of $\mathbf{M}$, while $\mathbf{L} = \mathbf{SV}^T$ is a low-rank representation matrix encoding the projection of the trajectory matrix onto this subspace.

This particular choice of dictionary $\mathbf{D}$ implies a specific relationship between the spectral characteristics of the trajectory matrix $\mathbf{M}$ and the low-rank representation matrix $\mathbf{L}$. To understand this relationship, we consider the spectral decomposition of each individual matrix in the form $\mathbf{D} = \mathbf{UG}_1\mathbf{R}^{-1}$ and $\mathbf{L} = \mathbf{RG}_2\mathbf{V}^*$ The matrices $\mathbf{U}, \mathbf{R}$ and $\mathbf{V}$ are unitary while $\mathbf{G}_1$ and $\mathbf{G}_2$ are diagonal matrices containing the singular values of the $\mathbf{D}$ and $\mathbf{L}$, respectively. The particular factorization permits us to utilize the product SVD [51, 52] and expresses the singular value decomposition of the product according to the expression $\mathbf{DL} = \mathbf{U}(\mathbf{G}_1\mathbf{G}_2)\mathbf{V}^*$, where the singular values of the matrix product are given by the product of the singular values of the corresponding matrices.

In this work, we consider orthogonal dictionaries, as opposed to overcomplete ones. Orthogonality of the dictionary guarantees that the vectors encoded in the dictionary span the low-dimensional subspace and therefore the representation of the measurements is possible. Furthermore, an orthonormal dictionary, such as the one considered in this work, is characterized by $\mathbf{G}_1 = \mathbf{I}$, leaving $\mathbf{G}_2$ responsible for the representation. We target exactly $\mathbf{G}_2$ in our problem formulation by seeking a low-rank representation matrix $\mathbf{L}$.

In our experimental results, we consider sets of training data associated with fully sampled time series from the first days of each experiment for generating the dictionaries. The subspace identified by the fully sampled data is used for the subsequent recovery of past measurements and prediction of future ones. Alternatively, the dictionary could be updated during the course of the SS-MC application via an incremental subspace learning method [53, 54]. We opted out from an incremental subspace learning since although it can potentially lead to better estimation, it is also associated with increased computational load and the higher probability of estimation drift and lower performance.

5.3 Networking aspects of SS-MC

In the context of IoT applications utilizing WSN infrastructures, communication can take place among nodes, but most typically between the nodes and the base station where data analytics are extracted. This communication can be supported (a) by a direct wireless link between the nodes and the sink/base station; (b) via appropriate paths that allow multi-hop communications; or (c) via more powerful cluster heads what forward the measurements to the base station.

For the multi-hop scheme, equal weight of each sample (democratic sampling) implies that no complicated processing needs to take place by the resource limited forwarding nodes. Furthermore, for high-performance WSNs, where point-to-point communication between nodes is available and processing capabilities are sufficient, nodes could perform reconstruction of a local neighborhood thus offering advantages similar to other distributed estimation schemes [55].

From a practical point-of-view, we argue that recovery and prediction of measurements from low sampling rates offer numerous advantages. First, it saves energy by reducing the number of samples that have to be acquired, processed, and communicated thus increasing the lifetime of the network. The proposed sampling scheme also reduces the frequency of sensor re-calibrations for sensors that perform complex signal acquisition, including chemical and biological sampling. As a result, higher quality measurements and therefore more reliable estimation of the field samples can be achieved. Furthermore, the method increases robustness to communication errors by estimating measurements included in lost or dropped packets, without the need for retransmission. Last, our scheme does not require explicit knowledge of node locations for the estimation of the missing measurements, since the incomplete measurement matrices and the corresponding trajectory matrices are indexed by the sensor id, thus allowing greater flexibility during deployment.

6 Experimental results

To evaluate the performance of the proposed low-rank reconstruction and prediction scheme, we consider real data from the Intel Berkeley Research Lab dataset[1] [56] and the SensorScope Grand St-Bernard dataset[2] [57]. The former dataset contains the recordings of 54 multimodal sensors located in an indoor environment over a 1-month period, while the latter contains multimodal measurements from 23 stations deployed at the Grand-St-Bernard pass between Switzerland and Italy.

In both cases, we analyze temperature measurements as an exemplary modality, while we exclude failed sensors from the recovery process. Unless stated otherwise, in all cases, we fix the SSA parameters, $K = 50$ and $L = 100$, and we train using a single day's worth of data while testing on the five consecutive ones. The threshold τ for the singular value thresholding operator is set to preserve 90 % of the signals' energy, while the parameter μ was set to 0.01 through a validation process, although the specific value had a minimal impact in performance.

To evaluate the performance, we consider three state-of-the-art methods and we compare them to the proposed SS-MC. More specifically, we evaluate the performance of the ADMM version of MC [44], the Knn-imputation [58], and the RegEM [12]. The reconstruction error is measured by the normalized mean squared error between the true $\mathbf{M}$ and the estimated $\mathbf{X}$ trajectory matrices given by $\frac{\sum \|\mathbf{M}-\mathbf{X}\|^2}{\sum \|\mathbf{M}\|^2}$.

6.1 Recovery with respect to measurement availability

The objective of this subsection is to present the recovery capabilities of the proposed SS-MC and state-of-the-art methods with respect to the availability of measurements, i.e., the sampling rate.

The two plots shown in Fig. 3 present the reconstruction error for the Intel-Berkeley data at 20 % (top) and 50 % (bottom) sampling rates, averaged over all sensing nodes. Naturally, one can see that increasing the sampling rate has a positive effect on all methods. Nevertheless, we also observe that not all sampling instances are equally difficult to estimate and that the reconstruction error exhibits a periodic trend across sampling instances. These variations are attributed to the significant changes in the environmental conditions due to the transition from nighttime to daytime.

Comparing the four methods, we observe that under all measurement availability scenarios, the proposed SS-MC scheme typically achieves the lowest reconstruction error and exhibits the most stable performance. The performance of SS-MC is closely followed, especially in low sampling rates, by RegEM which also exhibits a very stable performance, while on the other hand, MC and Knn-impute are more sensitive to the sampling instance, exhibiting a more erratic behavior.

To further illuminate the behavior of each method, we consider a large set of sampling instances and present the averaged recovery performance as a function of the sampling rate in Fig. 4 for Intel-Berkeley (top) and SensorScope (bottom) data. Regarding the performance on the Intel-Berkeley dataset, we observe that the proposed SS-MC and RegEM achieve comparable performance, much better than typical MC and Knn-impute. An interesting observation is that while SS-MC, RegEM, and Knn-impute all exhibit a monotonic reduction in reconstruction error at higher sampling rates, MC reaches a performance plateau around a 25 % sampling rate. This phenomenon is attributed to the rank constrains of MC leading to a low rank estimation which causes an incorrect estimation of missing measurements.

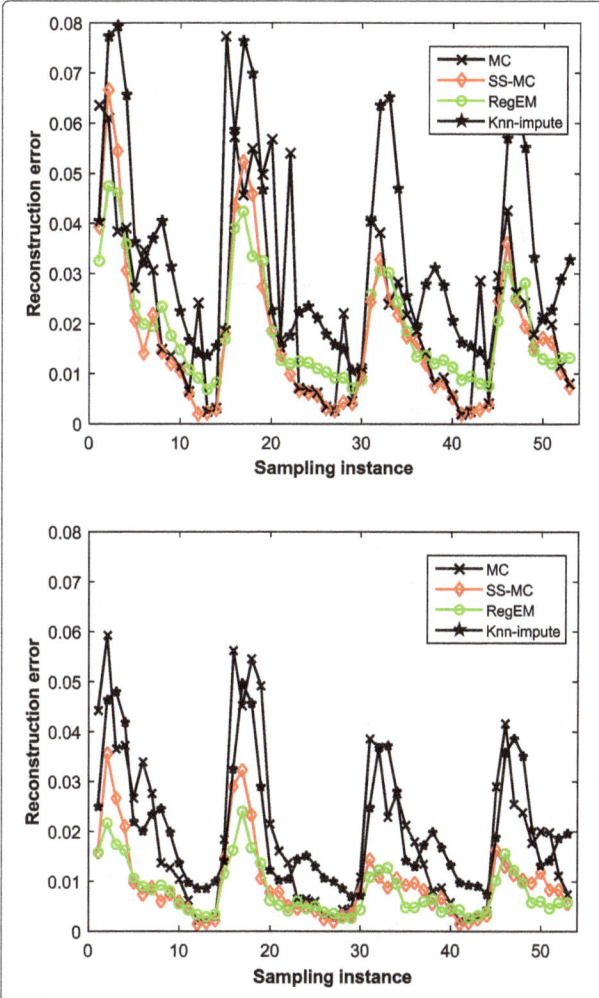

Fig. 3 Reconstruction error at 20 % (*top*) and 50 % (*bottom*) sampling rates for the Intel-Berkeley dataset

Regarding the performance on the SensorScope data, one can observe that in this case RegEM achieves a significantly better performance compared to the other methods, followed by MC at low sampling rates and SS-MC at large ones. Similar to the behavior observed for the Intel-Berkeley data, MC again reaches a performance plateau while the other methods achieve a monotonically reducing reconstruction error. Note that although RegEM achieves the lowest reconstruction error, it is also the most computationally demanding of the four methods.

6.2 Recovery from multiple sources

In this subsection, we investigate the recovery capabilities of the SS-MC and state-of-the-art method as a function of the number of sensors/sources that are simultaneously considered. Figure 5 presents the reconstruction error for the multiple source/sensor cases, where 2 (top), and 5 (bottom) sources from the Intel-Berkeley dataset are simultaneously considered. Comparing these results

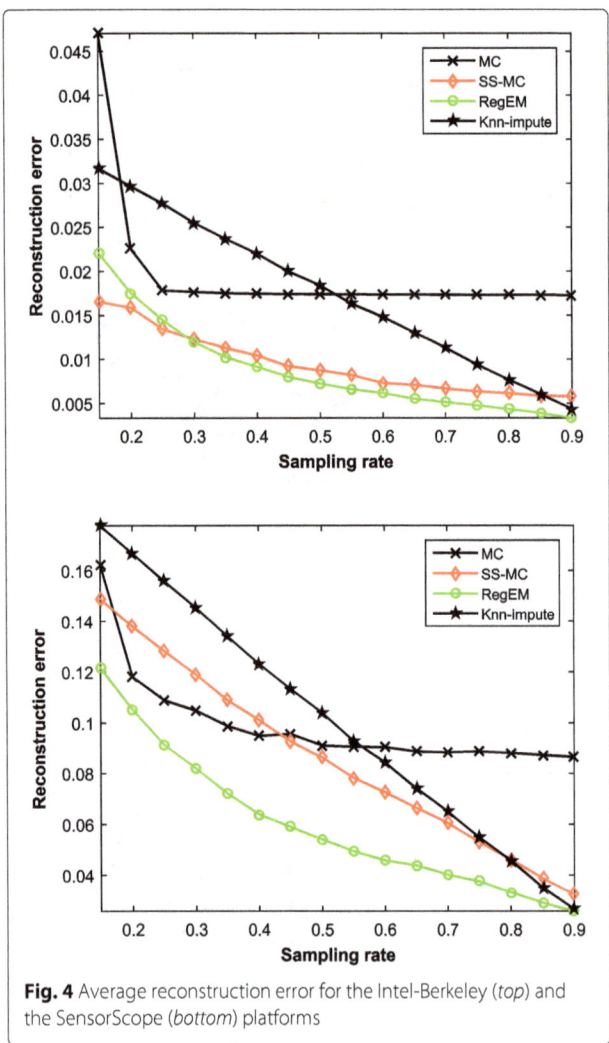

Fig. 4 Average reconstruction error for the Intel-Berkeley (*top*) and the SensorScope (*bottom*) platforms

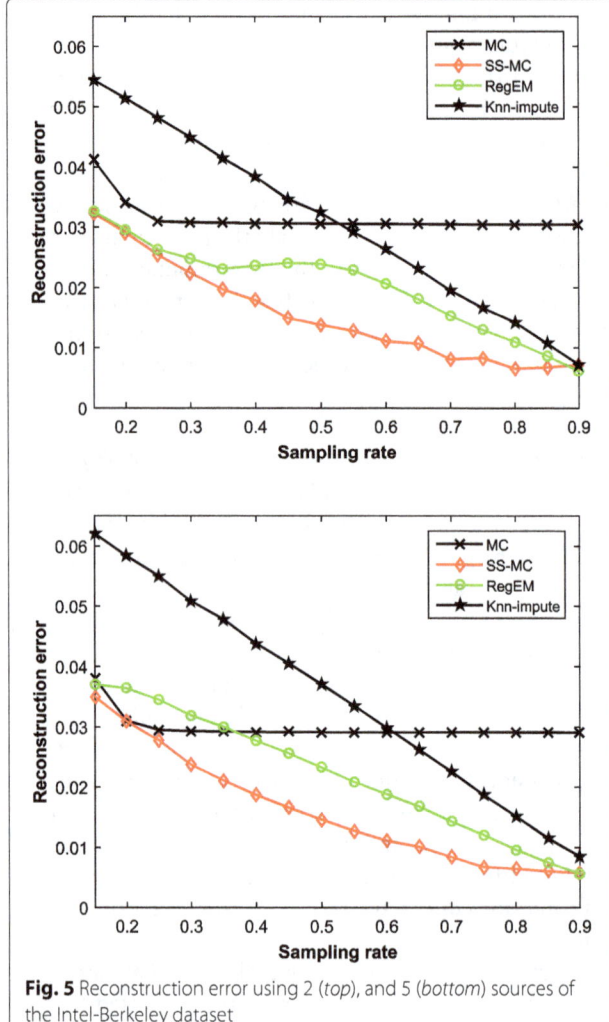

Fig. 5 Reconstruction error using 2 (*top*), and 5 (*bottom*) sources of the Intel-Berkeley dataset

with the results shown Fig. 4 (top), one can observe that increasing the number of sources that a method considers simultaneously can have a different effect for each method, although no method appears to be able to exploit the additional sources of data.

State-of-the-art methods, like Knn-impute and RegEM, not only appear to be unable to exploit the additional sources of data, but introducing the additional sources leads to an increase in reconstruction error for a given sampling rate. On the other hand, typical MC is unaffected by the different scenarios, exhibiting the same plateau in behavior regardless of the number of sources under consideration. Unlike the other methods, the proposed SS-MC is able to better handle the additional data. Although applying SS-MC with multiple sources of data does not lead to better performance, the proposed method is better in handling such complex data streams, offering the lowest reconstruction error among all methods considered.

The situation differs however for the SensorScope data shown in Fig. 6 for 2 (top) and 5 (bottom) sources, respectively. In this case, Knn-input appears to suffer a significant reduction in reconstruction quality due to the additional data sources, leading to a notable increase in reconstruction error compared to the single stream case. RegEM and typical MC also do not appear to benefit from the additional sources. In contrast to these methods, the proposed SS-MC achieves a more robust behavior leading to a significantly better behavior compared to the single source case. The improvement is more dramatic when moving from the single to two sources; however, introducing additional sources has a positive effect on recovery performance.

In general, for the state-of-the-art methods we consider, experimental results suggest that introducing multiple correlated sources does not necessarily aid in the recovery performance, while under different scenarios, the aggregation of multiple sources may also introduce prohibitively large communication overheads. On the other

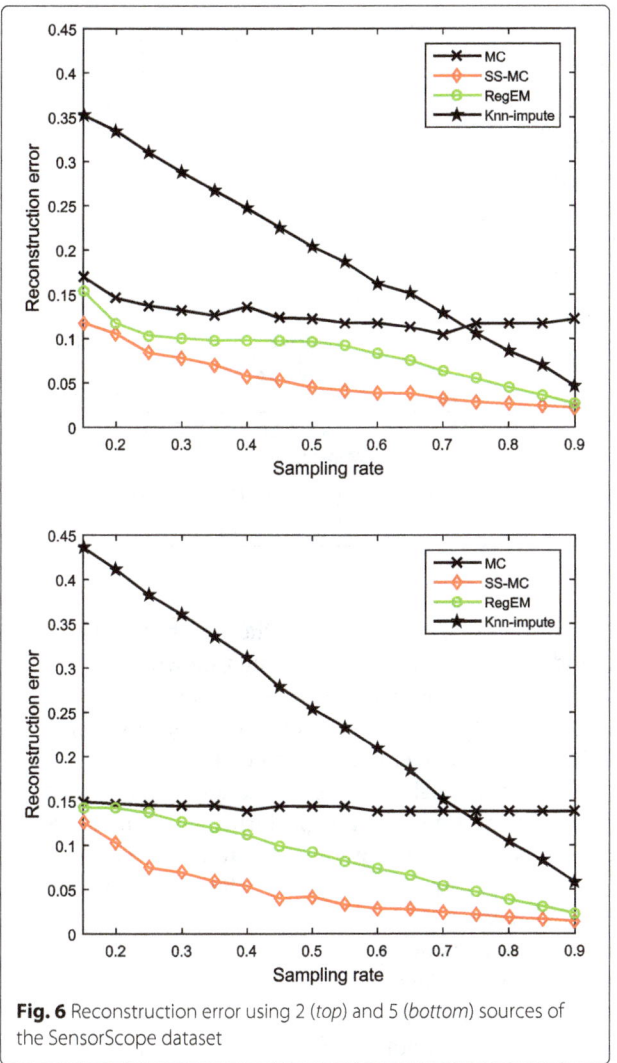

Fig. 6 Reconstruction error using 2 (*top*) and 5 (*bottom*) sources of the SensorScope dataset

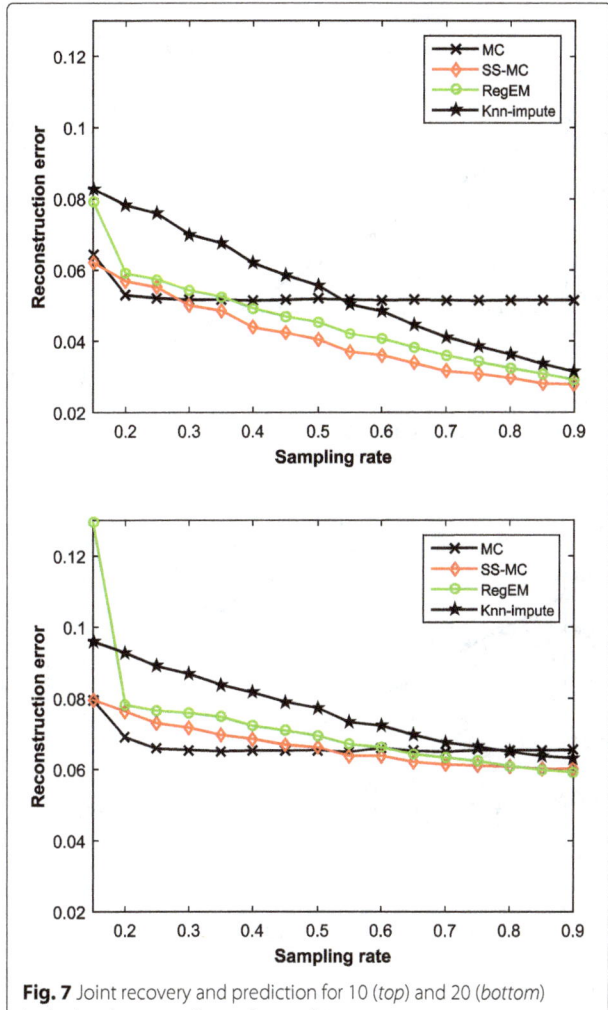

Fig. 7 Joint recovery and prediction for 10 (*top*) and 20 (*bottom*) look-ahead steps on SensorScope data

hand, the proposed SS-MC can smoothly transition from the single sensor/source case to multiple sensors/sources achieving compelling gains in certain scenarios.

6.3 Joint recovery and prediction

In this set of results, we consider the more challenging scenario where the method must simultaneously recover and predict future measurements. The results for the SensorScope data shown in Fig. 7 demonstrate the competitive performance of the proposed SS-MC method compared to state-of-the-art methods for both 10 (top) and 20 (bottom) look-ahead steps. The benefits of our method are more clearly shown for the short-term prediction (top) while for the long term, we observe a similar behavior for all methods. Naturally, the performance is significantly better for the short term compared to the long term; however, we observe that both the MC and the SS-MC approaches achieve a very stable performance

in both cases, suggesting that the low-rank regularization can provide strong benefits in this challenging scenario.

Figure 8 illustrates the recovery/estimation performance on the Intel-Berkeley data where we observe that the proposed SS-MC method achieves a dramatic reduction in reconstruction error, clearly surpassing the other methods in both short-term and long-term predictions. Similar to the SensorScope data, both MC and SS-MC achieve a very stable performance while SS-MC is much less affected by the increase in prediction horizon. Considering the results for both cases, we can conclude that SS-MC is an excellent choice for the challenging problem, achieving a very low prediction error even when only a small subset of measurements is available.

6.4 Performance with respect to computational resources

The results reported in the previous subsections assume that a single day's worth of data is utilized during the training phase where the dictionary **D** is obtained. Here, we

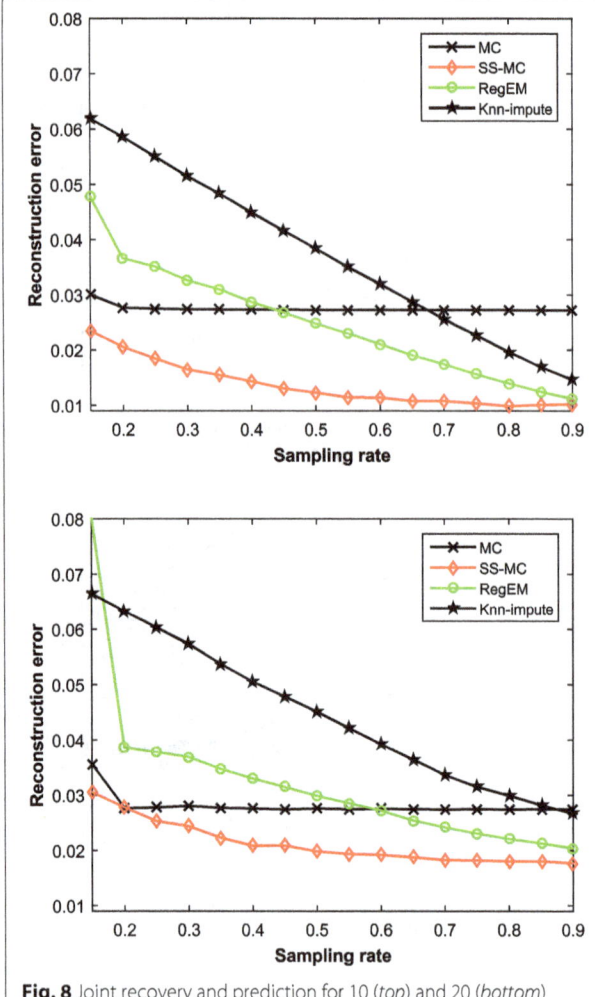

Fig. 8 Joint recovery and prediction for 10 (*top*) and 20 (*bottom*) look-ahead steps on Intel-Berkeley data

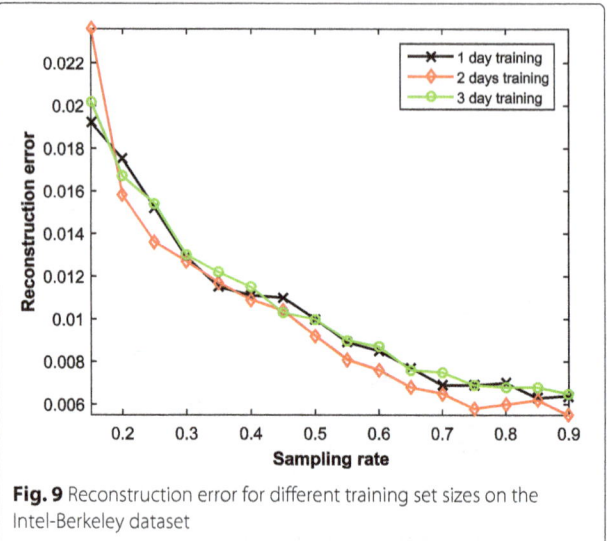

Fig. 9 Reconstruction error for different training set sizes on the Intel-Berkeley dataset

investigate the recovery capability of the proposed SS-MC method as a function of the amount of training data, i.e., the number of days used for training.

Figure 9 presents the reconstruction error for the Intel-Berkeley data using 1, 2, and 3 days of training data. The results clearly indicate that introducing more data from training has limited impact on the reconstruction performance. When one considers that the process of collecting fully sampled data can have a dramatic impact on the life time of the network, we can conclude that given a limited set of representative data suffices for SS-MC.

This aspect is critical since we assume that the training data is fully populated without any missing measurements. To achieve the acquisition of such training data requires extra care in terms of communication robustness as well as a larger energy consumption due to full sampling.

In addition to the amount of the training data that is required for a given performance, we also investigated the

SS-MC recovery as a function of the number of iterations and the sampling rate. The results shown in Fig. 10 demonstrate that the quality of the recovery is affected by the availability of the measurements where for larger sampling rates, a smaller number of iterations is required. Despite this relationship, however, we also observe that there is a clear limit on the performance gain above 50 iterations. This is the number of iterations we have assumed in our experiments unless the approximation error drops below 10^{-4}.

The requirements of Big Data processing mandate algorithm that can achieve high quality performance with minimal processing requirements. To better illustrate the computational requirements for each method, Table 1 presents the processing time (in seconds) for the proposed (SS-MC) and the three state-of-the-art methods under

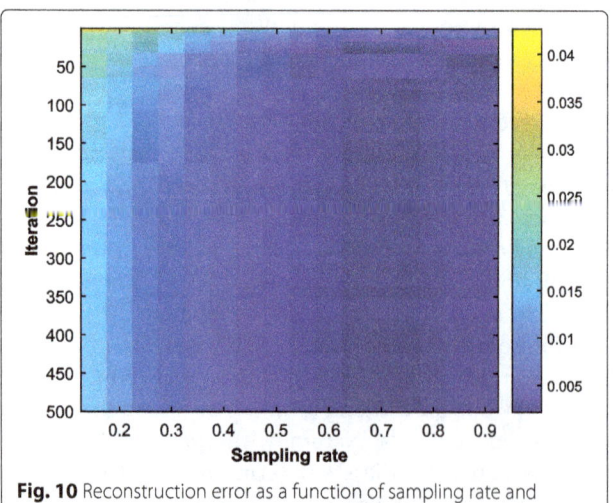

Fig. 10 Reconstruction error as a function of sampling rate and iteration number

Table 1 Computational time for different number of sensors and measurement availability

	25 %		50 %		75 %	
	1	5	1	5	1	5
SS-MC	0.188	0.950	0.137	0.719	0.087	0.358
MC	0.101	0.140	0.101	0.146	0.103	0.152
RegEM	0.092	0.137	0.098	0.407	0.154	1.194
Knn	0.153	0.866	0.102	0.632	0.051	0.275

different sampling rates when considering a single (1) or multiple (5) sources.

Table 1 clearly demonstrates the relationships of each method with respect to the sampling rate where we observe that for the proposed SS-MC method, increasing the sampling rate leads to lower processing time for both the single and the multiple source cases. On the other hand, MC requires a fixed processing time independently of the number of available measurements, while the effect of the number of sources is minimal. RegEM's processing time is increasing as the number of available measurements increase due to the inner mechanics of the algorithm which require multiple regression to take place. Last, the Knn-impute method exhibits a decrease in processing time with respect to the measurement availability and an increase associated with multiple sources. Overall, the proposed SS-MC exhibits a stable and predicable performance, achieving a very good trade-off between processing requirements and reconstruction quality.

7 Conclusions

Acquiring, transmitting, and processing Big Data presents numerous challenges due to the complexity and volume issues among others. The situation becomes even more complicated when one considers data sources associated with the Internet-of-Things paradigm, where component and architecture limitations, including processing capabilities, energy availability, and communication failures, must also be considered. In this work, we proposed a distributed sampling-centralized recovery scheme where due to various design choices and physical constraints, only a small subset of the entire set of measurements is collected during each sampling instance. The proposed SS-MC approach exploits the low-rank representation of appropriately generated trajectory matrices, when expressed in the subspace associated with dictionaries learned using training data, in order to recover missing measurements as well as predict future values. The recovery and prediction procedures are implemented via an efficient optimization based on the augmented Lagrange multipliers method. Experimental results on real data from the Intel-Berkeley and the SensorScope datasets validate the merits of the proposed scheme compared to state-of-the-art

methods like typical matrix completion, RegEM, and Knn-imputation, both in terms of pure reconstruction as well as in the demanding case of simultaneous recovery and prediction.

Endnotes
[1] http://db.csail.mit.edu/labdata/labdata.html.
[2] http://lcav.epfl.ch/page-86035-en.html.

Competing interests
The authors declare that they have no competing interests.

Acknowledgements
This work was founded by the DEDALE (contract no. 665044) within the H2020 Framework Program) of the EC. This work was also supported by the PETROMAKS Smart-Rig (grant 244205 /E30), SFI Offshore Mechatronics (grant 237896/O30), both from the Research Council of Norway, and the RFF Agder UiA CIEMCoE grant.

Author details
[1] Institute of Computer Science, Foundation for Research & Technology - Hellas (FORTH), Crete, Greece. [2] Lab Intelligent Signal Processing & Wireless Networks (WISENET), Department of Information and Communication Technologies, University of Agder, Grimstad, Norway. [3] Department of Computer Science, University of Crete, Crete, Greece.

References
1. K Slavakis, G Giannakis, G Mateos. IEEE Signal Proc. Mag. **31**(5), 18 (2014)
2. J Gubbi, R Buyya, S Marusic, M Palaniswami. Futur. Gener. Comput. Syst. **29**(7), 1645 (2013)
3. DB Rawat, JJ Rodrigues, I Stojmenovic, *Cyber-Physical Systems: From Theory to Practice.* (CRC Press, Boca Raton, 2015)
4. G Tzagkarakis, G Tsagkatakis, D Alonso, E Celada, C Asensio, A Panousopoulou, P Tsakalides, B Beferull-Lozano, in: *Cyber Physical Systems: From Theory to Practice.* (DB Rawat, J Rodrigues, I Stojmenovic, eds.) (CRC Press, USA, 2015)
5. F Sivrikaya, B Yener. IEEE Netw. **18**(4), 45 (2004)
6. B EJ Candès, Recht, Found. Comput. Math. **9**(6), 717 (2009)
7. T EJ Candès, IEEE Tao, Trans. Inf. Theory. **56**(5), 2053 (2010)
8. A Jindal, K Psounis. ACM Trans. Sens. Netw. (TOSN). **2**(4), 466 (2006)
9. GE Batista, MC Monard. Appl. Artif. Intell. **17**(5–6), 519 (2003)
10. N Cressie. Terra Nova. **4**(5), 613 (1992)
11. J Li, A Heap. Ecol. Informa. **6**(3), 228 (2011)
12. T Schneider. J. Clim. **14**(5), 853 (2001)
13. C Luo, F Wu, J Sun, C Chen, in *International conference on Mobile computing and networking.* (ACM, Beijing China, 2009), pp. 145–156
14. C Luo, F Wu, J Sun, CW Chen, IEEE Trans. Wirel. Commun. **9**(12), 3728 (2010)
15. A Fragkiadakis, I Askoxylakis, E Tragos, in: *International Workshop on Computer Aided Modeling and Design of Communication Links and Networks.* (IEEE, 2013), pp. 84–88
16. Z Xiong, A Liveris, S Cheng, IEEE Signal Proc. Mag. **21**(5), 80 (2004)
17. A Majumdar, RK Ward, Biomed. Signal Process. Control. **13**, 142 (2014)
18. A Shukla, A Majumdar, Biomed. Signal Process. Control. **18**, 174 (2015)
19. JJ Meng, W Yin, H Li, E Houssain, Z Han, in: *Acoustics Speech and Signal Processing (ICASSP) 2010 IEEE International Conference on.* (IEEE, Dallas, 2010), pp. 3114–3117
20. S Nikitaki, G Tsagkatakis, P Tsakalides, IEEE Trans. Mob. Comput. **14**(11), 2244 (2015)
21. S Nikitaki, G Tsagkatakis, P Tsakalides, in: *Signal Processing Conference (EUSIPCO) 2012 Proceedings of the 20th European.* (IEEE, Bucharest, 2012), pp. 195–199
22. PJ Shin, PE Larson, MA Ohliger, M Elad, JM Pauly, DB Vigneron, M Lustig, Magn. Reson. Med. **72**(4), 959 (2014)
23. G Tsagkatakis, P Tsakalides, in: *Machine Learning for Signal Processing (MLSP) 2012 IEEE International Workshop on.* (IEEE, Santander, 2012), pp. 1–6

24. A Majumdar, R Ward, in: *Data Compression Conference, 2010*. (IEEE, Snowbird, 2010), pp. 542–542

25. G Tsagkatakis, P Tsakalides, in: *Sensor Array and Multichannel Signal Processing Workshop (SAM)*. (IEEE, Hoboken, 2012), pp. 117–120

26. F Fazel, M Fazel, M Stojanovic, in: *Information Theory and Applications Workshop (ITA)*. (IEEE, San Diego, 2012), pp. 300–305

27. S Savvaki, G Tsagkatakis, P Tsakalides, in *ACM International Workshop on Cyber-Physical Systems for Smart Water Networks*. (ACM, New York, 2015)

28. A Majumdar, A Gogna, Sensors. **14**(9), 15729 (2014)

29. G Liu, Z Lin, S Yan, J Sun, Y Yu, Y Ma, IEEE Trans. Pattern Anal. Mach. Intell. **35**(1), 171 (2013)

30. E Elhamifar, R Vidal, IEEE Trans. Pattern Anal. Mach. Intell. **35**(11), 2765 (2013)

31. M Mardani, G Mateos, GB Giannakis, IEEE Trans. Signal Process. **63**(10), 2663 (2015)

32. N Golyandina. Singular Spectrum Analysis for time series (Springer Science & Business Media, New York, 2013)

33. G Tzagkarakis, M Papadopouli, P Tsakalides, in: *ACM Symposium on Modeling, analysis, and simulation of wireless and mobile systems*. (ACM, Chania, 2007), pp. 99–108

34. DH Schoellhamer, Geophys. Res. Lett. **28**(16), 3187 (2001)

35. D Kondrashov, M Ghil, Nonlinear Process. Geophys. **13**(2), 151 (2006)

36. N Golyandina, E Osipov, J. Stat. Plan. Infer. **137**(8), 2642 (2007)

37. K Patterson, H Hassani, S Heravi, A Zhigljavsky, J. App. Stat. **38**(10), 2183 (2011)

38. N Golyandina, D Stepanov, in: *5th St. Petersburg workshop on simulation*, vol. 293. (St. Petersburg State University, St. Petersburg, 2005), p. 298

39. N Golyandina, A Korobeynikov, A Shlemov, K Usevich. J. Stat. Softw. **67**(1), 1 (2015)

40. I Markovsky. Low rank approximation: algorithms, implementation, applications (Springer Science & Business Media, New York, 2011)

41. Y E Candès, Plan, Proc. IEEE. **98**(6), 925 (2010)

42. B Recht, M Fazel, P Parrilo, SIAM Rev. **52**(3), 471 (2010)

43. JF Cai, EJ Candès, Z Shen, SIAM J. Optim. **20**(4), 1956 (2010)

44. Z Lin, M Chen, Y Ma, *arXiv preprint 1009.5055*, (2010). http://arxiv.org/abs/1009.5055

45. DP Bertsekas. 1st edn. Constrained Optimization and Lagrange Multiplier Methods (Optimization and Neural Computation Series) (Athena Scientific, Nashua, 1996)

46. S Boyd, N Parikh, E Chu, B Peleato, J Eckstein, Found. Trends Mach Learn. **3**(1), 1 (2011)

47. Z Liu, L Vandenberghe, SIAM J. Matrix Anal. Appl. **31**(3), 1235 (2009)

48. M Grant, S Boyd, Y Ye (2008). Online accessiable: http://stanford.edu/~boyd/cvx. Accessed 1 Jan 2014

49. S Boyd, N Parikh, E Chu, B Peleato, J Eckstein, Distributed optimization and statistical learning via the alternating direction method of multipliers. Found. Trends® Mach. Learn. **3**(1), 1–122 (2011). Now Publishers Inc.

50. Luo ZQ, arXiv preprint arXiv:1208.3922 (2012). http://arxiv.org/abs/1208.3922

51. KV Fernando, S Hammarling. Linear algebra in signals, systems, and control (Boston, MA, 1986), pp. 128–140 (1988)

52. B De Moor, Signal Process. **25**(2), 135 (1991)

53. DA Ross, J Lim, RS Lin, MH Yang, Int. J. Comput. Vis. **77**(1–3), 125 (2008)

54. Y Li, Pattern Recogn. **37**(7), 1509 (2004)

55. ID Schizas, GB Giannakis, ZQ Luo, IEEE Trans. Signal Process. **55**(8), 4284 (2007)

56. S Madden, *Intel lab data, 2004*, (2012). http://db.csail.mit.edu/labdata/labdata.html

57. F Ingelrest, G Barrenetxea, G Schaefer, M Vetterli, O Couach, M Parlange, ACM Trans. Sens. Netw. (TOSN). **6**(2), 1 (2010)

58. O Troyanskaya, M Cantor, G Sherlock, P Brown, T Hastie, R Tibshirani, D Botstein, R Altman, Bioinformatics. **17**(6), 520 (2001)

13

Adaptive parameter particle CBMeMBer tracker for multiple maneuvering target tracking

Jinlong Yang[1,2*], Le Yang[1], Zhiyong Xiao[1] and Jianjun Liu[1]

Abstract

Cardinality-balanced multi-target multi-Bernoulli (CBMeMBer) filter has been demonstrated as a promising algorithm for multi-target tracking, and the multi-model (MM) method has been incorporated into the CBMeMBer filter to solve the problem of multiple maneuvering target tracking. However, it is difficult to construct a proper set of models due to the unknown maneuvering parameters of the targets. Moreover, the number of models may increase exponentially if more unknown parameters have to be taken into account to match the target motion modes, which may lead to prohibitive computational complexity. To address this aspect, this paper proposes to incorporate the adaptive parameter estimation (APE) method in the framework of CBMeMBer filters, so that the model with unknown maneuvering parameter can be modified adaptively by using the selected parameter particles. Moreover, a particle labeling technique is introduced in the proposed algorithm in order to obtain the individual target track, which results in the adaptive parameter particle filter CBMeMBer (APPF-CBMeMBer) tracker. Simulation results show that the proposed algorithm can effectively track multiple maneuvering targets with abruptly changing parameters and exhibit better robustness than those of the well-known MM-based approaches.

Keywords: Multi-target multi-Bernoulli filter, Adaptive parameter estimation (APE), Multi-model (MM) method, Multiple maneuvering target tracking

1 Introduction

In recent years, random finite set (RFS) [1–3] is an elegant formulation of the multi-target tracking (MTT) problem and has generated substantial interest due to the development of the probability hypothesis density (PHD) filter [2] and the cardinalized PHD (CPHD) filter [3]. The PHD and CPHD filters were proposed as Poisson RFS density approximations of the multi-target posterior, which can be used to estimate the target states by recursively computing the first-order moment of the multi-target posterior probability distribution. The existing closed-form solutions of PHD mainly include particle filter PHD (PF-PHD) [4, 5] and the Gaussian mixture PHD (GM-PHD) filter [6], which have opened the door to numerous novel extensions and applications as shown in [7–13]. Moreover,

being different from the PHD and CPHD filters, the multi-target multi-Bernoulli (MeMBer) [1] recursion was recently proposed by Mahler as a tractable approximation to the Bayesian multi-target recursion under low clutter density scenarios, which can achieve multi-target tracking by directly propagating the approximate posterior density of the targets. Unfortunately, the MeM-Ber filter has significant cardinality bias. To tackle this problem, the cardinality-balanced MeMBer (CBMeMBer) filter and its improved versions, such as the δ-generalized labeled multi-Bernoulli (δ-GLMB) and LMB filters, were developed in [14–18]. They eliminate the posterior cardinality bias by modifying the measurement-updated track parameters. These algorithms exhibit good MTT performance only when the model parameters are known precisely. In the presence of unknown measurement noise variances, clutter and detection probability, improved RFS filters capable of jointly estimating target states, and unknown parameters were proposed (see e.g., [9, 15, 19] and references therein). These methods

* Correspondence: yjlgedeng@163.com
[1]School of Internet of Things Engineering, Jiangnan University, Wuxi 214122, China
[2]Key Laboratory of Advanced Process Control for Light Industry (Ministry of Education), Wuxi 214122, China

would also suffer from performance degradation if the targets make maneuvers with unknown abruptly changing maneuvering parameters.

For maneuvering target tracking, the jump Markov system (JMS) has proved to be an effective method, which switches among a set of candidate models in a Markovian fashion [20, 21]. Pasha et al. introduced the linear JMS into the PHD filters and derived a closed-form solution for the PHD recursion in [22]. Furthermore, the unscented transform (UT) and the linear fractional transformation (LFT) are combined with the closed-form solution for the nonlinear jump Markov multi-target models in [23, 24]. In [25], a GM-PHD filter for jump Markov models is developed by employing the best-fitting Gaussian (BFG) approximation approach. These algorithms assume the Gaussianity of the PHD distribution, which may limit the scope of their applications. The multiple-model PHD (MM-PHD) filter and the MM-CPHD filter implemented using the sequential Monte Carlo (SMC) method were presented in [26, 27], and a corrected version, also known as the jump Markov multi-target Bayes filter, was later proposed in [28]. However, the MMP-CBMeMBer filter [29] has a higher accuracy than the MM-PHD filter due to the fact that the multi-Bernoulli-based method propagates the parameterized approximation to the posterior cardinality distribution. Most of the MM-based filters track multiple maneuvering targets through the interaction of multiple models, which is realized via combining estimates from different models according to their respective model likelihoods. The difficulty of applying them in tracking targets with abruptly changing parameters comes from the need to specify priorly the set of candidate models. The number of models may increase exponentially if more unknown parameters have to be taken into account to match the target motion modes, which may lead to prohibitive computational complexity.

In this work, we attempt to incorporate the adaptive parameter estimation (APE) technique into the framework of the CBMeMBer filter for addressing the problem of multiple maneuvering target tracking. The adaptive Liu and West (LW) filter is adopted to propagate the posterior marginal of the time-varying parameters as a mixture of multivariate Gaussian distributions [30–32]. The obtained adaptive parameter particle filter CBMeMBer (APPF-CBMeMBer) filter can track multiple maneuvering targets in the presence of unknown model parameters. Simulation results show that the proposed algorithm exhibits better robustness and improved tracking performance over the MMP-CBMeMBer algorithms. Furthermore, in order to obtain the individual target tracks, the particle labeling technique is introduced in the proposed algorithm.

The remainder of the paper is organized as follows. Section 2 formulates the problem of tracking a target in the presence of unknown model parameters. It also briefly reviews the APE technique and the CBMeMBer filter. Section 3 proposes the APPF-CBMeMBer algorithm with the closed-form solution and describes the track maintenance method. Simulation results are given in Section 4. Finally, conclusions are provided in Section 5.

2 Backgrounds
2.1 Formulation of the problem
The state-space models for tracking a single target moving on a two-dimensional plane are given by

$$x_{k+1} = \mathbf{F}x_k + \mathbf{G}v_k \tag{1}$$

$$y_k = h(x_k) + w_k \tag{2}$$

where $x_k = \begin{bmatrix} x_k, v_{x_k}, y_k, v_{y_k} \end{bmatrix}^{\mathrm{T}}$ denotes the target state at time k, (x_k, y_k) and (v_{x_k}, v_{y_k}) denote its position and velocity. $\mathbf{F}$ and $\mathbf{G}$ are the state transition matrices of the state vector and the process noise gain matrix. y_k is the measurement vector. v_k and w_k are the process noise and the measurement noise, respectively. They are independent with each other and modeled as zero-mean Gaussian random vectors with covariance Q_k and R_k, respectively.

In many practical applications, the state-space model in (1) and (2) may contain unknown parameters. For example, if the motion of a target follows a coordinated-turn (CT) model [26], the state transition matrix would become

$$\mathbf{F}(\omega) = \begin{bmatrix} 1 & \dfrac{\sin\omega T}{\omega} & 0 & -\dfrac{1-\cos\omega T}{\omega} \\ 0 & \cos\omega T & 0 & -\sin\omega T \\ 0 & \dfrac{1-\cos\omega T}{\omega} & 1 & \dfrac{\sin\omega T}{\omega} \\ 0 & \sin\omega T & 0 & \cos\omega T \end{bmatrix} \tag{3}$$

The maneuvering parameter (turn rate ω) may be unknown and time-varying. In this case, joint estimating the posterior distribution of the target state and the unknown maneuvering parameter from the measurements is needed.

Let θ_k be a time-varying parameter in the state-space model. The posterior probability density function (PDF) of the target state vector x_k and the unknown parameter vector θ_k conditioned on the measurements up to time k is, according to Bayes' rule,

$$p(x_k, \theta_k | y_{1:k}) = \frac{p(y_k | x_k, \theta_k)p(x_k, \theta_k | y_{1:k-1})}{\displaystyle\int p(y_k | x_k, \theta_k)p(x_k, \theta_k | y_{1:k-1})dx_k d\theta_k} \tag{4}$$

where $p(x_k, \theta_k | y_{1:k-1})$ is the predictive PDF and can be expressed as

$$p(\pmb{x}_k, \theta_k | \pmb{y}_{1:k-1}) = \int p(\pmb{x}_k | \pmb{x}_{k-1}, \theta_{k-1}) p(\pmb{x}_{k-1}, \theta_{k-1} | \pmb{y}_{1:k-1}) d\pmb{x}_{k-1} d\theta_{k-1} \tag{5}$$

Deriving exact recursive solutions for the posterior distribution $p(\pmb{x}_k, \theta_k | \pmb{y}_{1:k})$ from (4) and (5) is in general intractable and as a result, approximate solutions are usually resorted to. One such approach is the particle filter (PF) [4, 14, 31].

2.2 Adaptive parameter estimation
In [30, 31], the Liu and West (LW) filter was proposed for the joint identification of static parameters and the target states. In particular, the marginal posterior distribution of the unknown parameters is approximated and propagated using a mixture of multivariate Gaussian distributions. In [32], the particle learning technique was introduced into the LW filter. The obtained APE filter can handle both static and time-varying parameters.

The development of the APE method starts with factorizing the predicting PDF $p(\pmb{x}_k, \theta_k | \pmb{y}_{1:k-1})$ into

$$p(\pmb{x}_k, \theta_k | \pmb{y}_{1:k-1}) = p(\pmb{x}_k | \pmb{y}_{1:k-1}, \theta_k) p(\theta_k | \pmb{y}_{1:k-1}) \tag{6}$$

The predicting distribution $p(\theta_k | \pmb{y}_{1:k-1})$ of the time-varying parameter vector θ_k can be approximated via

$$p(\theta_k | \pmb{y}_{1:k-1}) \approx \begin{cases} \sum_{i=1}^{N} \omega_{k-1}^i N(\theta_k | m_{k-1}^i, h^2 V_{k-1}), & \text{with probability } 1-\beta \\ p_\theta(\theta_0), & \text{with probability } \beta \end{cases} \tag{7}$$

where $N(\theta_k | m_{k-1}^i, h^2 V_{k-1})$ is a Gaussian component with mean m_{k-1}^i and covariance V_{k-1}, and ω_{k-1}^i is the associated weight. Here, β is introduced to model the temporal evolution of θ_k. It is defined as the probability that θ_k is subject to an abrupt change at time k, or equivalently speaking, time instant k is a changepoint [32]. The time-varying vector θ_k is assumed to be piecewise constant between two neighboring changepoints. As shown in (7), if there is no abrupt change in θ_k, its predicting PDF follows a Gaussian mixture model of N components. The mean and covariance of each component are obtained by

$$m_{k-1}^i = \alpha \theta_{k-1}^i + (1-\alpha) \bar{\theta}_{k-1} \tag{8}$$

$$V_{k-1} = \sum_{i=1}^{N} \omega_{k-1}^i (\theta_{k-1}^i - \bar{\theta}_{k-1})(\theta_{k-1}^i - \bar{\theta}_{k-1})^T \tag{9}$$

where $\bar{\theta}_{k-1} = \sum_{i=1}^{N} \omega_{k-1}^i \theta_{k-1}^i$ is the minimum mean square error (MMSE) estimate of θ_{k-1} at time $k-1$, and $\alpha = \sqrt{1-h^2}$ is the shrinkage factor suggested in [33] to correct for the over-dispersion of the Gaussian mixture model.

In the case that time instant k is a changepoint, the predicting distribution of the time-varying vector θ_k will be reset to $p_\theta(\theta_0)$, its prior distribution.

With the predicting PDF given in (7), the APE filter utilizes the PF to produce an approximation of the posterior distribution $p(\pmb{x}_k, \theta_k | \pmb{y}_{1:k})$ in (4). Suppose at time $k-1$, the posterior distribution is represented by N particles $\{\pmb{x}_{k-1}^i, \theta_{k-1}^i\}_{i=1}^N$ with weights ω_{k-1}^i. At time k, each particle is given two weights [32]

$$\omega_{k,1}^i \propto p(\pmb{y}_k | \mu_k^i, \theta_{k-1}^i),$$
$$\text{where } \theta_k^i \sim N(\xi_k | m_{k-1}^i, h^2 V_{k-1}), \mu_k^i = E[\pmb{x}_k^i | \pmb{x}_{k-1}^i, \theta_{k-1}^i] \tag{10}$$

$$\omega_{k,2}^i \propto p(\pmb{y}_k | \mu_k^i, \gamma_k^i), \text{where } \gamma_k^i \sim p_\theta(\theta_0), \mu_k^i = E[\pmb{x}_k^i | \pmb{x}_{k-1}^i, \gamma_k^i] \tag{11}$$

which essentially leads to $2N$ particles. $\omega_{k,1}^i$ and $\omega_{k,2}^i$ correspond to the probability of the current measurement $\pmb{y}_k$ when there is no changepoint and when there is a changepoint. In the former case, the value of time-varying parameter vector θ_k^i is drawn from the Gaussian distribution $N(\xi_k | m_{k-1}^i, h^2 V_{k-1})$ while for the latter case, its value γ_k^i is produced using the prior distribution $p_\theta(\theta_0)$ (see also (7)). Resampling is then performed on the basis of the weights $(1-\beta)\omega_{k,1}^i$ and $\beta\omega_{k,2}^i$ to select N particles out of $2N$ particles and propagate them to generate the approximation of the posterior $p(\pmb{x}_k, \theta_k | \pmb{y}_{1:k})$ at time k. For more details on the APE filter for tracking a single maneuvering target, please refer to [32].

2.3 Cardinality-balanced MeMBer filter
The CBMeMBer filter was proposed in [14], which eliminates the posterior cardinality bias existed in the MeMBer filter [1] by modifying the measurement-updated tracks parameters. The CBMeMBer recursion is summarized as follows.

Prediction: Assume the posterior multi-target density at time $k-1$ can be presented by the multi-Bernoulli parameter set, i.e.,

$$\pi_{k-1} = \left\{ \left(r_{k-1}^{(i)}, p_{k-1}^{(i)} \right) \right\}_{i=1}^{M_{k-1}} \tag{12}$$

where $r_{k-1}^{(i)} \in (0, 1)$ and $p_{k-1}^{(i)}$ denote the existence probability and probability density of the i th Bernoulli component, respectively. M_{k-1} denotes the number of the posterior hypothesized tracks at time $k-1$.

Then, the predicted multi-target density $\pi_{k|k-1}$ can also be expressed by the multi-Bernoulli parameter set and is given by

$$\pi_{k|k-1} = \left\{ \left(r_{P,k|k-1}^{(i)}, p_{P,k|k-1}^{(i)} \right) \right\}_{i=1}^{M_{k-1}} \cup \left\{ \left(r_{\Gamma,k}^{(i)}, p_{\Gamma,k}^{(i)} \right) \right\}_{i=1}^{M_{\Gamma,k}}$$

$$(13)$$

where $\left\{ \left(r_{P,k|k-1}^{(i)}, p_{P,k|k-1}^{(i)} \right) \right\}_{i=1}^{M_{k-1}}$ and $\left\{ \left(r_{\Gamma,k}^{(i)}, p_{\Gamma,k}^{(i)} \right) \right\}_{i=1}^{M_{\Gamma,k}}$ denote the parameter sets of the multi-Bernoulli RFS of the surviving targets and the spontaneous births, respectively. M_{k-1} and $M_{\Gamma,k}$ denote the predicted hypothesized track number of the surviving targets and the spontaneous births, respectively.

$$r_{P,k|k-1}^{(i)} = r_{k-1}^{(i)} \left\langle p_{k-1}^{(i)}, p_{S,k} \right\rangle \qquad (14)$$

$$p_{P,k|k-1}^{(i)}(x) = \frac{\left\langle f_{k|k-1}(x|\cdot), p_{k-1}^{(i)} p_{S,k} \right\rangle}{\left\langle p_{k-1}^{(i)}, p_{S,k} \right\rangle} \qquad (15)$$

where $f_{k|k-1}(x|\cdot)$ denotes the single target transition density and $p_{S,k}$ denotes the target survival probability.

As can be seen from Eq. (13), in essence, the multi-Bernoulli parameter set for the predicted multi-target density $\pi_{k|k-1}$ is formed by the union of the multi-Bernoulli parameter sets for the surviving targets and the spontaneous births. The total number of predicted hypothesized tracks is $M_{k|k-1} = M_{k-1} + M_{\Gamma,k}$.

Update: Assume the predicted multi-target density at time k can be expressed by a known multi-Bernoulli parameter set as follows:

$$\pi_{k|k-1} = \left\{ \left(r_{k-1}^{(i)}, p_{k-1}^{(i)} \right) \right\}_{i=1}^{M_{k|k-1}} \qquad (16)$$

Then, the posterior multi-target density can be approximated by the union of the multi-Bernoulli parameter sets for the legacy tracks [the first term in Eq. (17)] and measurement-corrected tracks [the second term in Eq. (17)], i.e.,

$$\pi_k \approx \left\{ \left(r_{L,k}^{(i)}, p_{L,k}^{(i)} \right) \right\}_{i=1}^{M_{k|k-1}} \cup \left\{ \left(r_{U,k}^*(y), p_{U,k}^*(\cdot;y) \right) \right\}_{y \in Y_k} \qquad (17)$$

where

$$r_{L,k}^{(i)} = r_{k|k-1}^{(i)} \frac{1 - \left\langle p_{k|k-1}^{(i)}, p_{D,k} \right\rangle}{1 - r_{k|k-1}^{(i)} \left\langle p_{k|k-1}^{(i)}, p_{D,k} \right\rangle} \qquad (18)$$

$$p_{L,k}^{(i)} = p_{k|k-1}^{(i)}(x) \frac{1 - p_{D,k}(x)}{1 - \left\langle p_{k|k-1}^{(i)}, p_{D,k} \right\rangle} \qquad (19)$$

$$r_{U,k}^*(y) = \frac{\sum_{i=1}^{M_{k|k-1}} \frac{r_{k|k-1}^{(i)} \left(1 - r_{k|k-1}^{(i)} \right) \left\langle p_{k|k-1}^{(i)}, \psi_{k,y} \right\rangle}{\left(1 - r_{k|k-1}^{(i)} \left\langle p_{k|k-1}^{(i)}, p_{D,k} \right\rangle \right)^2}}{\kappa_k(y) + \sum_{i=1}^{M_{k|k-1}} \frac{r_{k|k-1}^{(i)} \left\langle p_{k|k-1}^{(i)}, \psi_{k,y} \right\rangle}{1 - r_{k|k-1}^{(i)} \left\langle p_{k|k-1}^{(i)}, p_{D,k} \right\rangle}}$$

$$(20)$$

$$p_{U,k}^*(x;y) = \frac{\sum_{i=1}^{M_{k|k-1}} \frac{r_{k|k-1}^{(i)}}{1 - r_{k|k-1}^{(i)}} p_{k|k-1}^{(i)}(x) \psi_{k,y}(x)}{\sum_{i=1}^{M_{k|k-1}} \frac{r_{k|k-1}^{(i)}}{1 - r_{k|k-1}^{(i)}} \left\langle p_{k|k-1}^{(i)} \psi_{k,y} \right\rangle} \qquad (21)$$

$$\psi_{k,y}(x) = p_k(y|x) p_{D,k}(x) \qquad (22)$$

$p_k(y|x)$ is the single target measurement likelihood, $p_{D,k}(x)$ is the target detection probability, Y_k is the measurement set, and $\kappa_k(y)$ is the intensity of clutter which follows the Poisson distribution. The total number of posterior hypothesized tracks is $M_k = M_{k|k-1} + |Y_k|$.

3 APPF-CBMeMBer tracker

3.1 Adaptive parameter particle filter

In this section, we propose the adaptive parameter particle filter to implement the CBMeMBer filter. Notice that the particles consist of the state and maneuvering parameter with associated weights. The detailed processes of particle implementation are described as follows.

(1) Prediction: Suppose that at time $k-1$, the posterior multi-target density is described as $\pi_{k-1} = \left\{ \left(r_{k-1}^{(i)}, p_{k-1}^{(i)}(x, \theta) \right) \right\}_{i=1}^{M_{k-1}}$, θ denotes the maneuvering parameter of a Bernoulli component. $p_{k-1}^{(i)}(x, \theta)$ is comprised of a set of weighted samples $\left\{ w_{k-1}^{(i,j)}, x_{k-1}^{(i,j)}, \theta_{k-1}^{(i,j)} \right\}_{j=1}^{L_{k-1}^{(i)}}$, i.e.,

$$p_{k-1}^{(i)}(x, \theta) = \sum_{j=1}^{L_{k-1}^{(i)}} w_{k-1}^{(i,j)} \delta \left(x - x_{k-1}^{(i,j)}, \theta - \theta_{k-1}^{(i,j)} \right) \qquad (23)$$

$L_{k-1}^{(i)}$ denotes the number of particles of the i th Bernoulli component. Then, the predicted multi-target density $\pi_{k|k-1}$ can be expressed as $\pi_{k|k-1} = \left\{ \left(r_{P,k|k-1}^{(i)}, p_{P,k|k-1}^{(i)}(x, \theta) \right) \right\}_{i=1}^{M_{k-1}} \cup \left\{ \left(r_{\Gamma,k}^{(i)}, p_{\Gamma,k}^{(i)}(x, \theta) \right) \right\}_{i=1}^{M_{\Gamma,k}}$.

(2) Parameter particle selection:

(2.1) Predict parameter particles $\left\{ w_{k-1}^{(i,j)}, x_{k-1}^{(i,j)}, \theta_{k-1}^{(i,j)} \right\}_{j=1}^{L_{k-1}^{(i)}}$, where $\theta_{k-1}^{(i,j)} \sim N \left(\cdot | \bar{\theta}_{k-1}^{(i)}, h^2 V_{k-1}^{(i)} \right)$, $\bar{\theta}_{k-1}^{(i)}$ and $V_{k-1}^{(i)}$ denote the mean and covariance of the maneuvering parameter of

ith component at time $k-1$ and can be obtained in the following step of parameter update (see Eq.(47)~(49)). Given important densities $q_k(\mathbf{x}_k|\mathbf{x}_{k-1}^{(i,j)}, \theta_{k-1}^{(i,j)}, Y_k)$, the steps of parameter particle prediction are as follows:

$$\mathbf{x}_{P,k|k-1}^{(i,j)} \sim q_k(\mathbf{x}_k|\mathbf{x}_{k-1}^{(i,j)}, \theta_{k-1}^{(i,j)}, Y_k), \quad i=1,...,M_{k-1}, j=1,...,L_{k-1}^{(i)}$$

(24)

$$\theta_{P,k|k-1}^{(i,j)} = \theta_{k-1}^{(i,j)}$$

(25)

$$\omega_{P,k|k-1}^{(i,j)} \propto \frac{f_{k|k-1}\left(\mathbf{x}_{P,k|k-1}^{(i,j)}|\mathbf{x}_{P,k-1}^{(i,j)}, \theta_{k-1}^{(i,j)}\right) p_{S,k}\left(\mathbf{x}_{k-1}^{(i,j)}\right)}{q_k\left(\mathbf{x}_k^{(i,j)}\Big|\mathbf{x}_{k-1}^{(i,j)}, \theta_{k-1}^{(i,j)}, Y_k\right)} w_{k-1}^{(i,j)}$$

(26)

At time k, each particle is given another weight which is proportional to the predictive likelihood corresponding to no changepoint parameter $\theta_{k-1}^{(i,j)}$, i.e.,

$$\omega_1^{(i,j)} \propto p\left(Y_k|\mathbf{x}_{P,k|k-1}^{(i,j)}, \theta_{k-1}^{(i,j)}\right)$$

(27)

(2.2) In order to obtain better parameter particles, produce new parameter particles with random parameter sampled from the initial distribution, i.e., $\left\{w_{k-1}^{(i,j)}, x_{k-1}^{(i,j)}, \gamma_k^{(i,j)}\right\}_{j=L_{k-1}^{(i)}+1}^{2L_{k-1}^{(i)}}$, where $\gamma_k^{(i,j)} \sim p_{\theta_0}(\cdot)$. The steps of new parameter particle prediction are as follows:

$$\mathbf{x}_{P,k|k-1}^{(i,j)} \sim q_k(\mathbf{x}_k|\mathbf{x}_{k-1}^{(i,j)}, \gamma_k^{(i,j)}, Y_k), \quad i=1,...,M_{k-1}, j=L_{k-1}^{(i)}+1,...,2L_{k-1}^{(i)}$$

(28)

$$\omega_{P,k|k-1}^{(i,j)} \propto \frac{f_{k|k-1}\left(\mathbf{x}_{P,k|k-1}^{(i,j)}|\mathbf{x}_{P,k-1}^{(i,j)}, \gamma_k^{(i,j)}\right) p_{S,k}\left(\mathbf{x}_{k-1}^{(i,j)}\right)}{q_k\left(\mathbf{x}_k^{(i,j)}\Big|\mathbf{x}_{k-1}^{(i,j)}, \gamma_k^{(i,j)}, Y_k\right)} w_{k-1}^{(i,j)}$$

(29)

At time k, each particle is also given another weight which is proportional to the predictive likelihood corresponding to changepoint parameter $\gamma_k^{(i,j)}$, i.e.,

$$\omega_2^{(i,j)} \propto p\left(Y_k|\mathbf{x}_{P,k|k-1}^{(i,j)}, \gamma_k^{(i,j)}\right)$$

(30)

(2.3) We then select $L_{k-1}^{(i)}$ particles out of the $2L_{k-1}^{(i)}$ obtained particles. Denote their indices as $l^j \in \left\{1,...,2L_{k-1}^{(i)}\right\}$, where $j=1,\cdots,L_{k-1}^{(i)}$, the selection processes are as follows.

(a) For $j=1,\boxtimes,L_{k-1}^{(i)}$, select indices l^j with probability $(1-\beta)\omega_1^{(i,j)}$ from $\left[1,...,L_{k-1}^{(i)}\right]$ and $\beta\omega_2^{(i,j)}$ from $\left[L_{k-1}^{(i)}+1,...,2L_{k-1}^{(i)}\right]$, where β is the probability that

an abrupt change occurred and it is assumed to be known.

(b) If $l^j \boxtimes \left\{1,...,L_{k-1}^{(i)}\right\}$, then update the time-varying parameter particles using $\theta_{p,k|k-1}^{(i,j)} = \theta_{p,k|k-1}^{(i,l^j)}$.

(c) If $l^j \boxtimes \left\{L_{k-1}^{(i)}+1,...,2L_{k-1}^{(i)}\right\}$, then set the time-varying parameter particles to be $\theta_{p,k|k-1}^{(i,j)} = \gamma_k^{(i,l^j)}$

(2.4) Relabel the selected particles with indices $j=1,\cdots,L_{k-1}^{(i)}$, i.e., $\mathbf{x}_{P,k|k-1}^{(i,j)} = \mathbf{x}_{P,k|k-1}^{(i,l^j)}$, $\omega_{P,k|k-1}^{(i,j)} = \omega_{P,k|k-1}^{(i,l^j)}$, and sample $L_{\Gamma,k}^{(i)}$ new-born particles from the proposal distribution $b_k(\cdot|\gamma_{\Gamma,k}^{(i,j)}, Y_k)$ via

$$\theta_{\Gamma,k}^{(i,j)} \sim p_{\theta_0}(\cdot), \quad i=1,...,M_{k-1}, j=1,...,L_{\Gamma,k}^{(i)}$$

(31)

$$\mathbf{x}_{\Gamma,k}^{(i,j)} \sim b_k(\cdot|\theta_{\Gamma,k}^{(i,j)}, Y_k)$$

(32)

$$\omega_{\Gamma,k}^{(i,j)} \propto \frac{p_{\Gamma,k}\left(\mathbf{x}_{\Gamma,k}^{(i,j)}\right)}{b_k\left(\mathbf{x}_{\Gamma,k}^{(i,j)}, \theta_{\Gamma,k}^{(i,j)}, Y_k\right)}$$

(33)

(3) *Update*: Assume the predicted multi-target density at time k is $\pi_{k|k-1} = \left\{\left(r_{k|k-1}^{(i)}, p_{k|k-1}^{(i)}(x,\theta)\right)\right\}_{i=1}^{M_{k|k-1}}$, where $M_{k|k-1} = M_{k-1} + M_{\Gamma,k|k-1}$. Each $p_{k|k-1}^{(i)}(x,\theta)$ is comprised of a set of weighted samples $\left\{w_{k|k-1}^{(i,j)}, x_{k|k-1}^{(i,j)}, \theta_{k|k-1}^{(i,j)}\right\}_{j=1}^{L_{k|k-1}^{(i)}}$, where $i=1,...,M_{k-1}$, i.e.,

$$p_{k|k-1}^{(i)}(x,\theta) = \sum_{j=1}^{L_{k|k-1}^{(i)}} w_{k|k-1}^{(i,j)} \delta\left(x-x_{k|k-1}^{(i,j)}, \theta-\theta_{k|k-1}^{(i,j)}\right)$$

(34)

Then, the updated multi-target density $\pi_k = \left\{\left(r_{L,k}^{(i)}, p_{L,k}^{(i)}(x,\theta)\right)\right\}_{i=1}^{M_{k|k-1}} \cup \left\{\left(r_{U,k}^*(y), p_{U,k}^*(x,\theta;y)\right)\right\}_{y \in Y_k}$ can be computed as follows:

$$r_{L,k}^{(i)} = r_{k|k-1}^{(i)} \frac{1-p_{L,k}^{(i)}(x,\theta)}{1-r_{k|k-1}^{(i)}p_{L,k}^{(i)}(x,\theta)}$$

(35)

$$p_{L,k}^{(i)}(x,\theta) = \sum_{j=1}^{L_{k|k-1}^{(i)}} \tilde{w}_{L,k}^{(i,j)} \delta\left(x-x_{k|k-1}^{(i,j)}, \theta-\theta_{k|k-1}^{(i,j)}\right)$$

(36)

$$r^*_{U,k}(y) = \frac{\sum_{i=1}^{M_{k|k-1}} \dfrac{r^{(i,)}_{k|k-1}\left(1-r^{(i)}_{k|k-1}\right)\rho^{(i)}_{U,k}(y)}{\left(1-r^{(i)}_{k|k-1}p^{(i)}_{L,k}(x,\theta)\right)^2}}{\kappa_k(y) + \sum_{i=1}^{M_{k|k-1}} \dfrac{r^{(i)}_{k|k-1}\rho^{(i)}_{U,k}(y)}{1-r^{(i)}_{k|k-1}p^{(i)}_{L,k}(x,\theta)}} \tag{37}$$

$$p^*_{U,k}(x,\theta;y) = \sum_{i=1}^{M_{k|k-1}}\sum_{j=1}^{L^{(i)}_{k|k-1}} \tilde{w}^{(i,j)}_{U,k}(y)\delta\left(x-x^{(i,j)}_{k|k-1},\theta-\theta^{(i,j)}_{k|k-1}\right) \tag{38}$$

where

$$\rho^{(i)}_{L,k}(x,\theta) = \sum_{j=1}^{L^{(i)}_{k|k-1}} w^{(i,j)}_{k|k-1}p_{D,k}\left(x^{(i,j)}_{k|k-1}\right)\delta\left(x-x^{(i,j)}_{k|k-1},\theta-\theta^{(i,j)}_{k|k-1}\right) \tag{39}$$

$$\tilde{w}^{(i,j)}_{L,k} = w^{(i,j)}_{L,k}/\sum_{j=1}^{L^{(i)}_{k|k-1}} w^{(i,j)}_{L,k} \tag{40}$$

$$w^{(i,j)}_{L,k} = w^{(i,j)}_{k|k-1}\left(1-p_{D,k}\left(x^{(i,j)}_{k|k-1}\right)\right) \tag{41}$$

$$\rho^{(i)}_{U,k}(y) = \sum_{j=1}^{L^{(i)}_{k|k-1}} w^{(i,j)}_{k|k-1}\psi_{k,y}\left(x^{(i,j)}_{k|k-1},\theta^{(i,j)}_{k|k-1}\right)\delta\left(x-x^{(i,j)}_{k|k-1},\theta-\theta^{(i,j)}_{k|k-1}\right) \tag{42}$$

$$\tilde{w}^{(i,j)}_{U,k}(y) = w^{(i,j)}_{U,k}(y)/\sum_{i=1}^{M_{k|k-1}}\sum_{j=1}^{L^{(i)}_{k|k-1}} w^{(i,j)}_{U,k}(y) \tag{43}$$

$$w^{(i,j)}_{U,k}(y) = w^{(i,j)}_{k|k-1}\frac{r^{(i)}_{k|k-1}\psi_{k,y}\left(x^{(i,j)}_{k|k-1},\theta^{(i,j)}_{k|k-1}\right)}{1-r^{(i)}_{k|k-1}} \tag{44}$$

$$\psi_{k,y}\left(x^{(i,j)}_{k|k-1},\theta^{(i,j)}_{k|k-1}\right) = p_k\left(y|x^{(i,j)}_{k|k-1},\theta^{(i,j)}_{k|k-1}\right)p_{D,k}\left(x^{(i,j)}_{k|k-1}\right) \tag{45}$$

$p_k\left(y|x^{(i,j)}_{k|k-1},\theta^{(i,j)}_{k|k-1}\right) = p_k\left(y|x^{(i,j)}_{k|k-1}\right)$ is the likelihood function.

(4) Resampling: To alleviate the effect of the particle degeneracy, the updated particle set $\left\{\tilde{w}^{(i,j)}_k,x^{(i,j)}_k,\theta^{(i,j)}_k\right\}_{j=1}^{L^{(i)}_{k|k-1}}$ is resampled to get $\left\{w^{(i,j)}_k,x^{(i,j)}_k,\theta^{(i,j)}_k\right\}_{j=1}^{L^{(i)}_k}$. The resampling step can effectively eliminate the particles with low weights and multiply the particles with high weights to focus on the important zones of the state space. The resampling process is similar to that of the CBMeMBer filter [14]. Notice that the number of the particles increases due to the spontaneous births in the prediction

and the averaging of the hypothesized tracks in the update. Therefore, the hypothesized tracks need to be pruned by discarding those with existence probabilities below a threshold η, which can reduce the number of particles effectively.

The posterior density of each Bernoulli component can be obtained by

$$p^{(i)}_k(x,\theta) = \sum_{j=1}^{L^{(i)}_k} w^{(i,j)}_k\delta\left(x-x^{(i,j)}_k,\theta-\theta^{(i,j)}_k\right) \tag{46}$$

(5) Parameter update: The mean and covariance of each component are obtained by

$$\bar{\theta}^{(i)}_k = \sum_{j=1}^{L^{(i)}_k} w^{(i,j)}_k\theta^{(i,j)}_k \tag{47}$$

$$V^{(i)}_k = \sum_{j=1}^{L^{(i)}_k} \omega^{(i,j)}_k\left(\theta^{(i,j)}_k-\bar{\theta}^{(i)}_k\right)\left(\theta^{(i,j)}_k-\bar{\theta}^{(i)}_k\right)^T \tag{48}$$

The parameter of each particle can be updated by

$$\theta^{(i,j)}_k = \alpha\theta^{(i,j)}_k + (1-\alpha)\bar{\theta}^{(i)}_k \tag{49}$$

where $\alpha = \sqrt{1-h^2}$ is the shrinkage factor suggested in [33] to correct for the over-dispersion of the Gaussian mixture model.

(6) State estimation: The estimated number of the targets is the cardinality mean, which can be obtained by

$$\hat{N}_k = \sum_{i=1}^{M_{k|k-1}} r^{(i)}_{L,k} + \sum_{y\in Y_k} r^*_{U,k}(y) \tag{50}$$

Individual state estimates can be obtained by calculating the means of the posterior densities of the hypothesized tracks with existence probabilities exceeding a given threshold (e.g., 0.5) [14], which is inexpensive and scales linearly with the number of hypothesized tracks.

3.2 Track maintenance

Since the MMP-CBMeMBer algorithm cannot give the tracks, the track maintenance algorithm is proposed by introducing the particle labeling method, which can effectively achieve the track continuity for the multiple maneuvering target tracking. The detailed process of the track maintenance is described as follows.

(1) Prediction: Suppose at time $k-1$ ($k\geq2$), the particle label of each multi-Bernoulli component can be described as

$$L_{k-1} = \left\{L^{(j)}_{k-1}\right\}_{j=1}^{J_{k-1}} = \left\{l^{(j)(1)}_{k-1},l^{(j)(2)}_{k-1},\cdots,l^{(j)(N_j)}_{k-1}\right\}_{j=1}^{J_{k-1}} \tag{51}$$

where J_{k-1} is the number of the Bernoulli components

at time $k-1$, and N_j denotes the number of the particles of the j th Bernoulli component.

The labels of the prediction Bernoulli components can be expressed as

$$L_{k|k-1} = L_{k-1} \cup L_\gamma \qquad (52)$$

where L_γ denotes the labels of the Bernoulli components of the spontaneous births and can be expressed as

$$L_\gamma = \left\{ L_\gamma^{(i)} \right\}_{i=1}^{J_\gamma} = \left\{ l_{k-1}^{(i)(1)}, l_{k-1}^{(i)(2)}, \cdots, l_{k-1}^{(i)(N_\gamma)} \right\}_{i=1}^{J_\gamma} \qquad (53)$$

where J_γ denotes the number of the Bernoulli components of the spontaneous births, and N_γ denotes the number of the particles of each Bernoulli component.

(2) *Update*: In the update state, there will appear $|Y_k|$ $+1$ Bernoulli components due to the measurements, where $|Y_k|$ denotes the number of the measurements. At time k, measurement-updated components are assigned the label of the predicted track, i.e., the label can be expressed as

$$L_{k|k} = L_{k|k-1} \cup L_{k|k-1}^1 \cup \cdots \cup L_{k|k-1}^{|Y_k|} \qquad (54)$$

where $L_{k|k-1}^n = L_{k|k-1}$, $n = 1, \cdots, |Z_k|$.

(3) *Resampling*: Resample each component of the $|Z_k| +$ 1 Bernoulli components. The resampling particles need to keep the same label as their father particles and the label of the remaining Bernoulli components are given as

$$L_k = L_k^1 \cup \cdots \cup L_k^{J_k} \qquad (55)$$

where J_k is the number of the remaining Bernoulli components at time k,

$$L_k^j = \left\{ L_k^{(j)(1)}, L_k^{(j)(2)}, \cdots, L_k^{(j)(N_j)} \right\}, j = 1, \cdots, J_k \qquad (56)$$

(4) *Track continuity*: Track continuity can be completed according to the particle labels by the data association technique [34], i.e., the track can be obtained by comparing the number of particles with the same labels in each component.

3.3 Simulations

In order to demonstrate the performance of the proposed APPF-CBMeMBer algorithm, a two-dimensional tracking example is simulated. The benchmark technique is the MMP-CBMeMBer algorithms [29]. In the considered scenario, the measurements are obtained at four stationary sensors located at $(0, 0)$ m, $(0, 1 \times 10^4)$ m, $(1 \times 10^4, 0)$ m, and $(1 \times 10^4, 1 \times 10^4)$ m. At time k, each sensor outputs the measured bearing of the received signal, which is given by

$$y_k^{S_i} = tan^{-1}\left(\frac{z_k - z_{S_i}}{x_k - x_{S_i}} \right) + w_k \qquad (57)$$

where (x_{S_i}, z_{S_i}) denotes the location of the i th sensor, $i = 1, 2, 3, 4$. w_k is the zero-mean Gaussian distributed measurement noise with variance $\sigma_w^2 = 1 \times 10^{-4} \text{rad}^2$.

There are four maneuvering targets. Targets 1 and 2 appear throughout the tracking process, and they are traveling from their initial positions $(-1 \times 10^3, 4 \times 10^3)$ m and $(1.4 \times 10^4, 1 \times 10^4)$ m. Target 3 is a spontaneous

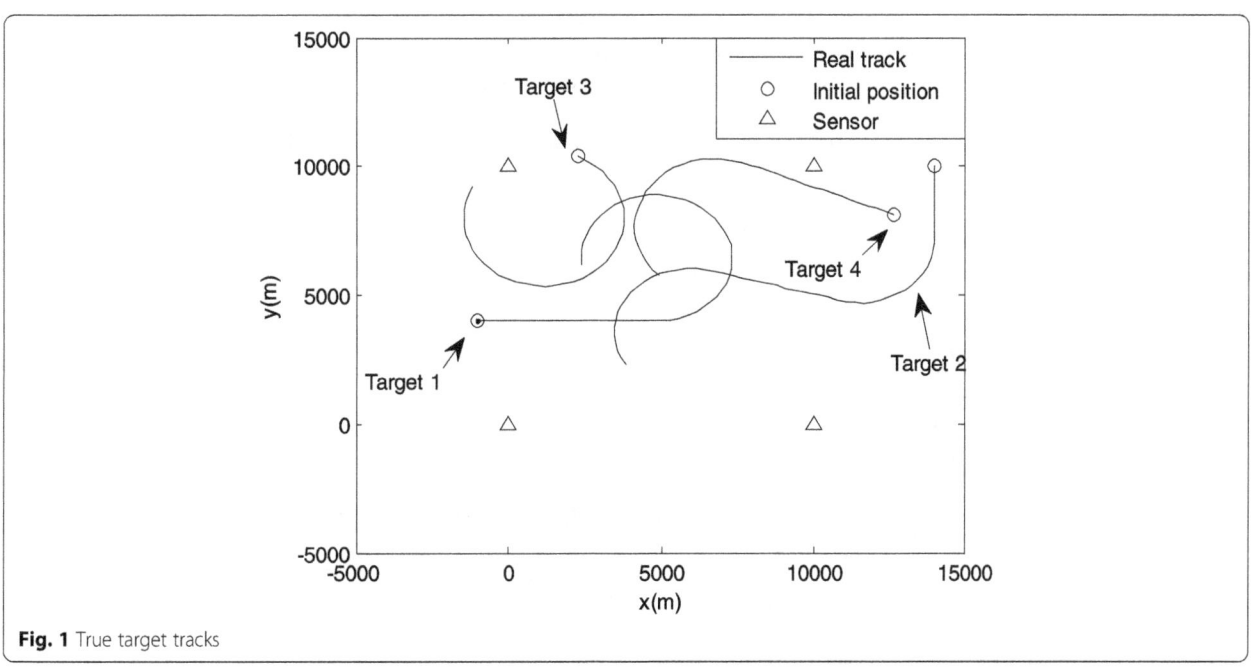

Fig. 1 True target tracks

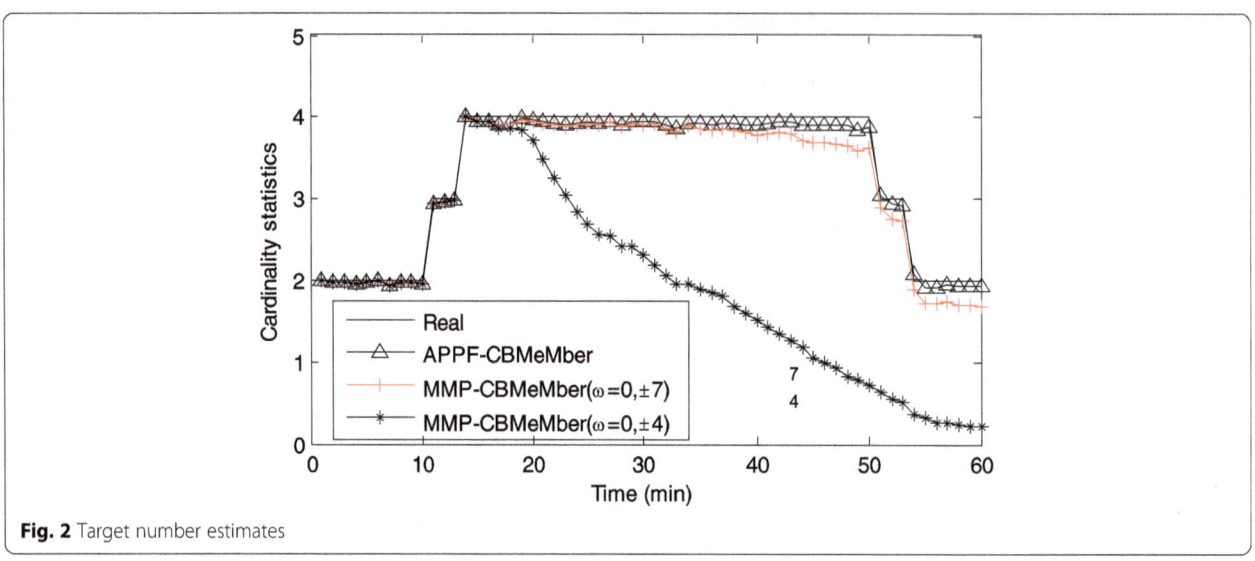

Fig. 2 Target number estimates

birth at 10th minute with initial position $(2 \times 10^3, 10.5 \times 10^3)$ m and disappears at 50th minute. Target 4 is a spontaneous birth at 13th minute with initial position $(1.3 \times 10^4, 8 \times 10^3)$ m and disappears at 53th minute. The true trajectories of the four targets are depicted in Fig. 1.

We model the birth process using a Poisson RFS with intensity

$$\Gamma_k^{(i)}(\mathbf{x}) = \sum_{i=1}^{3} 0.2 \mathcal{N}\left(\mathbf{x}; \mathbf{m}_\Gamma^{(i)}, \mathbf{P}_\Gamma^{(i)}\right) , \quad i = 1, 2, 3 \quad (58)$$

where $\mathbf{m}_\Gamma^{(1)} = (-1 \times 10^3\text{m}, 0\text{m/s}, 4 \times 10^3\text{m}, 0\text{m/s})$, $\mathbf{m}_\Gamma^{(2)} = (1.4 \times 10^4\text{m}, 0\text{m/s}, 1 \times 10^4\text{m}, 0\text{m/s})$, $\mathbf{m}_\Gamma^{(3)} = (2 \times 10^3\text{m}, 0\text{m/s}, 10.5 \times 10^3\text{m}, 0\text{m/s})$, $\mathbf{m}_\Gamma^{(4)} = (2 \times 10^3\text{m}, 0\text{m/s}, 10.5 \times 10^3\text{m}, 0\text{m/s})$, and $\mathbf{P}_\Gamma^{(1)} = \mathbf{P}_\Gamma^{(2)} = \mathbf{P}_\Gamma^{(3)} = \mathbf{P}_\Gamma^{(4)} = \text{diag}(400, 1, 400, 1)$. The clutter is modeled as a Poisson RFS with the mean rate $r = 10$ over the observation space.

The probabilities of the target survival and detection are $p_{S,k} = 0.99$ and $p_{D,k} = 0.98$.

To contrast the performance of different algorithms, two performance metrics are used. One is the statistics of the target number estimates. The other one is the optimal subpattern assignment (OSPA) distance [35], which is recently developed and defined as

$$d_p^{(c)}(X, Y) = \left(\frac{1}{n} \left(\min_{\pi \in \prod_n} \sum_{i=1}^{m} d^{(c)}\left(\mathbf{x}_i, \mathbf{y}_{\pi(i)}\right)^p + c^p(n-m) \right) \right)^{1/p}$$
(59)

where $X = \{\mathbf{x}_1, \cdots, \mathbf{x}_m\}$ and $Y = \{\mathbf{y}_1, \cdots, \mathbf{y}_n\}$ are arbitrary finite subsets, $1 \le p < \infty$, $c > 0$, $m, n \in N_o = \{0, 1, 2, \cdots\}$. If $m > n$, $\bar{d}_p^{(c)}(X, Y) = \bar{d}_p^{(c)}(Y, X)$. In the simulation, the parameters of OSPA distance are set as $p = 2$ and $c = 1000$.

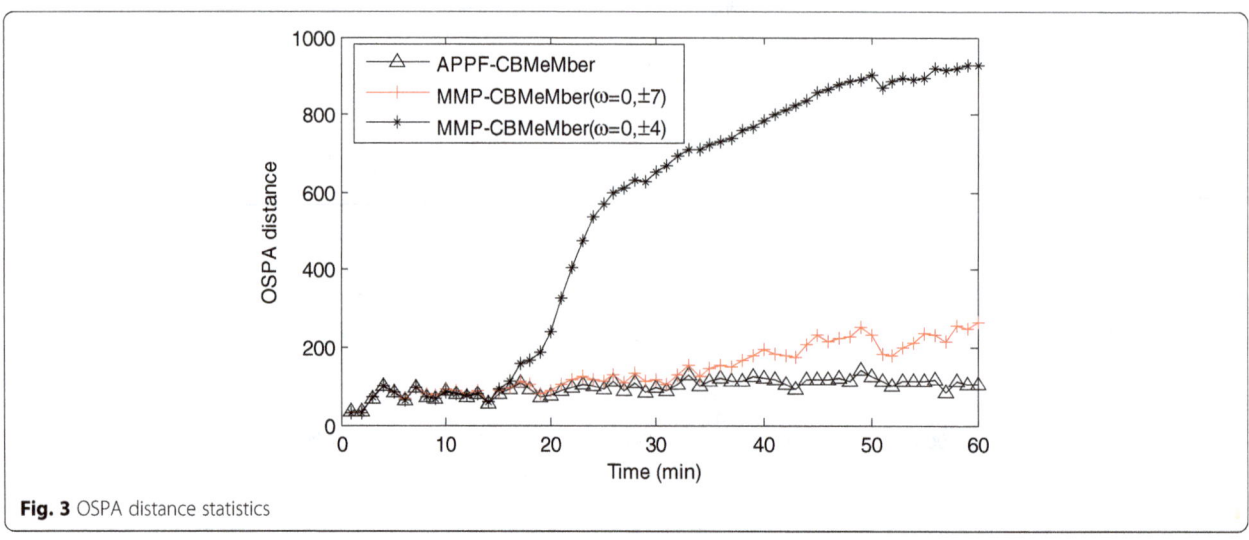

Fig. 3 OSPA distance statistics

The simulation results are obtained from Monte Carlo experiments of 200 ensemble runs.

In this experiment, we compare the performance of multiple abruptly maneuvering targets tracking with different algorithms. The turn rate ω is considered as unknown and time-varying model parameter for the proposed APPF-CBMeMBer algorithm and the MMP-CBMeMBer algorithm. Moreover, the MMP-CBMeMBer algorithm uses a constant velocity (CV) model and two coordinated-turn CT models, and the turn rate ω is assumed a known value priori. We compare the proposed APPF-CBMeMBer algorithm with the MMP-CBMeMBer ($\omega = 0, \pm 4$) algorithm and the MMP-CBMeMBer ($\omega = 0, \pm 7$) algorithm. It is noted that in the tracking scenario, the real turn rates are $\omega = \pm 7$. The simulation results for this experiment are shown in Figs. 2, 3, 4, 5, 6, and 7.

Figure 2 shows the averaged target number estimates of the APPF-CBMeMBer and MMP-CBMeMBer algorithms. It can be seen that the proposed APPF-CBMeMBer algorithm provides more accurate target number estimates than the benchmark MMP-CBMeMBer algorithm. The behind reason is that the proposed algorithm can effectively joint estimate the unknown model parameter ω which can be well matched with the motion model of each target. While for the MMP-CBMeMBer algorithm, the tracking accuracy depends on the matching degree of the prior designed multiple model sets with the real target motion models. Unfortunately, the prior parameters of the various models are unknown; thus, the models cannot be well matched with the real motion model of each target (such as $\omega = 0, \pm 4$). Moreover, although the MMP-CBMeMBer ($\omega = 0, \pm 7$) algorithm includes the real turn rates, the tracking accuracy is slightly lower than the proposed algorithm due to the disturbance of each model.

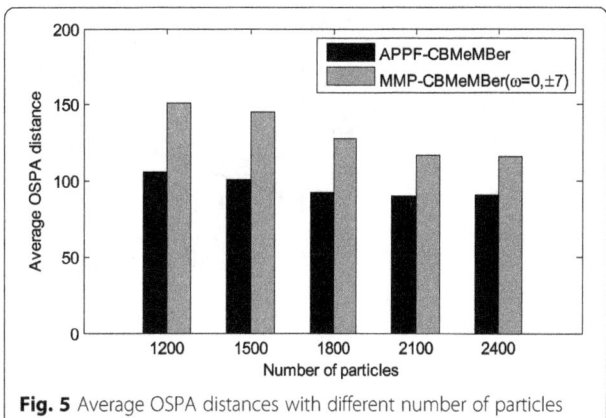
Fig. 5 Average OSPA distances with different number of particles

Figure 3 compares the OSPA distances of the two simulated algorithms, and it is clear that the proposed algorithm again outperforms the MMP-CBMeMBer algorithm. This is also due to the fact that the proposed method can adapt to the temporal evolution of the target maneuvering parameters via estimating them with the target states.

Figures 4 and 5 show the average RMSEs of target number estimates and the average OSPA distances under different number of particles. It is clear that the tracking accuracy of the two algorithms increase with the increase of the particle number. In addition, the accuracy of the APPF-CBMeMBer algorithm is always higher than that of the MMP-CBMeMBer algorithm. However, in Fig. 6, we can see that the proposed algorithm has a higher run time than the MMP-CBMeMBer algorithm, the reason is that the propose algorithm needs to select the parameter particles from the doubled particles, and the steps of parameter update occupy some time. However, for MM-based methods, the computational complexity may become prohibitive as the number of models needed would increase exponentially if more unknown parameters have to be taken into account to match the possible target motion modes.

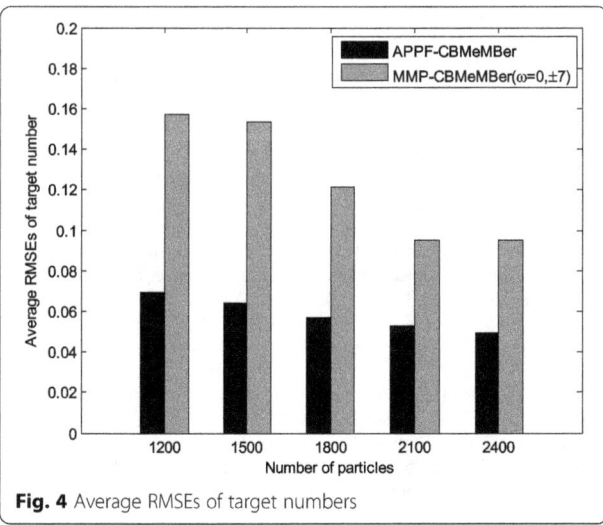
Fig. 4 Average RMSEs of target numbers

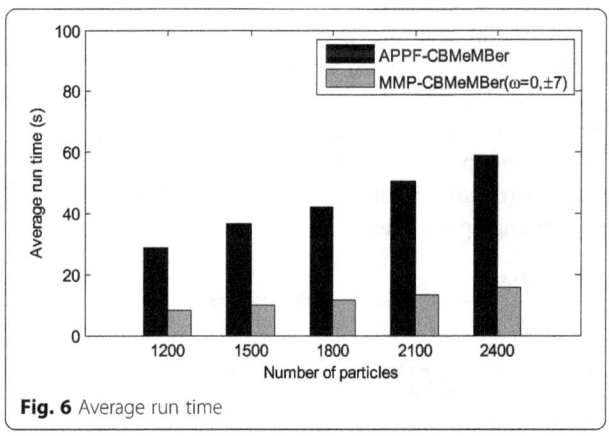
Fig. 6 Average run time

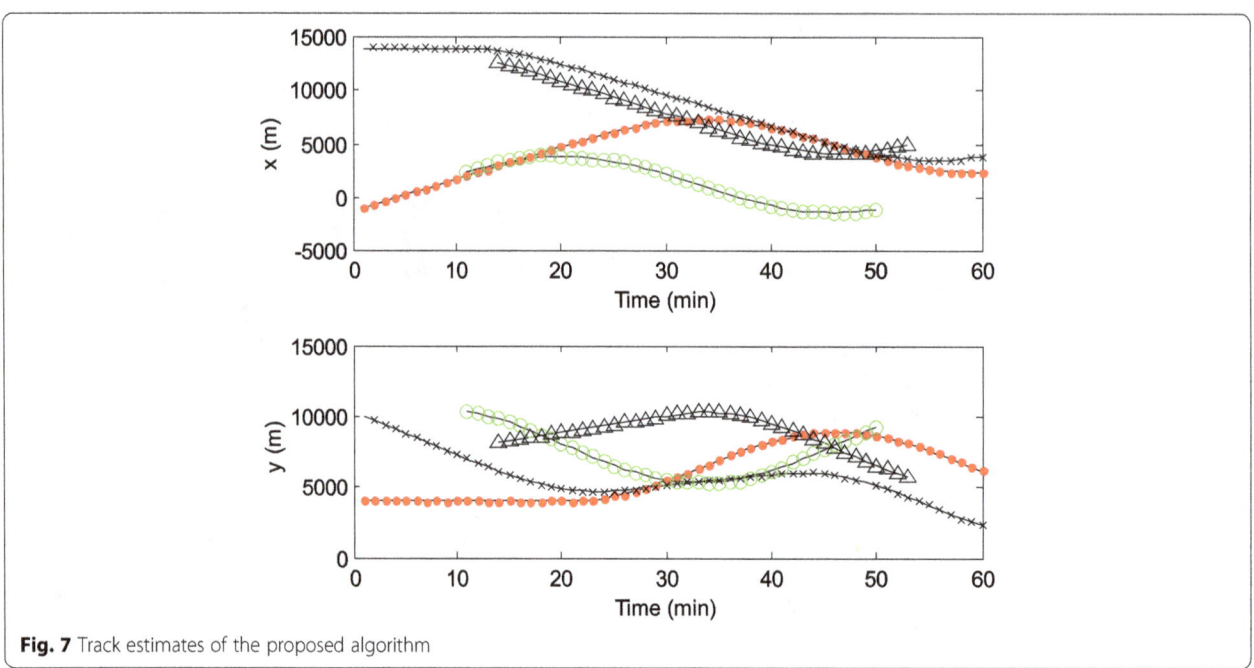

Fig. 7 Track estimates of the proposed algorithm

Figure 7 shows the track estimates of the proposed APPF-CBMeMBer algorithm. It is clear that the proposed algorithm has a good track maintenance performance due to its good performance in terms of the state estimates and their number estimates.

4 Conclusions

In this paper, we developed a new multiple maneuvering target tracker algorithm, referred to as the APPF-CBMeMBer tracker, to handle the presence of unknown and time-varying maneuvering parameter. In the proposed algorithm, the APE technique was incorporated to achieve online maneuvering parameter estimation, and the selected parameter particles were utilized to derive the approximation closed solution. Simulations showed that the newly proposed algorithm can offer higher tracking accuracy in the case of multiple maneuvering targets over the existing MMP-CBMeMBer algorithm. Furthermore, in order to obtain the individual target tracks, the particle labeling technique is introduced in the proposed algorithm.

In the future works, we shall consider introducing the APE technique into the LMB filter to obtain a better algorithm for tracking multiple targets with unknown but abruptly changing parameters.

Competing interests
The authors declare that they have no competing interests.

Acknowledgements
This paper is supported by the National Natural Science Foundation of China (Nos. 61305017, 61304264) and the Natural Science Foundation of Jiangsu Province (No. BK20130154).

References
1. R Mahler, *Statistical multisource-multitarget information fusion* (Artech House, Norwood, MA, 2007)
2. R Mahler, Multi-target Bayes filtering via first-order multi-target moments. IEEE Trans. Aerosp. Electron. Syst. **39**(4),1152–1178 (2003)
3. R Mahler, PHD filters of higher order in target number. IEEE Trans. Aerosp. Electron. Syst. **43**(4), 1523–1543 (2007)
4. BN Vo, S Singh, A Doucet, Sequential Monte Carlo methods for multi-target filtering with random finite sets. IEEE Trans. Aerosp. Electron. Syst. **41**(4),1224–1245 (2005)
5. D Clark, J Bell, Convergence results for the particle PHD filter. IEEE Trans. Signal Process. **54**(7), 2652–2661 (2006)
6. BN Vo, WK Ma, The Gaussian mixture probability hypothesis density filter. IEEE Trans. Signal Process. **54**(11),4091–4104 (2006)
7. BT Vo, BN Vo, A Cantoni, Analytic implementations of the cardinalized probability hypothesis density filter. IEEE Trans. Signal Process. **55**(7),3553–3567 (2007)
8. R Mahler, *Approximate multisensor CPHD and PHD filters*. Proceedings of the 13th International Conference on Information Fusion, Edinburgh, UK, 2010, pp. 1152–1178
9. R Mahler, BT Vo, BN Vo, CPHD filtering with unknown clutter rate and detection profile. IEEE Trans. Signal Process. **59**(8),3497–3513 (2011)
10. N Whiteley, S Singh, S Godsil, Auxiliary particle implementation of probability hypothesis density filter. IEEE Trans. Aerosp. Electron. Syst. **46**(3),1437–1454(2010)
11. M Yazdian-Dehkordi, Z Azimifar, MA Masnadi-Shirazi, Penalized Gaussian mixture probability hypothesis density filter for multiple target tracking. Signal Process. **92**(5),1230–1242 (2012)
12. ZX Liu, LJ Li, WX Xie et al., Sequential measurement-driven multi-target Bayesian filter. EURASIP Journal on Advances in Signal Processing **43**,1–9 (2015)
13. ZX Liu, LJ Li, WX Xie et al., Two implementations of marginal distribution Bayes filter for nonlinear Gaussian models. AEU - International Journal of Electronics and Communications **69**(9),1297–1304 (2015)
14. BT Vo, BN Vo, A Cantoni, The Cardinality balanced multi-target multi-Bernoulli filter and its implementations. IEEE Trans. Signal Process. **57**(2),409–423 (2009)

15. BT Vo, BN Vo, R Hoseinnezhad et al., Robust multi-Bernoulli filtering. IEEE J. Sel. Top. Sign. Proces. **7**(3),399–409 (2013)

16. BT Vo, BN Vo, Labeled random finite sets and multi-object conjugate priors. IEEE Trans. Signal Process. **61**(13),3460–3475 (2013)

17. S Reuter, BT Vo, BN Vo et al., The labeled multi-Bernoulli filter. IEEE Trans. Signal Process. **62**(12),3246–3260 (2014)

18. MA Beard, BT Vo, BN Vo, Bayesian multi-target tracking with merged measurements using labelled random finite sets. IEEE Trans. Signal Process. **63**(6),1433–1447 (2015)

19. JL Yang, HW Ge, An improved multi-target tracking algorithm based on CBMeMBer filter and variational Bayesian approximation. Signal Process. **93**(9),2510–2515 (2013)

20. ML Hernandez, B Ristic, A Farina et al., Performance measure for Markovian switching systems using best-fitting Gaussian distributions. IEEE Trans. Aerosp. Electron. Syst. **44**(2),724–747 (2008)

21. XR Li, VP Jilikov, Survey of maneuvering target tracking. Part V: multiple-model methods. IEEE Trans. Aerosp. Electron. Syst. **41**(4),1255–1321 (2005)

22. A Pasha, BN Vo, HD Tuan et al., *Closed-form PHD filtering for linear jump Markov models*. Proceedings of the 9th International Conference on Information Fusion, Florence, Italy, 2006

23. SA Pasha, BN Vo, HD Tuan et al., A Gaussian mixture PHD filter for jump Markov system models. IEEE Trans. Aerosp. Electron. Syst. **45**(3),919–936 (2009)

24. SA Pasha, HD Tuan, P Apkarian, *The LFT based PHD filter for nonlinear jump Markov models in multi-target tracking*. Proceedings of the IEEE Conference on Decision and Control, Shanghai, China, 2009, pp. 5478–5483

25. WL Li, YM Jia, Gaussian mixture PHD filter for jump Markov models based on best-fitting Gaussian approximation. Signal Process. **91**(4), 1036–1042 (2011)

26. K Punithakumar, T Kirubarajan, A Sinha, Multiple-model probability hypothesis density filter for tracking maneuvering targets. IEEE Trans. Aerosp. Electron. Syst. **44**(1), 87–98 (2008)

27. R Georgescu, P Willett, The multiple model CPHD tracker. IEEE Trans. Signal Process. **60**(4), 1741–1751 (2012)

28. R Mahler, *On multitarget jump-Markov filters*. Proceedings of the 15th International Conference on Information Fusion, Singapore, 2012, pp. 149–156

29. JL Yang, HB Ji, HW Ge, Multi-model particle cardinality-balanced multi-target multi-Bernoulli algorithm for multiple manoeuvring target tracking. IET Radar Sonar Navig. **7**(2),101–112 (2013)

30. J Liu, M West, Combined parameter and state estimation in simulation-based filtering, in *Proceedings of the sequential Monte Carlo Methods in practice, Springer, New York*, 2001, pp. 197–223

31. C Carvalho, M Johannes, H Lopes et al., Particle learning and smoothing. Stat. Sci. **25**(1),88–106 (2010)

32. C Nemeth, P Fearnhead, L Mihaylova, Sequential Monte Carlo methods for state and parameter estimation in abruptly changing environments. IEEE Trans. Signal Process. **62**(5),1245–1255 (2014)

33. M West, Approximating posterior distributions by mixture. J. R. Stat. Soc. **55**(2),409–422 (1993)

34. DE Clark, J Bell, Multi-target state estimation and track continuity for the particle PHD filter. IEEE Trans. Aerosp. Electron. Syst. **43**(4),1441–1452 (2007)

35. D Schuhmacher, BT Vo, BN Vo, A consistent metric for performance evaluation of multi-object filters. IEEE Trans. Signal Process. **56**(8),3447–3457 (2008)

14

Color interpolation algorithm for an RWB color filter array including double-exposed white channel

Ki Sun Song, Chul Hee Park, Jonghyun Kim and Moon Gi Kang[*]

Abstract

In this paper, we propose a color interpolation algorithm for a red-white-blue (RWB) color filter array (CFA) that uses a double exposed white channel instead of a single exposed green (G) channel. The double-exposed RWB CFA pattern, which captures two white channels at different exposure times simultaneously, improves the sensitivity and provides a solution for the rapid saturation problem of W channel although spatial resolution is degraded due to the lack of a suitable color interpolation algorithm. The proposed algorithm is designed and optimized for the double-exposed RWB CFA pattern. Two white channels are interpolated by using directional color difference information. The red and blue channels are interpolated by applying a guided filter that uses the interpolated white channel as a guided value. The proposed method resolves spatial resolution degradation, particularly in the horizontal direction, which is a challenging problem in the double-exposed RWB CFA pattern. Experimental results demonstrate that the proposed algorithm outperforms other color interpolation methods in terms of both objective and subjective criteria.

Keywords: Color interpolation, Double-exposed white channel, Directional color difference, Guided filter

1 Introduction

Most digital imaging devices use a color filter array (CFA) to reduce the cost and size of equipment instead of using three sensors and optical beam splitters. The Bayer CFA, which consists of primary colors such as red, green, and blue (R, G, and B) is a widely used CFA pattern [1]. Recently, methods using a new CFA have been studied to overcome the limited sensitivity of the Bayer CFA under low-light conditions. The reason for this is that the amount of absorbed light decreases due to the RGB color filters. When white (panchromatic) pixels (W) are used, sensors can absorb more light, thereby providing an advantage in terms of sensitivity [2–7]. Despite of the sensitivity improvement, various RGBW CFA patterns suffer from spatial resolution degradation. The reason for this degradation is that the sensor is composed of more color components than the Bayer CFA pattern. In spite of this drawback, some industries have developed image sensors using the RGBW CFA pattern to improve sensitivity [2, 7].

They have attempted to overcome the degradation of the spatial resolution by using new CI algorithms especially designed for their RGBW CFA pattern.

In order to overcome the problem of degradation in RGBW CFA, a pattern of RWB CFA that does not absorb G is proposed [8, 9]. The spatial resolution of this RWB CFA is similar to that of the Bayer CFA because the pattern of this RWB CFA is identical to the pattern of the Bayer CFA. Although spatial resolution is improved when RWB CFA uses W rather than G, however, it is impossible to guarantee color fidelity. The reason is that correct color information cannot be produced due to the lack of G. To maximize the advantage of this CFA without color degradation, two techniques are required. First, accurate G should be reconstructed based on the correlation among W, R, and B. Second, the image should be fused, for which a high dynamic range (HDR) reconstruction technique that combines highly sensitive W information with RGB information is needed, as shown in Fig. 1. If the above techniques are applied to images obtained through the RWB CFA (ideal HDR and G value reconstruction algorithms are applied), images without color degradation can be obtained while the spatial resolution is improved as

*Correspondence: mkang@yonsei.ac.kr
School of Electrical and Electronic Engineering, Yonsei University, 50
Yonsei-Ro, Seodaemun-Gu, 03722 Seoul, Republic of Korea

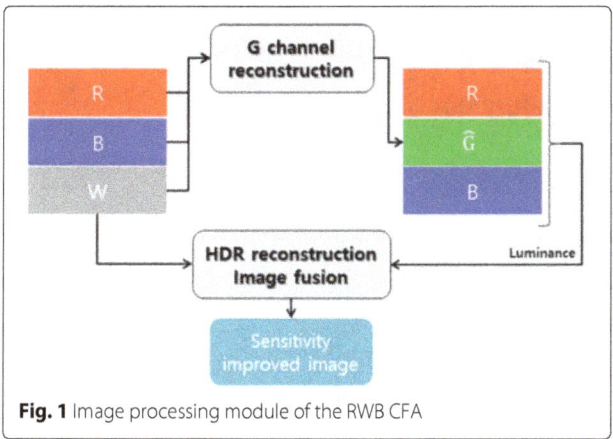

Fig. 1 Image processing module of the RWB CFA

compared with images obtained through the RGBW CFA. The sensitivity is also significantly improved in comparison to images obtained from the existing Bayer CFA, as shown in Fig. 2.

However, there is a problem in obtaining W. W channel saturates at a lower light level than the R, G, and B channels. That is, W is saturated faster than R, G, and B since W absorbs more light compared to R,G, and B. In the RGBW CFA, the saturation problem can be handled by using the HDR reconstruction scheme that combines W with luminance of RGB. On the contrary, the rapid saturation of W is a important problem in the RWB CFA. As shown in Fig. 3, rapid saturation of W occurs when obtaining W with R and B as is the case with the existing Bayer CFA. If W is saturated, G cannot be estimated accurately. In order to prevent the saturation of W, the image is captured with a shorter exposure time. Unfortunately, this leads to another problem, i.e., reduced signal-to-noise ratio (SNR) for R and B. To solve this issue, a new pattern of the RWB CFA that obtains two W values at different exposure times has been proposed [10]. The pattern of this CFA is designed as shown in Fig. 4. In spite of

the degradation of spatial resolution along the horizontal direction, R and B are placed in odd rows and W is placed in even rows to consider a readout method of complementary metal oxide semi-conductor (CMOS) image sensor and apply a dual sampling approach to the even rows. The dual sampling approach is the method of improving the sensitivity by adding a second column signal processing chain circuit and a capacitor bank to a conventional architecture of CMOS image sensor without modifying the data readout method [11]. In the CMOS image sensor, the sensor data is acquired by reading each row. Considering the row-readout process and using the dual sampling approach, it is possible to resolve the saturation problem of W and improve the sensitivity by using two W values at the same time. As well as the high SNR for R and B can be obtained. However, conventional color interpolation (CI) algorithms cannot be applied since the pattern of this RWB CFA is different from that of the widely used Bayer CFA. Consequently, the spatial resolution is degraded particularly in the horizontal direction as compared to the conventional RWB CFA pattern. In this paper, we propose a CI algorithm that improves the spatial resolution of images for this RWB CFA pattern. In the proposed algorithm, two sampled W channels captured with different exposure times are reconstructed as high resolution W channels using directional color difference information, and the sampled R and B are interpolated using a guided filter.

The rest of the paper is organized as follows. Section 2 provides the analysis of various CFA patterns. Section 3 describes the proposed algorithm in detail. Section 4 presents the experimental results, and Section 5 concludes the paper.

2 Analysis of CFA patterns

In 1976, Bayer proposed a CFA pattern to capture an image by using one sensor [1]. The Bayer pattern consists of 2×2 unit blocks that include two sampled G channels

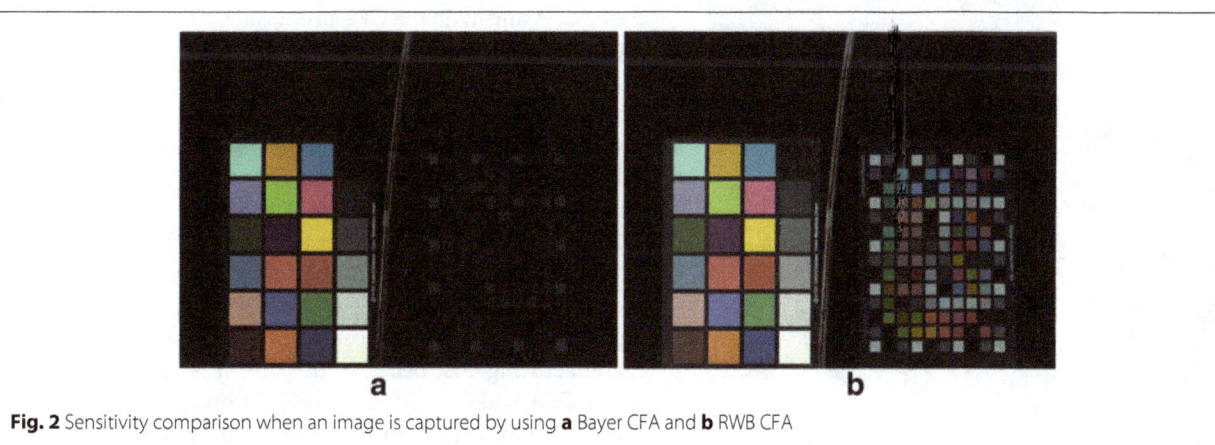

Fig. 2 Sensitivity comparison when an image is captured by using **a** Bayer CFA and **b** RWB CFA

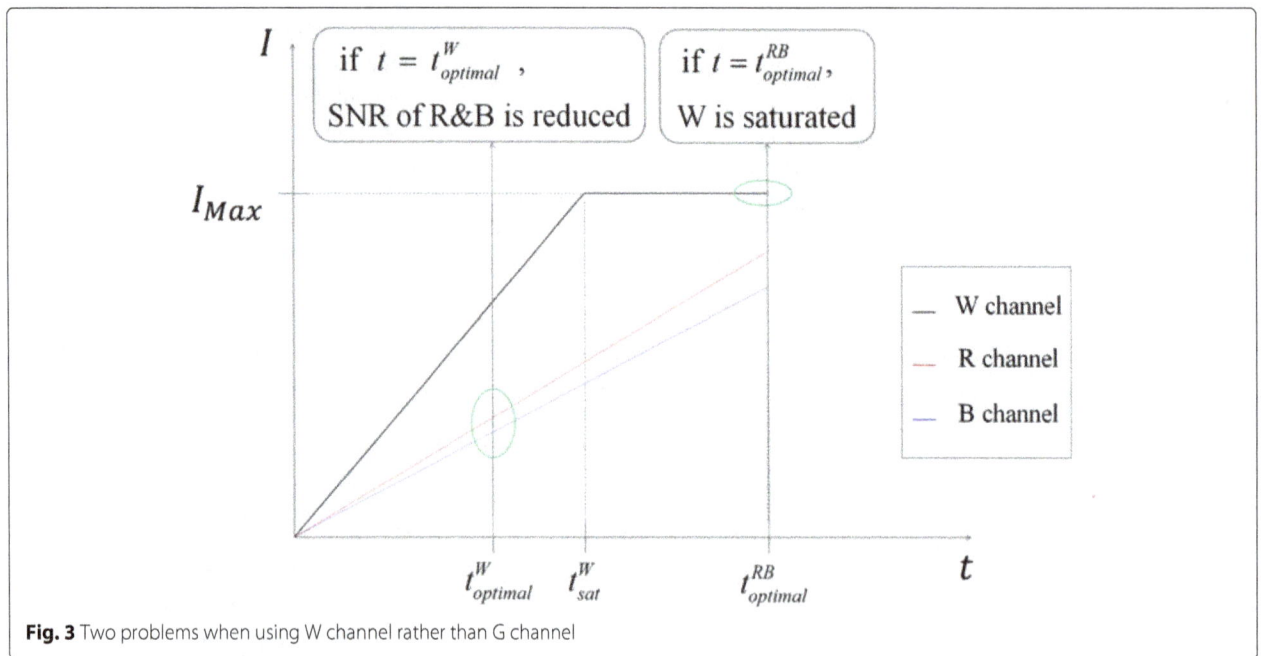

Fig. 3 Two problems when using W channel rather than G channel

diagonally, one sampled R channel, and one sampled B channel, as shown in Fig. 5a. The reason for this is as follows. Human eye is most sensitive to G and the luminance signal of the incident image is represented by G. In order to reconstruct full color channels from the sampled channels, many color interpolation (demosaicking) methods have been studied [12–18].

In order to overcome the limited sensitivity of the Bayer CFA pattern, new CFA patterns using W have been proposed in a number of patents and publications [2–6].

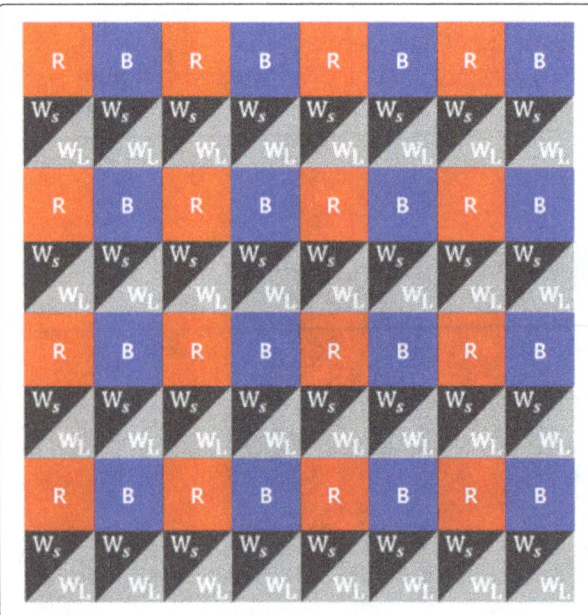

Fig. 4 RWB CFA pattern acquiring W at two exposure times [10]

Yamagami et al. were granted a patent for a new CFA pattern comprising RGB and W, as shown in Fig. 5b [2]. They acknowledged the abovementioned rapid saturation problem of W, and dealt with the problem by using CMY filters rather than RGB filters or a neutral density filter on the W. The drawback of this CFA is spatial resolution degradation caused by small proportions of R and B channels which occupy one-eighth of the sensor, respectively [19].

Gindele et al. granted a patent for a new CFA pattern using W with RGB in 2 × 2 unit blocks, as shown in Fig. 5c [3]. Sampling resolution of all the channels in this pattern are equal as a quarter of the sensor. It is possible to improve the spatial resolution as compared to the CFA of Yamagami et al. due to the increment of the sampling resolution for R and B. At the same time, improvement of the spatial resolution is limited because of the lack of a high-density channel. In comparison with other CFAs, G in the Bayer CFA and W in the CFA proposed by Yamagami et al. are high-density channels that occupy half of the sensor.

A new CFA pattern that consists of R, B, and W without G in 2 × 2 unit blocks was proposed in [8, 9], as shown in Fig. 5d. Since the pattern of this CFA is identical to the pattern of the Bayer CFA where G is replaced with W, the spatial resolution of this CFA is similar to that of the Bayer CFA. Recently, some industries have researched this CFA pattern with interest owing to some merits in terms of high resolution and high sensitivity. Further, this CFA pattern can be produced with minimal manufacturing cost because the pattern is similar to the Bayer CFA widely used today [20]. In order to obtain an accurate color image, they estimated G by using a color

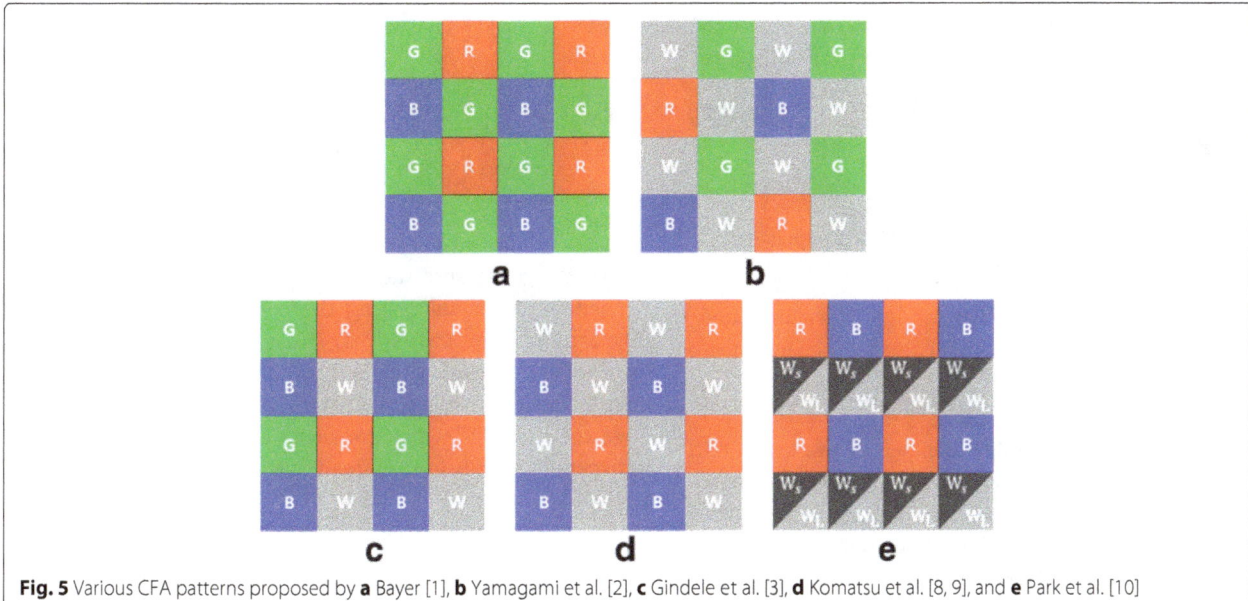

Fig. 5 Various CFA patterns proposed by **a** Bayer [1], **b** Yamagami et al. [2], **c** Gindele et al. [3], **d** Komatsu et al. [8, 9], and **e** Park et al. [10]

correction matrix. The drawback of this CFA pattern is that it cannot cope with the above mentioned rapid saturation problem of W. If saturation occurs at W, the accuracy of the estimated G decreases considerably because the estimation process is conducted two times. Since the real W value cannot be obtained, first, the real value of the saturated W which is larger than the maximum value is estimated. Then, the color correction matrix is used to estimate G. The results by 'two times' estimation have a larger error than the results by "one time" estimation because the relationship among R, G, B, and W are not accurately modeled.

Park et al. proposed an RWB CFA pattern using R, B, and double-exposed W instead of a single exposed W to solve the rapid saturation problem [10]. Two W values are obtained at different exposure times, and then fused to prevent saturation. This approach is similar to the HDR reconstruction algorithm. The pattern of this RWB CFA pattern is shown in Fig. 5e. R and B are placed sequentially one after the other in odd rows, and W is placed in even rows to consider the data readout method of the CMOS image sensor and apply the the dual sampling approach to W. Using the dual sampling approach, two values with different exposure times can be obtained without modifying the readout method in which the sensor data is acquired by scanning each row [11]. Applying the dual sampling approach to the even rows, double-exposed W are obtained to cope with the rapid saturation problem and improve the sensitivity. The obtained R and B show the high SNR since they are captured with optimal exposure time. The disadvantage of this CFA is the degradation of the spatial resolution along the horizontal direction than other CFAs. Since there are no W in

the odd rows, W located in vertical or diagonal directions except the horizontal direction are unavoidably referred for interpolating the missing W.

In this paper, we adopt the double exposed RWB CFA pattern proposed in [10]. Although this CFA pattern is disadvantageous with respect to the spatial resolution along the horizontal direction, we focus on the improvement of sensitivity and color fidelity by resolving the rapid saturation problem. In order to overcome the drawback of this CFA pattern, we propose a color interpolation method that reconstructs the four high resolution channels (long exposed W, short exposed W, R, and B) from the sampled low resolution channels. As a result, the images are improved in terms of spatial resolution, sensitivity, and color fidelity simultaneously compared with the results of other CFAs.

3 Proposed color interpolation algorithm

In this section, we describe the proposed method in detail. First, the missing W values are interpolated by calculating the color difference values for each direction, and then the estimated W values are updated to improve the spatial resolution by using subdivided directional color difference values. Second, the R and B values are estimated by using a guided filter that uses the interpolated W channel as a guided image. Then, the estimated R and B values are modified to obtain accurately estimated values by using an edge directional residual compensation method.

3.1 Two W channels interpolation

The proposed method calculates the directional color difference values for each direction similar to [16] and conducts a weighted summation of the obtained values according to the edge direction in order to interpolate the missing pixels. The existing method considers both vertical and horizontal directions. In the proposed method, the color difference for the vertical and diagonal directions are calculated as

$$\widetilde{\Delta}_V(i,j) = \begin{cases} \widetilde{W}_V(i,j) - R(i,j), & \text{if } W \text{ is interpolated,} \\ W(i,j) - \widetilde{R}_V(i,j), & \text{if } R \text{ is interpolated,} \end{cases}$$

$$\widetilde{\Delta}_{D1,D2}(i,j) = \begin{cases} \widetilde{W}_{D1,D2}(i,j) - R(i,j), & \text{if } W \text{ is interpolated,} \\ W(i,j) - \widetilde{R}_{D1,D2}(i,j), & \text{if } R \text{ is interpolated,} \end{cases}$$

$$(1)$$

where (i,j) denotes the index of pixel location. $\widetilde{\Delta}_V$ and $\widetilde{\Delta}_{D1,D2}$ denote the vertical and diagonal color difference values between the white and red channels, respectively. The direction of D_1 is from the upper right to the lower left and the direction of D_2 is from the upper left to the lower right. $\widetilde{W}$ and $\widetilde{R}$ represent the temporary estimated values of each channel according to the direction. They are calculated as

$$\widetilde{W}_V(i,j) = \frac{1}{2}\left(W(i-1,j) + W(i+1,j)\right)$$
$$+ \frac{1}{2}R(i,j) - \frac{1}{4}\left(R(i-2,j) + R(i+2,j)\right),$$

$$\widetilde{W}_{D1}(i,j) = \frac{1}{2}\left(W(i-1,j+1) + W(i+1,j-1)\right)$$
$$+ \frac{1}{2}R(i,j) - \frac{1}{4}\left(R(i-2,j+2) + R(i+2,j-2)\right),$$

$$\widetilde{W}_{D2}(i,j) = \frac{1}{2}\left(W(i-1,j-1) + W(i+1,j+1)\right)$$
$$+ \frac{1}{2}R(i,j) - \frac{1}{4}\left(R(i-2,j-2) + R(i+2,j+2)\right).$$

$$(2)$$

The equations are similar for calculating $\widetilde{R}_V$, $\widetilde{R}_{D1}$, and $\widetilde{R}_{D2}$. The reason for considering these directions rather than the horizontal direction is that it is impossible to calculate the color difference in the horizontal direction with this CFA pattern, whereas the calculation is possible in the diagonal direction.

In order to apply the dual sampling approach without modifying the readout method of CMOS image sensor, R and B are placed sequentially in odd rows and W is placed in even rows. However, this CFA pattern results in a zipper artifact at the horizontal edge owing to the energy difference between R and B. To remove this, the color difference value is modified by considering both R and B,

$$\widetilde{\Delta}_H(i,j) = \begin{cases} \widetilde{W}_V(i,j) - R^{hor}(i,j), & \text{if } W \text{ is interpolated,} \\ W(i,j) - \widetilde{R}_V^{hor}(i,j), & \text{if } R \text{ is interpolated,} \end{cases}$$

$$(3)$$

where

$$R^{hor}(i,j) = \frac{1}{2}R(i,j) + \frac{1}{4}\left(B(i,j-1) + B(i,j+1)\right),$$
$$\widetilde{R}_V^{hor}(i,j) = \frac{1}{2}\widetilde{R}_V(i,j) + \frac{1}{4}\left(\widetilde{B}_V(i,j-1) + \widetilde{B}_V(i,j+1)\right),$$

$$(4)$$

where $\widetilde{B}_V$ represents the temporary estimated value of B channel. After calculating the color difference values for each direction i.e., $\widetilde{\Delta}_V$, $\widetilde{\Delta}_{D1}$, $\widetilde{\Delta}_{D2}$, and $\widetilde{\Delta}_H$, the initial interpolated result is obtained by combining color difference values according to the edge direction. In horizontal edge regions, R^{hor} which considers the energy difference between R and B is used instead of R to prevent the zipper artifact,

$$\widehat{W}^{ini}(i,j) = \{w_V(R(i,j)+\widehat{\Delta}_V(i,j)) + w_{D1}(R(i,j)+\widehat{\Delta}_{D1}(i,j)) + w_{D2}(R(i,j) + \widehat{\Delta}_{D2}(i,j)) + w_H(R^{hor}(i,j)+\widehat{\Delta}_H(i,j))\}/\sum_{\mathbf{d}_4} w_{\mathbf{d}_4},$$

$$(5)$$

where $\mathbf{d}_4 = \{V, D1, D2, H\}$ and $w_{\mathbf{d}_4}$ represent the weight values for each direction. When calculating the horizontal direction of $\widehat{W}^{ini}$, $R^{hor} + \widehat{\Delta}_H$ should be used instead of $R + \widehat{\Delta}_H$ because $\widetilde{\Delta}_H$ is calculated using R^{hor} and B^{hor}. Since the scaling of each direction is matched, there are no artifacts in the interpolated image. $\widehat{\Delta}_{\mathbf{d}_4}$ represents the directional averaging values of the directional color difference values,

$$\widehat{\Delta}_V(i,j) = \mathbf{A}_V \otimes \widetilde{\Delta}_V(i-1:i+1,j),$$
$$\widehat{\Delta}_{D1}(i,j) = \mathbf{A}_{D1} \otimes \widetilde{\Delta}_{D1}(i-1:i+1,j-1:j+1),$$
$$\widehat{\Delta}_{D2}(i,j) = \mathbf{A}_{D2} \otimes \widetilde{\Delta}_{D2}(i-1:i+1,j-1:j+1),$$
$$\widehat{\Delta}_H(i,j) = \mathbf{A}_H \otimes \widetilde{\Delta}_H(i,j-1:j+1),$$

$$(6)$$

where $\otimes$ denotes element-wise matrix multiplication (often called the Hadamard product) and subsequent summation. Directional averaging matrices $\mathbf{A}_{\mathbf{d}_4}$ for each direction are as

$$\mathbf{A}_V = \begin{bmatrix} 1 & 2 & 1 \end{bmatrix}^T \cdot \frac{1}{4},$$

$$\mathbf{A}_{D1} = \begin{bmatrix} 0 & 0 & 1 \\ 0 & 2 & 0 \\ 1 & 0 & 0 \end{bmatrix} \cdot \frac{1}{4},$$

$$\mathbf{A}_{D2} = \begin{bmatrix} 1 & 0 & 0 \\ 0 & 2 & 0 \\ 0 & 0 & 1 \end{bmatrix} \cdot \frac{1}{4},$$

$$\mathbf{A}_H = \begin{bmatrix} 1 & 2 & 1 \end{bmatrix} \cdot \frac{1}{4}.$$

$$(7)$$

The directional weight values $w_{\mathbf{d}_4}$ in Eq. (5) for each direction are calculated as

$$
w_V = \left[\frac{1}{2}\left(|I(i,j) - I(i,j-1)| + |I(i,j) - I(i,j+1)|\right) \right.
$$
$$
\left. + \frac{1}{5}\sum_{k=-2}^{2}|I(i+k,j-1) - I(i+k,j+1)|\right]^{-2},
$$

$$
w_{D1} = \left[\frac{1}{2}\left(|I(i,j) - I(i-1,j-1)| + |I(i,j) - I(i+1,j+1)|\right) \right.
$$
$$
\left. + \frac{1}{5}\sum_{k=-2}^{2}|I(i-1,j-1+k) - I(i+1,j+1+k)|\right]^{-2},
$$

$$
w_{D2} = \left[\frac{1}{2}|I(i,j) - I(i-1,j+1)| + |I(i,j) - I(i+1,j-1)| \right.
$$
$$
\left. + \frac{1}{5}\sum_{k=-2}^{2}|I(i-1,j+1+k) - I(i+1,j-1+k)|\right]^{-2},
$$

$$
w_H = \left[\frac{1}{2}|I(i,j) - I(i-1,j)| + |I(i,j) - I(i+1,j)| \right.
$$
$$
\left. + \frac{1}{5}\sum_{k=-2}^{2}|I(i-1,j+k) - I(i+1,j+k)|\right]^{-2},
$$

$$\tag{8}$$

where I represents the input patterned image.

In order to further improve the resolution, the initial result is updated similar to [16]. After further subdividing the edge direction, the color difference values are calculated in eight directions i.e., $\mathbf{d}_8 = \{N, NE, E, SE, S, SW, W, NW\}$. The obtained values are again combined according to the edge direction to obtain the final interpolated result,

$$
\widehat{W}^{up}(i,j) = \sum_{\mathbf{d}_4}\frac{w_{\mathbf{d}_4}\widetilde{W}_{\mathbf{d}_4}^{up}(i,j)}{\sum w_{\mathbf{d}_4}}, \tag{9}
$$

where $\widetilde{W}_{\mathbf{d}_4}^{up}$ represents the updated W values for each $\mathbf{d}_4$ direction. They are calculated by arbitrating the directional color difference value $(\widehat{\Delta}_{\mathbf{d}_4})$ with updated directional color difference value $(\widehat{\Delta}_{\mathbf{d}_4}^{up})$ as

$$
\widetilde{W}_V^{up}(i,j) = R(i,j) + \{w_{up}\widehat{\Delta}_V^{up}(i,j) + (1 - w_{up})\widehat{\Delta}_V(i,j)\},
$$
$$
\widetilde{W}_{D1}^{up}(i,j) = R(i,j) + \{w_{up}\widehat{\Delta}_{D1}^{up}(i,j) + (1 - w_{up})\widehat{\Delta}_{D1}(i,j)\},
$$
$$
\widetilde{W}_{D2}^{up}(i,j) = R(i,j) + \{w_{up}\widehat{\Delta}_{D2}^{up}(i,j) + (1 - w_{up})\widehat{\Delta}_{D2}(i,j)\},
$$
$$
\widetilde{W}_H^{up}(i,j) = R^{hor}(i,j) + \{w_{up}\widehat{\Delta}_H^{up}(i,j) + (1 - w_{up})\widehat{\Delta}_H(i,j)\},
$$

$$\tag{10}$$

where w_{up} is the weight value which determines the ratio of the directional color difference value to the updated directional color difference value. This weight value is set to 0.7 in our experiments. The updated directional color difference values are calculated considering the subdivided directions ($\mathbf{d}_8$) as

$$
\widehat{\Delta}_V^{up}(i,j) = w_N\widehat{\Delta}_V(i-2,j) + w_S\widehat{\Delta}_V(i+2,j),
$$
$$
\widehat{\Delta}_{D1}^{up}(i,j) = w_{NE}\widehat{\Delta}_{D1}(i-2,j+2) + w_{SW}\widehat{\Delta}_{D1}(i+2,j-2),
$$
$$
\widehat{\Delta}_{D2}^{up}(i,j) = w_{NW}\widehat{\Delta}_{D2}(i-2,j-2) + w_{SE}\widehat{\Delta}_{D2}(i+2,j+2),
$$
$$
\widehat{\Delta}_H^{up}(i,j) = w_E\widehat{\Delta}_H(i,j-2) + w_W\widehat{\Delta}_H(i,j+2),
$$

$$\tag{11}$$

where

$$
w_N = \frac{1}{15}\sum_{k=-4}^{0}\sum_{l=-1}^{1}\frac{1}{|I(i+k,j+l-1) - |I(i+k,j+l+1)|^2},
$$
$$
w_{NE} = \frac{1}{15}\sum_{k=-4}^{0}\sum_{l=-1}^{1}\frac{1}{|I(i+k+l-1,j-k+1) - |I(i+k+l+1,j-k-1)|^2},
$$
$$
w_E = \frac{1}{15}\sum_{k=-1}^{1}\sum_{l=0}^{4}\frac{1}{|I(i+k-1,j+l) - |I(i+k+1,j+l)|^2},
$$
$$
w_{SE} = \frac{1}{15}\sum_{k=0}^{4}\sum_{l=-1}^{1}\frac{1}{|I(i+k+l+1,j+k+1) - |I(i+k+l-1,j+k-1)|^2},
$$
$$
w_S = \frac{1}{15}\sum_{k=0}^{4}\sum_{l=-1}^{1}\frac{1}{|I(i+k,j+l-1) - |I(i+k,j+l+1)|^2},
$$
$$
w_{SW} = \frac{1}{15}\sum_{k=0}^{4}\sum_{l=-1}^{1}\frac{1}{|I(i+k+l+1,j-k-1) - |I(i+k+l-1,j-k+1)|^2},
$$
$$
w_W = \frac{1}{15}\sum_{k=-1}^{1}\sum_{l=-4}^{0}\frac{1}{|I(i+k-1,j+l) - |I(i+k+1,j+l)|^2},
$$
$$
w_{NW} = \frac{1}{15}\sum_{k=-4}^{0}\sum_{l=-1}^{1}\frac{1}{|I(i+k+l-1,j+k-1) - |I(i+k+l+1,j+k+1)|^2}.
$$

$$\tag{12}$$

The double-exposed RWB CFA used in this paper obtains two W values that have different exposure times. Although the proposed interpolation algorithm can be applied to short exposed W, it cannot be applied to long exposed W. This is because the color difference value cannot be calculated due to the saturation. In the proposed method, the edge directional weighted summation of the temporary estimated W is set as the final interpolated value for long exposed W. The temporary estimated W for the vertical direction and both diagonal directions are obtained by using Eq. (2). In horizontal edge regions, spatial resolution is more improved when using $\widetilde{W}_V$ than using $\widetilde{W}_{D1}$ or $\widetilde{W}_{D2}$ because W values are not located in the horizontal direction. However, this results in the zipper artifact due to the energy difference between R and B. To remove this artifact, we propose a method to estimate the temporary W values for the horizontal edge ($\widetilde{W}_H$) by modifying $\widetilde{W}_V$ similar to Eq. (4),

$$
\begin{aligned}
\widetilde{W}_H\left(i,j\right) = &\frac{1}{2}\left(W\left(i-1,j\right)+W\left(i+1,j\right)\right) \\
&+\frac{1}{4}R\left(i,j\right)-\frac{1}{8}\left(R\left(i-2,j\right)+R\left(i+2,j\right)\right) \\
&+\frac{1}{8}B\left(i,j-1\right)-\frac{1}{16}\left(R\left(i-2,j-1\right)+R\left(i+2,j-1\right)\right) \\
&+\frac{1}{8}B\left(i,j+1\right)-\frac{1}{16}\left(R\left(i-2,j+1\right)+R\left(i+2,j+1\right)\right).
\end{aligned}
$$
(13)

The weighted values used in the edge directional weighted summation are identical to the weighted values used for short exposed W because they are determined for the same area,

$$
\widehat{W}_L\left(i,j\right)=\sum_{\mathbf{d}_4}\frac{w_{\mathbf{d}_4}\widetilde{W}_{\mathbf{d}_4}\left(i,j\right)}{\sum w_{\mathbf{d}_4}}.
$$
(14)

With the above mentioned interpolation method, the missing W values located at the sampled R pixels are calculated. The remaining missing W values located at the sampled B pixels are estimated in a similar manner.

3.2   R and B channels interpolation

The proposed method interpolates the sampled R and B channels into high resolution channels through a guided filter that uses the interpolated W channel as a guided value. The guided filter conducts the image filtering considering the guided value. If the image filtering is conducted using the guided filter approach, the filtered image has the characteristics of the guided image. Therefore, the guided filter approach demonstrates high performance particularly in edge-aware smoothing. Further, this approach is utilized in many image processing areas including detail enhancement, HDR compression, image matting/feathering, dehazing, and joint upsampling, [21]. In the proposed algorithm, the property of the guided filter, which conducts filtering by reflecting the characteristics of the guided value, is used. Since the interpolated W channel is a high resolution image, the sampled R and B channels can be efficiently interpolated by reflecting the characteristic of the W channel in terms of high resolution,

$$
\widetilde{R}\left(i,j\right)=a\cdot\widehat{W}^{up}\left(i,j\right)+b,
$$
(15)

where

$$
a=\frac{\frac{1}{N_{mask}}\left(\sum\limits_{i,j\in mask}\widehat{W}^{up}\left(i,j\right)R\left(i,j\right)\right)-\mu_{mask}^{\widehat{W}^{up}}\cdot\mu_{mask}^{R}}{\left[\sigma_{mask}^{\widehat{W}^{up}}\right]^2+\varepsilon},
$$

$$
b=\mu_{mask}^{R}-a\cdot\mu_{mask}^{\widehat{W}^{up}},
$$
(16)

where μ_{mask} and $\sigma_{mask}^{\widehat{W}^{up}}$ are the mean and standard deviation values of the estimated W or sampled R channels. N_{mask} is the number of pixels in the $n\times m$ mask. The size of the mask is set as 7×7 in our experiment, and the regularization parameter ε is set as 0.001. The real value of R is required for estimating the coefficients used in the guided filter. The input patterned image contains the sampled real value of R, allowing for an accurate estimation of the coefficient values. There is a difference between the initial estimated R values that are obtained by substituting the estimated coefficients into Eq. (15) and the real values of R at the location of sampled pixels,

$$
\lambda\left(i,j\right)=R\left(i,j\right)-\widetilde{R}\left(i,j\right),
$$
(17)

where $\widetilde{R}$ is the initial estimated value of R. λ represent the residual values which are calculated by the difference between the real R values and the initial interpolated R values. Such residuals occur also at the location of missing pixel. Thus, the initial interpolated R values should be compensated considering these residuals to improve the accuracy of the estimated R values [18]. In the proposed method, the final interpolated result of R $(\widehat{R})$ is obtained by using an edge directional residual compensation method as

$$
\widehat{R}\left(i,j\right)=\widetilde{R}\left(i,j\right)+\widehat{\lambda}\left(i,j\right),
$$
(18)

where $\widehat{\lambda}$ represents the edge directional residual values by considering the residual values at the surrounding sampled pixels. Because the pixels with real values neighboring a missing pixel are exclusively located in a diagonal direction, the diagonal residual value is first calculated as

$$
\begin{aligned}
\widehat{\lambda}\left(i,j\right)=&w_{NE}\lambda\left(i-1,j+1\right)+w_{SE}\lambda\left(i+1,j+1\right)\\
&+w_{SW}\lambda\left(i+1,j-1\right)+w_{NW}\lambda\left(i-1,j-1\right),
\end{aligned}
$$
(19)

where the weight values for each direction are calculated by using Eq. (12). The remaining missing pixels are interpolated by combining the vertical and horizontal residual values,

$$
\begin{aligned}
\widehat{\lambda}\left(i,j\right)=&w_N\lambda\left(i-1,j\right)+w_E\lambda\left(i,j+1\right)+w_S\lambda\left(i-1,j\right)\\
&+w_W\lambda\left(i,j-1\right).
\end{aligned}
$$
(20)

The B channel is interpolated similarly by using the guided filter with the reconstructed W value as the guidance value.

4   Experimental results

First, we tested the validity of the adoption of the double exposed CFA pattern. Various CFA patterns, as mentioned in Section 2, were considered in the experiments and compared in terms of the spatial resolution, sensitivity, and color fidelity. For this purpose, we designed experimental equipment for obtaining R, G, B, W, and double-exposed W in high resolution images, and then sampled these using each CFA pattern to obtain patterned images. All images were captured in our test room where the outside light could be blocked. Since other commercial products did not support this pattern, we built an imaging system using a single monochrome sensor and rotating filter wheel system including red, green, blue, and white (luminance) filters. Using this rotating filter wheel, capturing the high resolution image of each channel was possible. In our experiments, two W channels (W_L and W_s) were captured with different exposure times when using the W filter. The exposure time of each channel was determined as follows. First, the proper exposure time of R and B channels (t_{RB}) was set to prevent the saturation of those channels. Then, the exposure times of the long exposed W (t_{W_L}) and short exposed W (t_{W_s}) were set as $\frac{5}{6}t_{RB}$ and $\frac{1}{6}t_{RB}$, respectively. The ratio of exposure time between W_L and W_s was experimentally determined considering the sensitivity difference of each channel. Note that, the saturated region never occurs in the short exposed W (W_s). In the long exposed W (W_L), the generation of the saturated regions is not a concern. The brightness of the dark regions is important. The brighter the dark region, the higher SNR for the final resulting image. If object moving or handshaking occurs, long exposed W and short exposed W could capture different images. The problem of the moving object in the double exposed W mentioned previously was also problem in the R and B channels. In the images of the R and B channels, the object was captured with motion blur. The research regarding color interpolation including moving objects is beyond the scope of this paper. In our experiment, we assumed that the object did not move when capturing the images with t_{RB}.

In order to reconstruct the full color images, CI algorithms suitable for each CFA pattern were applied. In our experiments, multiscale gradients (MSG) based CI algorithm [16] was used for the patterns shown in Fig. 5a, d. For patterns shown in Fig. 5b, c, e, edge directional interpolation was utilized to generate the intermediate images with quincuncial patterns [22], and then a full color image was obtained by applying the MSG based CI algorithm. Then, G was estimated by using a color correction matrix in case of the RWB CFA, and the HDR reconstruction algorithm was applied if the image contained W.

In our experiments, the color correction matrix was used to estimate G from R, B, and W [9]. The matrix form of the method to estimate G is presented as

$$\mathbf{X} = \Phi^T \mathbf{Y}, \tag{21}$$

where $\mathbf{X} = \{R, G, B\}^T$ and $\mathbf{Y} = \{R, W, B\}^T$. Φ is the color correction matrix whose components are determined by considering the correlation among R, G, B, and

W. These components were changed in accordance with the color temperature and brightness. In our experiments, we experimentally determined the values of these components considering the correlation among R, G, B, and W for each experimental condition (brightness and kind of illumination). By using these values, the same color can be obtained regardless of the experimental conditions.

Exposure fusion method was used for the HDR reconstruction [23]. The weight values that determine the best parts in each image were calculated by multiplication of a set of quality measures. The final HDR image (L_{hdr}) was obtained by fusing the best parts of each image. The three images (interpolated short exposed W channel, interpolated long exposed W channel, and luminance image from RGB values) were used to improve the sensitivity.

$$L_{hdr} = \widehat{w}_s^{hdr}(i,j)\,\widehat{W}^{up}(i,j) + \widehat{w}_L^{hdr}(i,j)\,\widehat{W}_L(i,j) \\ + \widehat{w}_{lumi}^{hdr}(i,j)\,L(i,j), \tag{22}$$

where $\widehat{w}_s^{hdr}$, $\widehat{w}_L^{hdr}$, and $\widehat{w}_{lumi}^{hdr}$ represent the normalized HDR weight values for interpolated short exposed W ($\widehat{W}^{up}$), interpolated long exposed W ($\widehat{W}_L$), and luminance (L) value from RGB values, respectively. These values are calculated as

$$\widehat{w}_\mathbf{s}^{hdr}(i,j) = \left[\sum_\mathbf{s} w_\mathbf{s}^{hdr}(i,j)\right]^{-1} w_\mathbf{s}^{hdr}(i,j), \tag{23}$$

where $\mathbf{s} = \{\widehat{W}^{up}, \widehat{W}_L, L\}$. The HDR weight values for each image $w_\mathbf{s}^{hdr}$ are calculated as

$$w_\mathbf{s}^{hdr}(i,j) = C_\mathbf{s}^\alpha \times S_\mathbf{s}^\beta \times E_\mathbf{s}^\gamma, \tag{24}$$

where $C_\mathbf{s}$, $S_\mathbf{s}$, and $E_\mathbf{s}$ represent the values of quality measures for contrast, saturation, and well-exposure, respectively. α, β, and γ are the weight values of each measure. The detailed description for these quality measures are presented in [23].

Figure 6 shows the comparison results for a test image captured using each CFA pattern. The test image simultaneously includes a bright region (left side) and a dark region (right side). The processed results show that the sensitivity is improved when W is obtained. In Fig. 6b–f, the color checker board located on the right side was clearly visible. The average brightness of Fig. 6e was lower than that of the other results because of the shorter exposure time to prevent the saturation of W. If W is saturated, incorrect color information is produced due to the inaccurate estimation of G, as shown in Fig. 6d.

The enlarged parts, including letters, show the spatial resolution of each CFA pattern. Figure 6a, e shows higher spatial resolution than other results because the MSG based CI algorithm is designed for the Bayer CFA. In Fig. 6d, saturation occurs at the letters, and thus the spatial resolution is lower than that of Fig. 6e even though the CFA patterns are similar. Figure 6b, c shows that the

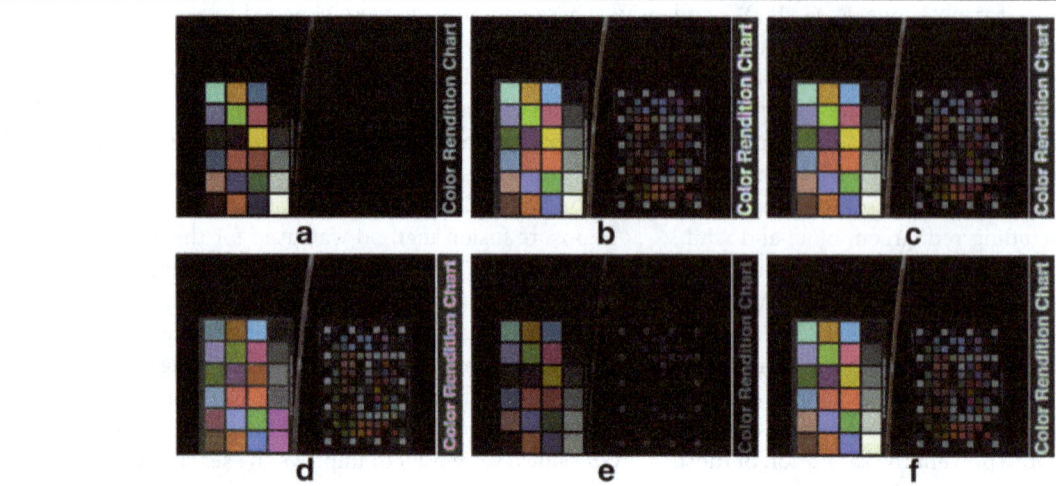

Fig. 6 Comparison of CI and HDR reconstruction results and the enlarged parts obtained by using **a** Bayer CFA, **b** RGBW CFA [2], **c** RGBW CFA [3], **d** RWB CFA with saturated *W* [8, 9], **e** RWB CFA with unsaturated W [8, 9], and **f** double-exposed RWB CFA [10]

lower spatial resolution due to the weakness of the CFA patterns. Figure 6f shows better spatial resolution than Fig. 6b, c. Although the double exposed CFA pattern may suffer from low resolution along horizontal direction due to the weakness of CFA pattern, the spatial resolution of this CFA pattern is not the worst. If an effective CI algorithm is applied, it is possible to improve the spatial resolution.

In Table 1, the performance of each CFA pattern is compared by using objective metrics for the test images captured by our experimental equipment. The color peak signal-to-noise ratio (CPSNR), brightness of each region (bright and dark regions), and angular error are used to measure the spatial resolution, sensitivity, and color fidelity, respectively. Comparing the CPSNR metric, the double-exposed RWB CFA pattern recorded a larger value than the RGBW CFA patterns. It should be improved further, since the Bayer CFA pattern provides a larger CPSNR value. In terms of the brightness values of bright and dark

regions, the CFA patterns with *W* recorded larger values compared with those of the Bayer CFA pattern. For the RWB CFA pattern with unsaturated W, however, the brightness value is smaller than that of other CFA patterns including W because of the shorter exposure time to avoid the saturation of W. By comparing the angular error values, it is shown that satisfactory color fidelity is obtained when using the double-exposed RWB CFA although the Bayer CFA pattern and RGBW CFA patterns provide smaller angular error than the double exposed RWB CFA pattern. If an accurate G estimation method is developed, the double exposed RWB CFA pattern will provide improved color fidelity compared with the results estimated by using a color correction matrix.

From above analysis, we adopted the double-exposed CFA pattern and proposed the CI algorithm to improve the spatial resolution. Despite spatial resolution degradation, the color fidelity was satisfactory and the sensitivity was greatly improved in comparison with the Bayer CFA pattern. If the spatial resolution is improved, it is possible to capture high-resolution and noise-free images in low light conditions because the sensitivity is significantly improved without color degradation.

Next, we tested the performance of the proposed algorithm. For this purpose, we obtained R and B, and the double exposed W in high resolution images by using our experimental equipment, and then sampled these images by using the CFA pattern proposed in [10] to obtain a patterned image. After applying the proposed method (PM) to the patterned image in order to reconstruct each color channel as a high resolution channel, we compared its values to the original values. We also compared the results of the PM to those of the conventional methods (CMs): a simple color interpolation algorithm (CM1) described in [10]; MSG based CI algorithm with

Table 1 Performance comparison of various CFA patterns

	Spatial resolution (CPSNR)	Sensitivity (Brightness)	Color fidelity (Angular error)
Bayer CFA [1]	35.27	62.22, 5.20	1.03
RGBW CFA [2]	19.93	94.65, 28.29	1.68
RGBW CFA [3]	20.65	94.67, 28.47	1.54
RWB CFA with saturated W [8, 9]	17.50	80.92, 28.55	8.73
RWB CFA with unsaturated W [8, 9]	27.60	45.38, 8.34	2.55
RWB CFA with double exposed W [10]	23.49	91.67, 27.87	2.42

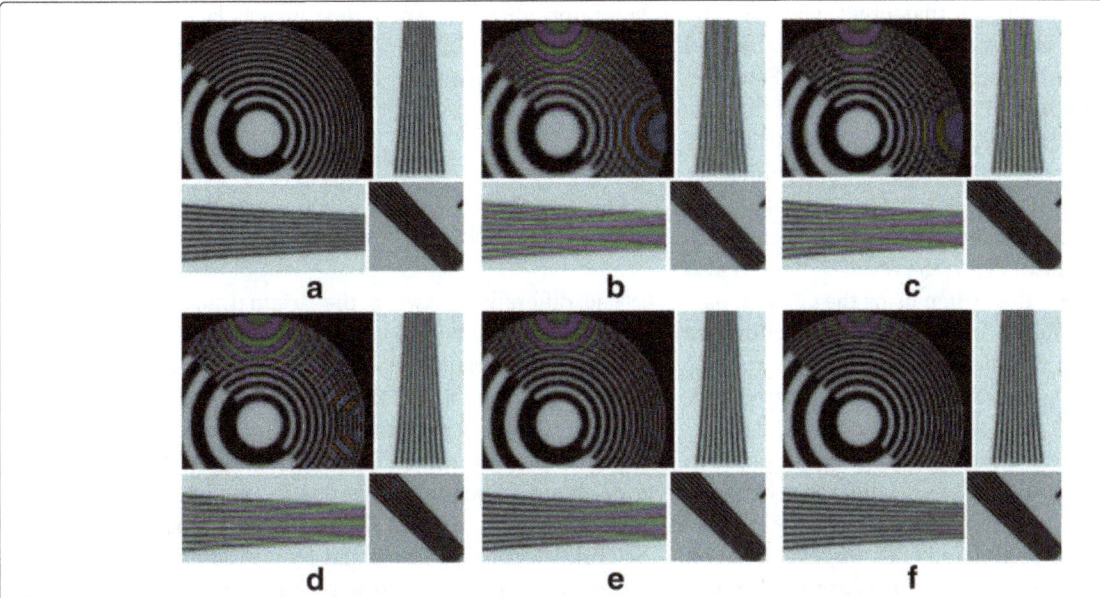

Fig. 7 Comparison of the various CI algorithms for color images including short exposed W: **a** original image, **b** CM1, **c** CM2, **d** CM3, **e** CM4, and **f** PM

intermediate quincuncial patterns (CM2) [16, 22]; an intra field deinterlacing algorithm (CM3) [24], which uses the characteristic of this CFA pattern with sampled W only in the even row; and a method (CM4) that modifies the algorithm for the existing Bayer CFA such that it is compatible with the double exposed RWB CFA pattern [17].

Figure 7 shows the result images which compare the original image with the reconstructed images by using the PM and CMs. Figure 7a shows the enlarged parts of the original image. Figure 7b shows the results when using CM1. It is clear that CM1 does not consider the edge direction at all. The ability of CM1 to improve the spatial resolution was very poor. In Fig. 7c, improvement of the spatial resolution is shown as compared with CM1 since

the edge directional interpolation for obtaining quincuncial patterns is used. Figure 7d shows the results when using CM3, where there is an obvious improvement to the spatial resolution compared with CM1, but is similar to CM2. This is because CM3 interpolated the W channel by estimating the edge direction. When estimating the edge direction, however, CM3 relied exclusively on pixel information corresponding to an even row. Consequently, the edge direction was falsely estimated in some areas, leading to incorrect interpolation results. With CM4, an interpolation kernel was applied for missing pixels considering the edge direction, the results for which are shown in Fig. 7e. Since the interpolation kernel was applied considering R and B, as well as W, the image was interpolated

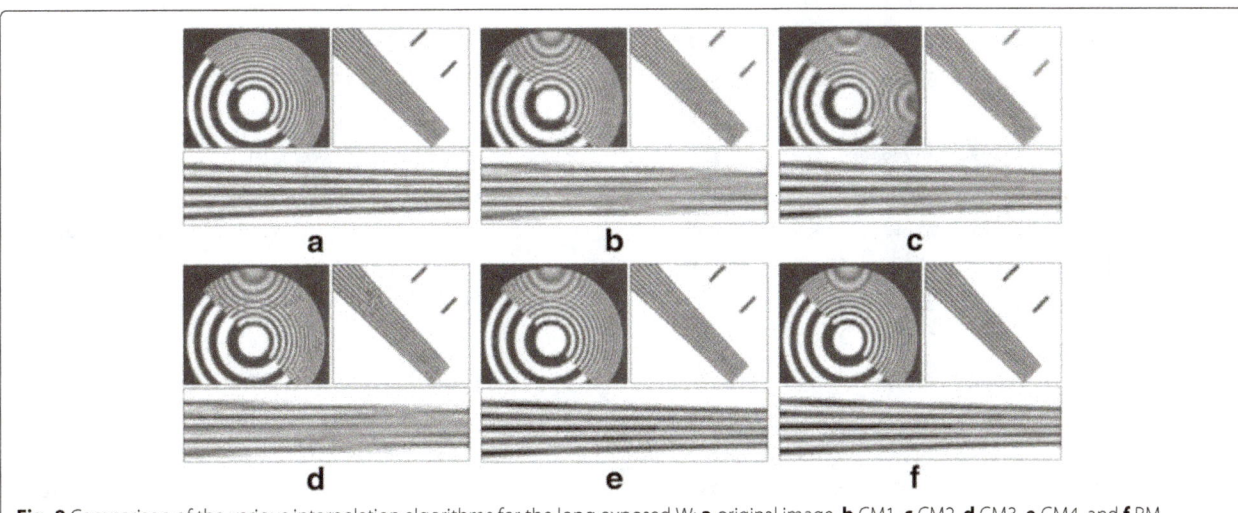

Fig. 8 Comparison of the various interpolation algorithms for the long exposed W: **a** original image, **b** CM1, **c** CM2, **d** CM3, **e** CM4, and **f** PM

while preserving the edge better than other methods. The biggest disadvantage of this method is poor resolution and zipper artifacts in the horizontal edge regions. Figure 7f presents the result of the PM. It is confirmed that the spatial resolution was further improved than when using the CMs. In particular, the spatial resolution is significantly improved near the horizontal edge.

The interpolation results for the long exposed W are shown in Fig. 8. Similar to Fig. 7, the result of the PM also shows higher resolution than when using the CMs. In particular, the spatial resolution of the diagonal direction was greatly improved and zipper artifacts located at regions near the horizontal edge were removed as compared with the results of the CMs. The reason is that the CMs interpolate the missing pixels without considering the pixels in the diagonal directions and the energy difference between R and B.

The performance of the PM and CMs was evaluated on a Kodak test set and the five test images captured in our laboratory. Because Kodak test images lack W channels, both exposures of W (W_s^{gen} and W_L^{gen}) were generated artificially for the test,

$$W_s^{gen}\left(i,j\right) = \frac{R\left(i,j\right) + G\left(i,j\right) + B\left(i,j\right)}{3} \times 0.5,$$

$$W_L^{gen}\left(i,j\right) = \frac{R\left(i,j\right) + G\left(i,j\right) + B\left(i,j\right)}{3} \times 1.5. \quad (25)$$

The difference between the original and interpolated images was determined by using two metrics: color peak signal-to-noise ratio (CPSNR) and S-CIELAB color difference (ΔE) formula [12, 25]. Generally, when calculating the CPSNR, three channels (R, G, and B) are used. In our experiments, this three-channel equation is modified to a four-channel equation since our results have four channels (R, B, long exposed W, and short exposed W). The S-CIELAB ΔE was calculated by averaging the ΔE values of two color images using long exposed W and short

Table 2 Comparison of objective evaluation: CPSNR

Image No.	CM1	CM2	CM3	CM4	PM
Captured img 1	33.714	35.178	36.533	38.175	**40.375**
2	32.685	34.580	35.064	37.491	**40.175**
3	35.769	37.226	37.502	39.528	**41.016**
4	29.218	31.417	32.153	**37.481**	37.414
5	33.483	34.825	35.265	37.959	**40.684**
Kodak test img 1	29.936	31.758	32.808	35.700	**38.715**
2	36.032	36.771	38.724	40.522	**41.661**
3	36.642	37.908	38.047	40.004	**40.191**
4	36.779	37.630	39.276	40.312	**42.234**
5	30.500	33.030	33.986	35.877	**38.737**
6	31.111	33.064	33.448	35.060	**38.433**
7	36.151	37.751	39.922	40.267	**42.658**
8	27.638	29.162	31.956	34.783	**36.001**
9	35.915	37.036	37.353	39.148	**41.917**
10	35.887	37.129	39.419	40.565	**42.869**
11	32.835	34.784	35.867	36.112	**39.333**
12	35.843	36.909	38.766	41.579	**43.724**
13	27.956	30.616	29.897	32.328	**35.165**
14	32.743	34.529	35.092	36.485	**39.065**
15	34.754	35.374	37.196	39.085	**41.519**
16	34.109	35.893	35.877	38.302	**40.759**
17	35.312	37.017	38.819	40.377	**43.665**
18	32.058	34.472	34.606	37.232	**40.675**
19	31.790	32.910	34.650	37.244	**40.070**
20	34.083	36.609	36.251	39.276	**42.327**
Avg.	33.318	34.943	35.939	38.032	**40.335**

Boldface indicates the best results in each metric among PM and CMs

Table 3 Comparison of objective evaluation: S-CIELAB ΔE

Image No.	CM1	CM2	CM3	CM4	PM
Captured img 1	13.500	6.854	9.068	6.792	**3.946**
2	16.642	8.882	13.194	9.884	**5.903**
3	9.466	5.324	6.804	4.580	**3.251**
4	18.804	13.493	15.592	13.947	**7.839**
5	14.044	7.825	9.947	7.468	**4.553**
Kodak test img 1	17.080	11.926	13.764	11.543	**7.604**
2	8.785	6.103	6.895	5.432	**3.578**
3	2.754	1.508	2.167	1.692	**1.054**
4	3.285	1.864	2.422	1.822	**1.314**
5	7.648	3.307	4.977	3.636	**1.862**
6	7.251	3.802	5.804	4.293	**2.644**
7	3.184	1.842	2.204	1.819	**1.112**
8	10.548	6.666	6.587	4.930	**2.712**
9	3.444	1.976	2.378	1.768	**1.105**
10	3.390	1.797	2.291	1.674	**1.031**
11	5.554	3.010	4.165	3.177	**1.808**
12	3.340	2.053	2.545	2.010	**1.228**
13	11.001	4.834	8.793	6.670	**3.494**
14	5.956	2.982	4.560	3.424	**2.226**
15	3.427	2.240	2.447	1.967	**1.334**
16	5.026	2.704	4.168	3.120	**1.873**
17	3.619	1.809	2.541	1.979	**1.143**
18	6.098	2.956	4.540	3.343	**2.193**
19	5.648	3.426	3.719	2.810	**1.547**
20	3.586	1.960	2.774	2.154	**1.283**
Avg.	5.409	2.921	3.953	2.972	**1.762**

Boldface indicates the best results in each metric among PM and CMs

exposed W rather than G. Table 2 shows the CPSNR comparison. The PM outperformed the other methods for 24 of the 25 test images. Likewise, the S-CIELAB ΔE results demonstrate that the PM had the best performance for all test images. The S-CIELAB ΔE values of each result are compared in Table 3.

5 Conclusions

In this paper, we proposed a color interpolation algorithm for the RWB CFA pattern that acquires W at different exposure times. The proposed algorithm interpolates two W channels by using directional color difference information, and then reconstructs R and B channels with a guided filter. The proposed method outperformed conventional methods, which was confirmed with objective metrics and a comparison using actual images. In future research, we intend to study G channel estimations and image fusion based on HDR reconstruction algorithms, and apply these methods to high resolution R,W, and B channels generated by using the proposed algorithm. Through these methods, an image with highly improved sensitivity compared to the existing Bayer patterned image can be obtained.

Competing interests

The authors declare that they have no competing interests.

Acknowledgements

This research was supported by the Basic Science Research Program through the National Research Foundation of Korea(NRF) funded by the Ministry of Science, ICT and future Planning(No. 2015R1A2A1A14000912).

References

1. BE Bayer, Color imaging array. Jul. 20. U.S. Patent 3,971,065 (1976)
2. T Yamagami, T Sasaki, A Suga, Image signal processing apparatus having a color filter with offset luminance filter elements. Jun. 21. U.S. Patent 5,323,233 (1994)
3. EB Gindele, AC Gallagher, Sparsely sampled image sensing device with color and luminance photosites. Nov. 5. U.S. Patent 6,476,865 (2002)
4. M Kumar, EO Morales, JE Adams, W Hao, in *Image Processing (ICIP), 2009 16th IEEE International Conference On*. New , digital camera sensor architecture for low light imaging (IEEE, Cairo, Egypt, 2009), pp. 2681–2684
5. J Wang, C Zhang, P Hao, in *Image Processing (ICIP), 2011 18th IEEE International Conference On*. New color filter arrays of high light sensitivity and high demosaicking performance (IEEE, Brussels, Belgium, 2011), pp. 3153–3156
6. JT Compton, JF Hamilton, Image sensor with improved light sensitivity. Mar. 20. U.S. Patent 8,139,130 (2012)
7. ON Semiconductor. TRUESENSE Sparse ColorFilter Pattern. Application Notes (AND9180/D), Sep. (2014). http://www.onsemi.com/pub_link/Collateral/AND9180-D.PDF
8. T Komatsu, T Saito, in *Proc. SPIE, Sensors and Camera Systems for Scientific, Industrial, and Digital Photography Applications IV*. Color image acquisition method using color filter arrays occupying overlapped color spaces, vol. 5017 (SPIE, Santa Clara, CA, USA, 2003), pp. 274–285
9. M Mlinar, B Keelan, Imaging systems with clear filter pixels. Sep. 19. U.S. Patent App. 13/736,768 (2013)
10. J Park, K Choe, J Cheon, G Han, A pseudo multiple capture cmos image sensor with rwb color filter array. J. Semicond. Technol. Sci. **6**, 270–274 (2006)
11. O Yadid-Pecht, ER Fossum, Wide intrascene dynamic range cmos aps using dual sampling. Electron Devices IEEE Trans. **44**(10), 1721–1723 (1997)
12. W Lu, Y-P Tan, Color filter array demosaicking: new method and performance measures. Image Process. IEEE Trans. **12**(10), 1194–1210 (2003)
13. BK Gunturk, J Glotzbach, Y Altunbasak, RW Schafer, RM Mersereau, Demosaicking: color filter array interpolation. Signal Process. Mag. IEEE. **22**(1), 44–54 (2005)
14. X Li, B Gunturk, L Zhang, in *Proc. SPIE, Visual Communications and Image Processing 2008*. Image demosaicing: a systematic survey, vol. 6822 (SPIE, San Jose, CA, USA, 2008), pp. 68221–6822115
15. D Menon, G Calvagno, Color image demosaicking: An overview. Image Commun. **26**(8-9), 518–533 (2011)
16. I Pekkucuksen, Y Altunbasak, Multiscale gradients-based color filter array interpolation. Image Process. IEEE Trans. **22**(1), 157–165 (2013)
17. S-L Chen, E-D Ma, Vlsi implementation of an adaptive edge-enhanced color interpolation processor for real-time video applications. Circ. Syst. Video Technol. IEEE Trans. **24**(11), 1982–1991 (2014)
18. L Wang, G Jeon, Bayer pattern cfa demosaicking based on multi-directional weighted interpolation and guided filter. Signal Process. Lett. IEEE. **22**(11), 2083–2087 (2015)
19. Y Li, P Hao, Z Lin, Color Filter Arrays: A Design Methodology. Research report (RR-08-03), Dept of Computer Science, Queen Mary, University of London (2008). https://core.ac.uk/download/files/145/21174373.pdf
20. Image Sensors World. Samsung Announces 8MP RWB ISOCELL Sensor (2015). http://image-sensors-world.blogspot.kr/2015/03/samsung-announces-8mp-rwb-isocell-sensor.html
21. K He, J Sun, X Tang, Guided image filtering. Pattern Anal. Mach. Intell. IEEE Trans. **35**(6), 1397–1409 (2013)
22. SW Park, MG Kang, Channel correlated refinement for color interpolation with quincuncial patterns containing the white channel. Digital Signal Process. **23**(5), 1363–1389 (2013)
23. T Mertens, J Kautz, F Van Reeth, in *Computer Graphics and Applications, 2007. PG '07. 15th Pacific Conference On*. Exposure fusion (IEEE, Maui, Hawaii, USA, 2007), pp. 382–390
24. MK Park, MG Kang, K Nam, SG Oh, New edge dependent deinterlacing algorithm based on horizontal edge pattern. Consum. Electron. IEEE Trans. **49**(4), 1508–1512 (2003)
25. RWG Hunt, MR Pointer, *Measuring Colour (4th Edition)*. (John Wiley Sons, Ltd, Chichester, UK, 2011)

15

Feature extraction of SAR scattering centers using M-RANSAC and STFRFT-based algorithm

Hui Sheng, Yesheng Gao[†*], Bingqi Zhu, Kaizhi Wang and Xingzhao Liu[†]

Abstract

This paper introduces a modified random sample consensus (M-RANSAC) and short-time fractional Fourier transform (STFRFT)-based algorithm for feature extraction of synthetic aperture radar (SAR) scattering centers. In this algorithm, the range migration curve (RMC) of a scattering center is formulated as a parametric model. By estimating these parameters, the backscattering envelope of scattering center, corresponding to the backscattering variation in synthetic aperture time, is extracted directly from a time-domain range-compressed signal. The estimated parameters can also reconstruct the geographical location and along-track velocity of scattering centers. Thus, even without knowing explicit knowledge of platform velocity and forming a SAR image, this algorithm is capable of realizing feature extraction. To estimate parameters scatter by scatter, M-RANSAC approach is proposed as an implementary method with iterative procedure. In the iterations, fitting precision indicator (FPI) works cooperatively with construction fitness coefficient (CFC) to determine the optimal parameters of different scattering centers. Adapting this method to more general cases, STFRFT is introduced to separate the overlapped trajectories of RMCs of scattering centers. The root mean squared errors (RMSEs) of parameter estimation are close to their Cramér-Rao lower bounds (CRLB). The effectiveness of feature extraction based on the devised algorithm is validated by both simulated and real SAR data.

Keywords: SAR, M-RANSAC, STFRFT, Parameter estimation, Feature extraction

1 Introduction

Feature extraction has confirmed its usage in synthetic aperture radar (SAR) target recognition and classification, where a given target is classified as a specific target type by feature matching over the known database [1–5]. In fact, the high-frequency scattering response of a target is well approximated as a sum of response from individual scattering centers [6]. The attributes of these scattering centers, including scattering mechanism, location, and velocity, are physically relevant to those of the target [7]. Thus, to characterize target properties, feature extraction of corresponding scattering centers is a meaningful approach.

Interested attributes for each scattering center generally include backscattering envelope, geographical location,

and the relative velocity between radar platform and scattering center. Backscattering envelope indicates the backscattering variation of a scattering center within synthetic aperture time. Illuminated by radar signals, some targets, like metallic surfaces, have a very directive backscattering pattern or can be sensitive only to a singular frequency (anisotropic scatters or dihedral corner reflectors). Oppositely, some targets like trihedral corner reflectors have isotropic patterns. It leads to a stable backscattering during the acquisition. Therefore, the backscattering envelope can be the feature of major concern to characterize target properties, especially when a wide-angle SAR is operated [8]. Moreover, the geographical location and relative velocity are equivalently important, since the location denotes the cross-track and along-track positions while the relative velocity reflects the along-track speed.

To extract the attributes of scattering centers, a family of time-frequency analysis (TFA) approaches has been devised. They use Wigner-Ville decomposition [9],

---

*Correspondence: ysgao@sjtu.edu.cn
†Contributed equally
School of Electronic Information and Electrical Engineering, Shanghai Jiao Tong University, 800 Dongchuan Road, 200240 Shanghai, China

wavelet transforms [10], and Fourier transform [8, 11] to realize feature extraction. Starting with spectrum of SAR imagery, these methods are constrained with knowing explicit knowledge of platform velocity and forming a SAR image first. Free from SAR image formation, another group of approaches can directly extract the feature from the spectrum of raw data. These methods rely on spectral estimation and include parametric [12–14], nonparametric [15–17], and semi-parametric approaches [18]. However, sometimes, the spectrum may wrap around azimuth frequency as a result of ambiguity [19]. Since the aforementioned methods start with the spectrum, it may degrade the effectiveness of feature extraction.

In this paper, we propose an innovative algorithm to realizes feature extraction. Starting with a time-domain range-compressed signal, this algorithm establishes its main contribution as the signal-level ambiguity-free feature extraction of scattering centers. The realization of feature extraction without knowing explicit knowledge of platform velocity and forming a SAR image provides additional novelty of this algorithm. The procedure of this algorithm is detailed as follows. First, a parametric model is presented to describe the range migration curve (RMC) of scattering center in a range-compressed signal of SAR raw data. Then, using the points extracted from the contour of the range-compressed signal, an modified random sample consensus (M-RANSAC)-based algorithm is developed to estimate the parameters scatter by scatter. Within the method, fitting precision indicator (FPI) works cooperatively with construction fitness coefficient (CFC) to determine the optimal parameters of different scattering centers through iterations. Given the estimated parameters, the backscattering envelopes can be extracted from the range-compressed signal. Along with the backscattering envelopes, geographical location and relative velocity can also be reconstructed. However, the performance of M-RANSAC-based algorithm may be degraded when the trajectories of RMCs are overlapped in the range-compressed signal. To guarantee the effectiveness in more general cases, a trajectories separation method based on STFRFT [20] is proposed, further improving the M-RANSAC-based algorithm in feature extraction.

This paper is organized as follows. Section 2 reviews the mathematical expression of received signal and models the RMC of scattering center. Section 3 describes the M-RANSAC-based algorithm for feature extraction of SAR scattering centers. Section 4 introduces a STFRFT-based trajectories separation method. An enhanced M-RANSAC algorithm embedded with this STFRFT-based method is also detailed in this section. Section 5 discusses the root mean squared error and Cramér-Rao bounds of the parameter estimation. Section 6 presents the experimental results to validate the performance of the algorithm in feature extraction and demonstrates the usage of extracted feature in target recognition and classification. In the end, Section 7 concludes this paper.

2 Mathematical model

The demodulated received signal is the superposition of those of multiple scattering centers, the expression can be written as:

$$s(\tau, \eta) = \sum_{i=1}^{M} \sigma_i (\eta - \zeta_i) w_r \left(\tau - \frac{2R_i(\eta)}{c} \right) w_a (\eta - \zeta_i)$$

$$\times \exp \left\{ -j \frac{4\pi R_i(\eta)}{\lambda} \right\} \exp \left\{ j\pi k_r \left(\tau - \frac{2R_i(\eta)}{c} \right)^2 \right\}$$

(1)

in which M is the number of overall scattering centers in the illuminated scene, τ and η represent the fast time and slow time, respectively, c is the speed of light, k_r stands for frequency modulation (FM) rate of the transmitted chirp signal, and λ is the carrier wavelength. w_r denotes the range envelope which is usually considered as a rectangle function for chirp signal, and w_a means the azimuth beam pattern which is normally a sin-squared function. ζ_i, R_i, and σ_i are defined as the ith scattering center's beam center time, the instantaneous slant range, and the complex backscattering envelope, respectively. After matched filtering in range direction, the range-compressed signal of (1) can be expressed as:

$$s_{rc}(\tau, \eta) = \sum_{i=1}^{M} \sigma_i(\eta - \zeta_i) p_r \left\{ \tau - \frac{2R_i(\eta)}{c} \right\} w_a(\eta - \zeta_i)$$

$$\times \exp \left\{ -j \frac{4\pi R_i(\eta)}{\lambda} \right\}.$$

(2)

Here, $p_r \left\{ \tau - \frac{2R_i(\eta)}{c} \right\}$ is a sinc function. For a single pulse, the peak locates at $\frac{2R_i(\eta)}{c}$. The locations of these peaks decide the trajectory of the RMC during synthetic aperture time. To implement RMC fitting, $R_i(\eta)$ which indicates the instantaneous slant range between antenna phase center (APC) and the scattering center should be well understood. As show in Fig. 1, $R_i(\eta)$ can be formulated as:

$$R_i(\eta) = \sqrt{R_{0i}^2 + V_{ri}^2(\eta - \eta_{0i})^2},$$

(3)

where the ith scattering center has the nearest slant range R_{0i} at the time η_{0i} and relative velocity V_{ri} between APC and itself. Consider some scattering centers may be moving target, $V_{ri} = V - v_{ai}$ may not be the same as platform speed V (see Fig. 1). To simplify the further derivation, $R_i(\eta)$ is approximated with Taylor's series. In squint mode, $R_i(\eta)$ should be expanded at ζ_i rather than η_{oi}. Let $\eta_{ci} =$

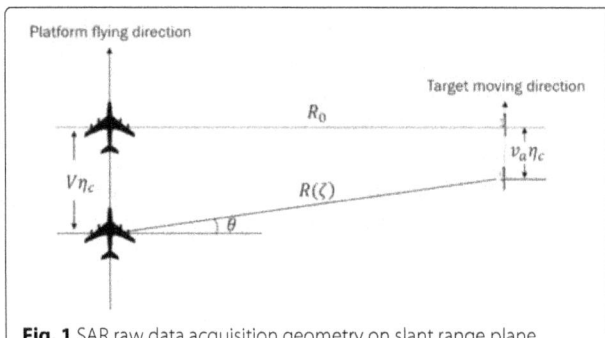

Fig. 1 SAR raw data acquisition geometry on slant range plane

$R_{0i} \tan \theta / V_{ri}$ be the offset between zero doppler time η_{oi} and beam center time ζ_i, which yields:

$$\zeta_i = \eta_{0i} - \eta_{ci}. \tag{4}$$

(3) is expanded by Taylor's series at ζ_i:

$$R_i(\eta) = R(\zeta_i) - V_{ri}\sin\theta(\eta - \zeta_i) + \frac{V_{ri}^2\cos^2\theta}{2R(\zeta_i)}(\eta - \zeta_i)^2. \tag{5}$$

Since we assume both the exposure time and the squint angle are moderate, the terms up to quadratic order in (5) are sufficient to model a RMC precisely.

Define a new coordinate:

$$\vec{\psi} = (X, Y)^T = \left(\frac{2\gamma R_i(\eta_n)}{c}, \eta_n \right), \tag{6}$$

in which the subscript n represents discrete sampling and γ scales fast time $2R_i(\eta_n)/c$ to a similar scale of magnitude of slow time η_n. Here, $\gamma = \text{PRI} \cdot f_s$ is decided by range sampling frequency f_s and pulse repetition interval PRI. For convenience, we let $\vartheta = c/2\gamma$. Together with (5), the discrete version of the ith scattering center's RMC can be modeled as:

$$X = A_i Y^2 + B_i Y + C_i \tag{7}$$

where

$$A_i = \frac{V_{ri}^2\cos^2\theta}{2\vartheta R_i(\zeta_i)}$$

$$B_i = \frac{-V_{ri}R_i(\zeta_i)\sin\theta - V_{ri}^2\zeta_i\cos^2\theta}{\vartheta R_i(\zeta_i)} \tag{8}$$

$$C_i = \frac{V_{ri}^2\zeta_i^2\cos^2\theta + 2R_i(\zeta_i)V_{ri}\zeta_i\sin\theta + 2R_i^2(\zeta_i)}{2\vartheta R_i(\zeta_i)}$$

In this proposed algorithm, $\vec{\mu} = \{A, B, C\}$, which parameterizes the RMC of an individual scattering center, is estimated scatter by scatter. Applying the estimated $\vec{\mu}$, the backscattering envelope σ can be extracted from range-compressed signal. Along with it, the geographical information R_0 and η_0 and the relative velocity V_r will be reconstructed. The process will be detailed in the next section.

3 M-RANSAC-based feature extraction algorithm

The proposed algorithm is an iterative method to estimate $\vec{\mu}$ of different scattering centers through fitting their RMCs. Then, the estimated parameters will be used to realize feature extraction . As shown in Fig. 2, this algorithm consists of two major steps: parameter estimation and feature extraction.

In the step of parameter estimation, the observed data is extracted from the contour of the range-compressed signal. It is a mix set of "inliers" and "outliers", indicating the trajectories of RMCs. The inliers can be explained by the parameter set $\vec{\mu}$ of current scattering center, while the outliers do not fit the model and may come from other scattering centers' RMCs or noise.

To separate the inliers from the outliers and obtain the current optimal fitting RMC with parameterized representation $\vec{\mu}$, RMC construction and performance measure are implemented iteratively in this algorithm. The iterative procedure of M-RANSAC-based approach continues until the points within observed data set are classified according to their corresponding RMCs, thus scattering centers. Along with the classified points, the overall number of scattering centers M and a set of $\vec{\mu}$ corresponding to different scattering centers are obtained.

Then, the step of feature extraction starts with these classified points and the estimated $\vec{\mu}$. The location and relative velocity of scattering centers can be directly reconstructed by $\vec{\mu}$. The backscattering scattering envelopes will be extracted from the range-compressed signal. Thus, feature extraction of M scattering centers are accomplished. The details of parameter estimation and feature extraction are summarized in the following subsections.

3.1 RMC construction with hypothetical inliers

Within a single iteration, a subset of observed data is randomly selected to construct a candidate RMC with the parametric representation $\vec{\mu}_c$. However, the observed data, which is directly extracted from the contour of the range-compressed signal, is sampled by sampling frequency f_s and pulse repetition frequency $1/\text{PRI}$. Thus, the original coordinate of the observed data $\vec{x} = (x, y)^T$ obviously differs from the new coordinate $\vec{\psi} = (X, Y)^T$ in (6). To locate the points of subset in the new coordinate system, a coordinate transformation should be processed first. The mapping relationship is expressed as:

$$\vec{\psi} = K\vec{x} + O, \tag{9}$$

where

$$K = \begin{bmatrix} \frac{c}{2\vartheta} & 0 \\ 0 & \text{PRI} \end{bmatrix} \quad \text{and} \quad O = \begin{bmatrix} \frac{R_s}{\vartheta} \\ \eta_s \end{bmatrix}. \tag{10}$$

Here, η_s and R_s are the minimum slow time and slant range of the given raw data, respectively.

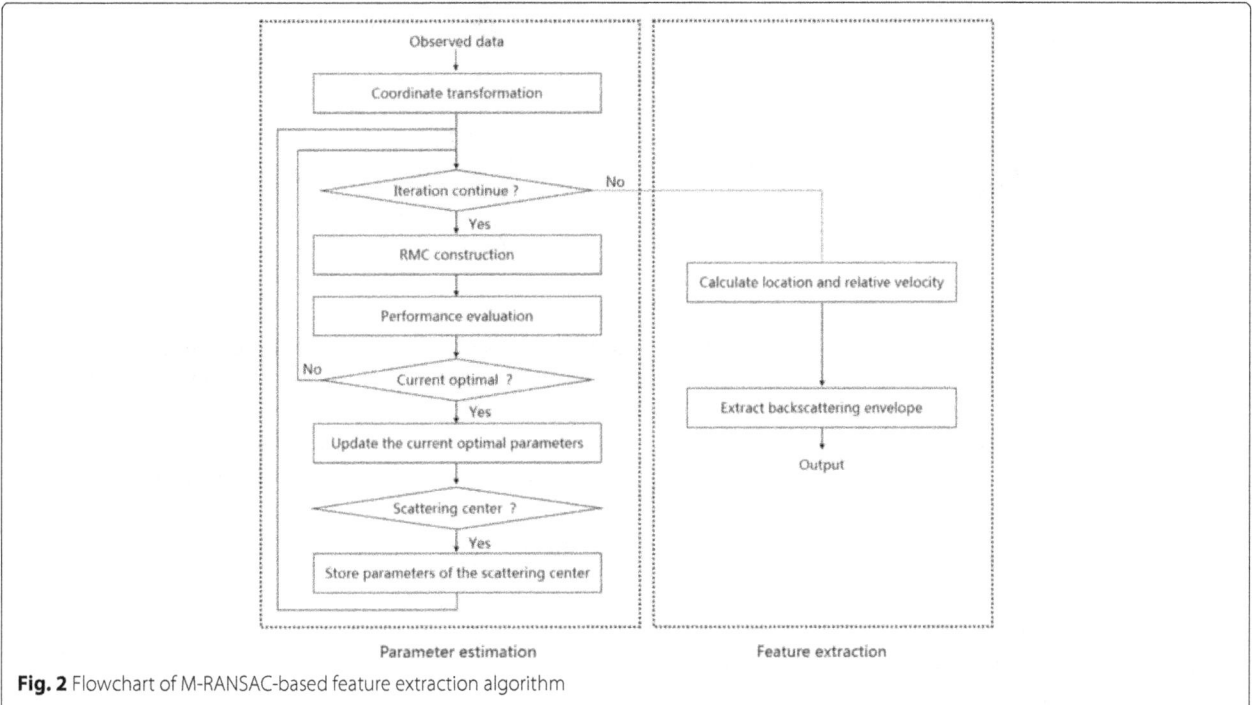

Fig. 2 Flowchart of M-RANSAC-based feature extraction algorithm

After coordinate transformation, the subset data in the new coordinate system is qualified for RMC construction. Since the degree of freedom (DOF) of (7) is three, a subset with $\vec{\psi}_1$, $\vec{\psi}_2$, and $\vec{\psi}_3$ is sufficient to calculate corresponding model parameters. $\vec{\mu}_c$ of this candidate RMC is therefore computed by:

$$A_c = \frac{X_{12}Y_{23} - X_{23}Y_{12}}{X_{23}Y_{12}^* - Y_{12}Y_{23}^*}$$

$$B_c = \frac{X_{23}Y_{12}^* - X_{12}Y_{23}^*}{Y_{23}Y_{12}^* - Y_{12}Y_{23}^*} \tag{11}$$

$$C_c = X_1 - A_c Y_1^2 - B_c Y_1$$

in which

$$X_{ij} = X_i - X_j \text{ and } X_{ij}^* = X_i^2 - X_j^2 \tag{12}$$

The accurate construction mainly depends on the accuracy of the selected points to solve (11). Only when $\vec{\psi}_1$, $\vec{\psi}_2$, and $\vec{\psi}_3$ come from the same RMC, this constructed $\vec{\mu}_c$ can be the parametric representation of a scattering center. However, the randomly chosen points might belong to different RMCs or be just noise points. Therefore, to assess the performance of this constructed RMC, a measure needs to be established in the iterations.

3.2 Performance measure establishment based on quadratic orthogonal distance

In this subsection, a double-measure system is developed to evaluate the performance of a candidate RMC. To deal with the situation that selected points come from different RMCs or are noise points, CFC, which denotes the number of points in observed data set can be explained by the candidate RMC with $\vec{\mu}_c$, is introduced. Another measure, called FPI, is proposed to assure that a more precise RMC will be chosen when two candidate RMCs share the same CFC.

To judge whether a point can be explained by the candidate RMC, quadratic orthogonal distance (QOD) between a point and a curve is developed. The qualification of a point is decided by its QOD to the RMC with a specific threshold value. Other than the least-squares distance, QOD is defined as the minimum connecting length from a point to the given curve, which is more precise in practical applications [21]. To obtain this distance, the geometric feature of RMC is further analyzed. After coordinate transformation, RMC can be considered as a parabola with the vertex at $\vec{\psi}_{ct} = \left(C - \frac{B^2}{4A}, -\frac{B}{2A}\right)$. According to (7), the mathematical expression of RMC can be written as:

$$X - \left(C - \frac{B^2}{4A}\right) = A\left(Y + \frac{B}{2A}\right)^2. \tag{13}$$

As shown in Fig. 3, $\vec{\psi}_g$ is a point in the observed data. To calculate the QOD between $\vec{\psi}_g$ and a candidate RMC parameterized with $\vec{\mu}_c$, $\vec{\psi}_p$, which is the closest projection of $\vec{\psi}_g$ on the RMC, should be located. Define $\vec{\psi}_g = \left(X_g, Y_g\right)$ and $\vec{\psi}_p = \left(X_p, Y_p\right)$. Since $\vec{\psi}_p$ locates on the RMC, an equation can be obtained:

$$f_1\left(X_p, Y_p\right) = A_c Y_p^2 + B_c Y_p + C_c - X_p = 0 \tag{14}$$

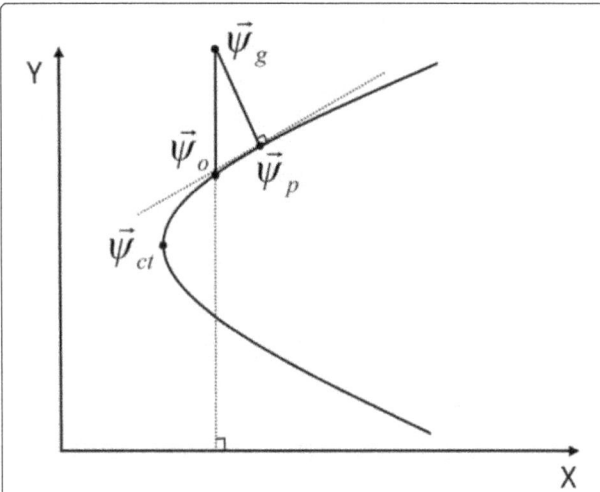

Fig. 3 Quadratic orthogonal distance to a curve

Moreover, the connecting line of $\vec{\psi}_g$ and $\vec{\psi}_p$ is perpendicular to the tangent line of the RMC at the point $\vec{\psi}_p$. This relationship can be formulated as:

$$\frac{dY}{dX} \cdot \frac{Y_g - Y_p}{X_g - X_p} = \frac{1}{2A_c Y_p + B_c} \cdot \frac{Y_g - Y_p}{X_g - X_p} = -1. \quad (15)$$

Rewriting the above equation, it yields:

$$f_2(X_p, Y_p) = (Y_g - Y_p) + (X_g - X_p)(2A_c Y_p + B_c) = 0. \quad (16)$$

Combining (14) and (16) into a quartic equation will result in maximum four solutions. Generally, the solution with the minimum geometric distance to $\vec{\psi}_g$ is chosen as the optimal projection $\vec{\psi}_p$. However, this numerical method is not stable. To optimize the calculation, Ahn [21] proposes a generalized Newton-Raphson method to locate the closest projection point. It is an efficient iterative method which converges quickly. Given the functions $f_1(X_p, Y_p)$ and $f_2(X_p, Y_p)$, we define the derivative matrix D, the current approximate result $\vec{\psi}_k$, and a more accurate approximation $\vec{\psi}_{k+1}$ to compute $\vec{\psi}_p$. The process of iterations can be presented as:

$$D\Delta\vec{\psi} = -f(\vec{\psi}_k)$$
$$\vec{\psi}_{k+1} = \vec{\psi}_k + \Delta\vec{\psi} \quad (17)$$

where

$$D = \begin{pmatrix} \frac{\partial f_1}{\partial X_p} & \frac{\partial f_1}{\partial Y_p} \\ \frac{\partial f_2}{\partial X_p} & \frac{\partial f_2}{\partial Y_p} \end{pmatrix} = \begin{pmatrix} -1 & 2A_c Y_p + B_c \\ -2A_c Y_p - B_c & -1 + 2A_c(X_g - X_p) \end{pmatrix} \quad (18)$$

An initial value $\vec{\psi}_0$ is given in Fig. 3, and its expression is:

$$\vec{\psi}_0 = \begin{cases} \begin{pmatrix} C_c - \frac{B_c^2}{4A_c} \\ -\frac{B_c}{2A_c} \end{pmatrix} & \text{if } X_g \le C_c - \frac{B_c^2}{4A_c} \\ \begin{pmatrix} X_g \\ \text{sign}\left(Y_g + \frac{B_c}{2A_c}\right)\sqrt{\frac{X_g + \frac{B_c^2}{4A_c} - C_c}{A_c}} - \frac{B_c}{2A_c} \end{pmatrix} & \text{if } X_g > C_c - \frac{B_c^2}{4A_c} \end{cases} \quad (19)$$

Equation 17 starts its iteration with the initial value $\vec{\psi}_0$ and ends when $|\Delta\vec{\psi}|$ is no more than a given threshold. During the iterations, parameters related to $\vec{\psi}_p$ in (18) are assigned with the value of the current iterative result $\vec{\psi}_k$, and the final closest projection $\vec{\psi}_p$ is set to be $\vec{\psi}_k$ when the iterations end. Therefore, we define the square Euclidean distance between $\vec{\psi}_p$ and $\vec{\psi}_g$ as the QOD of $\vec{\psi}_g$:

$$\text{rho} = |\vec{\psi}_g - \vec{\psi}_p|^2. \quad (20)$$

When rho of $\vec{\psi}_g$ stays no more than the given threshold rho_thr, this point is regarded as an inlier, otherwise an outlier. The overall number of inliers $N(\vec{\mu}_c)$ within the observed data is denominated as CFC. As shown in Fig. 4, this measure utilizes the number of inliers to define the fitting degree of the candidate RMC. To evaluate the degree of matching between the inliers and the candidate RMC, FPI is introduced as:

$$\chi(\mu) = -\sum_{k=1}^{N} \varepsilon(\text{rho\_thr} - \text{rho}_k) \cdot \text{rho}_k. \quad (21)$$

in which ε means unit step function. FPI, which is the negative overall QOD of inliers, is known as the accuracy of fitting. It works cooperatively with CFC to locate the optimal candidate RMC with the largest number of inliers and best fitting precision. Conventional RANSAC-based algorithm [22] only considers CFC as measure without applying weighting for inliers' QOD and the maximum likelihood estimation sample consensus (MLESAC)-based method [23] obtains the overall error with a computationally complicated process. They fail in either accuracy or efficiency. The double-measure system of CFC and FPI in this algorithm steps out of this dilemma and achieves a balance between precision and efficiency.

As shown in Fig. 5, the comparative tests are conducted to evaluate this double-measure system. In the test, each set of observed data contains 100 inliers and 200 outliers (see Fig. 5a). Those inliers can be explained by a RMC with the parameter set $\vec{\mu} = \{0.2, 0, 4\}$, while the outliers do not fit the RMC and surpass the QOD threshold rho_thr = 0.5 to this RMC. There exists 100 sets of observed data in all. Among them, the inliers are fixed while the outliers are randomly generated which may

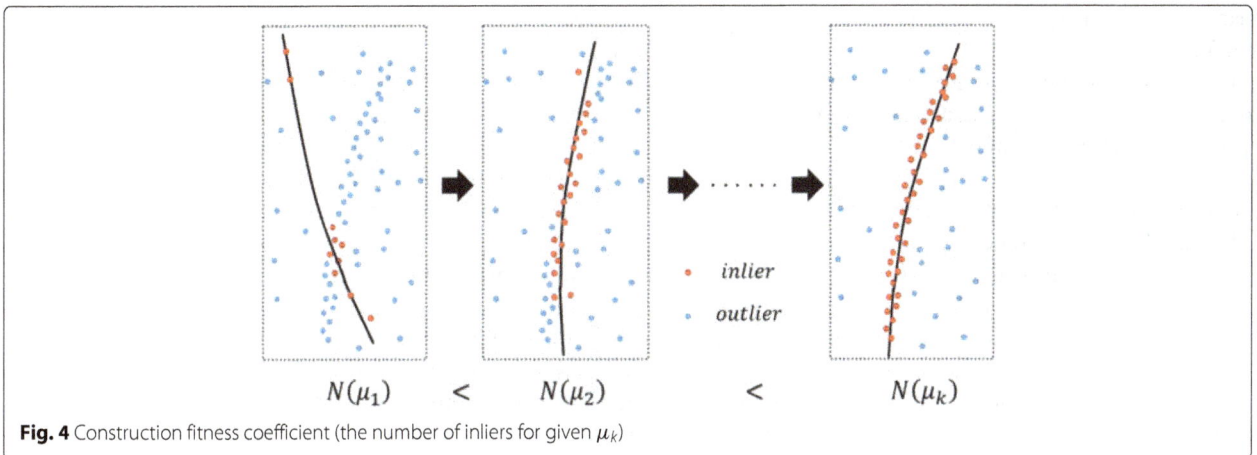

Fig. 4 Construction fitness coefficient (the number of inliers for given μ_k)

change from set to set. Based on the observed sets, 100 Monte-Carlo tests are conducted with iterative times 50, 100, 150 and 200. First, the average Euclidean norm of parameter set $\vec{\mu}$'s estimation error $\left\| \Delta \hat{\mu} \right\|$ versus iterative times are provided when the RMC is fitted by three different algorithms respectively. As shown in Fig. 5b, both the estimation accuracy and the convergence rate of MLESAC and M-RANSAC methods outperform those processed by RANSAC method. And, the estimation accuracy of the M-RANSAC method is surpassed by that of the MLESAC

method. However, the superiority of MLESAC's performance costs heavy computational burden. In this paper, the computational times on a desktop PC (i5-3210M CPU at 2.5 GHz and DDR3 RAM at 8 GB) corresponding to iterative times are listed in Fig. 5c. According to this figure, we can tell the computational efficiency of MLESAC is obviously surpassed by those of the M-RANSAC and RANSAC methods. Therefore, to balance the accuracy and efficiency, the application of M-RANSAC algorithm may be the optimal choice for RMC fitting, and the

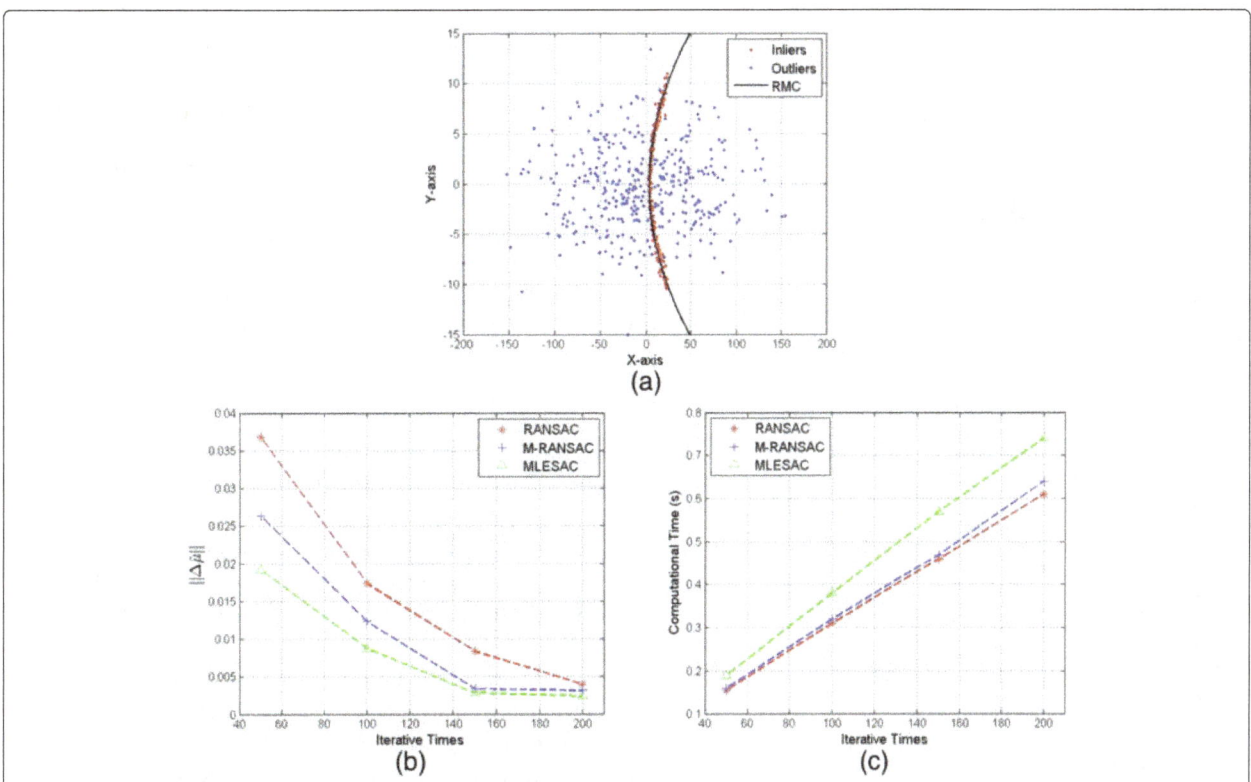

Fig. 5 Performance evaluation of the double-measure system in M-RANSAC algorithm. **a** The set of observed data. **b** Estimation accuracy compared with other measures. **c** Computational time compared with other measures

superior performance of this double-measure system is validated.

In the next subsection, M-RANSAC approach will integrate the measures and RMC construction into the iterative process of parameter estimation.

3.3 Iterative procedure of parameter estimation in M-RANSAC-based approach

The iterative process of proposed algorithm starts with a set of observed data D_set. It contains the points extracted from the contour of range compressed results and indicates the RMC trajectories of different scattering centers. This set also inevitably contains many noise points introduced by undesired background information. M-RANSAC-based algorithm is proposed to classify the groups of points corresponding to different scattering centers, get rid of existing noise, and realize parameter estimation of these RMCs simultaneously. The pseudodocode of this process is displayed in Fig. 6. It consists of iterations of two levels: point-level iterations and scatter-level iterations. In the point-level iterations, a parametric RMC with optimal CFC and FPI is located to explain a scattering center. Then, the scatter-level iterations will lock all these RMCs in a range-compressed signal scatter by scatter.

```
Given:
   D_set     – the set of observed data
   min, max  – the minimum and maximum iteration times
   m         – maximum non-updating times
   rho_thr   – the threshold QOD to define an inlier
   N_thr     – the minimum number of inliers to confirm a RMC
   CONT      – iteration proceeding factor
Output:
   M  – the number of scattering center
   set  – groups of classified points according to scattering centers they belong to
   setμ – the set containing the parametrized representation of fitted RMCs
while CONT==1 {
   Let M, iteration, and bestN = 0
   Let non_upd = infinity
   while iteration ≤ min || (iteration ≤ max && non_upd ≤ m){
      iteration+1, non_upd+1
      μ̄_c = RMC construction with 3 randomly selected points from D_set
      Calculate rho for every point in D_set to obtain an inlier data set set(μ̄_c)
         N(μ̄_c) = construction fitness coefficient with μ̄_c
         χ(μ̄_c) = fitting rank indicator with μ̄_c
      if (N(μ̄_c) ≥ N_thr && N(μ̄_c) > BestN ) || (N(μ̄_c) == BestN && χ(μ̄_c) > best χ){
         Define a current optimal fitting RMC, and update judgment conditions:
            non_upd = 0
            Bestμ   = μ̄_c
            BestN   = N(μ̄_c)
            Bestχ   = χ(μ̄_c)
            Bestset = set(μ̄_c)
            recalculate m
      }
   }
   CONT = 0
   if BestN ≥ N_thr {
      Confirm a new scattering center and store the relevant information:
         M+1
         setμ(M)  = Bestμ
         set(M)   = Bestset
      Iteration should continue to find another scattering center:
         CONT = 1
         clear Bestset from D_set
   }
}
return M, setμ and set
```

Fig. 6 The pseudocode of parameter estimation in M-RANSAC-based approach

To begin with, the minimum iterative times min, maximum iterative times max, the threshold QOD to define an inlier rho_thr and the threshold number of inliers to confirm a scattering center N_thr should be preestablished. What is more, we should initialize maximum non-updating times m to infinite, the number of scattering center M to zero, and the iteration proceeding factor CONT to 1.

A point-level iteration starts with randomly selecting three points from D_set to construct a candidate RMC. The $\vec{\mu}_c$ of this RMC is computed with (9) and (11). Then, according to subsection 3.2, the QOD between every point in the D_set and this candidate RMC are calculated and denoted by rho. The points whose rho stay no more than rho_thr are defined as inliers and stored in set($\vec{\mu}_c$). Then, CFC $N(\vec{\mu}_c)$ and FPI $\chi(\vec{\mu}_c)$ of this candidate RMC can be computed by (20) and (21).

This RMC can be regarded as the current optimal one in two cases. The CFC $N(\vec{\mu}_c)$ exceeds that of the former optimal $BestN$, or the FPI $\chi(\vec{\mu}_c)$ goes over that of former optimal $Best\chi$ under the circumstance that $N(\vec{\mu}_c)$ equals $BestN$. When the conditions are satisfied and the current optimal is renewed, not only $Best\mu$, $BestN$, $Best\chi$, and $Bestset$ are updated in line with the values of current optimal RMC but also the maximum non-updating times m will be recalculated. The point-level iteration stops when iterative times $iteration$ exceed max or non-updating times non_upd surpass m. An additional minimum iteration times min is used to remain the stability.

In scatter-level iterations, a new scattering center will be confirmed when the output of the point-level iterations $BestN$ goes beyond N_thr. At this time, the number of recovered scattering centers M is updated. $Best\mu$ and $Bestset$ are stored in $set\mu(M)$ and $set(M)$. The idea of CLEAN technique [24, 25] are taken, and the points in $Bestset$ will be subtracted from D_set. Another point-level iterations will be processed to locate the next RMC. Oppositely, if the point-level iteration fails to locate a scattering center, the remaining points in observed data set are considered as noise points. Thus, the scatter-level iterations stop by setting CONT $= 0$.

After the two-level iterations, the total number of scattering centers M is determined, the points of inliers are classified in set, and the parametric representation μ of scattering centers are estimated and saved in $set\mu$. These data will help to realize feature extraction for dominant scattering centers of the targets in the next subsection.

3.4 Feature extraction based on estimated parameters

To realize feature extraction, a vector $\vec{T}_i$ is established to cover the interested information of the ith scattering center, $\vec{T}_i = \{ R_{0i}, \eta_{0i}, V_{ri}, \sigma_i(\eta - \xi_i) \}$. In this vector set, R_{0i} and η_{0i} present the geographical location in both cross-track direction and along-track direction. V_{ri}, which is the relative velocity between the radar platform and scattering center, indicates the possible along-track speed of the scatter. They can be reconstructed using the estimated $\vec{\mu}_i$ in $set\mu(i)$:

$$R_{oi} = \frac{\vartheta \left(4A_iC_i - B_i^2\right)\cos\theta}{2A_i \left(2 - \tan^2\theta\right)}$$

$$\eta_{0i} = -\frac{B_i}{2A_i} + \frac{\sin 2\theta - 2\tan\theta}{4A_i}\sqrt{\frac{4A_iC_i - B_i^2}{2 - \tan^2\theta}} \qquad (22)$$

$$V_{ri} = \vartheta\sqrt{\frac{4A_iC_i - B_i^2}{2\cos^2\theta - \sin^2\theta}}.$$

In broadside case, $\sin\theta$ and $\tan\theta$ equal zero. (22) degrades to a simpler formula:

$$R_{0i} = \vartheta \left\{C_i - \frac{B_i^2}{4A_i}\right\}$$

$$\eta_{0i} = -\frac{B_i}{2A_i} \qquad (23)$$

$$V_{ri} = \vartheta\sqrt{2A_iC_i - \frac{B_i^2}{2}}.$$

Knowing R_{0i}, η_{0i}, and V_{ri}, compressed range envelope p_r, azimuth beam pattern w_a, and the phase information related to instantaneous slant range R_i are calculated. By locating inliers $set(i)$ in the range-compressed signal, the complex values along the RMC of scattering center can be extracted. According to (2), we obtain the complex backscattering envelope $\sigma_i(\eta - \zeta_i)$ by eliminating the influence of the aforementioned components in the extracted complex values. The vector set $\left\{\vec{T}_1, \vec{T}_2, \ldots \vec{T}_M\right\}$ are calculated scatter by scatter.

It is worth noting that, the process of M-RANSAC-based algorithm does not need the explicit parameters (e.g., platform velocity). However, for conventional methods of feature extraction based on SAR image formation, the platform velocity works as a crucial parameter of realizing range cell migration correction (RCMC) and azimuth matched filtering. Thus, compared with the conventional approach, M-RANSAC-based algorithm can be utilized in a more flexible way. Moreover, the proposed algorithm extracts the features directly from a range-compressed signal. Without forming SAR image, we may realize target recognition and classification directly in a signal level rather than in an image level.

When platform velocity is known and SAR image is formed, the backscattering envelope extracted by the proposed algorithm may classify targets which are similar in the gray-level SAR image. Moreover, given the platform velocity and the relative velocity between radar platform

and the dominant scattering center, the along-track velocity of target can be computed. Thus, even if SAR image is formed, this feature extraction algorithm may help us to better understand the target.

4 Trajectories separation based on STFRFT

Sometimes, RMC of one scattering center may overlap that of the other. This phenomenon is called trajectories overlapping in this paper. As shown in Fig. 7a, the trajectories of T1 and T2 are mixed after range compression (to clearly state the principle, the range curve are ignored under the low-resolution assumption). In this case, M-RANSAC-based algorithm may fail in extracting interested information of T1 and T2, respectively. To solve this problem, short-time fractional Fourier transform (STFRFT) is applied to separate the overlapped trajectories. With a spatial filtering using a rectangle window, the trajectories of different scattering centers will be separated in time-fractional frequency domain. The feature extraction can be successfully proceeded afterwards.

The phase component in (2) can be written as:

$$s_\phi(\eta) = \sum_{i=1}^{M} \exp\left\{j\phi_i + j2\pi f_i\eta - j\pi K_a\eta^2\right\} \qquad (24)$$

where,

$$\phi_i = -\frac{4\pi}{\lambda}\left\{R(\zeta_i) - V_{ri}\sin\theta\zeta_i\right\} - \frac{\pi V_{ri}^2\cos^2\theta}{R(\zeta_i)}\zeta_{ri}^2$$

$$f_i = \frac{2V_{ri}\sin\theta}{\lambda} + \frac{V_{ri}^2\cos^2\theta}{R(\zeta_i)}\zeta_i \qquad (25)$$

$$K_a = \frac{V_{ri}^2\cos^2\theta}{R(\zeta_i)} \approx \frac{V_{ref}^2\cos^2\theta}{R_{ref}}.$$

Here, K_a is considered as a constant. It is a reasonable assumption when the range span of processed data is moderate. f_i, related to the beam center time ζ_i, differs with the azimuth location of scattering center. Taking STFRFT of (24), it yields:

$$\mathrm{STF}_\phi(\eta, u) = \sum_{i=1}^{M} \exp\left\{j\phi_i\right\}\int_{-\infty}^{\infty} \exp\left\{j2\pi f_i t - j\pi K_a t^2\right\}$$
$$\times g(t - \eta)K_p(t, u)dt. \qquad (26)$$

where, to obtain the optimal 2D resolution [20], a Gaussian window $g(t)$ is used:

$$g(t) = \left(\pi\xi^2\right)^{-1/4}\exp\left(-\frac{t^2}{2\xi^2},\right) \qquad (27)$$

in which $\xi^2 = |\sin\alpha/K_a|$, and $K_p(t, u)$ denotes the kernel function of fractional Fourier transform (FRFT) with an expression:

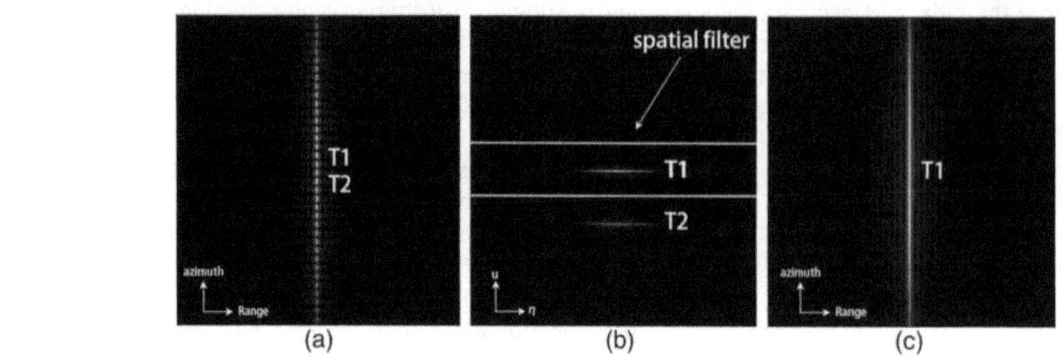

Fig. 7 Trajectories separation based on STFRFT. **a** Overlapped trajectories of T1 and T2. **b** Matched STFRFT and spatial filter of trajectories in selected range bin. **c** Separated trajectory of T1

$$K_p(t,u) = \begin{cases} \sqrt{\frac{1-j\cot\alpha}{2\pi}} \exp\left(j\frac{t^2+u^2}{2}\cot\alpha - jut\csc\alpha\right), \alpha \neq k\pi \\ \delta(u-t), \qquad \alpha = 2k\pi \\ \delta(u+t), \qquad \alpha = (2k+1)\pi \end{cases} \quad (28)$$

Based on the frequency shift property of STFRFT [20], (26) becomes:

$$\text{STF}_\phi(\eta,u) = \sum_{i=1}^{M} \exp\left\{j\phi_i + j2\pi uf_i\cos\alpha \right. \\ \left. -j2\pi f_i^2\sin\alpha\cos\alpha\right\}\text{STF}_{K_a}\left(\eta,u-2\pi f_i\sin\alpha\right) \quad (29)$$

where,

$$\text{STF}_{K_a}(\eta,u) = \int_{-\infty}^{\infty} \exp\left\{-j\pi K_a t^2\right\}g(t-\eta)K_p(t,u)dt. \quad (30)$$

Note that (30) is a matched STFRFT only when the second-order phase term is set to zero, thus, α is:

$$\alpha = \text{arc}\cot\left\{2\pi K_a \cdot N_a\text{PRI}^2\right\} \quad (31)$$

where, *PRI* is the pulse repetition interval and N_a denotes the length of azimuth sample. In the real application, $N_a\text{PRI}^2$ is used as the factor of coordinate transformation in digitalized computation [26, 27]. Back to (29), the STFRFT of an individual scattering center is decided by $\text{STF}_{K_a}(\eta,u)$. The matched STFRFT of (30) will locate the spectrogram line of a scattering center parallel to η. For multiple scattering centers, the shift $\Delta u = -2\pi f_i\sin\alpha$ along u axis in (29) separates their energy according to their different azimuth locations. As shown in Fig. 7b, a simple spatial filter using a rectangle window will separate the energy of one scattering center from the others. After inverse STFRFT, the trajectories of scattering centers with similar range position but different azimuth location are separated (e.g., Fig. 7c).

To realize trajectories separation in feature extaction, the STFRFT-based method is embedded in the aforementioned M-RANSAC-based approach. The processing steps can be summarized as follows:

1. Select the trajectory of an isolated scattering center and estimate K_a using the points extracted from it.
2. Calculate α based on (31).
3. Execute α-angle STFRFT for a range bin.
4. Implement spatial filtering using rectangle windows.
5. Realize trajectories separation with inverse STFRFT.
6. Repeat steps 3 to 5 until the last range bin is processed.
7. For every sub-patch range compressed signal, estimate M, $set\mu$, and set using M-RANSAC approach.
8. Use the estimated M, $set\mu$, and set to compute the vector set $\left\{\vec{T}_1, \vec{T}_2, \ldots \vec{T}_M\right\}$.

5 CRLB and RMSE of parameter estimation

The parameter estimation of $\vec{\mu}$ lays the foundation for feature extraction in this algorithm. In this section, Monte-Carlo tests are conducted to obtain the root mean squared errors (RMSEs) of the estimates. To evaluate the accuracy of estimation, these RMSEs of estimators compared their theoretical minimal errors, named Cramér-Rao lower bound (CRLB). We start this section with computing the CRLBs according to observation.

The observation can be derived from (2) and (7):

$$\chi_o[n] = \Gamma\left[n;\vec{\phi}_e\right] + \omega_0[n], n = 0, 1, 2, \ldots N_a - 1 \quad (32)$$

in which,

$$\Gamma[n;\vec{\phi}_e] = \sigma\{n\cdot\text{PRI}\}w_a\{n\cdot\text{PRI}\} \\ \exp\left\{-j\frac{4\pi\vartheta}{\lambda}\left(A\cdot\text{PRI}^2n^2 + B\cdot\text{PRI}\cdot n + C\right)\right\}. \quad (33)$$

Here, $\omega_0(n)$ denotes a complex white noise with zero mean and the variance of σ_0. The estimator vector $\vec{\phi}_e = [\hat{A}, \hat{B}, \hat{C}]$ contains three parameters waiting to be estimated.

According to [28], the Fisher information matrix of (32) can be calculated using the expression:

$$\left[I\left(\vec{\phi}_e \right) \right]_{ij} = \frac{1}{\sigma_o^2} \sum_{n=0}^{N_a-1} \frac{\partial \Gamma[n; \vec{\phi}_e]}{\partial [\vec{\phi}_e]_i} \frac{\partial \Gamma[n; \vec{\phi}_e]}{\partial [\vec{\phi}_e]_j} \quad (34)$$

The computation of its inverse is complicated and tedious, we omit the procedures of derivation and directly give the result:

$$\left[I(\vec{\phi}_e) \right]^{-1} = 3 \left(\frac{\sigma_0 \lambda}{4\pi \vartheta \Lambda} \right)^2$$
$$\begin{bmatrix} \frac{60}{\text{PRI}^4 \Theta_1} & \frac{60}{\text{PRI}^3 \Theta_2} & \frac{10}{\text{PRI}^2 \Theta_3} \\ \frac{60}{\text{PRI}^3 \Theta_2} & \frac{12(16N_a^2 - 30N_a + 11)}{\text{PRI}^2 \Theta_1} & \frac{-6(2N_a-1)}{\text{PRI}\Theta_3} \\ \frac{10}{\text{PRI}^2 \Theta_3} & \frac{-6(2N_a+1)}{\text{PRI}\Theta_3} & \frac{3N_a^2 - 3N_a + 2}{\Theta_3} \end{bmatrix}$$
$$(35)$$

where the Λ is the average amplitude of backscattering envelope $\sigma\{n \cdot PRI\}$ and azimuth beam pattern $w_a\{n \cdot PRI\}$. $\Theta_1 = N_a (N_a - 1)(N_a - 4)$, $\Theta_2 = N_a(-N_a^3 - N_a^2 + 4N_a + 4)$ and $\Theta_3 = N_a(N_a + 1)(N_a + 2)$. According to (35), the CRLBs of estimated parameters are the diagonal elements of inverse matrix. Thus, the CRLBs of $\hat{A}$, $\hat{B}$, and $\hat{C}$ are equal to $\left[I(\vec{\phi}_e) \right]^{-1}_{11}, \left[I(\vec{\phi}_e) \right]^{-1}_{22}$ and $\left[I(\vec{\phi}_e) \right]^{-1}_{33}$, respectively.

Then, 100 Monte-Carlo tests are conducted when signal-to-noise ratio (SNR) is $0, 2, 4, 6, 8$, and 10 dB, respectively. The experiments utilize the SAR simulation parameters listed in Table 1. To better exhibit the results, both RMSEs and CRLBs of the estimated parameters are expressed in decibels. As shown in Fig. 8, the RMSEs of the estimated values stay close to their CRLBs. Since CRLB is the theoretical lowest estimation error, we can conclude that the parameter estimation based on the proposed algorithm is accurate and effective. The precisely

estimated parameters guarantee the subsequent process of feature extraction.

6 Experimental results

To validate the performance of this algorithm, a series of experiments, named performance test, simulation test, and real data test, respectively, are presented in this section. In the performance test, we generate raw data of a single target when the broadside airborne SAR system operates. This raw data is added with various Gaussian white noise and then taken as the input of M-RANSAC algorithm to estimate the location and velocity of dominant scattering center. RMSE of the estimated parameters are listed corresponding to different input SNR and iterative times. Then, the scenario of multiple targets is considered in the simulation test. In the illuminated scene, three targets with different backscattering envelopes and along-track velocities are introduced. The features of dominant scattering centers are extracted from the generated data by M-RANSAC-based algorithm and compared with the theoretical ones. In addition, the potential usage of these extracted features in target recognition and classification are fully considered. In the end, the real data of RADARSAT-1 are processed to validate the performance of the proposed algorithm when the trajectories of targets are overlapped. The features of the dominant scattering centers, including locations, relative velocities, and backscattering envelopes, are extracted using both STFRFT-based trajectories separation and M-RANSAC-based feature extraction method. To verify the effectiveness of feature extraction, the reconstructed locations and velocities are compared with those obtained using conventional methods. To confirm the potential usage of these extracted features in target classification, the backscattering envelopes are used to further interpret the ships in English Bay which is located in the city of Vancouver, Canada (see Fig. 13a).

6.1 Performance test

In this subsection, raw data of a single target are simulated to evaluate the estimation accuracy. This target is a stationary one with the geographical location $R_0 = 7500$ m and $\eta_0 = 0.8717$ s. In the simulation, the system parameters are listed in the middle column of Table 1 and beam width in azimuth dimension is set to be 0.059 rad. The generated raw data are added with Gaussian white noise when the input SNR = $-10, -5, 0, 5$, and 10 dB. For each input SNR, we generate 150 sets of random Gaussian noise; thus, in total, 750 sets of observed raw data are obtained.

The M-RANSAC-based algorithm starts with those time-domain range compressed signals (see Fig. 9a). Its initial parameters are carefully designed. The scale factor

Table 1 System parameter of simulation and real data

Parameter	Simulation test	Real data experiment
Squint angle	0°	−1.584°
Signal bandwidth	150 MHz	30.111 MHz
Sample frequency	200 MHz	32.317 MHz
Pulse duration	5.12 μs	41.74 μs
PRI	1.7 ms	0.7956 ms
Platform velocity	153.3 m/s	7062 m/s
Range to scene center	7500 m	998263 m

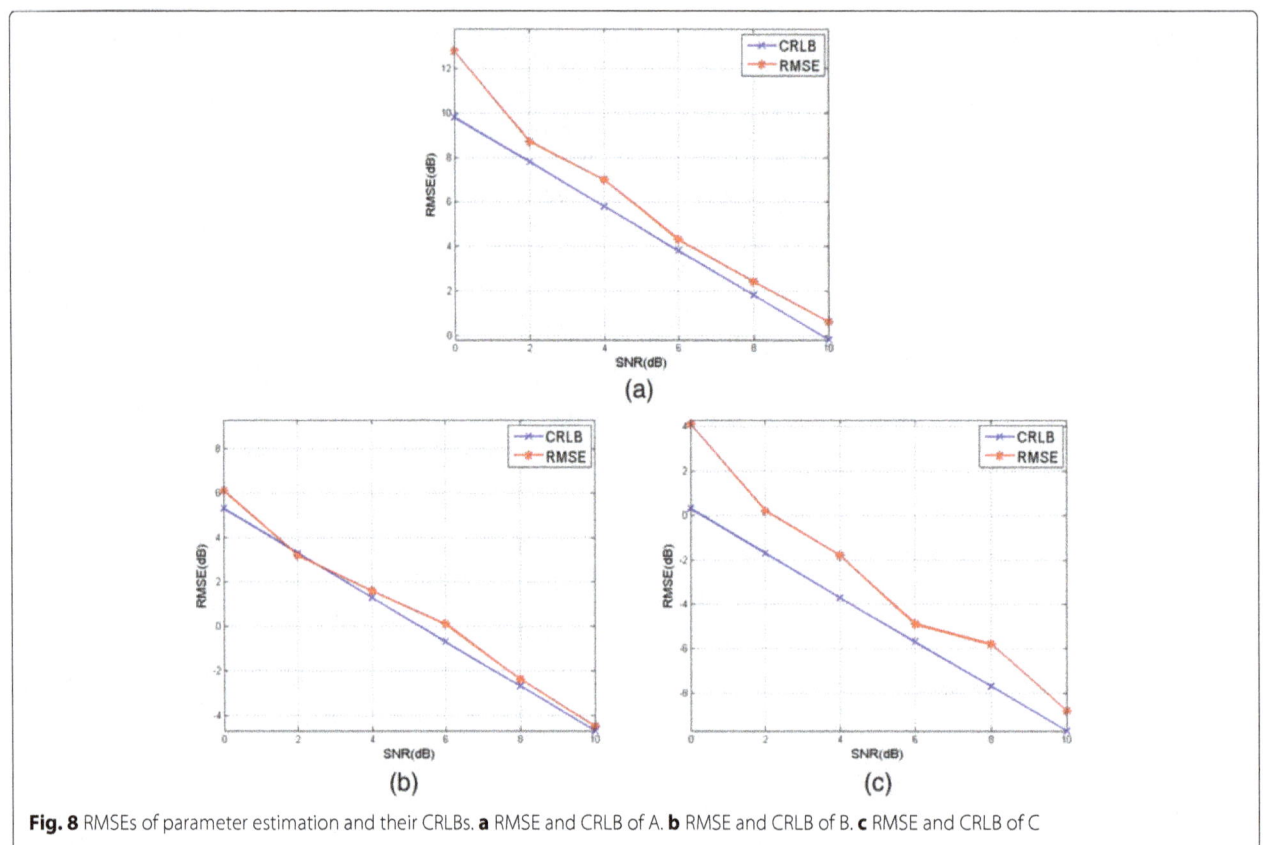

Fig. 8 RMSEs of parameter estimation and their CRLBs. **a** RMSE and CRLB of A. **b** RMSE and CRLB of B. **c** RMSE and CRLB of C

ϑ in (10) is set to be 450. The threshold of quadratic orthogonal distance rho_thr is 0.003 and the threshold number of inliers N_thr is $0.85N_a$. The overall iterative times are set to be a fixed number $T_{\text{iter}} = 50, 100, 150,$ and 500. After estimating $\vec{\mu}$ for the M dominant scattering centers and classifying inliers in *set*, the geographical locations and velocity information are reconstructed based on (23). For each input SNR, 150 sets of observed raw data will output 150 sets of $\left\{ \hat{R}_0, \hat{\eta}_0, \hat{V}_r \right\}$. Compared with the theoretical ones, RMSE of $\{R_0, \eta_0, V_r\}$ corresponding to different input SNR and iterative times are obtained and shown in Fig. 9b–d. According to this figure, several conclusions can be made:

(1) The estimation errors of location and relative velocity are quite limited especially when input SNR is 5 and 10 dB. Thus, We can expect a high estimation accuracy in a high-SNR case.

(2) The estimation accuracy may decrease along with the input SNR. The reasonable explanation is that a higher-level noise will impact the precision of inliers in a larger degree and thus decrease the estimation accuracy. In the performance test, this phenomenon becomes obvious when the low-SNR data is implemented.

(3) The estimation accuracy will be continuously enhanced with the increasing of iterative times until it converges. When input SNR is high, the estimation error converges fast. We can expect a high-precision output with a small number of iterative numbers. However, under low-SNR scenario, the estimation error converges slowly. max can be considerably increased to obtain relatively high-accuracy estimators. Unfortunately, there exists no SNR-related closed-form expression of max. The initial parameter max is an empirical parameter in this paper.

6.2 Simulation test

To further analyze this algorithm, the scenario of multiple targets is introduced in this subsection. According to the Li's research [12], most of man-made objects can be considered as the composition of trihedral corner reflectors and dihedral corner reflectors. Generally speaking, a trihedral corner reflector has a flat backscattering envelope during synthetic aperture time while a dihedral corner reflector has a variant one. Thus, in the simulation test, we simplify the types of targets in [8] to two: azimuth invariant target (e.g. Fig. 10a) and azimuth variant target (e.g. Fig. 10b), and their different backscattering envelopes

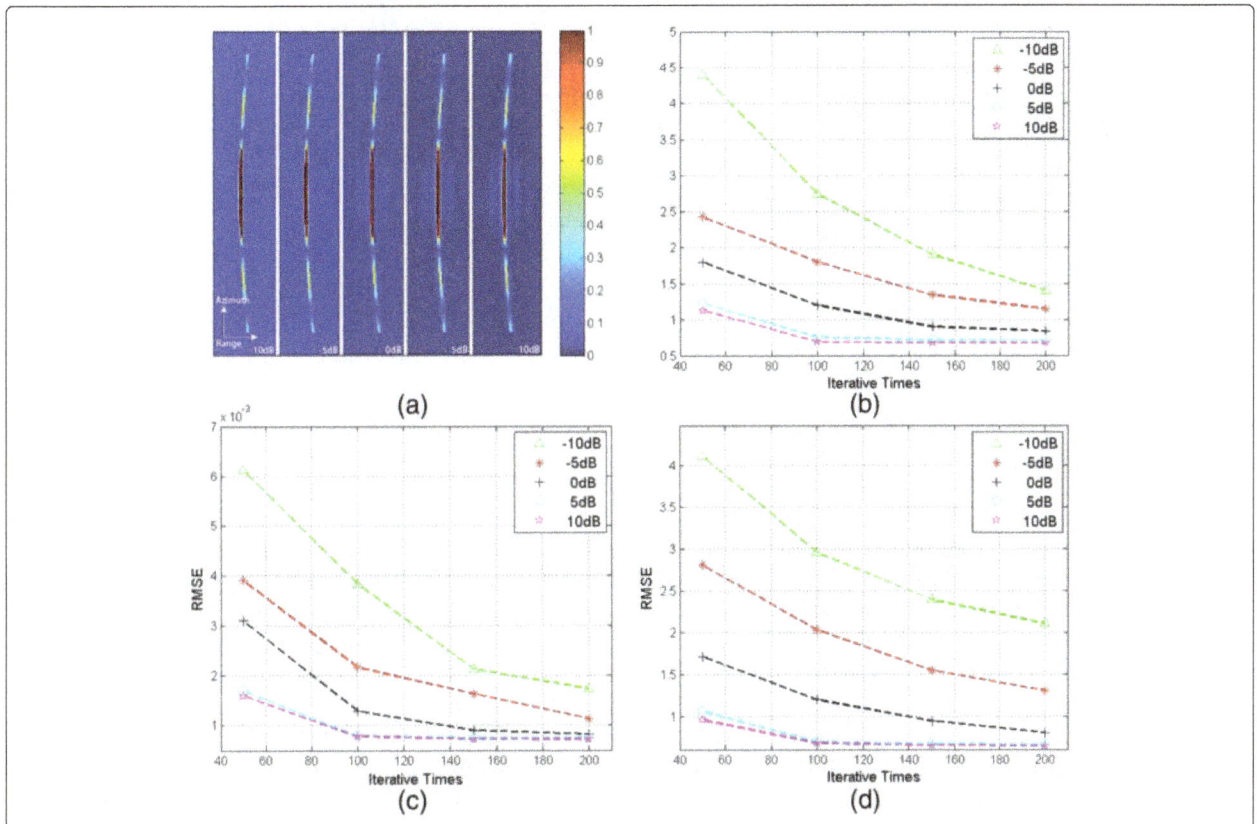

Fig. 9 Performance test. **a** Time-domain range compressed signal with different input SNR (range-compressed azimuth-time domain). **b** RMSE of R_0. **c** RMSE of η_0. **d** RMSE of V_r

are shown in Fig. 10c. In the illuminated scene, we have two azimuth invariant point targets: T1 and T2. The amplitude and phase of their theoretical backscattering envelopes are presented in Fig. 11a, b, respectively. T1 is set to be "brighter" than T2, which means it has a relatively higher backscattering coefficient or a larger radar cross section (RCS) [29]. We also have an azimuth variant point target T3, the maximum backscattering envelope of which stays lower than both T1 and T2 (see Fig. 11a).

Different from T1 and T2, T3 has an inconstant phase envelope during the synthetic aperture time (see Fig. 11b). In the simulation test, these targets are customized with the size of 2 m in range by 1 m in azimuth, and their geographical locations are shown in the columns R_0 and η_0 of Table 2. Moreover, both T1 and T3 are stationary targets while T2 is a moving target with a 5.5 m/s along-track velocity. The relative speed between radar platform and targets are listed in the column V_r of Table 2.

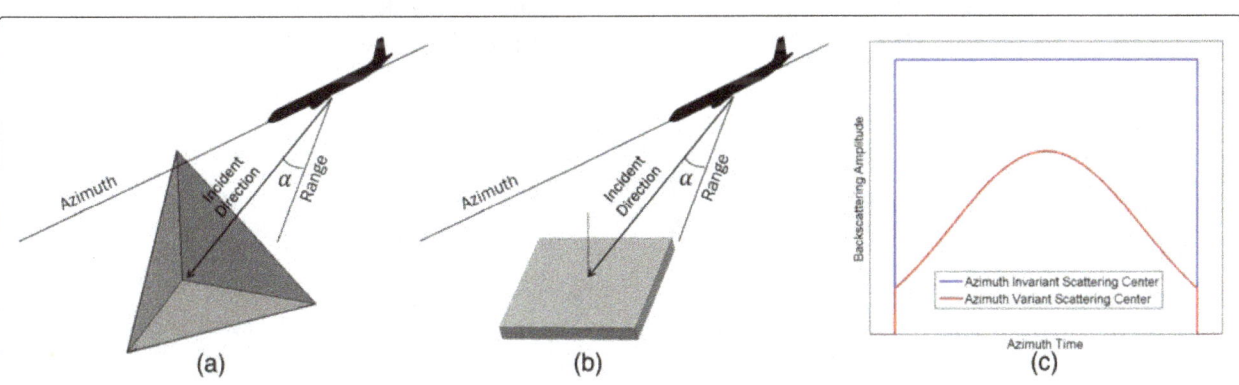

Fig. 10 Scattering centers with different backscattering envelopes. **a** Azimuth invariant scattering center. **b** Azimuth variant scattering center. **c** Backscattering envelopes of different scattering centers

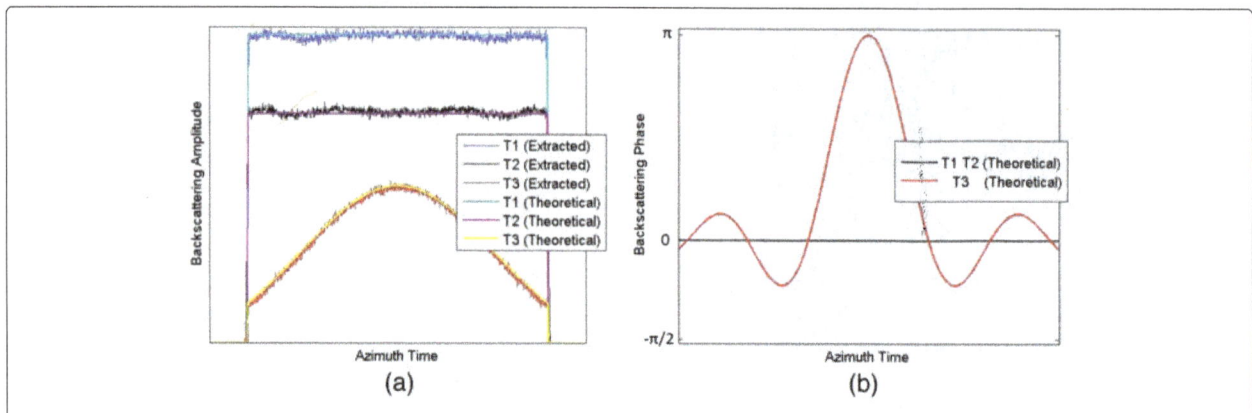

Fig. 11 Backscattering envelopes of targets. **a** Theoretical backscattering amplitude envelopes of targets and the extracted ones of dominant scattering centers. **b** Backscattering phase envelopes of targets

Using the SAR simulation parameters in Table 1, we generate the raw data of these three targets when SNR is 5 dB and beam width in azimuth dimension is 0.0785 rad. In this case, the Doppler spectrum of the range-compressed signal wraps around azimuth frequency as a result of ambiguity (see Fig. 12a). It may degrade the effectiveness of feature extraction algorithm starting with spectrum. To validate this assumption, a FFT-based time-frequency approach [8] and a complex spectral estimation algorithm called APES [16] are used to extract the spectral envelope of T3 respectively. As shown in Fig. 12b, the extracted results differ from the real data due to the impact of ambiguity. Thus, the proposed M-RANSAC-based feature extraction algorithm, which establishes itself as an ambiguity-free approach, is required in this case. The M-RANSAC-based algorithm starts with time-domain range compressed signal (see Fig. 12c). Its initial parameters are carefully designed. The scale factor ϑ in (10) is set to be 450. The lower bound of iterative times min is 30 and the upper bound max is 200, the threshold of quadratic orthogonal distance rho_thr is 0.003, and the threshold number of inliers N_thr is $0.85N_a$. After estimating $\vec{\mu}$ for the M dominant scattering centers and classifying inliers in *set*, the geographical locations and velocity information are reconstructed based on (23). As

shown in the columns $\hat{R}_0$, $\hat{\eta}_0$, and $\hat{V}_r$ of Table 2, the estimated errors are quite limited. Meanwhile, the backscattering envelopes are extracted and normalized. In Fig.11a, the extracted backscattering amplitude envelopes of dominant scattering centers match the theoretical ones of the corresponding targets.

Without knowing the explicit knowledge of platform velocity and forming a SAR image, the extracted backscattering envelopes can label T1 and T2 as azimuth invariant targets and T3 as an azimuth variant target. Thus, a rough target classification can be achieved. $\hat{R}_0$ and $\hat{\eta}_0$ present the geographical locations of dominant scattering centers. To visualize the extracted information, we map $\hat{R}_0$ and $\hat{\eta}_0$ of scattering centers into image domain. The amplitudes of them are obtained by averaging their backscattering envelopes. As shown in Fig. 12d, the image is free from the impact of sidelobes. Realizing the target classification and location, M-RANSAC-based algorithm help us to comprehend the targets without forming a SAR image.

When the explicit platform velocity is given, SAR image (see Fig. 12f) can be formed by chirp-scaling algorithm [30, 31]. Figure 12e presents the imaging result with eight times interpolation. In this image, T1 is well-focused while T2 and T3 are defocused. From the perspective of SAR image, T2 and T3 may be mistakenly classified

Table 2 Original and estimated parameters in both simulation test and real data test

Type	Scattering center	R_0 [m]	η_0 [s]	V_r [m/s]	$\hat{R}_0$ [m]	$\hat{\eta}_0$ [s]	$\hat{V}_r$ [m/s]
Simulation	T1	7500	0.8717	153.3	7500.7	0.8703	153.56
	T2	7462.5	0.8717	147.8	7462.2	0.8721	146.62
	T3	7537.5	0.8717	153.3	7538.7	0.8719	153.48
Real data	T1	997743	1.3143	7062.4	997744.3	1.3127	7062.9
	T2	997226	1.4202	7061.8	997226.8	1.4212	7061.9
	T3	997375	1.7073	7061.7	997374.1	1.7097	7061.2
	T4	997582	1.8665	7062.2	997582.5	1.8670	7061.7

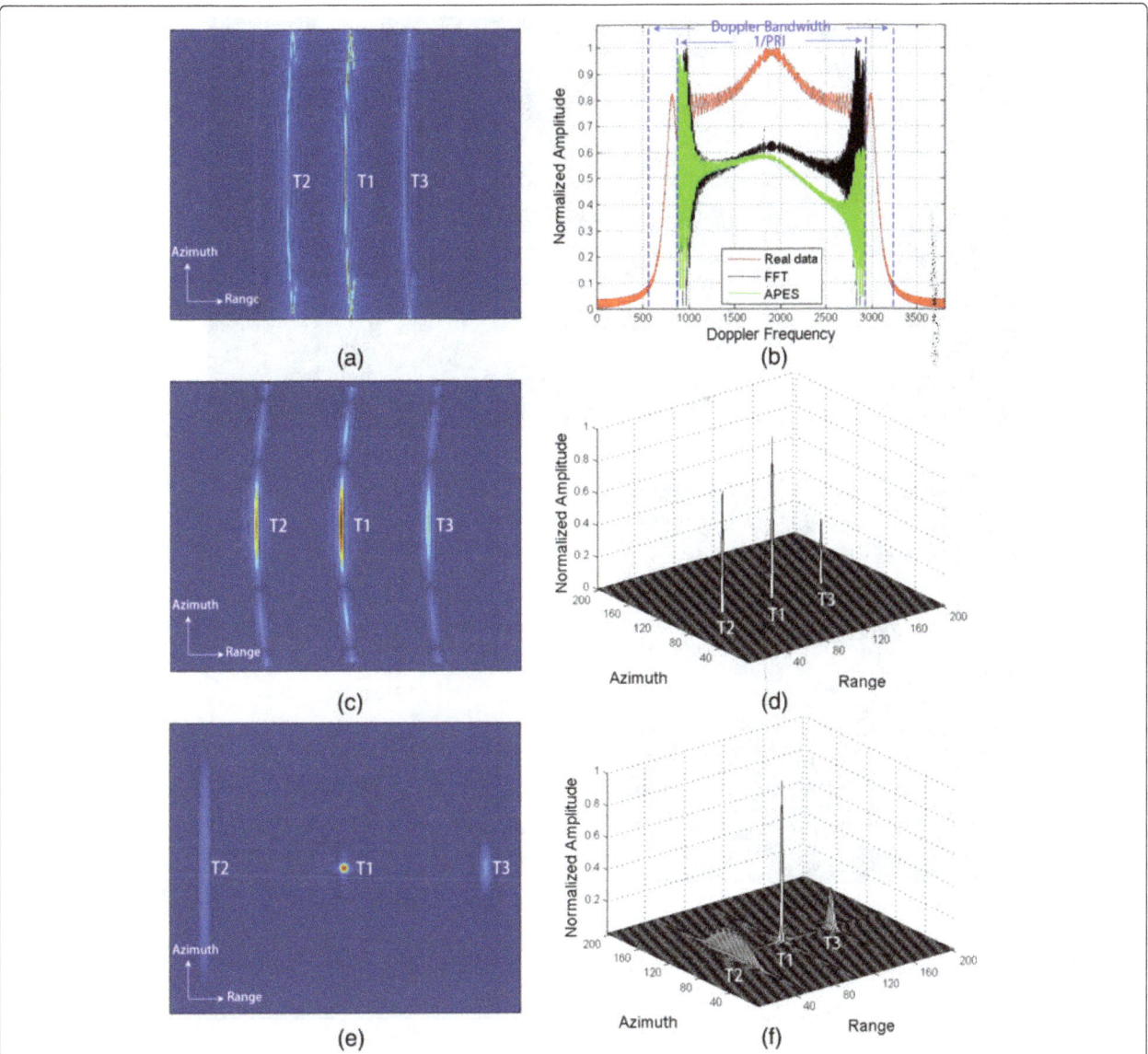

Fig. 12 Simulation test. **a** Doppler spectrum of a range-compressed signal (range-compressed Doppler domain). **b** Extracted spectral envelopes of T3 using FFT and APES. **c** Time-domain range-compressed signal (range-compressed azimuth-time domain). **d** SAR image formed by mapping the extracted feature. **e** SAR image with eight times interpolation formed by chirp-scaling algorithm. **f** SAR image formed by chirp-scaling algorithm

into the same type. However, defocus only indicates the mismatch of azimuth matched filter. It may result from either the along-track motion (T2) or the invariant azimuth envelope (T3). Since the two cases are hardly distinguished directly from SAR image, the importance of feature extraction is proved. The extracted envelopes of dominant scattering centers in Fig. 11a clearly reveals the backscattering feature of targets. Thus, we can label T1 and T2 as azimuth invariant targets and T3 an azimuth variant one. The backscattering envelope of target's dominant scattering center can be complementary to the SAR image in the application of target classification. Moreover, given the explicit platform velocity, we confirm T2 as a moving target according to column $\hat{V}_r$ of Table 2.

6.3 Real data test

In this subsection, RADARSAT-1 raw data included in the CD of [19] is applied in feature extraction. The key system parameters are listed in Table 1. As shown in Fig. 13a, the SAR image of English Bay is formed using the chirp-scaling algorithm. Then, the region of four ships, marked with a white rectangle, are truncated from this SAR image. This patch of complex image is converted to a range-compressed signal (use inverse chirp-scaling algorithm and range matched filter).

As shown in Fig. 13b, the trajectories of RMCs of the ship T2 and T3 are overlapped. Before extracting the feature scatter by scatter, the overlapped trajectories are separated using the STFRFT-based approach. First, K_a is

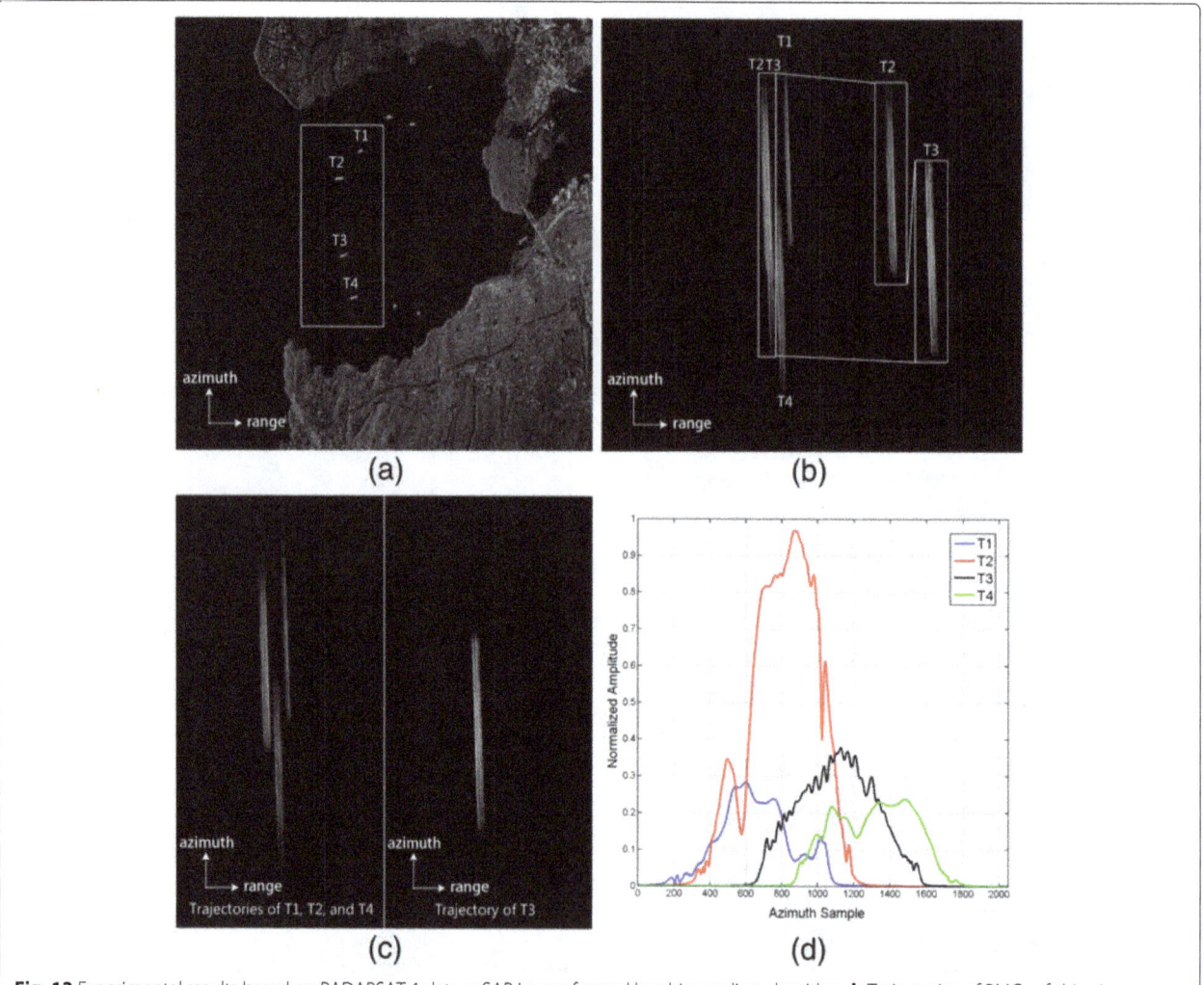

Fig. 13 Experimental results based on RADARSAT-1 data. **a** SAR image formed by chirp-scaling algorithm. **b** Trajectories of RMCs of ships in time-domain range compressed signal. **c** The result of trajectories separation based on STFRFT. **d** Extracted backscattering amplitude envelopes of dominant scattering centers

estimated using the dominant scattering center of isolated ship T1. Then, α is computed by (31). Using α-angle STFRFT and spatial filtering, the energy of T3 are separated from the whole range-compressed signal. After inverse STFRFT, two sub-patch results are obtained (see Fig. 13c).

The M-RASNAC-based feature extraction algorithm starts with the range-compressed signal in Fig. 13c. The scale factor $\vartheta = 5800$ in (10), the lower bound of iterative times min = 30, and the upper bound max = 300, the threshold of quadratic orthogonal distance rho_thr = 0.0016, and the threshold number of inliers $N\_thr = 0.85 \cdot N_a$. After estimating $\vec{\mu}$ for the dominant scattering centers of each ship, the geographical locations and velocity information are reconstructed based on (22) (see columns $\hat{R}_0$, $\hat{\eta}_0$, and $\hat{V}_r$ of Table 2). To verify the performance of parameters construction, we define the dominant scattering center of a ship as the point with maximum

amplitude in SAR image. Their geographical locations are listed in columns R_0 and η_0 of Table 2. The micro along-track velocity of dominant scattering centers are estimated using fractional Fourier transform (FRFT)-based method introduced in [32]. The relative velocity between radar platform and these scattering centers are then listed in column $\hat{V}_r$ of Table 2. Since two groups of data are in good agreements, the reconstructed errors are quite limited.

Then, to better understand the targets, the backscattering envelopes of dominant scattering centers are extracted from the range-compressed signal. As shown in Fig. 13d, T1 is much brighter than the others which may indicate the a relatively higher radar cross section (RCS) [29]. Moreover, T2 and T3 are azimuth variant while T1 and T4 are nearly azimuth invariant. It means the illuminated regions of T2 and T3 are more "flat" than those of T1 and T4.

7 Conclusions

An M-RANSAC and STFRFT-based technique is introduced to extract feature of SAR dominant scattering centers in this paper. Starting with the time-domain range-compressed signal, this algorithm provides an ambiguity-free signal-level approach. Meanwhile, this algorithm requires no explicit knowledge of platform velocity. It can conduct feature extraction without forming a SAR image. Within the extracted features, the backscattering envelope is promising to classify the target type in signal level, the geographical location indicates the target position relative to SAR platform, and the relative velocity denotes the along-track motion of illuminated target.

Experiments are conducted to illustrate the performance of this algorithm. In the tests, the estimation errors of location and relative velocity are quite limited when SNR is relatively high. The normalized extracted backscattering envelopes express their theoretical ones well. Moreover, these extracted features validate their usage in target recognition and classification. Without forming a SAR image, these extracted features will help us roughly understand and classify the illuminated targets. When SAR image is formed by conventional methods, the extracted backscattering envelopes can be complementary to SAR image in the application of target recognition and classification.

Competing interests
The authors declare that they have no competing interests.

Acknowledgements
This work has been supported by key project of the National Natural Science Foundation (NNSF) of China (nos.61132005). The authors also want to express gratitude to editors and anonymous reviewers who give the helpful comments and suggestions to his paper.

References

1. DE Dudgeon, RT Lacoss, An overview of automatic target recognition. Lincoln Lab J. **6**(1), 3–10 (1993)
2. GJ Owirka, SM Verbout, LM Novak, in *AeroSense'99*. Template-based SAR ATR performance using different image enhancement techniques (SPIE, Bellingham WA 98227-0010 USA, 1999), pp. 302–319. International Society for Optics and Photonics
3. Z Jianxiong, S Zhiguang, C Xiao, F Qiang, Automatic target recognition of SAR images based on global scattering center model. IEEE Trans Geosci Remote Sens. **49**(10), 3713–3729 (2011)
4. Y Chen, E Blasch, H Chen, T Qian, G Chen, in *SPIE Defense and Security Symposium*. Experimental feature-based SAR ATR performance evaluation under different operational conditions (SPIE, Bellingham WA 98227-0010 USA, 2008), pp. 69680–69680. International Society for Optics and Photonics
5. Y Huang, J Pei, J Yang, T Wang, H Yang, B Wang, Kernel generalized neighbor discriminant embedding for SAR automatic target recognition. EURASIP J Adv Signal Process. **2014**(1), 1–6 (2014)
6. JB Keller, Geometrical theory of diffraction. JOSA. **52**(2), 116–130 (1962)
7. LC Potter, RL Moses, Attributed scattering centers for SAR ATR. IEEE Trans Image Process. **6**(1), 79–91 (1997)
8. M Spigai, C Tison, J-C Souyris, Time-frequency analysis in high-resolution SAR imagery. IEEE Trans Geosci Remote Sens. **49**(7), 2699–2711 (2011)
9. VC Chen, H Ling, *Time-Frequency Transforms for Radar Imaging and Signal Analysis*. (Artech House, Norwood, 2001)
10. J-P Ovarlez, L Vignaud, J-C Castelli, M Tria, M Benidir, Analysis of SAR images by multidimensional wavelet transform. IEE Proc Radar Sonar Navig. **150**(4), 234–241 (2003)
11. G Lisini, C Tison, F Tupin, P Gamba, Feature fusion to improve road network extraction in high-resolution SAR images. IEEE Geosci Remote Sens Lett. **3**(2), 217–221 (2006)
12. Z-S Liu, J Li, in *Radar, Sonar and Navigation, IEE Proceedings-*. Feature extraction of SAR targets consisting of trihedral and dihedral corner reflectors, vol. 145 (IET, Michael Faraday House, Six Hills Way Stevenage, Herts, SG1 2AY, UK, 1998), pp. 161–172
13. Z Bi, J Li, Z-S Liu, Super resolution SAR imaging via parametric spectral estimation methods. IEEE Trans Aerosp Electron Syst. **35**(1), 267–281 (1999)
14. J Li, P Stoica, Efficient mixed-spectrum estimation with applications to target feature extraction. IEEE Trans Signal Process. **44**(2), 281–295 (1996)
15. EG Larsson, G Liu, P Stoica, J Li, High-resolution SAR imaging with angular diversity. IEEE Trans Aerosp Electron Syst. **37**(4), 1359–1372 (2001)
16. J Li, P Stoica, An adaptive filtering approach to spectral estimation and SAR imaging. IEEE Trans Signal Process. **44**(6), 1469–1484 (1996)
17. SR DeGraaf, Sidelobe reduction via adaptive FIR filtering in SAR imagery. IEEE Trans. Image Process. **3**(3), 292–301 (1994)
18. R Wu, J Li, Z Bi, P Stoica, SAR image formation via semiparametric spectral estimation. IEEE Trans Aerosp Electron Syst. **35**(4), 1318–1333 (1999)
19. IG Cumming, FH-c Wong, *Digital Processing of Synthetic Aperture Radar Data: Algorithms and Implementation*. (Artech House, Norwood, 2005)
20. R Tao, Y-L Li, Y Wang, Short-time fractional Fourier transform and its applications. IEEE Trans Signal Process. **58**(5), 2568–2580 (2010)
21. SJ Ahn, W Rauh, H-J Warnecke, Least-squares orthogonal distances fitting of circle, sphere, ellipse, hyperbola, and parabola. Pattern Recogn. **34**(12), 2283–2303 (2001)
22. YC Cheng, SC Lee, A new method for quadratic curve detection using K-RANSAC with acceleration techniques. Pattern Recogn. **28**(5), 663–682 (1995)
23. PH Torr, A Zisserman, MLESAC: a new robust estimator with application to estimating image geometry. Comput Vis Image Underst. **78**(1), 138–156 (2000)
24. J Tsao, BD Steinberg, Reduction of sidelobe and speckle artifacts in microwave imaging: the CLEAN technique. IEEE Trans. Antennas Propag. **36**(4), 543–556 (1988)
25. H Deng, Effective CLEAN algorithms for performance-enhanced detection of binary coding radar signals. IEEE Trans Signal Process. **52**(1), 72–78 (2004)
26. HM Ozaktas, O Arikan, MA Kutay, G Bozdagi, Digital computation of the fractional Fourier transform. IEEE Trans. Signal Process. **44**(9), 2141–2150 (1996)
27. S Chiu, Application of fractional fourier transform to moving target indication via along-track interferometry. EURASIP J Appl Signal Process. **2005**, 3293–3303 (2005)
28. TA Schonhoff, AA Giordano, *Detection and Estimation Theory and Its Applications*. (Pearson/Prentice Hall, New Jersey, 2006)
29. EF Knott, *Radar Cross Section Measurements*. (SciTech Publishing, Raleigh, 2006)
30. RK Raney, H Runge, R Bamler, IG Cumming, FH Wong, Precision SAR processing using chirp scaling. IEEE Trans Geosci Remote Sens. **32**(4), 786–799 (1994)
31. A Moreira, J Mittermayer, R Scheiber, Extended chirp scaling algorithm for air- and spaceborne SAR data processing in stripmap and ScanSAR imaging modes. IEEE Trans Geosci Remote Sens. **34**(5), 1123–1136 (1996)
32. J Yang, C Liu, Y Wang, Detection and imaging of ground moving targets with real SAR data. IEEE Trans Geosci Remote Sens. **53**, 920–932 (2015)

16

Performance of regression-based precoding for multi-user massive MIMO-OFDM systems

Ali Yazdan Panah[*†] [ID], Karthik Yogeeswaran[†] and Yael Maguire

Abstract

We study the performance of a single-cell massive multiple-input multiple-output orthogonal frequency-division multiplexing (MIMO-OFDM) system that uses linear precoding to serve multiple users on the same time-frequency resource. To minimize overhead, the channel estimates at the base station are obtained via comb-type pilot tones during the training phase of a time-division duplexing system. Polynomial regression is used to interpolate the channel estimates within each coherence block. We show how such regressors can be designed in an offline fashion without the need to obtain channel statistics at the base station, and we assess the downlink performance over a wide range of system parameters.

Keywords: MIMO, OFDM, Massive MIMO, Least squares, Interpolation, Channel estimation, Zero-forcing, Beamforming, Precoding

1 Introduction

Multi-user multiple-input multiple-output (MU-MIMO) systems with large number of base station antennas hold the promise of high throughput communications for emerging wireless deployment [1–4]. Using the notion of *spatial multiplexing*, the antenna array at the base station can serve a multiplicity of autonomous user terminals on the same time-frequency resource. This *spatial resource sharing policy* serves as an alternative not only to the need for costly spectrum licensing but also the costly procurement of additional base stations in conventional cell-shrinking strategies. While the benefits of spatial multiplexing may be fully realized when the number of base station antennas is equal to the number of scheduled user terminals, MU-MIMO systems with an excessively large number of antennas, also known as massive MIMO, have recently gained attention owing in part to the following benefits [5]:

- Massive MIMO can increase the throughput and simultaneously improve the radiated energy efficiency via energy focusing.
- Massive MIMO can be built with rather inexpensive components by replacing high-power (W) linear amplifiers with low-power (mW) counterparts.
- Massive MIMO can simplify the multiple-access layer (MAC) by scheduling the users on the entire band without the need for feedback[1].

Such benefits largely stem from asymptotic results on random matrix theory that illustrates how the effects of uncorrelated noise and small-scale fading are virtually eliminated (and the required transmitted energy per bit vanishes) as the number of antennas in a MIMO cell grows to infinity.

Massive MIMO systems are also versatile over a wide range of system parameters. For instance, the beamforming gain afforded by using a large number of transmit antennas may be used to overcome the large path loss associated with mmWave links in urban areas [6]. Alternatively, the beamforming gain may be harnessed at VHF/UHF frequencies to provide wide-coverage

---

*Correspondence: ayp@fb.com
†Equal contributors
Facebook Connectivity Lab, Facebook Inc., 1 Hacker Way, Menlo Park, CA 94025, USA

connectivity to rural areas of the world [7]. Given such promises, the practical and theoretical aspects of massive MIMO systems are actively under scrutiny for potential beyond-4G wireless communication deployments not only by standardization entities such as the 3rd Generation Partnership Project (3GPP) but also by many industrial base station and device manufacturers worldwide.

Coherent massive MIMO systems require channel state information (CSI) at the base station in order to compute linear precoder filters for the downlink and equalization filters for the uplink. Such systems are typically designed for a time-division duplexing (TDD) scheme where the uplink and downlink share the transmission bandwidth. This is primarily due to the fact that the CSI may be readily obtained in TDD mode when reciprocity is maintained in the signal path. For example, the base station may estimate the downlink (and uplink) channel using pilot symbols transmitted by the users during an uplink "training phase" [8]. The estimation of CSI is a well-studied area for MIMO [8], OFDM [9, 10], and MIMO-orthogonal frequency-division multiplexing (OFDM) [11] systems. For multi-user systems, the base station may use the estimated CSI obtained from uplink pilots to construct linear precoders (and equalizers). Fortunately, in the massive MIMO regime, the performance of such filters are known to be close to the optimal schemes. In this context, matched-filter (MF) and zero-forcing (ZF) are two popular linear filters [12]. The gains due to linear processing must be weighed by the increases in baseband computational complexity as a result of adding more antenna elements at the base station. For instance, MF and ZF equalization are known to have linear and cubic complexity, respectively, in the number of users. This may present a bottleneck given current hardware capabilities; hence, some researchers have devised suboptimal methods with reduced complexity such as the ordering scheme proposed in [13] for MF or the inversion-approximation for ZF proposed in [14]. The accuracy of these linear filters depend on the accuracy of the CSI on which they are obtained from.

Interpolating a reduced set of pilots is a popular method of estimating the CSI across the frequency band in single and multi-user MIMO-OFDM system (see, e.g., [9, 10, 15–17] and references therein). In this paper, we study the effects of *regression-based interpolation* of CSI and its effects on the accuracy of linear precoding in a downlink massive MIMO system. We propose polynomial regression as a way to interpolate the multiplexed pilots in the uplink into a single channel estimate over a block of bandwidth, i.e., over a coherence block. These regressors may be computed in an offline fashion without any knowledge of the channel. In Section 2, we formulate the problem and propose some notation and in Section 3, we

present numeric results. We make concluding remarks in Section 4.

Notation: Bold uppercase and lowercase letters represent matrices and vectors, respectively. $\mathbf{X}^*$, $\mathbf{X}^\mathrm{T}$, $\mathbf{X}^\mathrm{H}$, $\mathbf{X}^{-1}$, and $\mathbf{X}^+$ denote conjugate, transpose, conjugate-transpose, matrix inverse, and Moore-Penrose inverse of a matrix $\mathbf{X}$, respectively.

2   System model

We consider a linearly precoded MU-MIMO-OFDM system over N subcarriers with M antennas at the base station serving K single-antenna users. The system operates under a hardware-calibrated time division duplexing (TDD) scheme over a wireless channel with a coherence time of T_c seconds. This allows simultaneous uplink (users to base station) and downlink (base station to users) transmissions across a common frequency band.

2.1   Uplink pilot phase (training)

During the uplink pilot phase, the users transmit pilot symbols to the base station for the purposes of channel estimation and precoder/equalizer calculation. To minimize the overhead associated with pilot transmissions, we adopt a comb-type pilot arrangement where the pilot symbols are uniformly inserted into OFDM symbols during the uplink pilot phase. The pilot spacing in the frequency domain is chosen to be smaller than the coherence bandwidth of the channel which is approximated as $B_c = 0.02/\tau_\mathrm{rms}$, where τ_rms is the channel delay spread. As such, the channel estimation is processed on a per resource block (RB) basis, where a resource block is defined as a contiguous group of subcarriers spanning one coherence bandwidth (within the channel coherence time T_c). The pilot symbols are not precoded by the users and are instead transmitted in a multiple-access fashion. Figure 1 illustrates an example of an uplink pilot resource grid over one RB spanning 12 subcarriers with a total of 24 user-transmitting pilots across 6 OFDM symbols.

2.1.1   Least squares channel estimation

The OFDM channel between each base station antenna and each user can be estimated using the uplink pilots with a least squares (LS) method. With a sufficiently long cyclic prefix (CP) length, the received signal at symbol time t on antenna m at subcarrier n, from the kth user, at the base station is as follows:

$$y_m[t,n] = C_{m,k}[t,n]\, s_k[t,n] + \nu_m[t,n], \qquad (1)$$

where $C_{m,k}[t,n]$ is the channel frequency response, $s_k[t,n]$ is the transmitted (quadrature amplitude

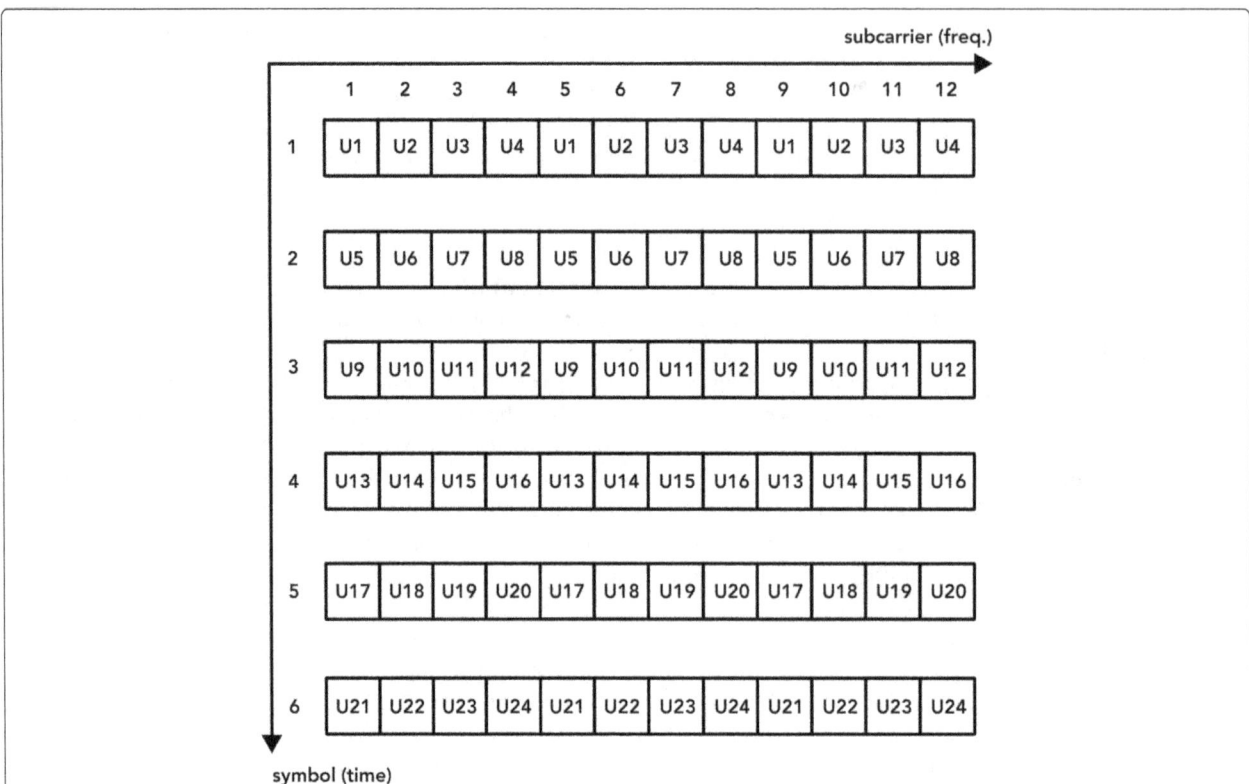

Fig. 1 Pilot tone allocation over one resource block (RB) consisting of 12 subcarriers and 6 symbols. Each user is allocated three pilot tones per RB, and this pattern is repeated over the frequency band

modulation (QAM)) pilot tone corresponding to user k, and $v_m[t, n]$ is additive white Gaussian noise (AWGN). Since the channel is assumed constant within an RB, we re-formulate the received signal of (1) to represent the rth RB:

$$y_m^{(r)}[t', n'] = C_{m,k}^{(r)}[t', n'] s_k^{(r)}[t', n'] + v_m^{(r)}[t', n']. \quad (2)$$

Here, t' and n' denote subsets of OFDM symbols and subcarriers, respectively, in which user k has transmitted a pilot tone within the rth RB. Let L denote the total number of pilot tones per user per RB.[2] For example, in Fig. 1 for user 1, we have $n' = \{1, 5, 9\}, t' = \{1\}$, and for user 2, we have $n' = \{2, 6, 10\}, t = \{1\}$, and for user 24, we have $n' = \{4, 8, 12\}, t' = \{6\}$, etc. In this case, for any user, we have $L = 3$.

The transmitted pilot tones are chosen from the unit-energy quadrature phase shift keying (QPSK) constellation space so that $\left| s_k^{(r)}[t', n'] \right|^2 = 1$. The noise is normal distributed: $v_m^{(r)}[t', n'] \sim \mathcal{CN}(0, 1)$ and i.i.d across m, t', and n', and the channel response $C_{m,k}^{(r)}[t', n']$ absorbs all

link budget parameters (such as path loss and thermal noise variance). The demodulated CSI on the pilot subcarriers for user k on RB r is as follows:

$$\widetilde{C}_{m,k}^{(r)}[t', n'] = y_m^{(r)}[t', n'] \left(s_k^{(r)}[t', n'] \right)^*. \quad (3)$$

2.2 Regressive interpolation

With the channel estimated on the pilot tones, the channel at other subcarriers may be computed via interpolation. In this paper, we use a *polynomial regression*-based approach formulated as a weighted average:

$$\widehat{C}_{m,k}^{(r)}[n] = \sum_{t', n'} \gamma_{m,k}^{(r)}[n, t', n'] \widetilde{C}_{m,k}^{(r)}[t', n'] = \boldsymbol{\gamma}_{m,k}^{(r)}[n] \widetilde{\mathbf{c}}_{m,k}^{(r)}, \quad (4)$$

where $\boldsymbol{\gamma}_{m,k}^{(r)}[n]$ is a row vector of length L of elements $\gamma_{m,k}^{(r)}[n, t', n']$ which are the interpolation weights associated with antenna m for user k for the rth RB for the n subcarrier, for all t', n'. We call $\boldsymbol{\gamma}_{m,k}^{(r)}[n]$ the *interpolation vector*. Also, $\widetilde{\mathbf{c}}_{m,k}^{(r)}$ is a column vector of length L of the values $\widetilde{C}_{m,k}^{(r)}[t', n']$ for all t', n'. The interpolation vector may be computed from a polynomial regression of order p, represented by the vector $\mathbf{x}_{m,k}^{<p>} = [x_0, x_1, \ldots, x_p]^T$ satisfying:

$$\mathbf{V}_{m,k}^{<p>} \, \mathbf{x}_{m,k}^{<p>} = \widetilde{\mathbf{c}}_{m,k}^{(r)}, \tag{5}$$

where $\mathbf{V}_{m,k}^{<p>}$ is the $L \times (p + 1)$ *Vandermonde matrix*. For $0 \leq p < L$, the solution to this linear system of equations is

$$\mathbf{x}_{m,k}^{<p>} = \left(\mathbf{V}_{m,k}^{<p>}\right)^+ \widetilde{\mathbf{c}}_{m,k}^{(r)}, \tag{6}$$

where

$$\left(\mathbf{V}_{m,k}^{<p>}\right)^+ = \left(\left(\mathbf{V}_{m,k}^{<p>}\right)^T \mathbf{V}_{m,k}^{<p>}\right)^{-1} \left(\mathbf{V}_{m,k}^{<p>}\right)^T \tag{7}$$

is the Moore-Penrose pseudo-inverse of $\mathbf{V}_{m,k}^{<p>}$. With $\mathbf{x}_{m,k}^{<p>}$ in hand, the channel estimate at any subcarrier n in the RB is simply an evaluation on the polynomial function: $n^0 x_0 + n^1 x_1 + n^2 x_2 + \ldots + n^p x_p$. Defining $\mathbf{d}[n] = [1, n, n^2, \ldots, n^p]$, in vector form, we have

$$\widehat{C}_{m,k}^{(r)}[n] = \mathbf{d}[n] \left(\mathbf{V}_{m,k}^{<p>}\right)^+ \widetilde{\mathbf{c}}_{m,k}^{(r)}, \tag{8}$$

Comparing the left-hand side of (8) with (4), we see that

$$\boldsymbol{\gamma}_{m,k}^{(r)}[n] = \mathbf{d}[n] \left(\mathbf{V}_{m,k}^{<p>}\right)^+. \tag{9}$$

Some further simplification is possible in (9) since the Vandermonde matrix and interpolation vector do not depend on the antenna index m (as seen in Fig. 1, the pilot subcarrier locations are fixed for any m). Also, the interpolation vector does not depend on the RB index since the pilot subcarrier location pattern is identical across RBs. For a system with N' subcarriers per RB, we have

$$\boldsymbol{\gamma}_k[n] = \mathbf{d}[n] \left(\mathbf{V}_k^{<p>}\right)^+, n = 1, 2, \ldots, N' \tag{10}$$

As a result, the interpolated CSI across all subcarriers (in any RB) in (4) can be rewritten as follows:

$$\widehat{C}_{m,k}^{(r)}[n] = \boldsymbol{\gamma}_k[n] \, \widetilde{\mathbf{c}}_{m,k}^{(r)}. \tag{11}$$

Finally, it should be noted that while suboptimal by design, the polynomial interpolation method described above may present some advantages compared to the well-known linear minimum mean square error (LMMSE) channel interpolators of [9, 10]. For example, the polynomial interpolators are both channel model and channel signal-to-noise ratio (SNR) independent. Moreover, the per-RB-based processing nature of the polynomial interpolation method may lead to computational savings since for N total subcarriers and N' subcarriers per RB, the LMMSE method requires inversion of *complex-valued* matrices of size $\frac{N}{N'}L$, while the polynomial interpolators require inversion of *real-valued* Vandermonde matrices of size p where $p < L \leq \frac{N}{N'}L$.

2.3 Downlink precoding

During the downlink phase, the base station transmits precoded data to the users. Let the vector $\mathbf{s}[n] = [s_1[n], s_2[n], \ldots, s_K[n]]$ represent the QAM symbols intended for the user terminals at subcarrier n and $\mathbf{v}[n] \sim \mathcal{CN}(\mathbf{0}, \mathbf{I}_K)$ be AWGN at the user terminals. Similar to (1), the received signal at the users may be modeled by the $K \times 1$ vector $\mathbf{y}[n]$ as follows:

$$\mathbf{y}[n] = \mathbf{C}[n] \, \mathbf{F}[n] \, \mathbf{s}[n] + \mathbf{v}[n], \tag{12}$$

where $\mathbf{C}[n]$ is the $K \times M$ downlink MIMO channel from the base station to the user terminals that absorbs the link budget parameters (such as path loss and noise variance) and also transmit power constraint of the base station. The elements of the channel matrix are estimated during the uplink pilot phase and are given by (8). $\mathbf{F}[n] = \left[\mathbf{f}_{n,1}^T, \mathbf{f}_{n,2}^T, \ldots, \mathbf{f}_{n,K}^T\right]$ is the $M \times K$ precoding matrix at subcarrier n so that $\mathbf{f}_{n,k}$ is the precoding vector allocated to user k by the base station for subcarrier n. We consider ZF precoding in this paper:

$$\mathbf{F}_{\mathrm{ZF}}[n] = \widehat{\mathbf{C}}[n]^H \left(\widehat{\mathbf{C}}[n] \, \widehat{\mathbf{C}}[n]^H\right)^{-1}, \tag{13}$$

where the elements of $\widehat{\mathbf{C}}[n]$ are obtained using polynomial regression via (11).

3 Numeric results

In this section, we assess the performance of the regression-based linear precoding described in Section 2 using a system level simulator with Monte Carlo simulations. We consider a single-cell multi-user MIMO-OFDM system with $N = 256$ subcarriers of which 180 subcarriers are used for data and control signals. Each RB consists of 12 contiguous subcarriers. The base station serves $K = 24$ users using $M \geq K$ antennas. The channel between each base station antenna and each user is modeled as a tapped-delay line with an effective delay spread of τ_{rms}. The UL pilot transmission phase consists of 6 OFDM symbols with QPSK pilot symbols multiplexed for 24 users as in Fig. 1. The pilot phase is followed by DL data transmissions with QPSK symbols. The DL transmit powers, path loss, link budgets, and noise variance are such that the SNR for each user is identical.

The channel frequency response estimates are computed using (8) using polynomial regressors of the order $p = 0, 1, 2$. The interpolation vectors $\boldsymbol{\gamma}_k[n]$ may be computed offline and selected from the rows of base matrices $\boldsymbol{\Gamma}_k^{<p>}$, where the subscripts denote the user indices corresponding to Fig. 1. We elaborate on this idea using an example below.

Example 1. In Fig. 1, for user 1, we have $n' = \{1, 5, 9\}, t' = \{1\}$, meaning there are $L = 3$ pilot tones per RB allocated to this user. For $p = 2$, the 3×3 Vandermonde matrix in (10) and its inverse can be computed as follows:

$$\mathbf{V}_1^{<p=2>} = \begin{bmatrix} 1 & 1 & 1 \\ 1 & 5 & 25 \\ 1 & 9 & 81 \end{bmatrix}, \left(\mathbf{V}_1^{<p=2>} \right)^{-1}$$

$$= \begin{bmatrix} +1.4063 & -0.5625 & +0.1563 \\ -0.4375 & +0.6250 & -0.1875 \\ +0.0313 & -0.0625 & +0.0313 \end{bmatrix}.$$

Since each RB is defined as 12 subcarriers in Fig. 1, the length of three interpolation vectors $\boldsymbol{\gamma}_1[n]$ can be computed for any $n = 1, 2, \ldots, 12$ via (10) and the inverse Vandermonde matrix above. For example, $\boldsymbol{\gamma}_1[1]$ and $\boldsymbol{\gamma}_1[2]$ are computed as follows:

$$\boldsymbol{\gamma}_1[1] = [1, 1, 1] \begin{bmatrix} +1.4063 & -0.5625 & +0.1563 \\ -0.4375 & +0.6250 & -0.1875 \\ +0.0313 & -0.0625 & +0.0313 \end{bmatrix}$$

$$= [1.0000, 0.0000, 0.0000],$$

$$\boldsymbol{\gamma}_1[2] = [1, 2, 4] \begin{bmatrix} +1.4063 & -0.5625 & +0.1563 \\ -0.4375 & +0.6250 & -0.1875 \\ +0.0313 & -0.0625 & +0.0313 \end{bmatrix}$$

$$= [0.6563, 0.4375, -0.0937],$$

and similarly for $\boldsymbol{\gamma}_1[3], \boldsymbol{\gamma}_1[4], \ldots, \boldsymbol{\gamma}_1[12]$. The interpolation vectors may be collected in the $N' \times L$ *base matrix*:

$$\boldsymbol{\Gamma}_1^{<p=2>} = \left[\boldsymbol{\gamma}_1[1]^T, \boldsymbol{\gamma}_1[2]^T, \ldots, \boldsymbol{\gamma}_1[12]^T \right]^T$$

$$= \begin{bmatrix} +1.0000 & +0.0000 & +0.0000 \\ +0.6563 & +0.4375 & -0.0937 \\ +0.3750 & +0.7500 & -0.1250 \\ +0.1562 & +0.9375 & -0.0938 \\ -0.0000 & +1.0000 & -0.0000 \\ -0.0938 & +0.9375 & +0.1562 \\ -0.1250 & +0.7500 & +0.3750 \\ -0.0938 & +0.4375 & +0.6563 \\ +0.0000 & -0.0000 & +1.0000 \\ +0.1563 & -0.5625 & +1.4062 \\ +0.3750 & -1.2500 & +1.8750 \\ +0.6563 & -2.0625 & +2.4062 \end{bmatrix}.$$

Finally, noting that some users have identical pilot allocation locations (e.g., users $1, 5, 9, 13, 17$, and 21 in Fig. 1), the base matrices are identical over such user sets. For completeness, these matrices are computed below for regression orders $p = 0, 1, 2$:

$$\boldsymbol{\Gamma}_{1,2,3,4,5\ldots,24}^{<p=0>} = \begin{bmatrix} +0.3333 & +0.3333 & +0.3333 \\ +0.3333 & +0.3333 & +0.3333 \\ +0.3333 & +0.3333 & +0.3333 \\ +0.3333 & +0.3333 & +0.3333 \\ +0.3333 & +0.3333 & +0.3333 \\ +0.3333 & +0.3333 & +0.3333 \\ +0.3333 & +0.3333 & +0.3333 \\ +0.3333 & +0.3333 & +0.3333 \\ +0.3333 & +0.3333 & +0.3333 \\ +0.3333 & +0.3333 & +0.3333 \\ +0.3333 & +0.3333 & +0.3333 \\ +0.3333 & +0.3333 & +0.3333 \end{bmatrix},$$

$$\boldsymbol{\Gamma}_{1,5,9,13,17,21}^{<p=1>} = \begin{bmatrix} +0.8333 & +0.3333 & -0.1667 \\ +0.7083 & +0.3333 & -0.0417 \\ +0.5833 & +0.3333 & +0.0833 \\ +0.4583 & +0.3333 & +0.2083 \\ +0.3333 & +0.3333 & +0.3333 \\ +0.2083 & +0.3333 & +0.4583 \\ +0.0833 & +0.3333 & +0.5833 \\ -0.0417 & +0.3333 & +0.7083 \\ -0.1667 & +0.3333 & +0.8333 \\ -0.2917 & +0.3333 & +0.9583 \\ -0.4167 & +0.3333 & +1.0833 \\ -0.5417 & +0.3333 & +1.2083 \end{bmatrix},$$

$$\boldsymbol{\Gamma}_{2,6,10,14,18,22}^{<p=1>} = \begin{bmatrix} +0.9583 & +0.3333 & -0.2917 \\ +0.8333 & +0.3333 & -0.1667 \\ +0.7083 & +0.3333 & -0.0417 \\ +0.5833 & +0.3333 & +0.0833 \\ +0.4583 & +0.3333 & +0.2083 \\ +0.3333 & +0.3333 & +0.3333 \\ +0.2083 & +0.3333 & +0.4583 \\ +0.0833 & +0.3333 & +0.5833 \\ -0.0417 & +0.3333 & +0.7083 \\ -0.1667 & +0.3333 & +0.8333 \\ -0.2917 & +0.3333 & +0.9583 \\ -0.4167 & +0.3333 & +1.0833 \end{bmatrix},$$

$$\Gamma^{<p=1>}_{3,7,11,15,19,23} = \begin{bmatrix} +1.0833 & +0.3333 & -0.4167 \\ +0.9583 & +0.3333 & -0.2917 \\ +0.8333 & +0.3333 & -0.1667 \\ +0.7083 & +0.3333 & -0.0417 \\ +0.5833 & +0.3333 & +0.0833 \\ +0.4583 & +0.3333 & +0.2083 \\ +0.3333 & +0.3333 & +0.3333 \\ +0.2083 & +0.3333 & +0.4583 \\ +0.0833 & +0.3333 & +0.5833 \\ -0.0417 & +0.3333 & +0.7083 \\ -0.1667 & +0.3333 & +0.8333 \\ -0.2917 & +0.3333 & +0.9583 \end{bmatrix},$$

$$\Gamma^{<p=2>}_{3,7,11,15,19,23} = \begin{bmatrix} +1.8750 & -1.2500 & +0.3750 \\ +1.4062 & -0.5625 & +0.1562 \\ +1.0000 & +0.0000 & -0.0000 \\ +0.6563 & +0.4375 & -0.0937 \\ +0.3750 & +0.7500 & -0.1250 \\ +0.1563 & +0.9375 & -0.0937 \\ +0.0000 & +1.0000 & +0.0000 \\ -0.0937 & +0.9375 & +0.1563 \\ -0.1250 & +0.7500 & +0.3750 \\ -0.0938 & +0.4375 & +0.6563 \\ -0.0000 & +0.0000 & +1.0000 \\ +0.1562 & -0.5625 & +1.4062 \end{bmatrix},$$

$$\Gamma^{<p=1>}_{4,8,12,16,20,24} = \begin{bmatrix} +1.2083 & +0.3333 & -0.5417 \\ +1.0833 & +0.3333 & -0.4167 \\ +0.9583 & +0.3333 & -0.2917 \\ +0.8333 & +0.3333 & -0.1667 \\ +0.7083 & +0.3333 & -0.0417 \\ +0.5833 & +0.3333 & +0.0833 \\ +0.4583 & +0.3333 & +0.2083 \\ +0.3333 & +0.3333 & +0.3333 \\ +0.2083 & +0.3333 & +0.4583 \\ +0.0833 & +0.3333 & +0.5833 \\ -0.0417 & +0.3333 & +0.7083 \\ -0.1667 & +0.3333 & +0.8333 \end{bmatrix}.$$

$$\Gamma^{<p=2>}_{4,8,12,16,20,24} = \begin{bmatrix} +2.4062 & -2.0625 & +0.6562 \\ +1.8750 & -1.2500 & +0.3750 \\ +1.4062 & -0.5625 & +0.1563 \\ +1.0000 & -0.0000 & +0.0000 \\ +0.6563 & +0.4375 & -0.0937 \\ +0.3750 & +0.7500 & -0.1250 \\ +0.1562 & +0.9375 & -0.0938 \\ -0.0000 & +1.0000 & -0.0000 \\ -0.0938 & +0.9375 & +0.1563 \\ -0.1250 & +0.7500 & +0.3750 \\ -0.0938 & +0.4375 & +0.6563 \\ -0.0000 & -0.0000 & +1.0000 \end{bmatrix}.$$

$$\Gamma^{<p=2>}_{1,5,9,13,17,21} = \begin{bmatrix} +1.0000 & +0.0000 & +0.0000 \\ +0.6563 & +0.4375 & -0.0937 \\ +0.3750 & +0.7500 & -0.1250 \\ +0.1562 & +0.9375 & -0.0938 \\ -0.0000 & +1.0000 & -0.0000 \\ -0.0938 & +0.9375 & +0.1562 \\ -0.1250 & +0.7500 & +0.3750 \\ -0.0938 & +0.4375 & +0.6563 \\ +0.0000 & -0.0000 & +1.0000 \\ +0.1563 & -0.5625 & +1.4062 \\ +0.3750 & -1.2500 & +1.8750 \\ +0.6563 & -2.0625 & +2.4062 \end{bmatrix},$$

$$\Gamma^{<p=2>}_{2,6,10,14,18,22} = \begin{bmatrix} +1.4062 & -0.5625 & +0.1562 \\ +1.0000 & +0.0000 & -0.0000 \\ +0.6563 & +0.4375 & -0.0937 \\ +0.3750 & +0.7500 & -0.1250 \\ +0.1563 & +0.9375 & -0.0937 \\ +0.0000 & +1.0000 & +0.0000 \\ -0.0937 & +0.9375 & +0.1563 \\ -0.1250 & +0.7500 & +0.3750 \\ -0.0937 & +0.4375 & +0.6563 \\ -0.0000 & +0.0000 & +1.0000 \\ +0.1562 & -0.5625 & +1.4062 \\ +0.3750 & -1.2500 & +1.8750 \end{bmatrix},$$

3.1 Performance vs. SNR: interpolation accuracy

The polynomial regression order p determines the interpolation matrices used to compute the channel estimates over the frequency band. The selection of the regression order depends on (a) the quality of the channel estimates on the pilot tones, i.e., the SNR and (b) the channel variability, i.e., the delay spread τ_{rms}. It is shown in [15] that in high channel noise, higher-order interpolation may be affected more adversely than lower-order interpolation. In Figs. 2 and 3, we confirm this observation for the proposed polynomial regressors by plotting the normalized channel estimation mean square error (NMSE) and the error vector magnitude (EVM) versus SNR. We plot results for both flat fading (Rayleigh), i.e., $\tau_{rms} = 0$, and a frequency selective channels with $\tau_{rms} = 0.104\ \mu s$. As a baseline for the EVM curves, we include results from a genie-aided system which computes the ZF precoding matrices on each subcarrier using perfect CSI; the genie-aided system does not suffer from the effects of thermal or interpolation noise. The interpolation noise floor is evident for the frequency selective channel at high SNR for the non-genie-aided approaches. Also, Fig. 2 shows how at low SNR, the zero-order-hold regressor, i.e., $p = 0$, performs best since it minimizes noise amplification while at high SNR $p = 2$ performs best by more accurately capturing the channel variation. In summary, for practical SNR ranges for QPSK (e.g., < 15 dB), the performance of the proposed

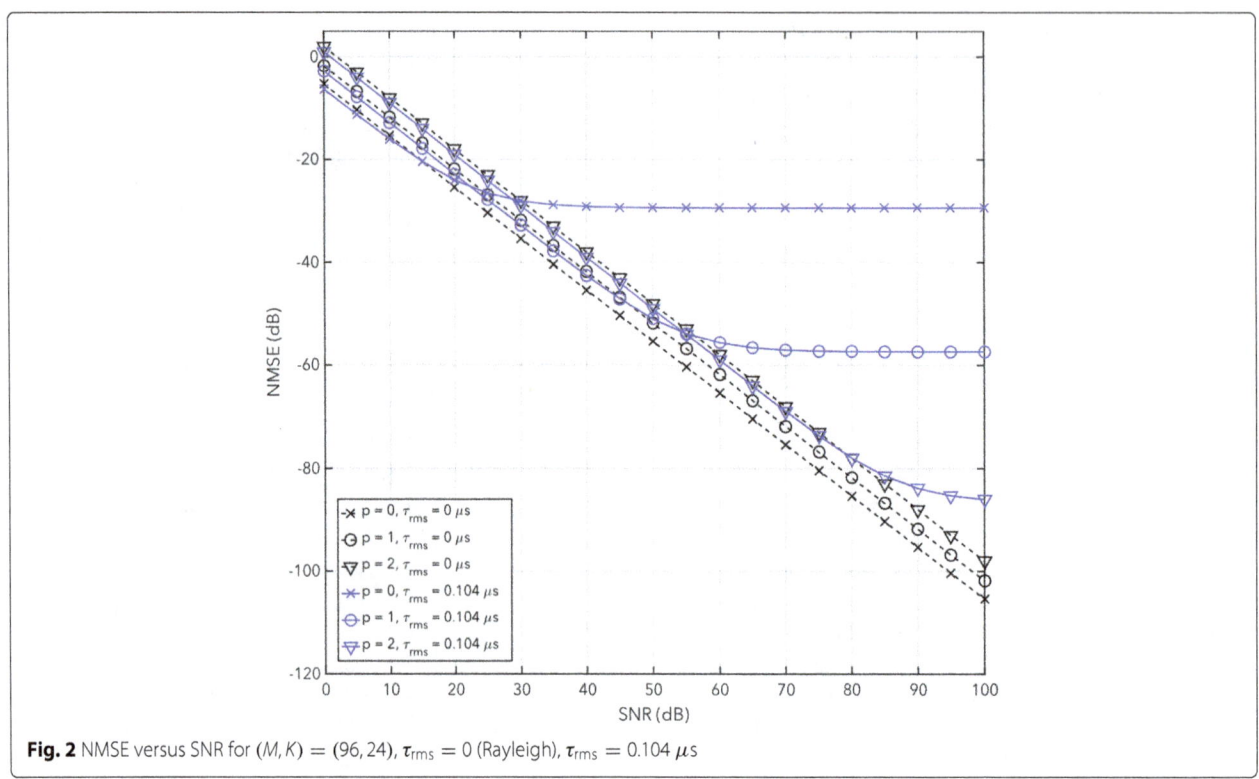

Fig. 2 NMSE versus SNR for $(M, K) = (96, 24)$, $\tau_{\text{rms}} = 0$ (Rayleigh), $\tau_{\text{rms}} = 0.104 \, \mu s$

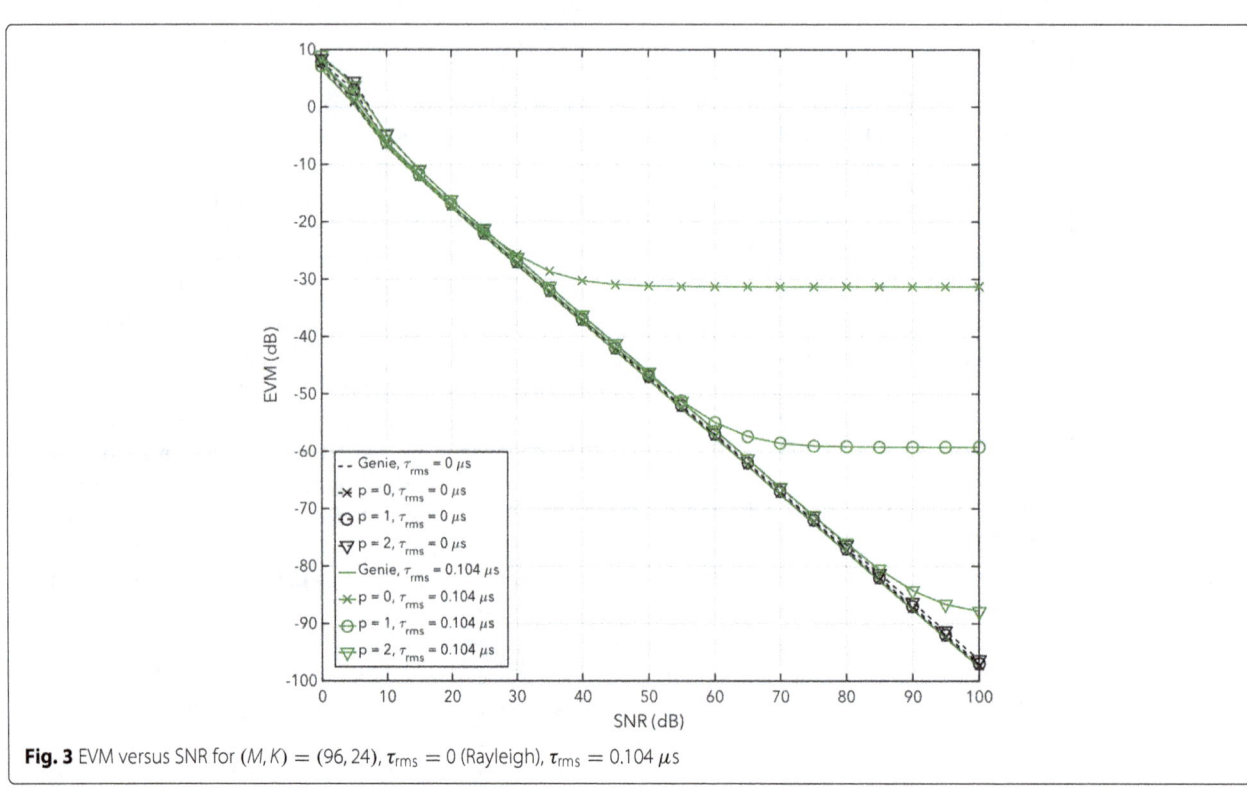

Fig. 3 EVM versus SNR for $(M, K) = (96, 24)$, $\tau_{\text{rms}} = 0$ (Rayleigh), $\tau_{\text{rms}} = 0.104 \, \mu s$

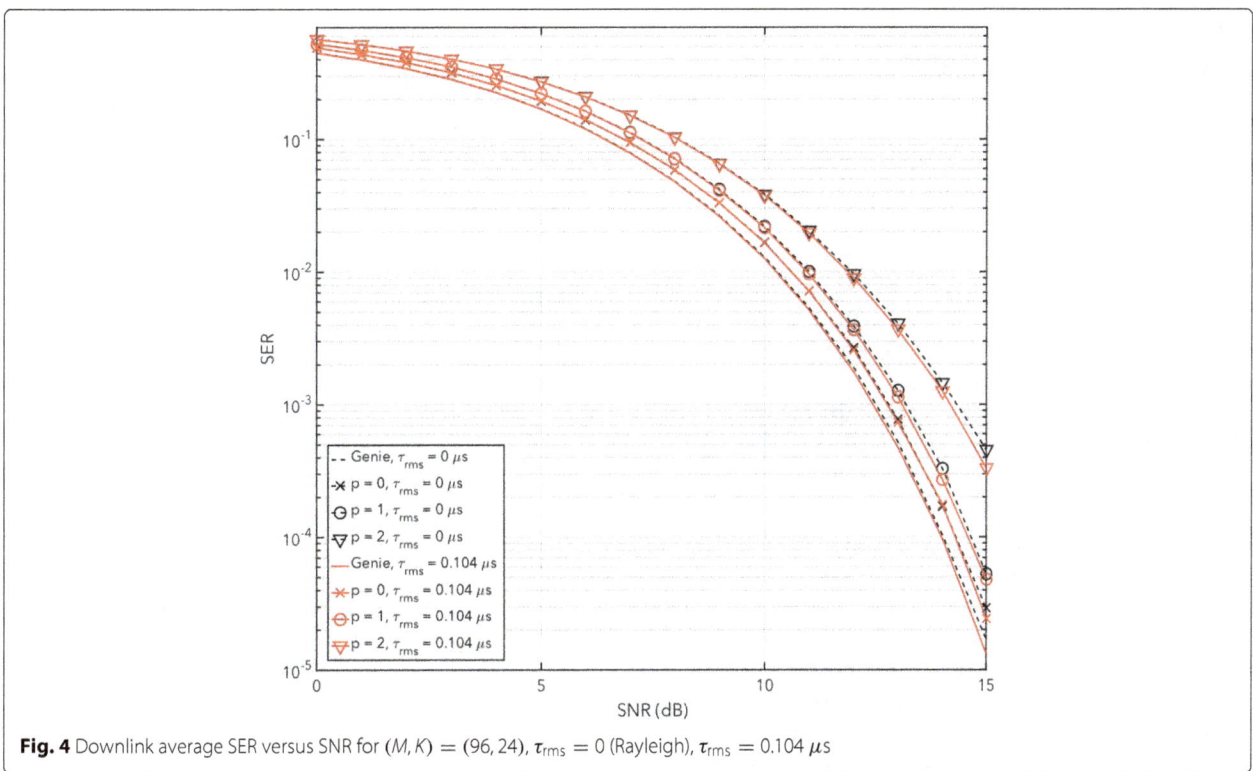

Fig. 4 Downlink average SER versus SNR for $(M, K) = (96, 24)$, $\tau_{rms} = 0$ (Rayleigh), $\tau_{rms} = 0.104 \, \mu s$

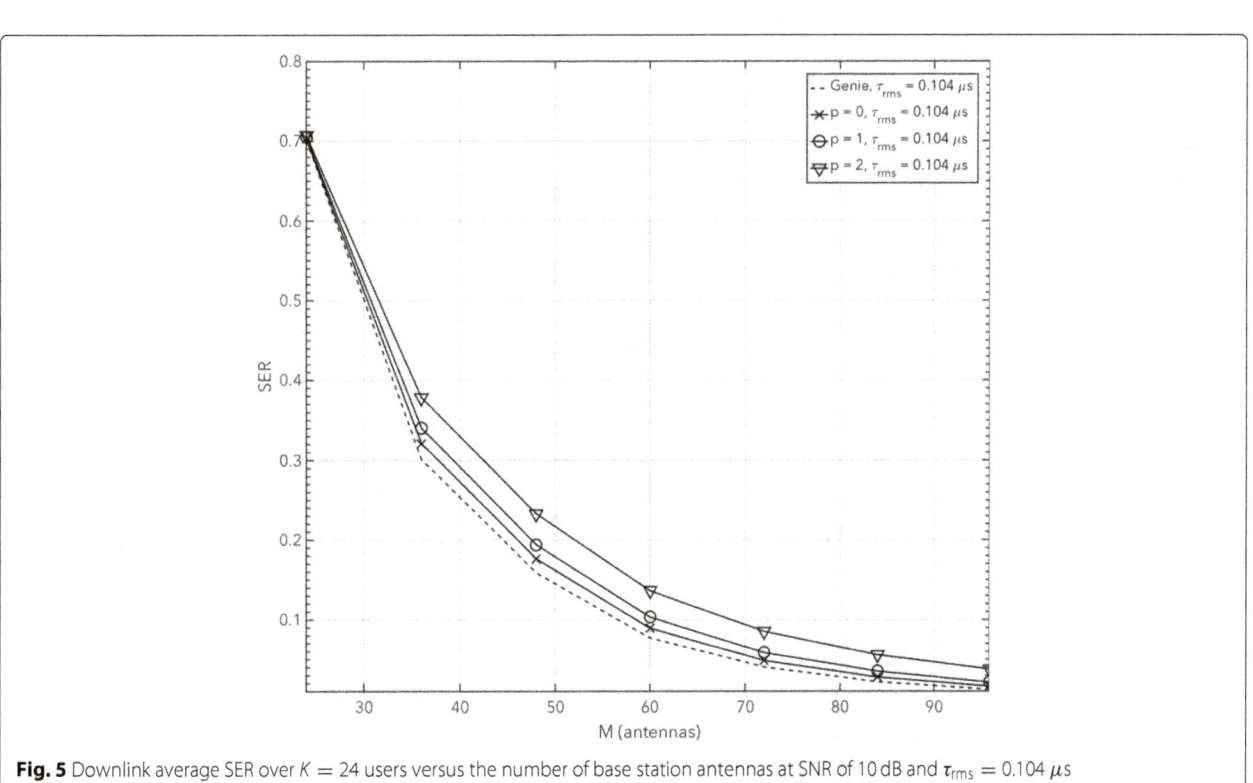

Fig. 5 Downlink average SER over $K = 24$ users versus the number of base station antennas at SNR of 10 dB and $\tau_{rms} = 0.104 \, \mu s$

methods are close to the genie-aided system, even for low-order regression models. Figure 4 shows the corresponding average symbol error-rate (SER) performance confirming this observation.

3.2 Performance vs. M: the massive MIMO effect

To serve K users, the base stations need to be equipped with at least $M = K$ antennas.[3] However, owing to larger array gain and "favorable propagation," the performance can improve by adding more antennas to the base station. To confirm this observation, in Fig. 5, we plot the simulated downlink SER versus the number of base stations antennas. The SNR is fixed at 10 dB for all the data points, and the channel delay spread is $\tau_{rms} = 0.104\ \mu s$. We compare results for polynomial regression vectors of orders $p = 0, 1, 2$. The results show that a zero-order-hold interpolator ($p = 0$) performs best and is within 6 dB of the genie-aided system when M is large.

4 Conclusions

In this correspondence, we assessed the performance of regression-based linear precoding in the downlink of a multi-user massive MIMO-OFDM system. Simple linear polynomial regressors were used to reduce multiple channel estimates over the resource blocks. These regressors do not depend on the channel statistics and can be computed in an offline manner. Simulations showed that for practical SNR ranges, the performance of the proposed methods are close to the genie-aided system, even for low-order regression selections. Moreover, the order of the regressor vectors may be adapted to the channel conditions to obtain optimal performance.

Endnotes

[1] This is true for time-division duplexing (TDD) where channel reciprocity holds.

[2] Assumed to be equal for all users over any RB.

[3] Otherwise, the ZF precoder matrix in (13) does not exist.

Competing interests
The authors declare that they have no competing interests.

Acknowledgements
The authors would like to thank members of the Connectivity Lab at Facebook for their valuable input during the course of this project.

References
1. TL Marzetta, Noncooperative cellular wireless with unlimited numbers of base station antennas. IEEE Trans. Wireless Commun. **9**(11), 3590–3600 (2010)
2. H Yang, TL Marzetta, Performance of conjugate and zero-forcing beamforming in large-scale antenna systems. IEEE J. Selected Areas Commun. **31**(31), 172–179 (2013)
3. F Rusek, D Persson, BK Lau, EG Larsson, TL Marzetta, O Edfors, F Tufvesson, Scaling up MIMO: opportunities and challenges with very large arrays. IEEE Signal Process. Mag. **30**(1), 40–60 (2013)
4. Y-H Nam, BL Ng, K Sayana, Y Li, J Zhang, Y Kim, J Lee, Full-dimension MIMO (FD-MIMO) for next generation cellular technology. IEEE Commun. Mag. **51**(6), 172–179 (2013)
5. O Edfors, F Tufvesson, Massive MIMO for next generation wireless systems. IEEE Commun. Mag. **52**, 187 (2014)
6. AL Swindlehurst, E Ayanoglu, P Heydari, F Capolino, Millimeter-wave massive MIMO: the next wireless revolution? IEEE Commun. Mag. **52**, 57 (2014)
7. H Suzuki, R Kendall, K Anderson, A Grancea, D Humphrey, J Pathikulangara, K Bengston, J Matthews, C Russell, in *Proceedings Int. Symp. on Commun. And Inform. Tech. (ISCIT)*. Highly spectrally efficient Ngara Rural Wireless Broadband Access Demonstrator (IEEE, Gold Coast QLD, 2012), pp. 914–919
8. M Biguesh, AB Gershman, Training-based MIMO channel estimation: a study of estimator tradeoffs and optimal training signals. IEEE Trans. Signal Process. **54**(3), 884–893 (2006)
9. S Coleri, M Ergen, A Puri, A Bahai, Channel estimation techniques based on pilot arrangement in OFDM systems. IEEE Trans. Broadcast. **48**(3), 223–229 (2002)
10. H Arslan, et al., Channel estimation for wireless OFDM systems. IEEE Surv. Tutorials. **9**(2), 18–48 (2007)
11. H Minn, N Al-Dhahir, Optimal training signals for MIMO OFDM channel estimation. IEEE Trans. Wireless Commun. **5**(5), 1158–1168 (2006)
12. M Vu, A Paulraj, MIMO wireless linear precoding. IEEE Signal Process. Mag. **24**(5), 86–105 (2007)
13. K Alnajjar, PJ Smith, GK Woodward, et al., in *2014 Communications Theory Workshop (AusCTW)*. Low complexity V-BLAST for massive MIMO (IEEE, Sydney, NSW, 2014), pp. 22–26
14. M Wu, B Yin, G Wang, C Dick, JR Cavallaro, C Studer, Large-scale MIMO detection for 3GPP LTE: algorithms and FPGA implementations. IEEE J. Selected Topics Signal Process. **8**(5), 916–929 (2014)
15. K-C Hung, DW Lin, Pilot-aided multi-carrier channel estimation via MMSE linear phase-shifted polynomial interpolation. IEEE Trans. Wireless Commun. **9**(8), 2539–2549 (2010)
16. X Wang, K Liu, in *IEEE Global Telecommunications Conference*. OFDM channel estimation based on time-frequency polynomial model of fading multi-path channels, vol. 1 (IEEE, San Antonio, TX, 2001), pp. 212–216
17. H Tang, KY Lau, RW Brodersen, in *IEEE Global Telecommunications Conference*. Interpolation-based maximum likelihood channel estimation using OFDM pilot symbols, vol. 2 (IEEE, Taipei, Taiwan, 2002), pp. 1860–1864

Permissions

All chapters in this book were first published in JASP, by Springer; hereby published with permission under the Creative Commons Attribution License or equivalent. Every chapter published in this book has been scrutinized by our experts. Their significance has been extensively debated. The topics covered herein carry significant findings which will fuel the growth of the discipline. They may even be implemented as practical applications or may be referred to as a beginning point for another development.

The contributors of this book come from diverse backgrounds, making this book a truly international effort. This book will bring forth new frontiers with its revolutionizing research information and detailed analysis of the nascent developments around the world.

We would like to thank all the contributing authors for lending their expertise to make the book truly unique. They have played a crucial role in the development of this book. Without their invaluable contributions this book wouldn't have been possible. They have made vital efforts to compile up to date information on the varied aspects of this subject to make this book a valuable addition to the collection of many professionals and students.

This book was conceptualized with the vision of imparting up-to-date information and advanced data in this field. To ensure the same, a matchless editorial board was set up. Every individual on the board went through rigorous rounds of assessment to prove their worth. After which they invested a large part of their time researching and compiling the most relevant data for our readers.

The editorial board has been involved in producing this book since its inception. They have spent rigorous hours researching and exploring the diverse topics which have resulted in the successful publishing of this book. They have passed on their knowledge of decades through this book. To expedite this challenging task, the publisher supported the team at every step. A small team of assistant editors was also appointed to further simplify the editing procedure and attain best results for the readers.

Apart from the editorial board, the designing team has also invested a significant amount of their time in understanding the subject and creating the most relevant covers. They scrutinized every image to scout for the most suitable representation of the subject and create an appropriate cover for the book.

The publishing team has been an ardent support to the editorial, designing and production team. Their endless efforts to recruit the best for this project, has resulted in the accomplishment of this book. They are a veteran in the field of academics and their pool of knowledge is as vast as their experience in printing. Their expertise and guidance has proved useful at every step. Their uncompromising quality standards have made this book an exceptional effort. Their encouragement from time to time has been an inspiration for everyone.

The publisher and the editorial board hope that this book will prove to be a valuable piece of knowledge for researchers, students, practitioners and scholars across the globe.

List of Contributors

Hsiao-feng (Francis) Lu
Department of Electrical and Computer Engineering, National Chiao Tung University, ED726, 1001 University Rd., 300 Hsinchu, Taiwan

Amaro Barreal, David Karpuk and Camilla Hollanti
Department of Mathematics and Systems Analysis, Aalto University, P.O. Box 11100, FI-00076 AALTO (Espoo), Finland

Alina Mirza, Ayesha Zeb and Shahzad Amin Sheikh
Department of Electrical Engineering, College of Electrical and Mechanical Engineering, National University of Sciences and Technology, Islamabad, Pakistan

Venkata Pathuri Bhuvana
Institute of Networked and Embedded Systems, Alpen-Adria-Universität, Klagenfurt, Austria
Department of Marine Engineering, Electrical, Electronics, and Telecommunications, University of Genova, Genova, Italy

Melanie Schranz, Bernhard Rinner and Andrea M. Tonello
Institute of Networked and Embedded Systems, Alpen-Adria-Universität, Klagenfurt, Austria

Carlo S. Regazzoni
Department of Marine Engineering, Electrical, Electronics, and Telecommunications, University of Genova, Genova, Italy

Mario Huemer
Institute of Signal Processing, Johannes Kepler University, Linz, Austria

Jan Østergaard
Department of Electronic Systems, Aalborg University, Fredrik Bajers Vej 7b, Aalborg, Denmark

Daniel Quevedo
Department of Electrical Engineering (EIM-E), Paderborn University, Paderborn, Germany

Guohua Ren and Ioannis D. Schizas
Department of Electrical Engineering, University of Texas at Arlington, Arlington, TX, USA

Vasileios Maroulas
Department of Math, University of Tennessee at Knoxville, Knoxville, TN, USA

S. M. Ali Tayaranian Hosseini and Hamidreza Amindavar
Amirkabir University of Technology, Tehran, Iran

James A. Ritcey
University of Washington, Seattle, WA, USA

Qi Min and Yingping Huang
School of Optical-Electrical and Computer Engineering, University of Shanghai for Science and Technology, Shanghai 200093, China

Naveen Kumar and Shrikanth S. Naryanan
Signal Analysis and Interpretation Laboratory (SAIL), Electrical Engineering, University of Southern California, Los Angeles, CA, USA

Fatemeh Fazel and Milica Stojanovic
Department of Electrical and Computer Engineering, Northeastern University, Boston, MA, USA

Hui Wen, Weixin Xie, Jihong Pei and Lixin Guan
ATR Key Lab of National Defense, Shenzhen University, 518060 Shenzhen, China

Joep de Groot
Signal Processing Systems group, Department of Electrical Engineering, Eindhoven University of Technology, 5600 MB, Eindhoven, The Netherlands
Genkey Solutions B.V., High Tech Campus 69, 5656 AG Eindhoven, The Netherlands

Jean-Paul Linnartz
Signal Processing Systems group, Department of Electrical Engineering, Eindhoven University of Technology, 5600 MB, Eindhoven, The Netherlands

Boris Škorić
Security and Embedded Networked Systems group, Department of Mathematics and Computer Science, Eindhoven University of Technology, 5600 MB Eindhoven, The Netherlands

Niels de Vreede
Discrete Mathematics group, Department of Mathematics and Computer Science, Eindhoven University of Technology, 5600 MB Eindhoven, The Netherlands

Elvin Isufi and Geert Leus
Faculty of Electical Engineering, Mathematics and Computer Science, Delft University of Technology, 2628 CD, Delft, The Netherlands

Henry Dol
Acoustics and Sonar Dept., TNO, 2597 AK, The Hague, The Netherlands

Grigorios Tsagkatakis
Institute of Computer Science, Foundation for Research & Technology - Hellas (FORTH), Crete, Greece

Baltasar Beferull-Lozano
Lab Intelligent Signal Processing & Wireless Networks (WISENET), Department of Information and Communication Technologies, University of Agder, Grimstad, Norway

Panagiotis Tsakalides
Institute of Computer Science, Foundation for Research & Technology - Hellas (FORTH), Crete, Greece
Department of Computer Science, University of Crete, Crete, Greece

Le Yang, Zhiyong Xiao and Jianjun Liu
School of Internet of Things Engineering, Jiangnan University, Wuxi 214122, China

Jinlong Yang
School of Internet of Things Engineering, Jiangnan University, Wuxi 214122, China
Key Laboratory of Advanced Process Control for Light Industry (Ministry of Education), Wuxi 214122, China

Ki Sun Song, Chul Hee Park, Jonghyun Kim and Moon Gi Kang
School of Electrical and Electronic Engineering, Yonsei University, 50 Yonsei-Ro, Seodaemun-Gu, 03722 Seoul, Republic of Korea

Hui Sheng, Yesheng Gao, Bingqi Zhu, Kaizhi Wang and Xingzhao Liu
School of Electronic Information and Electrical Engineering, Shanghai Jiao Tong University, 800 Dongchuan Road, 200240 Shanghai, China

Ali Yazdan Panah, Karthik Yogeeswaran and Yael Maguire
Facebook Connectivity Lab, Facebook Inc, 1 Hacker Way, Menlo Park, CA94025, USA

Index